Soil Biology

Volume 46

Series Editor
Ajit Varma, Amity Institute of Microbial Technology,
Amity University Uttar Pradesh, Noida, UP, India

More information about this series at http://www.springer.com/series/5138

Mukesh K. Meghvansi • Ajit Varma
Editors

Organic Amendments and Soil Suppressiveness in Plant Disease Management

Springer

Editors
Mukesh K. Meghvansi
Ministry of Defence
Defence R&D Organisation
Defence Research Laboratory
Tezpur
Assam
India

Ajit Varma
Amity University Uttar Pradesh
Amity Institute of Microbial Technology
Noida
Uttar Pradesh
India

ISSN 1613-3382 ISSN 2196-4831 (electronic)
Soil Biology
ISBN 978-3-319-23074-0 ISBN 978-3-319-23075-7 (eBook)
DOI 10.1007/978-3-319-23075-7

Library of Congress Control Number: 2015955001

Springer Cham Heidelberg New York Dordrecht London

Printed on acid-free paper

Springer International Publishing AG Switzerland is part of Springer Science+Business Media (www.springer.com)

Preface

In view of the rising public concerns about economic and ecological consequences of agricultural chemicals, the emphasis on crop improvement strategies has gradually been shifting from chemical to non-chemical approaches for sustainable agriculture. Soil amendment is one such approach that can play a significant role in building up soil fertility and improving soil health for sustainable agriculture. Various research reports have convincingly established the role of organic amendments in improving plant growth, health, and yield. In addition, organic amendments contribute to enhancing soil suppressiveness.

Soil suppressiveness is often attributed to activity of soil microorganisms or microbial metabolites. However, physicochemical properties of soil, including pH, organic matter, and clay content, can also contribute to the suppression of plant diseases directly or indirectly through their influence on soil microbial activity. It is therefore important to know the influence of soil physicochemical properties on disease suppression. Although one set of physicochemical attributes of soil considered as suppressive for a disease may be conducive for other one. It is therefore equally important to understand the physicochemical characteristics of soil which are unfavourable to the specific disease development. It has also been established that some of the soil-borne plant diseases can be effectively managed through organic amendments. It is, therefore, equally imperative to understand the relationship between organic amendments and soil suppressiveness. Despite being a very significant area from the view point of plant disease management through sustainable means, literature is scanty on the topic. The main objective of the present volume *Organic Amendments and Soil Suppressiveness in Plant Disease Management* is to make efforts to fill this gap by synthesising the literature on various aspects of organic amendments and soil suppressiveness in order to utilise potential of these phenomena more effectively and efficiently in sustainable agriculture.

The present volume has four parts with a total of 25 chapters. Part I deals with general paradigms and mechanisms of soil suppressiveness, comprising eight chapters. Parts II and III focus on concepts in plant disease management involving microbial soil suppressiveness and organic amendments, respectively. Part IV elaborates various combinatorial approaches in plant disease management. Each chapter in these parts provides an overview of the topic, current knowledge and recent developments, conclusions, and directions for future research following an in-depth and critical analysis of the literature.

In Chap. 1, George. M. Kariuki, Lilian K. Muriuki, and Emma M. Kibiro discuss how suppressive soils affect or influence plant pathogens' suppression in the soil and how they contribute to agricultural productivity. Chaney C. G. St. Martin in Chap. 2 provides a detailed account of current knowledge on enhancing soil suppressiveness using compost and compost tea, along with predictors and mechanisms of disease suppression and factors affecting the efficacy of compost and compost tea. Furthermore, the potential application of molecular tools for better understanding the relationship between microbial properties of compost and compost tea and soil suppressiveness is highlighted and core areas for research identified in Chap. 2. In Chap. 3, D. P. Singh reviews the information on research done on soils and crop health of rice–wheat system under conservation agriculture. Agronomic strategies for developing disease-suppressive soils for improved soil and plant health and productivity as well as for environmental benefits are discussed in Chap. 4 by R. S. Yadav, Jitendra Panwar, H. N. Meena, P. P. Thirumalaisamy, and R. L. Meena. In Chap. 5, Prashant P. Jambhulkar, Mahaveer Sharma, Dilip Lakshman, and Pratibha Sharma discuss natural mechanisms of soil suppressiveness against diseases caused by *Fusarium*, *Rhizoctonia*, *Pythium*, and *Phytophthora*. The pea footrot disease symptoms and assessment, molecular basis of pea footrot disease, and the potential role of agricultural soil health indices in pea footrot disease suppressiveness are discussed by Ebimieowei Etebu in Chap. 6. Subsequently, Chap. 7 contributed by Phatu W. Mashela, Zakheleni P. Dube, and Kgabo M. Pofu provides the dosage model as an alternative strategy in managing plant parasitic nematodes with specific reference to addressing efficacy, phytotoxicity, and inconsistent result issues of phytonematicides. Chapter 8 by Silvana Pompeia Val-Moraes focuses on recent progress towards unravelling the microbial basis of suppressive soils. In Chap. 9, Mona Kilany, Essam H. Ibrahim, Saad Al Amry, Sulaiman Al Roman, and Sazada Siddiqui present recent advances and findings regarding the role of beneficial microbes in the pythium damping-off disease suppression and the biological aspects highlighting the mechanisms of action of biocontrol process. Interaction of rhizobia with soil suppressiveness factors has been discussed at length by Kim Reilly in Chap. 10. In subsequent chapter, an overview of the biocontrol potential of opportunistic as well as AM fungi on the growth and improvement of various crop plants and population of plant parasitic nematodes in different pathosystems has been provided by Mohd. Sayeed Akhtar, Jitendra Panwar, Siti Nor Akmar Abdullah, and Yasmeen Siddiqui. This chapter also focuses on the cost-effective technologies used for the mass propagation of opportunistic fungi and AM fungi and their ample application in the expansion of practical control system desired for the sustainable agricultural practices. In Chap. 12, different aspects of microbial soil suppressiveness and their impact on wilt disease have been discussed in detail by M. K. Mahatma and L. Mahatma. Chapter 13 by Erin Rosskopf, Paula Serrano-Pérez, Jason Hong,

Utsala Shrestha, María del Carmen Rodríguez-Molina, Kendall Martin, Nancy Kokalis-Burelle, Carol Shennan, Joji Muramoto, and David Butler summarises the research that has been conducted on anaerobic soil disinfestations (ASD) around the world and to suggest research areas that are of interest and importance for the future. Topics of their discussion also include the impact that amendment choice and temperature have on generating anaerobic conditions; how the process of ASD changes soil chemistry; changes in the microbial community as a result of ASD and the role microbes play in anaerobicity; and what is currently known about creating a disease-suppressive soil using this method. Chapter 14 by Yasmeen Siddiqui, Yuvarani Naidu, and Asgar Ali highlights the potentiality of harnessing microbial diversity utilising compost and compost teas for mitigation of fungal diseases of fruits and vegetables in an eco-friendly manner. Yurdagul Simsek-Ersahin in Chap. 15 provides an overview of the current understanding of the influence of vermicompost products, solid or liquefied, on *fusarium* diseases. In Chap. 16, Christel Baum, Bettina Eichler-Löbermann, and Katarzyna Hrynkiewicz provide an overview on the causal agents of suppression of fusarium wilt evaluating the quality of different organic amendments. Further it aims to facilitate a selection and optimisation of the use of organic amendments in the arable management by reviewing the actual state of knowledge. In Chap. 17, Sazada Siddiqui, Saad Alamri, Sulaiman Alrumman, Mukesh K. Meghvansi, K. K. Chaudhary, Mona Kilany, and Kamal Prasad discuss the role of micronutrients, which can lead to a less disease-favourable environment and increase host plant resistance. The chapter carries out a critical analysis of various factors responsible for the suppression of certain plant fungal diseases due to micronutrients and determines key areas where sincere research efforts are still needed to develop strategies for manipulating micronutrient application in such a way that it could be more efficiently utilised in managing soil-borne plant fungal diseases. L. Grantina-Ievina, V. Nikolajeva, N. Rostoks, I. Skrabule, L. Zarina, A. Pogulis, and G. Ievinsh in Chap. 18 provide an analysis of the impact of organic amendments, i.e. green manure and vermicompost on the soil microorganisms and plant growth and health in conditions of organic agriculture of Northern temperate climate. In Chap. 19, Henok Kurabachew discusses the impact of silicon amendment on suppression of bacterial wilt caused by *Ralstonia solanacearum* in Solanaceous crops. In Chap. 20, various facets of suppression of soil-borne plant pathogens by cruciferous residues have been discussed by Ritu Mawar and Satish Lodha. In Chap. 21, Santiago Larregla del Palacio, María del Mar Guerrero Díaz, Sorkunde Mendarte Azkue, and Alfredo Lacasa Plasencia critically review the mechanisms involved in disease suppression and the organic amendment management strategies for the control of protected pepper crops' soil-borne diseases and soil fatigue. Chapter 22 by David Ruano-Rosa and Jesús Mercado-Blanco provides a brief overview on research efforts devoted to the use of biological control agents (BCAs) and organic amendments (OAs) against soil-borne diseases within integrated disease management strategies. More specifically, this chapter focuses on the ad hoc combination of BCAs and OAs and discuss aspects such as how these approaches may influence soil microbial communities or the suitability of using OAs as carriers to develop more stable and effective formulations of BCAs. Chapter 23 by Mohammad Haneef Khan, M. K. Meghvansi, Rajeev Gupta, K. K. Chaudhary, Kamal Prasad, Sazada Siddiqui, Vijay Veer, and Ajit Varma highlights the potential of individual and combined approach of vermiwash and AM fungi with a particular emphasis on understanding the possible underlying molecular mechanisms involved in the suppression of plant diseases. Chapter 24 by Massimo Pugliese, Giovanna Gilardi, Angelo Garibaldi, and Maria Lodovica Gullino focuses on the use of organic amendments, compost in particular, and soil suppressiveness for the management of diseases of vegetable and ornamental crops. In Chap. 25, a study conducted by Yohichi Matsubara, Jia Liu, and Tomohiro Okada on suppression of *fusarium* crown rot and the changes in free amino acid contents in mycorrhizal asparagus plants with NaCl treatment is discussed in order to clarify the mechanisms of disease tolerance.

The editors would like to express sincere gratitude to all the contributors for submitting their work and timely responding to all the post-submission editorial queries. We have received numerous insightful and constructive inputs from the researchers all across the world on this subject while editing this book for which we are sincerely grateful to them. Dr. Mukesh K. Meghvansi takes this opportunity to express his deep sense of gratitude to Dr. Vijay Veer, Director, Defence Research Laboratory, Tezpur, for his constant support, encouragement, and guidance. Dr. Meghvansi wishes to thank Mrs. Manju Meghvansi (wife) and Miss Lakshita Meghvansi (daughter) for their unconditional love, patience, understanding, and moral support while editing this volume. Last but not the least, we thank all the staff members of Springer Heidelberg, especially Dr. Jutta Lindenborn, project coordinator (Springer Books—Life Sciences and Biomedicine), for their critical evaluation, constant support, and encouragement.

Assam, India — Mukesh K. Meghvansi
Uttar Pradesh, India — Ajit Varma

Contents

Part I
Soil Suppressiveness: Paradigms and Mechanisms

Chapter 1
The Impact of Suppressive Soils on Plant Pathogens and Agricultural Productivity

George M. Kariuki, Lilian K. Muriuki, and Emma M. Kibiro

1.1 Introduction

Soil is a key element of agricultural production, which comprises of complex blend of organic and inorganic matter, including different species, the majority of which have not been described. A number of the organisms are pests that result in important crop losses as others carry out environmental activities such as aeration, biological pest control, drainage, and water and nutrient cycling. Soil is the foundation of sustainable agriculture and provides the physical support upon which majority of other human activities rely on (Singh 2013).

Agricultural soils that are suppressive to soilborne plant pathogens exist all over the world (Weller et al. 2002), and the biological basis of suppressiveness has been depicted for majority of the soils. The suppressive soil concept was initially introduced by Menzies (1959) who used the term in the description of the soils that inhibited *Streptomyces* potato scab (Weller et al. 2002). Suppressive soils have been referred to as soils in which there cannot be establishment or persistence of pathogen (Shurtleff and Averre 1997), there can be establishment of the pathogen but it causes little or no damage, or there can be establishment of the pathogen and development of disease but the disease is less significant, even though the pathogen may persist in the soil or soils in which some diseases are inhibited because of the presence, in the soil, of microbes that act antagonistically against the pathogen or pathogens (Baker and Cook 1974). On the contrary, conducive soils are ones in which disease occurs readily. Pathogen suppression is termed as the inhibition of saprophytic survival or growth of the pathogen in the soil, while disease

G.M. Kariuki (✉) • L.K. Muriuki • E.M. Kibiro
Department of Agricultural Science and Technology, Kenyatta University, P.O. Box 43844-00100, Nairobi, Kenya
e-mail: kariuki.george@ku.ac.ke; gmmkariuki@yahoo.com; lilnyix@gmail.com; emakip2005@yahoo.com

M.K. Meghvansi, A. Varma (eds.), *Organic Amendments and Soil Suppressiveness in Plant Disease Management*, Soil Biology 46,
DOI 10.1007/978-3-319-23075-7_1

suppression is the inhibition of the parasitic growth of the pathogen (Simon and Sivasithamparam 1989).

Soil suppressiveness is associated with the level of fertility, types, and numbers of soil organisms, as well as nature of the soil texture and drainage. Mechanisms through which soilborne pathogens are affected by rhizosphere microorganisms have been keyed out and include consumption of pathogen stimulatory compounds, antibiotic compound production, direct parasitism, (micro)nutrients competition, as well as production of lytic enzymes (Lugtenberg and Kamilova 2009). Suppressive soils are essential in agricultural production since severity or occurrence of disease is less than expected for the dominating environment or in comparison to that in surrounding soil that reciprocally results in higher crop yields. Suppressive soils are the best natural examples in which the natural microflora efficaciously offers protection to plants against pathogens. Various pathogens for which suppressive soils have been demonstrated include fungi such as *Pythium splendens* (Kao and Ko 1983), *Fusarium oxysporum* (Alabouvette et al. 1993), *Gaeumannomyces graminis* var. *tritici* (Hornby 1998), *Aphanomyces euteiches* (Persson et al. 1999), *Phytophthora cinnamomi* (Ko and Shiroma 1989), *Thielaviopsis basicola* (Stutz et al. 1986), *Phytophthora infestans* (Andrivon 1994), *Pythium ultimum* (Martin and Hancock 1986), *Plasmodiophora brassicae* (Murakami et al. 2000), and *Rhizoctonia solani* (Lucas et al. 1993); nematodes such as *Heterodera schachtii*, *H. avenae*, *Criconemella xenoplax*, and *Meloidogyne* spp.; and bacteria such as *Ralstonia solanacearum* and *Streptomyces scabies* (Haas and Défago 2005).

The main objective of this chapter is to describe how suppressive soils affect or influence plant pathogen suppression in the soil and how they contribute to agricultural productivity. We have discussed different types of soil suppressiveness and factors that influence them. Different types of suppressive soils which include fungi-suppressive soils, bacteria-suppressive soils, and nematode-suppressive soils have also been discussed highlighting the contribution of these types of soils to agricultural productivity.

1.2 Impact of Soil Health on Agriculture

Soil health is termed as the soil's capacity to function as a critical living system, within ecosystem and land-use boundaries, in order to sustain productivity of animals and plants, enhance or maintain air and water quality, as well as enhance animal and plant health. Soil health is critical to crop production. Since it is fragile and finite, soil is an important resource that needs special care from its users. Majority of crop and soil management systems today are not sustainable. On the one hand, overutilization of fertilizer has resulted in nitrogen deposition, which is a threat to the sustainability of an approximated 70 % of nature (Hettelingh et al. 2008). On the other hand, in most regions of sub-Saharan Africa, the underutilization of fertilizer entails that soil nutrients exported together with crops fail to be replenished, resulting in the degradation of soil, as well as decrease in yields. The aim of sustainable agriculture is to meet the demands of the present with no compromise of the productive potential for the following generations. Rational soil use practices ought to permit environmentally and economically sustainable yields that will just be attained with the recovery or maintenance of the soil health.

Different management and land uses impact the soil and the production systems' sustainability. Tillage systems grounded on disking and plowing in the tropical area lead to the reduction in soil organic matter, as well as rise in the process of erosion. This causes physical, chemical, and biological modifications in the characteristics of the soil, which promote the reliance on external inputs and accordingly enhance costs of production, resulting in environmental effects. Less impacting cropping systems, on the other hand, depend more on biological processes for sustainability (Kaschuk et al. 2010). Sustainable ecosystems, whether agricultural or natural, depend on the nutrient flux across trophic levels that are primarily intermediated by microorganisms and soil fauna (Chen et al. 2003). The microbial community and soil fauna are regarded as critical in any ecosystem through soil organic matter decomposition, cycling of nutrients, and affecting the soil's physical and chemical characteristics, with direct impacts on sustainability and soil fertility.

1.3 Types of Soil Suppressiveness

1.3.1 General Suppressiveness

General suppressiveness is termed as the widespread but confined ability of soils to inhibit the activity or growth of soilborne pathogens. It can as well be termed as nonspecific antagonism or biological buffering (Weller et al. 2002). General suppression is associated with the soil's total microbial biomass that engages in a competition with the pathogen for resources or results in suppression via more direct types of antagonism. It is frequently promoted by some agronomic practices, the addition of organic matter, or the accumulation of soil fertility (Rovira and Wildermuth 1981) all of which can enhance soil microbial activity. General suppression results from several organisms and cannot be transferred between soils (Rovira and Wildermuth 1981). Typically, in suppressive soils, inhibition is caused by the accumulative impacts of complex relationships between the pathogen and other factors. Soil suppressiveness has been ascribed to either or combination of biotic and abiotic factors, and it differs from a single pathogen to another (Weller et al. 2002). These factors cause antagonism against pathogens either through production of antibiotics, competition for food, secretion of lytic enzymes, or via direct parasitizing of the pathogens suppressing them from surpassing the levels of economic threshold. A number of the antagonistic microorganisms, which are known to raise suppressiveness in soils, include fungi, for instance, *Penicillium*,

Sporidesmium, and *Trichoderma* spp., or bacteria belonging to the genera *Streptomyces*, *Pseudomonas*, and *Bacillus* (Rani and Sudini 2013).

Components of suppressive soils added to conducive soil can decrease the amount of disease through introduction of microorganisms that are antagonistic to the pathogen. Amendment of soil with soil containing a strain of *Streptomyces* spp. that is antagonistic to the cause of potato scab has been demonstrated to cause significant reduction in potato scab. *Phytophthora* root rot of papaya was managed by planting papaya seedlings in suppressive soil put in holes in the orchard soil that was infested with the *P. palmivora* (Rani and Sudini 2013).

Planting the same crop, continuously, in a conducive soil results in raised microbial populations that are antagonistic to a number of pathogens. For instance, continuous wheat cultivation has been demonstrated to result in a decrease in take-all of wheat. Continuous watermelon cropping also permits the accumulation of antagonistic *Fusarium* spp. associated with the one causing watermelon Fusarium wilt that leads to a decrease in Fusarium wilt. Such was as well demonstrated in the accumulation of soil suppressiveness to root-knot nematodes of peanut in Florida, USA (Kariuki and Dickson 2007).

1.3.2 Specific Suppressiveness

This form of soil suppression arises from a direct inhibition of a known pathogen by one organism. There are incidences where an agent of biological control is introduced into the soil for the specific reduction in occurrence of the disease. Specific suppressiveness owes its activity to the impacts of individual or select groups of microorganisms. Specific suppressiveness can be transferred to conducive soil with small portions of soil, and this makes the nature of soil suppressiveness to be considered as biological (Shipton et al. 1973; Kariuki and Dickson 2007). Transferability of suppressiveness indicates specific soil suppressiveness against nematodes that are parasitic (Kerry 1988), especially when the antagonists of the nematode are not culturable or are not known. Based on a report by Mankau (1975), greenhouse soil amendment, which has been steam-sterilized, with soil infested with *Pasteuria penetrans* led to *Meloidogyne incognita* suppression by the obligate parasitic nematode. The transfer of 20 to 53-μm fraction of soil obtained from northern Europe, which comprised of *Nematophthora gynophila*, to *Heterodera avenae*-infested South Australia soils led to infection of the nematodes by fungi (Stirling and Kerry 1983).

The approach of soil transfer is particularly helpful when the active organisms have not yet been keyed out. For instance, soil transferability exhibited the biological nature for peach orchard soil that is *Criconemella xenoplax*-suppressive when 5 % of the orchard soil, which was not steamed, was blended into the peach orchard soil that had been subjected to steaming (Kluepfel et al. 1993). This form of transfer in *H. schachtii*-suppressive soil was attained in a field experiment with 1 and 10 % suppressive soil, while in the greenhouse it was achieved with as little as 0.1 %

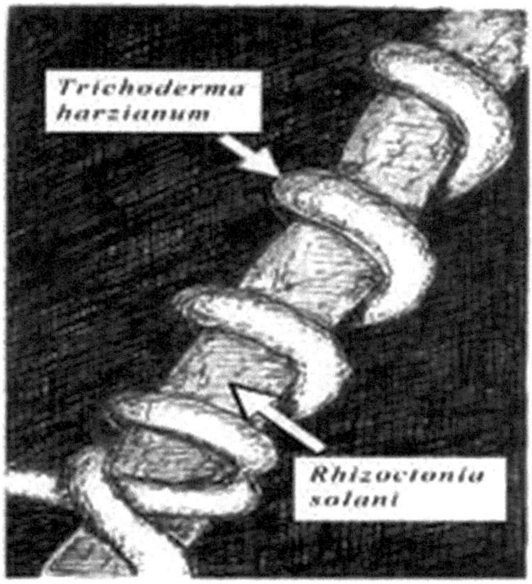

Fig. 1.1 Hyphae of *Trichoderma* spp. wrapped around the pathogenic fungus *Rhizoctonia* (*Source*: Hamid 2011)

suppressive soil to conducive (Westphal and Becker 2001). Soil suppressiveness onset was monitored, in the field experiment, with an infective J2 bioassay in field plots. Suppressiveness rose in the plots subjected to 10 % transfer in initially conducive plots more quickly than in the plots with 1 % transfer. It was as well indistinguishable from the suppressive control following a shorter incubation in the higher soil treatment of soil amendment than in the lower one. This observation offered additional proof of the nematode suppression's biological nature. There is occurrence of less root rot, as well as corresponding feeding sites' loss when test soil is being diluted. However, when a suppressive soil is being diluted, impacts on the reproduction of nematode are still measurable. For instance, in *H. schachtii*-suppressive soil, the number of eggs for each cyst was about 40, while there were nearly 120 in conducive soil (Westphal and Becker 2001) (Fig. 1.1).

1.4 Impact of Abiotic Factors on Soil Suppressiveness

The pH of the soil, level of calcium, nitrogen form, and the availability of other nutrients in the soil are important in soil suppressiveness playing key functions in the management of diseases. Sufficient crop nutrition renders plants more resistant to or tolerant of disease. The status of nutrients of the soil, as well as the application of certain amendments and fertilizers, can significantly impact the environment of the pathogen. For instance, in potato scab, the disease has more severity in soils with levels of pH of more than 5.2, while the disease is significantly inhibited with levels less than 5.2. Sulfur and ammonium nitrogen sources also lower the severity and occurrence of potato scab since they lower the soil pH, rendering it unconducive for the establishment of the pathogen. On the contrary, practices such as soil liming enhance disease severity. Soilborne diseases resulting from *Pythium* spp., for instance, damping off in peanuts, wheat, beans, soybeans, peppers, peas, sugar beets, tomatoes, snap dragons, as well as onions, cannot be managed by calcium availability in the soil. It has been reported that amendment of the soil with calcium and adding alfalfa meal to raise microbial populations significantly reduced damping off in cucumbers. Sufficient calcium levels have as well been found to lower crucifers' clubroot. The disease is suppressed in neutral to marginally alkaline soils with a pH of 6.7–7.2 (Campbell and Greathead 1990). In crops such as melons, cottons, tomatoes, as well as a number of ornamentals, sufficient levels of calcium, and soil pH increase, have been demonstrated to lower *Fusarium* spp. infestation levels (Jones et al. 1989) resulting in raised yields.

Nitrogen fertilizers have been demonstrated to inhibit tomato's Fusarium wilt since they have the tendency of raising the pH levels around the root zone, unlike the ammonia fertilizers, which enhance severity of the disease. Studies on tomatoes have demonstrated that the application of nitrate nitrogen in high-pH soil leads to even better wilt control (Woltz and Jones 1973). Levels of *Fusarium* disease have been shown to reduce by the use of calcium nitrate in comparison to ammonium nitrate. Nevertheless, ammoniacal nitrogen uptake has been shown to enhance plant manganese uptake, as well as reducing take-all disease. The same findings were attained in *Verticillium* wilt in potatoes, as well as corn stalk rot (Hamid 2011).

Adding potassium into the soil results in disease suppressiveness, as well as increase in yields. It was revealed that high levels of potassium lowered incidences of Fusarium wilt in tomatoes, as well as *Verticillium* wilt in cotton (Foster and Walker 1947). It was shown that cotton soils containing between 200 and 300 pounds of potassium for each acre had plants with between 22 and 62 % leaf infections, whereas levels of soil test of more than 300 pounds for each acre had an infection rate of between zero and 30 % (Obrien-Wray 1995). Amendment of agricultural soils, as well as soilless growing media with organic matter, enhances natural soil suppressiveness against soilborne pathogens, provides plant nutrients, and enhances biological and physicochemical features (Veeken et al. 2005; Janvier et al. 2007). Reciprocally, the quality of the soil also impacts plant health, as well as crop production.

There has been effective application of compost in high-value crops, in the nursery industry, as well as in mixtures of potting soil for root rot diseases' control. Successful suppression of disease by use of compost has been less common in soils than in potting mixtures. These have significant implications for management of soil and nutrient, as well as plant health and management of pests. In a research carried out at the University of Florida, field experiments demonstrated disease inhibition effects of compost and sewage sludge, subjected to heat treatment, on southern peas and snap beans (Ozores-Hampton et al. 1994). The compost used at 36 or 72 tons for each acre and the sludge at 0.67 and 1.33 tons for each acre produced larger beans and 25 % higher yields at the two rates of application than those from regions without compost application. In regions treated with sludge, the disease was decreased but nearly gotten rid of where compost had been used. In the portion of field where compost was not used, leaf death and leaf wilting were, however, pronounced.

Fig. 1.2 Compost-amended soils vs. unamended soils (*Source*: Matthew Ayres, SARDO, November 2007; with permission)

There has been demonstration of modification of the chemical, physical, and biological features of soil, by use of manure, which can indirectly or directly affect crop infection and the survival of the pathogen. Scheuerell and colleagues (2005) found that *Pythium* spp. suppression was linked to volatilization of ammonia from manure amendments. Similarly, Conn and Lazarovits (1999) reported that the application of liquid swine manure lowered the occurrence of wilt, as well as common scab in potato fields. It also lowered the number of plant-parasitic nematodes for a period of 3 years following a single use. A significant reduction in root disease of the red stele strawberry was as well observed in fields treated with steer/poultry and dairy manure compost, comparative to control (Millner et al. 2004). The difference between plants growing in compost-amended soils and unamended soils is illustrated in Fig. 1.2.

1.5 Effect of Beneficial Organisms in Disease Suppression and Plant Health

Several commercial products comprising of beneficial, disease-suppressive organisms such as *Flavobacterium* spp., *Trichoderma* spp., *Gliocladium* spp., *Streptomycetes* spp., *Pseudomonas* spp., and *Bacillus* spp. have been reported. These

Fig. 1.3 Pepper plants on the *right* treated with *Bacillus subtilis* and *Bacillus amyloliquefaciens* strains compared to untreated check (*Source*: Ayres et al. 2007; with permission)

products are applied in various ways through seed treatments, compost inoculation, soil inoculation, and soil drenches. These products have plant growth-promoting rhizobacteria (PGPR) that colonize plant roots and induce plant growth and/or decrease plant disease occurrence (Burkett Cadena et al. 2008). These PGPR serve as plant growth stimulators, aggressive colonizers, and as biocontrol. In PGPR present in soil acting as bio-fertilizers, promotion of plant growth prevails. This is ascribed to a number of processes, which include fixation of nitrogen, solubilization of phosphate, as well as the production of volatile growth stimulants and phytohormones. Other PGPR in the soil serve as biopesticides whereby the aspect of biocontrol is most conspicuous. *Bacillus* and *Pseudomonas* spp. are the key PGPR, which serve as antagonists of known root pathogens. Root-colonizing plant-beneficial fungi present naturally in the soil are as well significant in offering protection to plants against root pathogens. Among these are nonpathogenic *Trichoderma* and *Fusarium* spp. that engage in a symbiotic, rather than a parasitic, association with plants. These nonpathogenic strains result in raised growth, as well as plant vigor observed with the use of PGPR. For instance, pepper plants on the right received a treatment *Bacillus subtilis* strain GBO3 together with *B. amyloliquefaciens* (strain IN937a) as compared to the untreated ones on the left in Fig. 1.3.

1.6 Fungi-Suppressive Soils

Soilborne fungal pathogens of plants, one of the key factors restricting the agroecosystem productivity, are frequently hard to control via conventional techniques, for instance, the application of synthetic fungicides and host cultivars that are resistant. The absence of dependable chemical controls, the incidence of pathogen resistance to fungicide, and the circumvention or breakdown of host resistance by the populations of pathogens are among the reasons behind attempts to develop new control measures of diseases (McDonald and Linde 2002). The withdrawal from the markets of methyl bromide, the most effectual soil fumigant globally, has enhanced the need for alternate methods of control (Martin 2003). The search for highly efficient alternatives that have low costs and less environmental effect is a test for eco-sustainable contemporary agriculture. The application of organic amendments, for instance, green manure composts, peats, and animal manure, has been suggested, for biological and conventional agricultural systems, to enhance the structure of soil and fertility (Conklin et al. 2002; Cavigelli and Thien 2003) and reduce the occurrence of disease resulting from soilborne pathogens (Litterick et al. 2004; Noble and Coventry 2005). The introduction of fungicides, disease-resistant varieties, and synthetic organic fertilizers has permitted farmers to break the connection between soil fertility and organic amendments (Hoitink and Boehm 1999). Consequently, organic materials, for instance, manure and crop residues from necessary resources, turned into solid wastes. Following the decline in the organic input, organic matter in the soil reduced with time, soil fertility reduced, and a huge number of soilborne diseases extended in agroecosystems (Hoitink and Boehm 1999; Bailey and Lazarovits 2003). Fungi-suppressive soils enhance growth of plants unlike in the non-suppressive soils.

1.6.1 **Fusarium *Wilt-Suppressive Soils***

Fusarium wilt is a soilborne plant disease that occurs globally and is caused by *Fusarium oxysporum*, a plant pathogenic fungus. Fusarium wilt is linked to significant losses in yield in several crops, and its sufficient as well as sustainable control is yet to be achieved. Soil suppressiveness to Fusarium wilt was initially reported by Atkinson (1892) and more research has been undertaken. The suppressiveness is specific just to Fusarium wilts. Fungal and bacterial genera demonstrated to have soil suppressiveness against Fusarium wilt are nonpathogenic *F. oxysporum*, *Pseudomonas* spp., *Bacillus* spp., *Alcaligenes* spp., *Actinomycetes*, and *Trichoderma* spp.

Wilt-suppressive soils restrict the severity or occurrence of wilts of a number of plant species that lead to higher yields.

1.6.2 Take-All-Decline Suppressive Soils

Take-all decline (TAD), which is caused by the fungus *Gaeumannomyces graminis* var. *tritici*, is an important wheat root disease globally. Take-all decline is one of the most studied types of soil suppressiveness. It needs a susceptible host's monoculture, *G. graminis* var. *tritici*, as well as at least a single severe take-all outbreak. TAD can be termed as the spontaneous reduction in the severity and occurrence of

Fig. 1.4 Take-all decline disease increases and then declines with years of monoculture (*Source*: Berendsen et al. 2012)

take-all, which takes place with monoculture of wheat or other host crops that are susceptible following one or more severe eruptions of the disease (Simon and Sivasithamparam 1989). This form of suppressiveness may be lowered or gotten rid of through breaking of the monoculture with a crop that is not a host (Cook 1981), although a field with a long TAD history may regain suppressiveness when barley or wheat is again grown. In an experiment, fluorescent *Pseudomonas* spp. from the wheat rhizosphere grown in Moses Lake and Quincy TAD soils were compared to *Pseudomonas* spp. from wheat roots grown in conducive soils from Mt. Vernon and Lind. Every soil was diluted with fumed lind virgin soil and then adjusted with take-all inoculum. During the second wheat cropping, take-all was inhibited in mixtures with TAD and not conducive soils. All the roots had equal population densities of fluorescent species of Pseudomonas repressive to *G. graminis* var. tritici in vitro. There were significantly more on roots from mixtures with Moses Lake and Quincy TAD soils than on roots from mixtures of conducive soil. Moreover, fluorescent pseudomonads from TAD soils offered significantly better protection against take-all than pseudomonads from conducive soils, when applied as wheat seed treatments (Fig. 1.4).

1.7 Induction of Suppressiveness to Apple Replant Disease

Apple replant disease can be described as the poor apple tree growth, which occurs following replanting on a site that was antecedently cropped to apples. It results from a complex of fungi, which include *Rhizoctonia solani*, *Cylindrocarpon destructans*, *Pythium* spp., and *Phytophthora cactorum* (Mazzola 1998). Soils, which have not gone through cultivation of apple, are suppressive to replant disease. Orchard soils turn increasingly more conducive to monoculture replant disease. This phenomenon was demonstrated by Mazzola (1998) when he brought in *R. solani* inoculum into soils from orchard blocks in their first to fifth years of growth and from close noncultivated regions. Growth of apple seedling was considerably lowered in soils from the third-, fourth-, and fifth-year blocks in comparison to noncultivated soil growth or in first- and second-year block soil. There was a rise in the populations of decline pathogens obtained from the roots of the seedlings. There was also a reduction in populations of *Pseudomonas putida* and *Burkholderia cepacia*. *B. cepacia* secretes multiple antibiotics and has biocontrol activity against soilborne pathogens, which include *Pythium* spp. and *R. solani* (Parke and Gurian-Sherman 2001). *P. putida* isolates from these soils also acted antagonistically against *Rhizoctonia* and *Pythium* spp., though as they reduced in populations in the orchard soil, *P. fluorescents* boar C and *P. syringe* isolates became dominant.

1.8 Nematode-Suppressive Soils

Nematode-suppressive soils can be termed as the ecosystems in which an increase in population of a plant-parasitic nematode is less than in a conducive soil in spite of the presence of a virulent pathogen, a susceptible host, as well as conducive environmental conditions (Stirling 1991). Soils that are specifically suppressive against nematodes that are parasitic to plants are of interest in the definition of the mechanisms, which control population density. Suppressive soils preclude establishment and causing of a disease by nematodes. They as well decrease severity of the disease following initial damage of nematode in an uninterrupted culturing of a host. An array of nonspecific and specific soil treatments, followed by a target nematode infestation, has been used to key out nematode-suppressive soils. Soil transfer tests, baiting approaches, and biocidal treatments together with plant-parasitic nematode observations in the susceptible host plants' root zone have enhanced the apprehension of nematode-suppressive soils.

The utility of nematode-suppressive soils, in the study of biological control of nematodes that are parasitic to plants, is broadly accepted (Stirling 1991). It is thought that enhanced biological control mechanisms exploitation will, to a large extent, gain from a thorough apprehension of natural mechanisms of regulating population densities of nematodes in the soils. Nematode-suppressive soils, even though understood poorly, frequently comprise an array of nematode antagonistic microorganisms (Kerry 1990). Nematode-suppressive soils frequently are initially known or surmised when the nematode's population densities reduce following initial establishment (Gair et al. 1969) or when their populations stay significantly less in a number of fields than in others within the same area with the same histories of soil and crop (Carris et al. 1989; Westphal and Becker 1999). Suppressive soils are frequently linked to a susceptible host's monoculture (Gair et al. 1969; Westphal and Becker 1999; Noel and Wax 2003). Nevertheless, monoculture does not constantly result in a nematode-suppressive soil (Carris et al. 1989). Field observations of soil suspected to be nematode-suppressive have to be affirmed. Greenhouse experiments have been developed and applied in the characterization of suppressive soils for several soilborne diseases (Mazzola 2002; Weller et al. 2002), and several similar methods can be employed to key out nematode-suppressive soils (Kerry 1988). Nematode-suppressive soils occur globally, but just a limited number of them have been exhibited to be biological in nature (Kerry 1988; Crump 1989).

1.8.1 **Heterodera avenae**-*Suppressive Soils*

In an experiment carried out by Gair and colleagues (1969), *H. avenae* population densities and other plant-parasitic nematodes were followed under cereal monoculture for a number of growing periods, and it was demonstrated that population densities reduced after initial high population densities. Typically, the population densities of nematodes rose initially prior to reducing to low levels. Formaldehyde drenches of soils in which nematode decline had occurred and cropping of a susceptible host resulted in increased nematode reproduction in comparison to non-treated controls (Williams 1969). This prelude observation of lower population densities in natural soil that was not treated resulted in elaborate studies of organisms that contribute to the suppression of nematode and finally to the keying out of *Verticillium chlamydosporium* and *Nematophthora gynophila* as microorganisms mainly responsible for maintaining population densities of nematodes below the destruction threshold (Kerry and Crump 1980). Decrease in population densities in the soil that has not been disturbed after inoculation with several life phases backs the claim for soil suppressiveness. For instance, *H. schachtii*-infested California soil supported just low numbers of the sugar beet cyst nematode under uninterrupted host plants' cropping (Westphal 1998).

1.9 Bacteria-Suppressive Soils

1.9.1 Potato Scab Decline

Common scab is an important potato disease, which is caused by *Streptomyces scabies*, as well as other species of *Streptomyces* (Loria et al. 1997). The pathogenic strains secrete thaxtomins, phytotoxins that stimulate signs of scab when used in tubers with the lack of the pathogen. Thaxtomin nonproducers are nonpathogenic

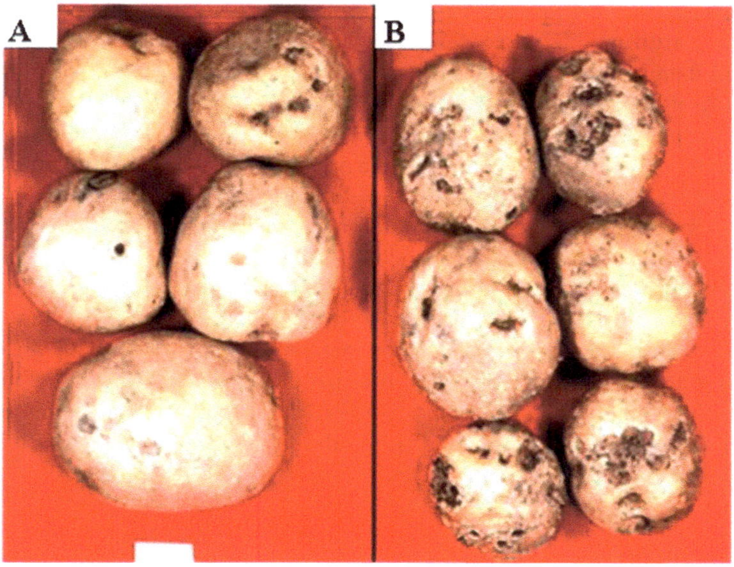

Fig. 1.5 Biological control of potato scab caused by the bacterium *Streptomyces scabies* with a suppressive strain of another *Streptomyces* spp. (**a**) Tubers harvested from soil treated with the biocontrol agent. (**b**) Tubers harvested from soil not amended with the biocontrol agent (*Source*: Agrios 2005)

(Loria et al. 1997). In a field observation, potatoes growing in old irrigated field, which had been used for many years for potato production, were nearly free of potato scab (Menzies 1959). In fields where monoculture production of potato was tried, scab took place uniformly on potatoes from new fields though it was not apparent on potatoes grown in the old fields. Potato scab has gone down with monoculture of potato in other regions that produce potatoes. A diverse *Streptomyces* isolates' collection from scab-free potatoes growing in the suppressive soil secreted antibiotics that were suppressive to *S. scabies* in vitro, and the strains of the pathogen were seen to have less inhibition than the strains that were suppressive against other isolates, regardless of pathogenicity (Liu et al. 1996). The reduction of the potato scab associated with monoculture is, thus, a clear demonstration of suppressive soils, and a rise in production, as well as yields, has been exhibited in infested fields (Fig. 1.5).

1.9.2 Bacterial Wilt-Suppressive Soils

Bacterial wilt is a disease caused by *Ralstonia solanacearum* and it is linked to high yield losses. Soils that are suppressive to bacterial wilt have been depicted. For instance, Islam and Toyota (2004) did a comparison of three soil types, which included chemical fertilized (CF) soil that had been amended, for 14 years, with chemical fertilizers; CF+FYM soil that had been amended, for 14 years, with farmyard manure and chemical fertilizers; and FYM soil that had been amended with farmyard manure for 14 years. This comparison was aimed at evaluating the level of suppressiveness of tomato bacterial wilt by the soil. Over 70 % of tomato plants were shown to have wilt symptom in the CF-FYM and CF soil following 30 cultivation days, while below 10 % of tomato plant wilted in the FYM soil. It was demonstrated that tomato bacterial wilt was inhibited in the poultry and FYM-added soil because of higher activity of microorganisms. It has also been demonstrated that amendments of soil can aid in reduction in the incidence of bacterial wilt and rise in yield. The best results, though, were from the mixture of inorganic and organic fertilizers, when potassium was added with an organic nitrogen source (Lemaga et al. 2001). Another demonstration is that the pig slurry addition significantly reduced the *R. solanacearum* population and decreased numbers of infected as well as diseased plants in the soil suppressiveness tests (Gorissen et al. 2004).

1.10 Biological Control Potential and Soil Suppressiveness

The bacterium *Pasteuria penetrans* has been demonstrated to inhibit populations of root-knot nematode effectively, in field, as well as in microplot trials (Freitas et al. 2000; Weibelzahl-Fulton et al. 1996). The *P. penetrans* role in inhibition of plant-parasitic nematodes has been tried on several crops, largely in greenhouse pots (Chen and Dickson 1998). *P. penetrans* inhibited *Meloidogyne* spp. on tomato, eggplant, tobacco, wheat, soybean, hairy vetch, bean, cucumber, peanut, rye, chicken pea, pepper, kiwi, brinjal, mung, grape, and okra. *Pasteuria* spp. isolates have been shown to inhibit *H. avenae* and *H. zeae* on bermudagrass turf (Giblin-Davis et al. 1990), *H. elachista* on rice, as well as *H. cajani* on cowpea (Singh and Dhawan 1994).

There exist only a few documented reports on soils that are suppressive against plant nematodes with majority regarding fungal antagonist (Gair et al. 1969). In the past few decades, there have been more reports concerning suppressive soils infested with huge numbers of *P. penetrans* (Stirling and Kerry 1983). Baker and Cook (1974) defined soil suppressiveness against soilborne disease as the inhospitability of some soils to a number of plant pathogens in a manner that either the pathogen is not able to establish, it establishes but does not produce disease, or it establishes and produces disease initially and decrease with extended crop culture. Nematode-suppressive soils are widely available, but just a few examples have been exhibited to have a biological nature (Kerry 1988; Crump 1989). Suppressive soils are linked to a susceptible host's monoculture (Gair et al. 1969; Westphal and Becker 1999; Noel and Wax 2003). Specifically soils suppressive against plant-parasitic nematodes are important in the definition of the mechanisms that control population density (Westphal 2005). A number of bacteria and fungi, for instance, some *Fusarium* spp., *Verticillium* spp., as well as *P. penetrans*, have wide range of hosts including both cyst and various root-knot nematodes (Davies et al. 2001).

1.11 Transfer of Suppressiveness

Transferability is a significant feature of biological soil suppressiveness to soilborne plant pathogens (Baker and Cook 1974). Transferability of suppressiveness is a demonstration of specific soil inhibition against plant-parasitic nematodes (Kerry 1988), especially when the nematode antagonists are not known or are not culturable. In diseases that are soilborne, specific suppressiveness can be transferred to conducive soils by the use of small portions of the soil. This observation indicates that suppressiveness is biological in nature (Menzies 1959). Amendments of biocidally treated, disease-conducive substrates or soils with between 1 and 10 % disease-suppressive soils have been demonstrated to transfer suppressiveness to diseases (Andrade et al. 1994; Wiseman et al. 1996). Even though there have been a few intensive studies on nematode-suppressive soils, soil suppressiveness transferability against plant-parasitic nematodes has not been given much attention.

1.12 Effect of Chemical Nematicide on *Pasteuria penetrans* Suppressiveness

The application of *P. penetrans* as a biological control agent together with other management practices, particularly nematicides, is of interest (Freitas et al. 1997). Infection of *M. javanica* by *P. penetrans* following an in vitro treatment was reported to withstand the effects of nematicides DBCP and 1,3-D (Stirling 1984). A synergistic decrease of root galling by *M. javanica* with aldicarb or carbofuran combined with *P. penetrans* has as well been exhibited (Brown and Nordmeyer 1985). This can be attributed to the stimulation of the nematode movement by the low carbamate nematicide concentration, which oriented the nematode toward the host roots. Hence, the possibility of nematode contact with endospores of the bacteria was raised. High concentrations of carbamate nematicides and organophosphates are known to reduce the mobility of nematodes; thereby, the most possible explanation of lowered infection was the reduced probability of contact between endospores and the nematodes.

1.13 Effects of Cropping System and Nematode Density on *Pasteuria penetrans* Suppressiveness

The *P. penetrans* endospore abundance has been demonstrated to be highest in monoculture of peanut and intermediate in two bahiagrass rotations, as well as one rotation of cotton (Timper et al. 2001). While studying the long-term *P. penetrans* persistence and suppressiveness against *M. arenaria* race 1, Cetintas and Dickson (2005) reported that J2 with endospores had the highest percentage in weed fallow (87 %), bahiagrass followed (63 %), and rhizomal peanut (53 %). In a field microplot trial, Oostendorp et al. (1990) demonstrated that the number of plots with infections of *P. penetrans* spore numbers attaching to J2 in the soil were raised continuously for 3 years and were under the influence of the cropping sequence. Differences in *M. arenaria* numbers in plots with no *P. penetrans* among three sequences of cropping were seen only during the spring of every year but not in autumn. This proposed that the summer crop, peanut, strongly influenced the population density of the nematode than the winter cover crops. Similar observations were made by Kariuki et al. (2010) where *P. Penetrans* was transferred from a suppressive field site to microplots located at the University of Florida, Gainesville, and thereafter evaluations done to determine the effect of two summer crops with different cycles.

It has been exhibited that with the introduction of *P. penetrans* into a soil with high *M. arenaria* densities, the bacterium amplifies to suppressive levels in 3 years (Oostendorp et al. 1990) or less if more endospore densities are added (Chen et al. 1996). Peanut can be an ideal crop for use in amplification of *P. penetrans* to suppressive densities since it grows in hot climate and is a comparatively long-season crop. These two conditions prefer *P. penetrans* development (Serracin et al. 1997). The harvesting methods for peanut also aid in the spread of the endospores since it involves digging the plants, drying on the surface of the soil, and then combining of the pods leaving behind the residues of the roots (Dickson and De Waele 2005). In order to sustain soil suppressiveness caused by *P. penetrans*, this bacterium needs some amplification in the soil (Cetintas and Dickson 2005). The downward dispersal of endospores with percolating water could result in depletion of *P. penetrans* endospores from the top 20 to 25 cm of the soil if they are not being continuously amplified in this zone (Cetintas and Dickson 2005). This may require that nematode population densities be maintained at low levels to maintain suppressiveness (Cetintas and Dickson 2005).

1.14 Conclusion

Soil is an essential component for sustainable agricultural production, and it also supports other major human activities. The ability of the soil to become suppressive to plant pathogens is of importance as it contributes to plant health which leads to high agricultural production. This is so because soil suppressiveness is associated with soil fertility and occurrence of beneficial organisms in the soil. Suppressive soils inhibit occurrence of diseases and it also decreases the level of disease severity in plants. Proper crop nutrition is an important aspect in agricultural production as it gives the soil disease-suppressive properties, and therefore crops planted in these soils tend to become either tolerant or more resistant to diseases. The addition of organic amendments in soils is also important as it contributes to disease suppressiveness and improves the plant nutrients. Organic amendments also enhance the plants' biological and physiochemical features which affect the rate of crop infection and also the survival of plant pathogens. Occurrence of beneficial organisms in the soil either naturally or through induction/inoculation also leads to suppressive soils. Most of these organisms colonize the plant roots which makes them resistant to harmful pathogens and/or induces plant growth which decreases the rate of occurrence of plant diseases and decreases the severity rate, and this in return leads to the increase in yields and production of agricultural crops. In general, suppressive soils lead to disease suppression in crops which then leads to increased agricultural productivity.

References

Agrios GN (2005) Plant pathology, 5th edn. Elsevier, New York

Alabouvette C, Lemanceau P, Steinberg C (1993) Recent advances in the biological control of *Fusarium* wilts. Pestic Sci 37(4):365–373

Andrade OA, Mathre DE, Sands DC (1994) Suppression of *Gaeumannomyces graminis* var. *tritici* in Montana soils and its transferability between soils. Soil Biol Biochem 26(3):397–402

Andrivon D (1994) Dynamics of the survival and infectivity to potato tubers of sporangia of *Phytophthora infestans* in three different soils. Soil Biol Biochem 26(8):945–952

Atkinson GF (1892) Some diseases of cotton. Albama Agric Exp Station Bull 41:19–65

Ayres M, Ballard R, Wicks T, Barnett S, Ophel-Keller K (2007) Suppression of soil-borne plant disease using compost. Trial 2(15/11):06

Bailey KL, Lazarovits G (2003) Suppressing soil-borne diseases with residue management and organic amendments. Soil Till Res 72(2):169–180

Baker KF, Cook RJ (1974) Biological control of plant pathogens. Freeman, San Francisco

Berendsen RL, Pieterse CM, Bakker PA (2012) The rhizosphere microbiome and plant health. Trends Plant Sci 17(8):478–486

Brown SM, Nordmeyer D (1985) Synergistic reduction in root galling by *Meloidogyne javanica* with *Pasteuria penetrans* and nematicides. Revue de Nématol 8(3):285–286

Burkett Cadena M, Kokalis Burelle N, Lawrence KS, Van Santen E, Kloepper JW (2008) Suppressiveness of root-knot nematodes mediated by rhizobacteria. Biol Control 47(1):55–59

Campbell RN, Greathead AS (1990) Control of clubroot of crucifers by liming. In: Engelhard AW (ed) Management of diseases with macro- and microelements. American Phytopathology Society, St. Paul, pp 90–101

Carris LM, Glawe DA, Smyth CA, Edwards DI (1989) Fungi associated with populations of *Heterodera glycines* in two Illinois soybean fields. Mycologia 81:66–75

Cavigelli MA, Thien SJ (2003) Phosphorus bioavailability following incorporation of green manure crops. Soil Sci Soc Am J 67(4):1186–1194

Cetintas R, Dickson DW (2005) Distribution and downward movement of *Pasteuria penetrans* in field soil. J Nematol 37(2):55

Chen ZX, Dickson DW (1998) Review of *Pasteuria penetrans*: biology, ecology, and biological control potential. J Nematol 30(3):313

Chen ZX, Dickson DW, McSorley R, Mitchell DJ, Hewlett TE (1996) Suppression of *Meloidogyne arenaria* race 1 by soil application of endospores of *Pasteuria penetrans*. J Nematol 28(2):159

Chen G, Zhu H, Zhang Y (2003) Soil microbial activities and carbon and nitrogen fixation. Res Microbiol 154:393–398

Conklin AE, Erich MS, Liebman M, Lambert DG, Halteman WA (2002) Effects of red clover (*Trifolium pratense*) green manure and compost soil amendments on wild mustard (*Brassica kaber*) growth and incidence of disease. Plant Soil 238(2):245–256

Conn KL, Lazarovits G (1999) Impact of animal manures on verticillium wilt, potato scab, and soil microbial populations. Can J Plant Pathol 21(1):81–92

Cook RJ (1981) The influence of rotation crops on take-all decline phenomenon. Phytopathology 71(2):189–192

Crump DH (1989) Interaction of cyst nematodes with their natural antagonists. Aspects Appl Biol 22:135–140

Davies KG, Fargette M, Balla G, Daudi A, Duponnois R, Gowen SR, Trudgill DL (2001) Cuticle heterogeneity as exhibited by Pasteuria spore attachment is not linked to the phylogeny of parthenogenetic root-knot nematodes (*Meloidogyne* spp.). Parasitology 122(01):111–120

Dickson DW, De Waele D (2005) Nematode parasites of peanut. In: Luc M, Sikora RA, Bridge J (eds) Plant parasitic nematodes in subtropical and tropical agriculture. CABI Publishing, Wallingford, pp 393–436

Foster RE, Walker JC (1947) Predisposition of tomato to *Fusarium* wilt. J Agric Res 74:165–185

Freitas LG, Mitchell DJ, Dickson DW (1997) Temperature effects on the attachment of *Pasteuria penetrans* endospores to *Meloidogyne arenaria* race 1. J Nematol 29(4):547

Freitas LG, Dickson DW, Mitchell DJ, McSorley R (2000) Suppression of *Meloidogyne arenaria* by *Pasteuria penetrans* in the field. Nematol Bras 24(2):147–156

Gair R, Mathias PL, Harvey PN (1969) Studies of cereal nematode populations and cereal yields under continuous or intensive culture. Ann Appl Biol 63(3):503–512

Giblin-Davis RM, McDaniel LL, Bilz FG (1990) Isolates of the *Pasteuria penetrans* group from phytoparasitic nematodes in bermudagrass turf. J Nematol 22(45):750

Gorissen A, Van Overbeek LS, Van Elsas JD (2004) Pig slurry reduces the survival of *Ralstonia solanacearum* biovar 2 in soil. Can J Microbiol 50(8):587–593

Haas D, Défago G (2005) Biological control of soil-borne pathogens by fluorescent pseudomonads. Nat Rev Microbiol 3(4):307–319

Hamid K (2011) Crop rotations for managing soil-borne plant diseases. Afr J Food Sci Technol 2(1):001–009

Hettelingh JP, Slootweg J, Posch M (2008) Critical load, dynamic modeling and impact assessment in Europe: CCE status report 2008. Netherlands Environmental Assessment Agency, The Netherlands

Hoitink HA, Boehm MJ (1999) Biocontrol within the context of soil microbial communities: a substrate-dependent phenomenon. Annu Rev Phytopathol 37(1):427–446

Hornby D (1998) Take-all disease of cereals: a regional perspective. CAB International, Wallingford

Islam TM, Toyota K (2004) Suppression of bacterial wilt of tomato by *Ralstonia solanacearum* by incorporation of composts in soil and possible mechanisms. Microbes Environ 19(1):53–60
Janvier C, Villeneuve F, Alabouvette C, Edel-Hermann V, Mateille T, Steinberg C (2007) Soil health through soil disease suppression: which strategy from descriptors to indicators? Soil Biol Biochem 39(1):1–23
Jones JP, Engelhard AW, Woltz SS (1989) Management of *Fusarium* wilt of vegetables and ornamentals by macro- and microelement nutrition. In: Engelhard AW (ed) Soilborne plant pathogens: management of diseases with macro- and microelements. American Phytopathological Society, St. Paul, p 227
Kao CW, Ko WH (1983) Nature of suppression of *Pythium splendens* in a pasture soil in South Kohala, Hawaii. Phytopathology 73:1284–1289
Kariuki GM, Dickson DW (2007) Transfer and Development of *Pasteuria penetrans*. J Nematol 39(1):55
Kariuki GM, Kimenju JW, Dickson DW (2010) Influence of different crops and nematode densities on multiplication and abundance of a suppressive bacterium (*Pasteuria penetrans*). East Afr Agric Forest J 77(1):21–29
Kaschuk G, Alberton O, Hungria M (2010) Three decades of soil microbial biomass studies in Brazilian ecosystems: lessons learned about soil quality and indications for improving sustainability. Soil Biol Biochem 42:1–13
Kerry BR (1988) Fungal parasites of cyst nematodes. Agric Ecosyst Environ 24(1):93–305
Kerry BR (1990) An assessment of progress toward microbial control of plant-parasitic nematodes. J Nematol 22(45):621
Kerry BR, Crump DH (1980) Two fungi parasitic on females of cyst nematodes (*Heterodera* spp.). Trans Br Mycol Soc 74(1):19–125
Kluepfel DA, McInnis TM, Zehr EI (1993) Involvement of root-colonizing bacteria in peach orchard soils suppressive of the nematode *Criconemella xenoplax*. Phytopathology 83: 1240–1245
Ko WH, Shiroma SS (1989) Distribution of *Phytophthora cinnamomi*-suppressive soil in nature. J Phytopathol 127(1):75–80
Lemaga B, Kanzikwera R, Kakuhenzire R, Hakiza J, Manzi G (2001) The effect of crop rotation on bacterial wilt incidence and potato tuber yield. Afr Crop Sci J 9(1):257–266
Litterick AM, Harrier L, Wallace P, Watson CA, Wood M (2004) The role of uncomposted materials, composts, manures, and compost extracts in reducing pest and disease incidence and severity in sustainable temperate agricultural and horticultural crop production. A review. Crit Rev Plant Sci 23(6):453
Liu D, Anderson NA, Kinkel LL (1996) Selection and characterization of strains of Streptomyces suppressive to the potato scab pathogen. Can J Microbiol 42(5):487–502
Loria R, Bukhalid RA, Fry BA, King RR (1997) Plant pathogenicity in the genus *Streptomyces*. Plant Dis 81(8):836–846
Lucas P, Smiley RW, Collins HP (1993) Decline of rhizoctonia root rot on wheat in soils infested with *Rhizoctonia solani* AG-8. Phytopathology 83:260–265
Lugtenberg B, Kamilova F (2009) Plant-growth-promoting rhizobacteria. Annu Rev Microbiol 63: 541–556
Mankau R (1975) *Bacillus penetrans* n. comb. Causing a virulent disease of plant-parasitic nematodes. J Invertebr Pathol 26(3):333–339
Martin FN (2003) Development of alternative strategies for management of soilborne pathogens currently controlled with methyl bromide. Annu Rev Phytopathol 41(1):325–350
Martin FN, Hancock JG (1986) Association of chemical and biological factors in soils suppressive to *Pythium ultimum*. Phytopathology 76(11):1221–1231
Mazzola M (1998) Elucidation of the microbial complex having a causal role in the development of apple replant disease in Washington. Phytopathology 88:930–938
Mazzola M (2002) Mechanisms of natural soil suppressiveness to soilborne diseases. Antonie van Leeuwenhoek 81(1–4):557–564

McDonald BA, Linde C (2002) The population genetics of plant pathogens and breeding strategies for durable resistance. Euphytica 124(2):163–180

Menzies JD (1959) Occurrence and transfer of a biological factor in soil that suppresses potato scab. Phytopathology 49(10):648–652

Millner PD, Ringer CE, Maas JL (2004) Suppression of strawberry root disease with animal manure composts. Compost Sci Util 12(4):298–307

Murakami H, Tsushima S, Shishido Y (2000) Soil suppressiveness to clubroot disease of Chinese cabbage caused by *Plasmodiophora brassicae*. Soil Biol Biochem 32(11):1637–1642

Noble R, Coventry E (2005) Suppression of soil-borne plant diseases with composts: a review. Biocontrol Sci Technol 15(1):3–20

Noel GR, Wax LM (2003) Population dynamics of *Heterodera glycines* in conventional tillage and no-tillage soybean/corn cropping systems. J Nematol 35(1):104

Obrien-Wray K (1995) Potassium clobbers verticillium wilt. Soybean Digest, 38

Oostendorp M, Dickson DW, Mitchell DJ (1990) Host range and ecology of isolates of *Pasteuria* spp. from the southeastern United States. J Nematol 22(4):525

Ozores-Hampton M, Bryan H, McMillian R Jr (1994) Suppressing disease in field crops. BioCycle 35:60–61

Parke JL, Gurian-Sherman D (2001) Diversity of the *Burkholderia cepacia* complex and implications for risk assessment of biological control strains. Annu Rev Phytopathol 39(1):225–258

Persson L, Larsson-Wikström M, Gerhardson B (1999) Assessment of soil suppressiveness to aphanomyces root rot of pea. Plant Dis 83(12):1108–1112

Rani VD, Sudini H (2013) Management of soilborne diseases in crop plants: an overview. Int J Plant Anim Environ Sci 3(4):156–164

Rovira AD, Wildermuth GB (1981) The nature and mechanism of suppression. In: Asher MJC, Shipton PJ (eds) Biology and control of take all. Academic, New York, pp 385–415

Scheuerell SJ, Sullivan DM, Mahaffee WF (2005) Suppression of seedling damping-off caused by *Pythium ultimum*, *P. irregulare* and *Rhizoctonia solani* in container media amended with a diverse range of Pacific Northwest compost sources. Phytopathology 95(3):306–315

Serracin M, Schuerger AC, Dickson DW, Weingartner DP (1997) Temperature-dependent development of *Pasteuria penetrans* in *Meloidogyne arenaria*. J Nematol 29(2):228

Shipton PJ, Cook RJ, Sitton JW (1973) Occurrence and transfer of a biological factor in soil that suppresses take-all of wheat in eastern Washington. Phytopathology 63(4):511–517

Shurtleff MC, Averre CW (1997) Glossary of plant-pathological terms. APS Press, St. Paul, MN

Simon A, Sivasithamparam K (1989) Pathogen-suppression: a case study in biological suppression of *Gaeumannomyces graminis* var. *tritici* in soil. Soil Biol Biochem 21(3):331–337

Singh KP (2013) Suppressive soils in plant disease management. Centre of Advanced Faculty Training in Plant Pathology

Singh B, Dhawan SC (1994) Effect of *Pasteuria penetrans* on the penetration and multiplication of *Heterodera cajani* in *Vigna unguiculata* roots. Nematol Mediterr 22(2):159–161

Stirling GR (1984) Biological control of *Meloidogyne javanica* with *Bacillus penetrans*. Phytopathology 74(1):55–60

Stirling GR (1991) Biological control of plant parasitic nematodes: progress, problems, and prospects. CAB International, Wallingford

Stirling GR, Kerry BR (1983) Antagonists of the cereal cyst nematode Heterodera avenae Woll. in Australian soils. Anim Prod Sci 23(122):318–324

Stutz EW, Défago G, Kern H (1986) Naturally occurring fluorescent pseudomonads involved in suppression of black root rot of tobacco. Phytopathology 76(2):181–185

Timper P, Minton NA, Johnson AW, Brenneman TB, Culbreath AK, Burton GW, Gascho GJ (2001) Influence of cropping systems on stem rot (*Sclerotium rolfsii*), *Meloidogyne arenaria*, and the nematode antagonist *Pasteuria penetrans* in peanut. Plant Dis 85(7):767–772

Veeken AH, Blok WJ, Curci F, Coenen GC, Termorshuizen AJ, Hamelers HV (2005) Improving quality of composted biowaste to enhance disease suppressiveness of compost-amended, peat-based potting mixes. Soil Biol Biochem 37(11):2131–2140

Weibelzahl-Fulton E, Dickson DW, Whitty EB (1996) Suppression of *Meloidogyne incognita* and *M. javanica* by *Pasteuria penetrans* in field soil. J Nematol 28(1):43
Weller DM, Raaijmakers JM, Gardener BB, Thomashow LS (2002) Microbial populations responsible for specific soil suppressiveness to plant pathogens. Annu Rev Phytopathol 40(1):309–348
Westphal A (1998) Soil suppressiveness against *Heterodera schachtii* in a California field. University of California, Riverside
Westphal A (2005) Detection and description of soils with specific nematode suppressiveness. J Nematol 37(1):121
Westphal A, Becker JO (1999) Biological suppression and natural population decline of *Heterodera schachtii* in a California field. Phytopathology 89(5):434–440
Westphal A, Becker JO (2001) Components of soil suppressiveness against *Heterodera schachtii*. Soil Biol Biochem 33(1):9–16
Williams TD (1969) The effects of formalin, nabam, irrigation and nitrogen on *Heterodera avenae* Woll., *Ophiobolus graminis* Sacc. and the growth of spring wheat. Ann Appl Biol 64(2): 325–334
Wiseman BM, Neate SM, Keller KO, Smith SE (1996) Suppression of *Rhizoctonia solani* anastomosis group 8 in Australia and its biological nature. Soil Biol Biochem 28(6):727–732
Woltz SS, Jones JP (1973) Tomato *Fusarium* wilt control by adjustments in soil fertility. Proc Fla State Hort Soc 86:157–159

Chapter 2
Enhancing Soil Suppressiveness Using Compost and Compost Tea

Chaney C.G. St. Martin

2.1 Introduction

Enhancing soil suppressiveness using compost and compost tea represents an alternative biocontrol approach to the conventional paradigm of plant disease control, one that is based on the use of several microorganisms at the same time to control one or many pathogens rather than the conventional use of one active ingredient or microbial agent to target one or multiple pathogens. Inclusive in this paradigm shift in disease control are (1) mixing of several known types of biocontrol agents (BCAs) with diverse modes of action or that colonise different ecological niches (Siddiqui and Shaukat 2002), (2) the enhancement of resident populations existing on or around the plant (Mazzola 2007), and (3) the introduction of partially or uncharacterised microbial communities usually with no known activity (Litterick et al. 2004). Compost and compost tea used as biocontrol agents fall under the latter group of strategies in this paradigm shift.

Although research on compost and compost tea has been conducted for decades, there is now increasing interest in their possible role in developing suppressive soils and managing plant diseases. This interest has primarily arisen due to increasing demand for organically produced foods (Dimitri and Greene 2000) and concerns by the public over the use and potential negative impacts of synthetic pesticides on human health and environment.

St. Martin (2013) argued that the theoretical basis for the effectiveness of compost and compost tea in suppressing phytopathogens is their ability to alter the microbial profile and activity of the rhizosphere and/or soil as a whole. However, it is highly debatable whether compost tea alters the microbiota of the

C.C.G. St. Martin (✉)
Department of Life Sciences, The University of the West Indies, St. Augustine, Republic of Trinidad and Tobago
e-mail: cstmartin@hotmail.com

M.K. Meghvansi, A. Varma (eds.), *Organic Amendments and Soil Suppressiveness in Plant Disease Management*, Soil Biology 46,
DOI 10.1007/978-3-319-23075-7_2

rhizosphere and/or soil as a whole (Scheuerell and Mahaffee 2006; Larkin 2008). More so, there is no consensus on whether the suppressive effects of compost and compost tea satisfy the demonstration and measurement criteria of classic soil suppressiveness (Baker and Chet 1982). That is, a "natural reduction" in pathogen population levels and plant disease incidence, which is distinct from the decrease that occurs in monoculture of certain susceptible crops, and is often presumed to be long-standing (Hornby 1983).

Nonetheless, several field studies have shown that the compost and liquid preparations such as compost tea made from compost can suppress various phytopathogens and plant diseases (Pera and Filippi 1987; Joshi et al. 2009; Van Schoor et al. 2009; Zaccardelli et al. 2011). This means that at least one member of the microbial community of the soil, i.e. the pathogen, was affected by the application of compost or compost tea. Therefore, the application of compost or compost tea to soils either (1) made conditions more favourable for the development of resident antagonists, in which case the resulting effects can be categorised as induced suppression (Baker and Cook 1974), or (2) did not stimulate resident antagonists but added antagonists to the soil, in which case the resulting positive effects can be categorised as introduced suppression (Hornby 1983). In this context, the positive effects of compost and compost tea satisfy the more inclusive criteria of suppressive soils, that is, soils in which disease severity or incidence remains low, in spite of the presence of a pathogen, a susceptible host plant and climatic conditions favourable for disease development (Baker and Cook 1974). More so, because compost and compost tea have the potential to directly and indirectly affect the physico-chemical and biological properties of soils, they can be viewed as tools, which can be used to enhance or develop disease-suppressive soils (Trankner 1992; Stone et al. 2004). In this regard, the major impediments to the use of compost and compost tea have been the less than desirable and inconsistent levels of plant disease suppression in various cropping systems. Despite the plethora of studies done to date, our understanding of, and research into, compost and compost tea is at an early evolutionary stage, particularly as it relates to predicting disease suppression levels under field conditions.

The objectives of this chapter are to summarise current knowledge on enhancing soil suppressiveness using compost and compost tea. Predictors and mechanisms of disease suppression are discussed and factors affecting the efficacy of compost and compost tea are highlighted. Furthermore, the potential application of molecular tools for better understanding the relationship between microbial properties of compost and compost tea and soil suppressiveness is highlighted, and core areas for research are identified.

2.2 Definitions and Standards

Composting is the controlled, microbial aerobic decomposition and stabilisation of organic substrates, under conditions that allow the generation of high temperatures by thermophilic microbes, to obtain an end product that is stable and free of pathogens and viable weed seeds and can be used in plant culture (St. Martin and Brathwaite 2012). The end product, which is a solid particulate extracted during the maturation and curing phase, is termed compost (Litterick and Wood 2009). Compost tea is defined as filtered products of compost brewed in water (Litterick et al. 2004) and brewing, a steeping process of compost in any solvent (usually water), which lasts for more than one hour (NOSB 2004). Various other definitions have been provided for compost and compost tea in the literature. However, the definitions used in this chapter seem more succinctly technical and representative of attempts made to standardise meanings to facilitate greater clarity of research progress on disease suppression using compost and compost tea. In this light, terms such as compost-water extracts (CWE), which are used in many studies, have been recategorised as either aerated compost tea (ACT) or non-aerated compost tea (NCT) in accordance with definitions presented in the Compost Tea Task Force Report (NOSB 2004). ACTs refer to products where the compost-water extract is actively aerated during the brewing process, and NCTs are products where the compost-water extract is not aerated or receives minimal aeration only at the initial mixing stage of the brewing process (Litterick and Wood 2009). Compost-water extracts are filtered products of compost mixed primarily with water (or any solvent) but not brewed or held for more than one hour before use (Scheuerell and Mahaffee 2002; NOSB 2004). Scheuerell and Mahaffee (2002) and NOSB (2004) can be consulted for a more thorough review of these terms and others and, likewise, St. Martin and Brathwaite (2012) and Scheuerell and Mahaffee (2002) for detailed reviews on compost and compost tea production methods, practices and technologies.

2.3 Suppression of Phytopathogens and Diseases

2.3.1 Soilborne Phytopathogens and Diseases

2.3.1.1 Compost

An increasing body of evidence shows that soils amended with compost can partly or wholly suppress soilborne phytopathogens and plant diseases (Dickerson 1999; Fuchs 2002; Tilston et al. 2005). Most of the research efforts on enhancing soil suppressiveness using compost have focused primarily on root and soilborne pathogens including *Rhizoctonia*, *Pythium*, *Phytophthora* and *Fusarium* spp. For example, Fuchs (2002) found that after 5 years, the receptivity of soils applied annually with 10 tons/ha of compost to *Pythium ultimum* or *Rhizoctonia solani* was lower compared to soil not amended with compost. More so, the suppressive effects of compost were clearly observed 1 year after compost application, particularly in more intensively worked and cultivated fields. Similarly, Tilston et al. (2005) found that soils amended with green waste compost at a rate of 150 Mg ha^{-1} significantly suppressed take-all (*Gaeumannomyces graminis* var. *tritici*). However, residual or cumulative effects of compost application on the disease suppression were not detectable within the duration of the field trials. Escuadra and Amemiya (2008) reported that Fusarium wilt (*Fusarium oxysporum* f. sp. *spinaciae*) suppression in spinach was not evident during the first cultivation. However, notably higher disease suppressiveness was conferred by compost mixes applied before every two croppings compared to those applied only at planting.

Notwithstanding the absence or presence of residual, cumulative or delayed suppressive effects, most studies show that where >50 % disease control was recorded, compost was applied at a rate of at least 100 tons/ha (Coventry et al. 2006; Zaccardelli et al. 2011). Such high application rates exceed the allowable limit of 30 tons/ha for most green composts and 20–30 tons of green or food-derived compost per hectare set for nitrate vulnerable zones (NVZs) (Council Directive 91/676/EEC 1991). Moreover, these rates are also potentially hazardous to the environment, particularly with reference to groundwater and surface water pollution and the conveyance of heavy metals to the soil.

Juxtaposed against the potential environmental hazard of high compost application rates is the issue of the repeatability of disease suppression. This relates to the difficulty in replicating and standardising compost quality across production batches and differences in climate, soil type, crop production practices and/or experimental protocols used in the field. To date, this has been one of the major limitations in recommending compost as an input for enhancing soil suppressiveness in commercial crop production.

In this light, composts have been shown to have neutral and negative effects on phytopathogens and disease suppression. For example, Merriman (1976) found that after 245 days, tomato compost applied to sandy clay loam soil at a rate of 17.5 tons/ha significantly increased the mean number of viable sclerotia of *Sclerotinia sclerotiorum*, the causal agent of white mould. Likewise, Pera and Filippi (1987) reported that poplar bark compost applied to field plots of carnation plants at a rate of 15 % or 30 % (w/w of 20 cm of topsoil) did not suppress Fusarium wilt (*F. oxysporum* f. sp. *dianthi*). Similar results were reported in studies which evaluated various compost types against *Fusarium* blight (*Microdochium nivale*) (Pratt 2003), root rot (Kim et al. 1997; Rangarajan et al. 2001) and cavity rots (Coventry et al. 2005). In contrast, Dickerson (1999) found that sewage sludge compost applied at 48 tons/ha significantly suppressed root rot (*Phytophthora capsici* L.) of chile peppers, whereas rates of 72 tons/ha or higher enhanced the severity of root rot. More complex trends of the effect of compost on soil suppressiveness have been reported. For example, Abbasi et al. (2002) observed that compost showed a significant suppressive effect only in the year with a higher disease level. Similarly, Lodha et al. (2002) reported that the incidence of dry root rot (*M. phaseolina*) of cluster bean differed significantly between years, with disease suppression with compost being greater in the year with higher disease levels.

Currently, fewer direct comparisons are being made between the level of disease suppression achieved through the use of composts and that achieved using standard fungicide treatments (Litterick and Wood 2009). However, data from such comparisons are critical in rationalising the comparative advantages of using compost in combination with or rather than other control methods such as synthetic pesticides. In this regard, Asirifi et al. (1994) found that the application of the fungicide vinclozolin had a significant but lower suppressive efficacy than lucerne hay compost against *Sclerotinia* rot (*S. sclerotiorum*) in lettuce. In contrast, Coventry et al. (2006) reported that onion waste compost was as effective as a standard fungicide treatment (tebuconazole) in reducing onion white rot (*Sclerotinia cepivorum*).

2.3.1.2 Compost Tea

Research on enhancing soil suppressiveness against soilborne diseases using compost tea in open-field system is limited. Even rarer are studies on the residual, cumulative effects and comparative field evaluations of NCT and ACT against soilborne diseases. From this standpoint, the argument for compost tea as an input for enhancing soil suppressiveness is weaker than that of compost, particularly as compost tea has a lower capacity than compost to serve as a substantial carbon or nutrient source for introduced or resident soil microorganisms. Nonetheless, compost teas have been shown to suppress soilborne diseases in various crops and field conditions (Manandhar and Yami 2008; Joshi et al. 2009; Islam et al. 2014). For example, Manandhar and Yami (2008) found that aerated and non-aerated compost and vermicompost teas significantly suppressed foot rot (*F. moniliforme*) in rice. Similar results were reported in studies, which evaluated various compost tea types against bacterial wilt (*Ralstonia solanacearum*) (Islam et al. 2014), stem canker (*Rhizoctonia solani*) (Islam et al. 2013b), apple replant disease (Van Schoor et al. 2009) and dollar spot (*Sclerotinia homoeocarpa*) (Hsiang and Tian 2007). In contrast, Kelloway (2012) reported that the efficacy of the mink compost tea in controlling dollar spot disease was site specific and variable, with only one location showing significant control. In one of the few field studies to investigate the combinatory effects of compost tea and compost, Joshi et al. (2009) found that poultry manure, *Lantana camara* and *Urtica* spp. composts and fermented extracts made using these composts, significantly suppressed root rot (*R. solani*) in French bean (*Phaseolus vulgaris* L.) over two growing seasons. More so, the suppression levels of these treatments were similar with seeds treated with the chemical fungicide, carbendazim. Similarly, Larkin (2008) investigated the relative effects of biological amendments and crop rotations on soilborne diseases and found that soil applied with ACT and the combination of ACT with a mixture of beneficial microorganisms reduced stem canker (*R. solani*) and common scab (*Streptomyces scabiei*) on Irish potato tubers in the 2-year barley/ryegrass but not in the barley/clover rotations.

2.3.2 *Foliar and Fruit Phytopathogens and Diseases*

2.3.2.1 Compost

Field studies on the use of compost to enhance soil suppressiveness against foliar and fruit (aerial) phytopathogens and diseases are limited. However, the majority of published works show that composts suppress foliar diseases under field conditions, mainly by inducing plant defences (Zhang et al. 1996; Stone et al. 2003; Vallad et al. 2003). For example, Stone et al. (2003) found that the amendment of soil with paper mill residue compost (PMRC) at a rate of 78.4 Mg/ha resulted in the suppression of brown spot (*Pseudomonas syringae pv. syringae*) and anthracnose (*Colletotrichum lindemuthianum*) in snap bean and angular leaf spot (*P. syringae pv. lachrymans*) in cucumber. Similarly, Vallad et al. (2003) reported that bacterial speck (*P. syringae pv. tomato*) in tomato was suppressed with the application of PMRC or PMRC + bark composts to the soil at a rate of 78.4 Mg/ha. In contrast, Abbasi et al. (2002) found that the application of yard waste compost to soil at a rate of 12–15 tons/ha did not result in the suppression of anthracnose in tomato. However, applied at 24–30 tons/ha, yard waste compost significantly reduced the severity of anthracnose in tomato. Conversely, Stone et al. (2003) reported that soil amended with PMRC + bark composts at a rate of 38.1 or 78.4 Mg/ha had no effect on the severity of anthracnose or angular leaf spot of cucumber.

2.3.2.2 Compost Tea

Unlike compost, the majority of field studies conducted with compost tea have focused on suppressing aerial phytopathogens and diseases. Though the majority of these field studies show that compost tea can suppress aerial phytopathogens and diseases, the suppressive effect is often attributed to changes in the phyllosphere rather than the rhizosphere. The work done by Islam et al. (2013a) is one of the only published field study, which has evaluated the suppressive effects of compost tea applied as a soil drench against a foliar disease. Islam et al. (2013a) found that compost tea significantly suppressed the severity of late blight (*Phytophthora infestans*) in tomato and potato. They suggested that suppression was associated with the positive effects of compost tea on soil microbial communities as it relates to increasing the diversity and populations of beneficial microorganisms on root surfaces and the activation of plant defence pathways in host plants. Similar claims have been reported in controlled studies; however, further studies with similar objectives are needed to corroborate such findings.

2.4 Predictors of the Suppressive Capacity of Compost and Compost Tea

Although not fully understood, the predictors of the suppressive capacity of compost and compost tea have generally been linked to live microorganisms, since soil suppressiveness against various pathogens has been reduced or lost with the application of sterilised compost or compost tea (Serra-Wittling et al. 1996; Bonanomi et al. 2010). To this end, the presence, population density, diversity, activity, composition and function of microbes in compost and compost tea have been discussed as single or interrelated biological factors associated with the development of disease-suppressive soils.

Pal and Gardener (2006) noted that the microbes that contribute most to disease control are most likely competitive saprophytes and facultative hyperparasites and plant symbionts. Generally, these microbes, which are at low trophic levels, can survive on dead plant matter and are able to colonise and express biological control activities while growing on plant tissues (van Bruggen and Termorskuizen 2003; Pal and Gardener 2006). Avirulent species such as strains of *F. oxysporum* binucleate *Rhizoctonia*-like fungi, which are phylogenetically very similar to phytopathogens, also contribute significantly to disease control. In this light, though other genera are involved, bacteria in the genera *Bacillus*, fluorescent *Pseudomonas, Serratia* and *Streptomyces* and fungi in the genera *Penicillium*, *Trichoderma* and *Gliocladium* are generally regarded as the main microbes responsible for the suppressive effects of compost and compost tea (Phae et al. 1990; Hoitink and Fahy 1986; Litterick et al. 2004). As such, most studies have focused almost exclusively on bacterial and fungal consortia with little focus on specific fungal types such as yeasts or other microbes including protozoa and beneficial nematodes, as live agents responsible for the disease-suppressive effects of compost and compost tea. However, a recent study by St. Martin et al. (2012) highlighted the possible role of yeast present in ACTs in suppressing the growth of *P. ultimum.* Viruses have not generally been considered as agents responsible or related to the disease suppression resulting from compost and compost tea application. However, a study by Heringa et al. (2010), which found that five-strain bacteriophage mixture isolated from sewage effluent and applied to dairy manure compost reduced the population of *Salmonella enterica*, may illustrate the potential role of viruses in disease suppression with compost and compost tea. Though important, Hoitink and Fahy (1986) noted that the mere presence of known or suspected antagonists in the compost or compost tea does not ensure disease suppression.

In this regard, microbial population metrics of compost and compost tea have been evaluated as predictors of disease suppression. However, it is difficult to draw meaningful conclusions from the results of these studies. For example, Craft and Nelson (1996) reported that recoverable microbial populations, particularly of fungi and actinomycetes, were generally higher in suppressive than non-suppressive composts. However, Stockwell et al. (1994) reported that though no clear statistical relationships between bacterial populations and disease suppression were observed in their study, other reports indicate many of the bacteria and actinomycetes recovered from suppressive composts were suppressive to *P. graminicola* in laboratory bioassays. In a similar context, a review paper by Scheuerell and Mahaffee (2002) indicated that disease-suppressive compost teas had total bacterial populations ranging from 10^7 to 10^{10} CFU/ml. In contrast, Pane et al. (2012) found that compost tea with total bacterial count of lower than 10^{-3} CFU/ml inhibited *Alternaria alternata*, *B. cinerea* and *Pyrenochaeta lycopersici*.

In view of these seemingly contrasting findings, St. Martin et al. (2012) suggested that an examination of the population metrics of specific microorganisms rather than total microbial populations or types may prove to be more reliable in rationalising the efficacy between aerated and non-aerated compost teas. Borrero et al. (2004) found that the microbes in composts that were involved in suppression of Fusarium wilt in tomato were cellulolytic and oligotrophic actinomycetes and fungi. They also reported a strong negative correlation between Fusarium wilt severity and the ratios of cellulolytic actinomycetes/cellulolytic bacteria, oligotrophic bacteria/copiotrophic bacteria and oligotrophic actinomycetes/oligotrophic bacteria. To this end, in a meta-analytical review article, Bonanomi et al. (2010) reported that total culturable bacteria, fluorescent pseudomonads and *Trichoderma* populations were most useful in predicting disease suppressiveness of organic soil amendments against soilborne plant diseases. However, the authors cautioned that though total cultural bacteria is an important characteristic, it should not be considered in isolation to be a reliable predictor of disease suppression, either in relation to organic matter types or different pathogen species. With the exception of *Fusarium* spp., total cultural fungi were considered a poor predictor of disease suppression. Bonanomi et al. (2010) also reported that in some cases, the negative effects of composts and crop residues on disease suppression could be explained by the application of partially colonised organic materials that enhance the microbial population but also pathogen saprophytic activity.

Owing to the lack of a significant relationship between the level of pathogen inhibition and the abundance of culturable bacteria or fungi (after 24-h incubation) in ACT, Palmer et al. (2010) concluded that microbial diversity, more than abundance of culturable bacteria and fungi, was a main factor contributing to the suppression of disease by compost tea. Similarly, Nitta (1991) and Postma et al. (2008) all reported positive relationships between microbial diversity of compost and general disease suppression for various pathogens. In contrast, Borrero et al. (2004) reported that higher microbial diversity could not explain the suppression of Fusarium wilt (*F. oxysporum* f. sp. *lycopersici*) of tomato in plant growth media containing or not containing compost. More so, unlike the results of Nitta (1991), Borrero et al. (2004) found that lower diversity was not associated with conduciveness to Fusarium wilt. Though important, Borrero et al. (2004) cautioned that microbial diversity should not be regarded as a reliable predictor of disease suppression unless examined in the context of corresponding microbial activity and biomass.

In this light, Chen et al. (1988) reported a high positive correlation between microbial activity in a compost-amended medium and induction of damping-off

(*P. ultimum*) suppression. Conversely, Erhart et al. (1999) found that microbial activity was positively correlated to damping-off incidence. To this end, Bonanomi et al. (2010) and Chen et al. (1988) concluded that microbial activity is indicative of suppressiveness only when the plant growth substrate itself is not stimulatory to population development of the pathogen. Investigations on the effect of compost tea application on the enzymatic (e.g. microbial activity and substrate respiration) and microbiological (fluorescent pseudomonads and *Trichoderma* populations) properties of soil and their relationship to disease suppression are lacking and therefore needed.

With regard to microbial community and functions, Boehm et al. (1997) concluded that a shift in the microbial community composition from Gram-negative bacteria, which generally have antagonistic ability, to Gram-positive bacteria, which are less able to antagonise soilborne pathogens, reduces the suppressive capacity of compost. McKellar and Nelson (2003) found that bacteria and *Actinobacteria* capable of metabolising fatty acids (linoleic acid) reduced sporangium germination of *P. ultimum*, which resulted in induced suppression of *Pythium* damping off in cotton. Fuchs (2002) noted that the significant negative correlation between more intensively worked and cultivated fields and disease receptivity was likely due to a greater disturbance of the biological equilibrium in these fields compared to fields that were not as intensively worked or cultivated. However, the term "biological equilibrium", which can imply functional relationships among microorganisms, was not clearly defined by Fuchs (2002). It is not uncommon to find the use of such ambiguously defined terms, which implies some microbial functional relational offered as an explanation for the success or failure of disease control using compost or compost tea. This highlights the need for further research on the quantitative relationships between microbial abundance, diversity, functions and disease-suppressive efficacy of compost and compost tea. More so, a better understanding of mechanism of suppression will serve as an important proxy for developing more accurate predictors of the suppressive capacity of compost and compost tea under field conditions.

2.5 Mechanisms of Suppression of Compost and Compost Tea

Six mechanisms of suppression, which are related to the biotic or abiotic characteristics of compost and compost tea, have been identified: (1) competition for carbon and nutrients (such as Fe) by beneficial microorganisms, (2) production of antibiotics or other compounds that is toxic to phytopathogens, (3) hyperparasitism or predation of phytopathogens by lytic bacteria and fungi, (4) activation of disease-resistance genes in plants by the compost and compost tea microflora, (5) improved plant nutrition and vigour due to microbes and (6) physico-chemical properties of compost and compost tea that are directly toxic to phytopathogens, improve nutritional status of crops or induce disease resistance (Hoitink and Boehm 1999; Mehta et al. 2014). According to Hadar and Papadopoulou (2012), the first three mechanisms target the pathogen directly and reduce its survival and capacity to invade the plant, whereas the subsequent two act indirectly via the plant and affect disease progression in the host plant. The last mechanism shows features of both direct and indirect pathogen and disease suppression. Most researchers have explored each mechanism separately. However, it is likely that several mechanisms may be functioning simultaneously in the suppression of diseases. To date, microbiostasis (competition for growth resources and/or antibiosis) and hyperparasitism/predation have been identified as the principal mechanisms by which phytopathogens are suppressed (St. Martin and Brathwaite 2012; Scheuerell and Mahaffee 2002).

2.5.1 Microbiostasis

In the context of soil suppressiveness, microbiostasis refers to the process of inhibiting the growth, reproduction and multiplication of pathogens but not killing them (Ko 1982). It is mainly caused by nutrient deprivation imposed by microbial activity (Ko 1982), i.e. competition, or by antibiosis, which refers to the release of specific and/or non-toxic specific metabolites or antibiotics by one organism that directly suppresses the activity of pathogens (Litterick and Wood 2009). Suppression by microbiostasis seems to be more effective against pathogens with propagules <200 μm diam. including coliforms, *Phytophthora* and *Pythium* spp. (Hoitink and Ramos 2008).

2.5.2 Competition

Chen et al. (1987) noted that disease suppression based on competition could be related to microbial metabolic activities and is controlled by the availability and rate of utilisation of nutrients and energy sources. In this light, Sivan and Chet (1989) reported that some microorganisms reduce the disease incidence by limiting iron availability for pathogens through the production of low-molecular-weight ferric-specific ligands (siderophores) under iron-limiting conditions. Pantelides et al. (2009) reported that the main mechanism of action of the nonpathogenic *F. oxysporum* against *V. dahliae* was the competition for space or nutrients on the root surface of host plants. Likewise, Serra-Wittling et al. (1996) concluded that the suppression of Fusarium wilt was due to microbial nutrient competition, involving the total microflora of the soil and compost.

2.5.3 Antibiosis

Strains of *Bacillus subtilis* and other *Bacillus* spp., *Gliocladium virens*, *Enterobacter*, *Trichoderma harzianum* and *Pseudomonas* spp., which have been found in compost and compost tea, are known to produce antibiotics or enzymes that can inhibit growth germination and multiplication of many phytopathogens (Brinton and Droffner 1995). For example, Vinale et al. (2009) found that harzianic acid, a metabolite produced by a *T. harzianum* strain, displayed antibiotic activity against *Pythium irregulare*, *Sclerotia sclerotiorum* and *R. solani*. Chitinolytic enzymes produced by *Enterobacter* strains were also reported to be antagonistic to *R. solani* (Chernin et al. 1995), as was "gliotoxin" isolated from *Gliocladium virens* to *P. ultimum* (Lumsden et al. 1992).

2.5.4 Hyperparasitism/Predation

Microbial hyperparasitism refers to the phenomena in which pathogens are colonised by specific phylogenetically unrelated microorganisms resulting in lysis or death (Hoitink et al. 1997). In contrast, microbial predation refers to interactions in which pathogens are killed usually through phagocytosis (Matz et al. 2007). Microbial predation is pathogen non-specific, and disease suppression levels are usually less predictable than with microbial hyperparasitism (Pal and Gardener 2006).

In contrast to microbiostasis, hyperparasitism has generally been observed with phytopathogens with propagules of >200 μm diam. and in 20 % of uninoculated composts (Hoitink et al. 1996; Hoitink and Ramos 2008). According to Hoitink et al. (1996), parasitism is affected by the organic matter decomposition level and the presence of glucose and other soluble nutrients, which repress the production and effect of lytic enzymes used to kill pathogens. Hoitink et al. (1997) postulated that a similar relationship between organic matter decomposition levels and the production of antibiotics might exist. For example, in compost consisting of fresh bark, *Trichoderma* spp. including *T. hamatum* and *T. harzianum,* which produce many lytic enzymes, do not directly attack the phytopathogen, *R. solani*. However, as composting progresses, lower concentrations of readily available cellulose and glucose activate the chitinase genes of *Trichoderma* spp., producing chitinase to parasitise *R. solani* (Kwok et al. 1987; Benítez et al. 2004). Conversely, *Penicillium* spp. were the predominant hyperparasites recovered from sclerotia of *Sclerotium rolfsii*, in a high-sugar and low-cellulose-composted grape pomace (Hadar and Gorodecki 1991). It is possible for a pathogen to be hyperparasitised by several fungal species. For example, Kiss (2003) reported that together, *Acremonium alternate*, *Acrodontium crateriforme*, *Ampelomyces quisqualis*, *Cladosporium oxysporum* and *G. virens* have the capacity to parasitise powdery mildew pathogens.

With regard to liquid extract of compost, El-Masry et al. (2002) concluded that the presence of clear inhibition zones between compost-water extracts (CWE) from several composts and pathogenic fungi, the absence of antibiotics or siderophores in CWE and the presence of protease, chitinase, lipase and β-1,3-glucanase (cell wall-degrading enzymes) in the CW indicated a possible role for mycoparasitism. Similarly, Benhamou and Chet (1997) concluded that the marked alteration of the (beta)-1,3-glucan component of the *Pythium* cell wall suggested that (beta)-1,3-glucanases played a key role in the interaction between *T. harzianum* and *P. ultimum*.

2.5.5 Induced Resistance

Plant disease suppression with compost and compost tea through the induction of plant host defences was believed to be a fairly rare and variable phenomenon (Hadar and Papadopoulou 2012). However, it has been shown that this phenomenon is more common than previously thought (Zhang et al. 1998; Khan et al. 2004; Ntougias et al. 2008; Sang et al. 2010). Microbes present in compost and compost tea or extracts have been reported to induce plant host defences in the presence of soilborne and foliar pathogens (Zhang et al. 1998; Wei et al. 1991). Such inductions, which are described as being local and/or systemic in nature, are dependent on the type, source and amount of stimuli (Keen 1990). In this regard, two forms of induced plant resistance have been identified: systemic acquired resistance (SAR) and induced systemic resistance (ISR). In both SAR and ISR, plant defences are preconditioned by prior infection or treatment that results in resistance (or tolerance) against subsequent challenge by a pathogen or parasite (Vallad and Goodman 2004). However, SAR and ISR can be differentiated based on the nature of the elicitor and the regulatory pathways involved. SAR is induced by the exposure of root or foliar tissues to biotic or abiotic elicitors, is associated with the accumulation of pathogenesis-related (PR) proteins and is dependent on the phytohormone salicylate (Vallad and Goodman 2004), whereas ISR is induced by the exposure of roots to specific strains of plant growth-promoting *Rhizobacteria* (PGPR), is not associated with the accumulation of PR proteins and is independent of salicylate but dependent on the phytohormones ethylene and jasmonate (Vallad and Goodman 2004). Moreover, as demonstrated by their reliance on a functional version of the gene NPR1 in *Arabidopsis thaliana*, SAR and ISR are intertwined molecularly (Vallad and Goodman 2004).

Kavroulakis et al. (2006) found that the expression of certain PR genes in the roots of tomato plants grown in suppressive compost increased, even in the absence of any pathogen. They therefore concluded that the expression of PR genes may be triggered by the microflora of the compost or could be associated with abiotic characteristics of the compost. Using the split-root technique, Zhang et al. (1996) found that peroxidase activity, a putative marker of SAR in cucumber, was significantly enhanced in plants grown in the compost-amended mixes. They concluded that the interaction between compost and the pathogen appears to be a critical factor for rapid activation of SAR-associated gene expression in cucumber plants grown in compost mix. Similar findings have been reported for compost tea or extracts and microorganisms isolated from compost. For example, based on the increased concentration of inducible resistance-related compounds including peroxidase, phenol oxidase and phenylalanine ammonia lyase activities, Siddiqui et al. (2008) concluded that induced host resistance was stimulated in okra plants treated with non-sterilised and filter-sterilised compost teas. Likewise, Sang and Kim (2011) attributed the suppressive-effect compost-water extract against anthracnose in cucumber and pepper to a compost mediated ISR property. Hoitink et al. (2006) and Horst et al. (2005) reported that *Trichoderma* spp. isolated from compost triggered system resistance effect in host plants against *Phytophthora* spp. and *Botrytis cinerea*, respectively. *Trichoderma* spp., which are also known for their mycoparasitic and antibiosis effects, are also widely studied their ISR effects (Hoitink et al. 2006; Khan et al. 2004).

2.6 Improved Plant Nutrition and Vigour Due to Microbes

Compost and compost tea have been reported to contain plant growth-promoting *Rhizobacteria* (PGPR) and endophytes, which are known to improve plant growth and vigour (Scheuerell and Mahaffee 2002; Insam et al. 2002; Castaño et al. 2011). As such, even when the composts are not directly suppressive to phytopathogens, plant growth and vigour may be stimulated or induced by increased nutrient uptake. The resulting effect may be plants that are more resistant or tolerant to pathogen attack. Some Gram-negative bacteria species from the genera such as *Pseudomonas*; Gram-positive bacteria species from the genera *Bacillus, Paenibacillus* and actinomycetes; as well as arbuscular mycorrhizal fungi (AMF) species have been reported to be involved in such indirect mechanisms of phytopathogen and disease control. *Pseudomonas fluorescens*, which is the most studied species within the genus *Pseudomonas,* stimulate plant growth by suppressing deleterious rhizosphere microorganisms (Bouizgarne 2013), facilitating nutrient uptake from soil (De Weger et al. 1986) or by producing plant growth-promoting substances (Ryu et al. 2005). In contrast, species of *Paenibacillus* have been shown to induce plant growth by fixing atmospheric nitrogen (von der Weid et al. 2002) and producing auxins (Da Mota et al. 2008) and cytokinin (Timmusk et al. 1999). Moreover, certain *Bacillus,* actinomycetes and AMF species are reported to stimulate plant growth by increasing the uptake of soluble phosphorus (El-Tarabily 2008; Deepa et al. 2010).

Microorganisms including AMF and strains of *Pythium oligandrum* have also been shown to induce anatomical and morphological changes in root systems (Pharand et al. 2002; Atkinson et al. 1994), alter rhizosphere profiles (Meyer and Linderman 1986) and increase host tolerance to pathogen attack by compensating for the loss of root biomass or function caused by pathogens (Cordier et al. 1996).

However, the significance of these findings as it relates to plant protection or a mechanism of biocontrol has not yet been sufficiently considered or evaluated. Nonetheless, mature compost has been inoculated with some of these microbial species including *T. hamatum*, *Chryseobacterium gleum* and *B. subtilis* and non-pathogenic strains of *F. oxysporum* to improve disease-suppressive efficacy. Generally, results show small but significant increases in the suppressive effect of mature compost inoculated with suspected or known BCA or beneficial microorganisms (Coventry et al. 2006; Ryckeboer 2001). The effectiveness of microbial inocula to improve the suppressive effects of compost is dependent on the capacity of the substrate to support microbial growth and activity (Cotxarrera et al. 2002; Dukare et al. 2011; Hoitink and Fahy 1986).

2.7 Physico-chemical Properties of Compost and Compost Tea

While important, microbiological properties per se do not fully explain the capacity of compost and compost tea to enhance soil suppressiveness. Physico-chemical properties of compost and compost tea may protect plants against various diseases through direct toxicity, improved nutritional status or SAR. For example, Spencer and Benson (1982) and Hoitink and Fahy (1986) found that the ability of compost to suppress diseases caused by pathogens, to which free water is important for asexual multiplication, was dependent on the ability of compost to raise the air capacity of a substrate above 15 %. Cronin et al. (1996) and Sang et al. (2010) concluded that the suppressive effects of fermented compost extracts were not biological in nature since sterilising or micron filtering extracts did not significantly affect the results. They both suggested that suppression was likely due to presence or activity of heat-stable chemical compounds. However, without the identification of these specific heat-stable chemical compounds, and the use of molecular tools to elucidate the community structure and functional role of microbes in compost extracts, it is unclear whether this heat-stable chemical factor was produced by microorganisms.

Nonetheless, disease-suppressive effects have been attributed to organic and inorganic compounds present in compost or compost tea or released by microorganisms inhabiting these inputs. Humic, phenolics, bioactive compounds and volatile fatty acids (VFAs) have often been suggested as organic compounds, which play an important role in disease suppression with compost and compost tea. For example, Pascual et al. (2002) found that compost and its humic fractions significantly reduced *P. ultimum* populations in soil and the number of root lesions on pea plants. Tenuta et al. (2002) demonstrated that under acidic conditions (pH 4.75) non-ionised forms of VFAs from liquid swine manure were toxic to microsclerotia of *Verticillium dahliae* Kleb., the causal agent of *Verticillium* wilt in potato. However, the mechanism by which VFAs are toxic to *V. dahliae* is unknown.

Nonmicrobial inorganic compounds, such as aluminium and nitrogen from N-rich organic matter decomposition, can also affect pathogens (Fichtner et al. 2004; Lazarovits et al. 2005). Fichtner et al. (2004) reported an aluminium-mediated suppression of *Phytophthora parasitica* in a potting medium containing 20 % composted swine waste. Fichtner et al. (2004) noted that both abiotic and biotic suppression may have occurred, but at different times. They therefore concluded that aluminium amendments may be effective at protecting the plant before beneficial microbial populations reach a threshold necessary for suppression, if exchangeable aluminium levels of the medium are >2 μM Al g^{-1}. High nitrogen levels and high ammonium-to-nitrogen ratios have been reported to enhance Fusarium wilt incidence, and severity has been reported by several researchers (Woltz and Jones 1981; Hoitink et al. 1987; Borrero et al. 2012). The suppressive capacity of compost and compost tea is also affected by the pH and electrical conductivity of these inputs and of the soil (Spencer and Benson 1981; Jones et al. 1991; Hoitink et al. 1996; Cotxarrera et al. 2002). Hoitink et al. (1996) reported that highly saline compost (>10 dS/m) enhanced *Pythium* and *Phytophthora* diseases unless they are applied months ahead of planting to allow for leaching. In contrast, Pane et al. (2011) found a negative correlation between salinity of compost-amended substrates and damping off (*Sclerotinia minor*) in *Lepidium sativum*. MacDonald (1982) and Al-Sadi et al. (2010) reported that high salinity levels do not inhibit mycelial growth of *P. ultimum* but negatively affects plants, making them more susceptible to attack by the pathogen. Hoitink et al. (1996) noted that the pH of compost affects its potential to be colonised by beneficial bacteria. At pH values of <5.0, the growth, reproduction and multiplication of bacterial biocontrol agents are generally inhibited (Hoitink et al. 1991). Compost pH also affects the availability of macro-and micronutrients for plant uptake or pathogen use, which in turn affects disease incidence and severity. A classic example of this is the use of the pH of plant growth substrates as a chemical environmental index for Fusarium wilt in tomato (Woltz and Jones 1981; Jones et al. 1991). The unavailability of micronutrients such as Cu, Fe and Zn, at substrate pH values of ≥7.5, can limit growth, sporulation and pathogenicity of *F. oxysporum* (Jones et al. 1993). Furthermore, the low availability of Fe can induce siderophore production and microbial competition for Fe (Alabouvette 1999). Fusarium wilt severity tends to be higher at substrate pH values of 5–7, which are most favourable for the growth and survival of pathogenic *Fusarium* species (Oritsejafor 1986). In contrast, Blaker and MacDonald (1983) showed that low pH (≤4.5) reduced sporangium formation, zoospore release and motility of *Phytophthora cinnamomi*, a causal agent of root rot and dieback in many plant species. As such, pine bark compost with pH values of 4.4–4.5 has been used as a substrate or substrate component to suppress *Phytophthora*, *Rhizoctonia* and *Pythium* root rot diseases (Spring et al. 1980; Nelson et al. 1983).

Plant-based composts are suspected to contain compounds that mimic chemical signals from the root or shoot exudates of host plants (Mehta et al. 2014). These chemical signals are suspected to be important in host identification and triggering germination of pathogens before they come in contact with host plants

(Chen et al. 1988). Mehta et al. (2014) noted that the mimicry of these chemical signals may explain why Yogev et al. (2006) found that pathogen activity and proliferation is reduced with plant-based compost, even in the absence of a host. Mehta et al. (2014) termed this type of suppression as "ineffective pathogen proliferation" and categorised it as separate from more conventional mechanisms or factors. Further research is needed to clearly demonstrate this type of suppression and to support the need to distinguish it from more established mechanisms. This is particularly important since generally, a pathogen propagule does not significantly proliferate in the absence of a host (Lockwood 1990).

To this end, the six biological control mechanisms of compost and compost tea have been classified into two broad classes: "general" and "specific" suppression (Cook and Baker 1983). General refers to suppression that can be attributed to the activities of many different types of microorganisms, which result in the generation of a hostile environment for the development many pathogens or diseases (Hoitink and Boehm 1999). General suppression is linked to both abiotic and biotic substrate characteristics (Baker and Cook 1974). In contrast, specific suppression is attributed to the presence and/or activity of one or a few microorganisms. To this end, competition and production of antibiotics are mostly involved in general suppression effects, whereas predation, parasitism and activation of disease resistance are more often manifested by specific microorganisms (Hadar and Papadopoulou 2012; Cook and Baker 1983). It is likely that two broad classes of suppression are not mutually exclusive. In fact, Bonilla et al. (2012) hypothesised that in most cases, suppressive soils owe their activity to a combination of general and specific suppression. However, most researchers have concluded that compost and compost tea suppress phytopathogens and diseases through general rather than specific mechanisms. Unfortunately, the disease-suppressive effects resulting from general mechanisms are not easily transferable from one medium to another.

Regardless of the mechanisms, i.e. general or specific, the degree of disease suppression observed with the application of compost or compost tea to soils can vary greatly or be short lived. The duration of suppressiveness and degree of efficacy of compost and compost tea depend on many production, application and soil factors including the feedstock types composted; the composting or compost tea brewing process; the use of nutrient or microbial amendments; the rate, time and frequency of application; and the physical, chemical and biological characteristics of the soil. Some of these factors, e.g. organic matter decomposition level and compost maturity, were highlighted or briefly discussed in the previous subsections. For more comprehensive and detailed discussions of these factors, the works of St. Martin (2014), St. Martin and Ramsubhag (2014) and Scheuerell and Mahaffee (2002) can be consulted.

2.8 Conclusions and Future Outlook

Despite their limited use, compost and compost tea have much potential as tools for enhancing the suppressive capacity of soils. These potentially low-cost and environmentally benign alternatives to chemical fungicides have been shown to suppress many fungal and bacterial pathogens through similar mechanisms and processes ascribed to naturally suppressive soils. However, less than desirable and inconsistent levels of disease suppression achieved with compost and compost tea effectively limited their use as tools for enhancing soil suppressiveness in conventional cropping systems. Conversely, the use of compost and compost tea as tools for enhancing soil suppressiveness in organic crop production system is deemed important to producers who have limited disease control options (Mahaffee and Scheuerell 2006; St. Martin and Brathwaite 2012). This widespread use limitation appears to be related to the complexity and dynamisms of microbial ecological processes involved in the production and application of compost and compost tea. As such, it appears that increasing our understanding of the microbial ecology of the compost-soil-plant interactions may assist in improving the consistency and efficacy of disease suppression for particular compost types, pathogens, crops and soil and environmental conditions. To achieve this, a systems biology approach, which addresses the complex, simultaneous and dynamic interactions of variable communities that affect plant health, is needed (Lazarovits 2014). However, thus far, there have been limited applications of these techniques in the study aimed at enhancing soil suppressiveness using composts and compost tea.

St. Martin (2014), Mehta et al. (2014) and Mazzola (2004) reviewed the potential use of molecular-based methods, including polymerase chain reaction (PCR) combined with techniques such as DNA sequencing, denaturing gradient gel electrophoresis (DGGE), temperature gradient gel electrophoresis (TGGE) and terminal restriction fragment length polymorphism (T-RFLP); real-time PCR; DNA arrays; and metagenomic libraries for assessing community structure and function as it relates to disease suppression with compost or compost tea.

Pang et al. (2009) demonstrated that metagenomic analysis could be used to identify and characterise a novel endoglucanase enzyme from compost soils. Kim et al. (2010) used metagenomic libraries to characterise a novel family VIII alkaline esterase from a pig manure-mushroom waste compost. It is expected that much more studies using the "omics" technologies such as metagenomics, metatranscriptomics and metabolomics and next-generation sequencing will be completed in the near future. Such studies will provide a tremendous opportunity for elucidating the microbial and metabolic dynamics associated with suppressiveness of compost. In turn, this will provide the basis for developing or optimising production and application protocols for consistently suppressive compost and compost tea. As with microbial biocontrol in general, research and developmental work on compost and compost tea should focus on the ecology of plant-associated microbes, the application of antagonistic microbial strains/inoculant strategies, discovering novel strains and mechanisms of action, and practical integration of these findings into agricultural systems remain (Pal and Gardener 2006). More specifically, the long-term residual and cumulative effects of compost and compost tea soil amendments on general and specific suppression should be investigated. This research should be done on various soil types using low compost application rates.

Despite such research, inducing general or specific suppression in soils using compost or compost tea might not be sufficient or possible to achieve commercially viable disease control in many disease and cropping systems (Stone et al. 2004). In such cases, other strategies or combinations of strategies such as the use of crop rotation, cover and rotation crops, tillage and inputs including plant genetic resources and amendments will be necessary.

References

Abbasi P, Al-Dahmani J, Sahin F, Hoitink H, Miller S (2002) Effect of compost amendments on disease severity and yield of tomato in conventional and organic production systems. Plant Dis 86(2):156–161

Alabouvette C (1999) Fusarium wilt suppressive soils: an example of disease-suppressive soils. Australas Plant Pathol 28(1):57–64

Al-Sadi AM, Al-Habsi N, Al-Kiyumi K, Al-Said FA, Al-Rawahy SA, Ahmed M, Deadman ML, Hussain N (2010) Effect of salinity on growth, reproduction and pectolytic enzyme production by Pythium aphanidermatum: a serious soil borne pathogen of vegetable crops in Oman. In: Ahmed M, Al-Rawahy SA, Hussain N (eds) A monograph on management of salt-affected soils and water for sustainable agriculture. Sultan Qaboos University, Muscat, pp 95–98

Asirifi K, Morgan WC, Parbery D (1994) Suppression of sclerotinia soft rot of lettuce with organic soil amendments. Anim Prod Sci 34(1):131–136

Atkinson D, Berta G, Hooker JE (1994) Impact of mycorrhizal colonisation on root architecture, root longevity and the formation of growth regulators. In: Gianinazzi S, Schuepp H (eds) Impact of arbuscular mycorrhizas on sustainable agriculture and natural ecosystems. Springer, Berlin, pp 89–99

Baker R, Chet I (1982) Induction of suppressiveness. In: Schneider RW (ed) Suppressive soils and plant diseases. American Phytopathological Society, St. Paul, MN, pp 35–50

Baker K, Cook RJ (1974) Biological control of plant pathogens. WH Freeman and Company, San Francisco, CA

Benhamou N, Chet I (1997) Cellular and molecular mechanisms involved in the interaction between *Trichoderma harzianum* and *Pythium ultimum*. Appl Environ Microbiol 63 (5):2095–2099

Benítez T, Ricón AM, Limón MC, Codón AC (2004) Biocontrol mechanisms of *Trichoderma* strains. Int Microbiol 7:249–260

Blaker N, MacDonald J (1983) Influence of container medium pH on sporangium formation, zoospore release, and infection of rhododendron by *Phytophthora cinnamomi*. Plant Dis 67 (3):259–263

Boehm MJ, Wu T, Stone AG, Kraakman B, Iannotti DA, Wilson GE, Madden LV, Hoitink H (1997) Cross-polarized magic-angle spinning (sup13) c nuclear magnetic resonance spectroscopic characterization of soil organic matter relative to culturable bacterial species composition and sustained biological control of pythium root rot. Appl Environ Microbiol 63 (1):162–168

Bonanomi G, Antignani V, Capodilupo M, Scala F (2010) Identifying the characteristics of organic soil amendments that suppress soilborne plant diseases. Soil Biol Biochem 42 (2):136–144

Bonilla N, Gutiérrez-Barranquero JA, Vicente AD, Cazorla FM (2012) Enhancing soil quality and plant health through suppressive organic amendments. Diversity 4(4):475–491

Borrero C, Trillas MI, Ordovás J, Tello JC, Avilés M (2004) Predictive factors for the suppression of fusarium wilt of tomato in plant growth media. Phytopathology 94(10):1094–1101

Borrero C, Trillas M, Delgado A, Avilés M (2012) Effect of ammonium/nitrate ratio in nutrient solution on control of fusarium wilt of tomato by Trichoderma asperellum T34. Plant Pathol 61 (1):132–139

Bouizgarne B (2013) Bacteria for plant growth promotion and disease management. In: Maheshwari DK (ed) Bacteria in agrobiology: disease management. Springer, Berlin, pp 15–47

Brinton WF, Droffner M (1995) The control of plant pathogenic fungi by use of compost teas. Biodynamics 197:12–15

Castaño R, Borrero C, Avilés M (2011) Organic matter fractions by SP-MAS 13C NMR and microbial communities involved in the suppression of fusarium wilt in organic growth media. Biol Control 58(3):286–293

Chen W, Hoitink H, Schmitthenner A (1987) Factors affecting suppression of pythium damping-off in container media amended with composts. Phytopathology 77(5):755–760

Chen W, Hoitink H, Madden L (1988) Microbial activity and biomass in container media for predicting suppressiveness to damping-off caused by *Pythium ultimum*. Phytopathology 78 (11):1447–1450

Chernin L, Ismailov Z, Haran S, Chet I (1995) Chitinolytic *Enterobacter agglomerans* antagonistic to fungal plant pathogens. Appl Environ Microbiol 61(5):1720–1726

Cook RJ, Baker KF (1983) The nature and practice of biological control of plant pathogens. American Phytopathological Society, St. Paul, MN

Cordier C, Gianinazzi S, Gianinazzi-Pearson V (1996) Colonisation patterns of root tissues by *Phytophthora nicotianae* var. *parasitica* related to reduced disease in mycorrhizal tomato. Plant Soil 185(2):223–232

Cotxarrera L, Trillas-Gay M, Steinberg C, Alabouvette C (2002) Use of sewage sludge compost and *Trichoderma asperellum* isolates to suppress fusarium wilt of tomato. Soil Biol Biochem 34(4):467–476

Council Directive 91/676/EEC (1991) Protection of waters against pollution caused by nitrates from agricultural sources. Off J L 375:1–8

Coventry E, Noble R, Whipps JM (2005) Recycling horticultural wastes to produce pathogen suppressant composts for sustainable vegetable crop production. Final Report EU Project QLRT-2000-01458 RECOVEG

Coventry E, Noble R, Mead A, Marin F, Perez J, Whipps J (2006) Allium white rot suppression with composts and *Trichoderma viride* in relation to sclerotia viability. Phytopathology 96 (9):1009–1020

Craft CM, Nelson EB (1996) Microbial properties of composts that suppress damping-off and root rot of creeping bentgrass caused by *Pythium graminicola*. Appl Environ Microbiol 62 (5):1550–1557

Cronin M, Yohalem D, Harris R, Andrews J (1996) Putative mechanism and dynamics of inhibition of the apple scab pathogen *Venturia inaequalis* by compost extracts. Soil Biol Biochem 28(9):1241–1249

Da Mota FF, Gomes EA, Seldin L (2008) Auxin production and detection of the gene coding for the auxin efflux carrier (AEC) protein in Paenibacillus polymyxa. J Microbiol 46(3):257–264

De Weger L, Van Boxtel R, Van der Burg B, Gruters R, Geels F, Schippers B, Lugtenberg B (1986) Siderophores and outer membrane proteins of antagonistic, plant-growth-stimulating, root-colonizing *Pseudomonas* spp. J Bacteriol 165(2):585–594

Deepa C, Dastager SG, Pandey A (2010) Plant growth-promoting activity in newly isolated *Bacillus thioparus* (NII-0902) from Western ghat forest, India. World J Microbiol Biotechnol 26(12):2277–2283

Dickerson GW (1999) Damping-off and root rot. BioCycle 40:62–63

Dimitri C, Greene C (2000) Recent growth patterns in the US organic foods market. Agriculture Information Bulletin, vol 777. USDA Economic Research Service, Washington, DC

Dukare AS, Prasanna R, Chandra Dubey S, Nain L, Chaudhary V, Singh R, Saxena AK (2011) Evaluating novel microbe amended composts as biocontrol agents in tomato. Crop Prot 30 (4):436–442

El-Masry M, Khalil A, Hassouna M, Ibrahim H (2002) In situ and in vitro suppressive effect of agricultural composts and their water extracts on some phytopathogenic fungi. World J Microbiol Biotechnol 18(6):551–558

El-Tarabily KA (2008) Promotion of tomato (*Lycopersicon esculentum* Mill.) plant growth by rhizosphere competent 1-aminocyclopropane-1-carboxylic acid deaminase-producing streptomycete actinomycetes. Plant Soil 308(1–2):161–174

Erhart E, Burian K, Hartl W, Stich K (1999) Suppression of *Pythium ultimum* by biowaste composts in relation to compost microbial biomass, activity and content of phenolic compounds. J Phytopathol 147(5):299–305

Escuadra GME, Amemiya Y (2008) Suppression of Fusarium wilt of spinach with compost amendments. J Gen Plant Pathol 74(4):267–274

Fichtner E, Benson D, Diab H, Shew H (2004) Abiotic and biological suppression of *Phytophthora parasitica* in a horticultural medium containing composted swine waste. Phytopathology 94 (7):780–788

Fuchs J (2002) Practical use of quality compost for plant health and vitality improvement. In: Insam H, Riddech N, Klammer S (eds) Microbiology of composting. Springer, Heidelberg, pp 435–444

Hadar Y, Gorodecki B (1991) Suppression of germination of sclerotia of *Sclerotium rolfsii* in compost. Soil Biol Biochem 23(3):303

Hadar Y, Papadopoulou KK (2012) Suppressive composts: microbial ecology links between abiotic environments and healthy plants. Annu Rev Phytopathol 50:133–153

Heringa SD, Kim J, Jiang X, Doyle M, Erickson M (2010) Use of a mixture of bacteriophages for biological control of *Salmonella enterica* strains in compost. Appl Environ Microbiol 76 (15):5327–5332

Hoitink H, Boehm M (1999) Biocontrol within the context of soil microbial communities: a substrate-dependent phenomenon. Annu Rev Phytopathol 37(1):427–446

Hoitink HAJ, Fahy PC (1986) Basis for the control of soilborne plant pathogens with composts. Annu Rev Phytopathol 24(1):93–114

Hoitink H, Ramos L (2008) Impacts of composts on soil and plant health. Ohio State University. http://plantpath.osu.edu/people-and-programs/faculty-directory/emeritus/hoitink-harry-a-j/hoitink-ramos-acorbat2008-3.doc. Accessed 26 June 2011

Hoitink H, Daughtrey M, Tayama H (1987) Control of cyclamen fusarium wilt—a preliminary report. Ohio Florists Assoc Bull 693:1–3

Hoitink H, Inbar Y, Boehm M (1991) Status of compost-amended potting mixes naturally suppressive to soilborne diseases of floricultural crops. Plant Dis 75(9):869

Hoitink H, Stone A, Grebus M (1996) Suppression of plant disease by composts. In: De Bertoldi M (ed) The science of composting, vol 4. Chapman and Hall, London, pp 373–381

Hoitink H, Stone A, Han D (1997) Suppression of plant diseases by composts. HortSci 32 (2):184–187

Hoitink HA, Madden LV, Dorrance AE (2006) Systemic resistance induced by *Trichoderma* spp.: interactions between the host, the pathogen, the biocontrol agent, and soil organic matter quality. Phytopathology 96(2):186–189

Hornby D (1983) Suppressive soils. Annu Rev Phytopathol 21(1):65–85

Horst LE, Locke J, Krause CR, McMahon RW, Madden LV, Hoitink HAJ (2005) Suppression of *Botrytis* Blight of *Begonia* by *Trichoderma hamatum* 382 in peat and compost-amended potting mixes. Plant Dis 89(11):1195–1200. doi:10.1094/pd-89-1195

Hsiang T, Tian L (2007) Compost tea for control of dollar spot. Annual research report 2007, University of Guelph Ontario

Insam H, Riddech N, Klammer S (2002) Microbiology of composting. Springer, Heidelberg
Islam MR, Mondal C, Hossain I, Meah MB (2013a) Organic management: an alternative to control late blight of potato and tomato caused by *Phytophthora infestans*. Int J Theor Appl Sci 5 (2):32–42
Islam R, Mondal C, Hossain I, Meah MB (2013b) Compost tea and poultry litter extract: alternative organic management approaches for stem canker of potato caused by *Rhizoctonia solani*. J Agric Sci 5(10):261–272
Islam M, Mondal C, Hossain I, Meah M (2014) Compost tea as soil drench: an alternative approach to control bacterial wilt in brinjal. Arch Phytopathol Plant Prot 47(12):1475–1488
Jones JB, Jones JP, Stall RE, Zitter TA (1991) Compendium of tomato diseases. American Phytopathological Society, St. Paul, MN
Jones JP, Engelhard AW, Woltz S (1993) Management of Fusarium wilt of vegetables and ornamentals by macro- and micro-element nutrition. In: Engelhard WA (ed) Soilborne plant pathogens: management of diseases with macro- and microelements. The American Phytopathological Society, St. Paul, MN, pp 18–32
Joshi D, Hooda K, Bhatt J, Mina B, Gupta H (2009) Suppressive effects of composts on soil-borne and foliar diseases of French bean in the field in the western Indian Himalayas. Crop Prot 28 (7):608–615
Kavroulakis N, Papadopoulou KK, Ntougias S, Zervakis GI, Ehaliotis C (2006) Cytological and other aspects of pathogenesis-related gene expression in tomato plants grown on a suppressive compost. Ann Bot 98(3):555–564
Keen N (1990) Gene-for-gene complementarity in plant-pathogen interactions. Annu Rev Genet 24(1):447–463
Kelloway S (2012) Compost tea for the management of dollar spot (*Sclerotinia homoeocarpa*) on turfgrass. Dalhousie University, Halifax, NS
Khan J, Ooka J, Miller S, Madden L, Hoitink H (2004) Systemic resistance induced by *Trichoderma hamatum* 382 in cucumber against phytophthora crown rot and leaf blight. Plant Dis 88(3):280–286
Kim K, Nemec S, Musson G (1997) Effects of composts and soil amendments on soil microflora and phytophthora root and crown rot of bell pepper. Crop Prot 16(2):165–172
Kim YH, Kwon EJ, Kim SK, Jeong YS, Kim J, Yun HD, Kim H (2010) Molecular cloning and characterization of a novel family VIII alkaline esterase from a compost metagenomic library. Biochem Biophys Res Commun 393(1):45–49
Kiss L (2003) A review of fungal antagonists of powdery mildews and their potential as biocontrol agents. Pest Manag Sci 59(4):475–483
Ko W (1982) Effects of nutritional factors on chemical and soil microbiostasis. Research series, vol 012. University of Hawaii, Hilo, Hawaii
Kwok O, Fahy P, Hoitink HAJ, Kuter G (1987) Interactions between bacteria and *Trichoderma hamatum* in suppression of rhizoctonia damping-off in bark compost media. Phytopathology 77(8):1206–1212
Larkin RP (2008) Relative effects of biological amendments and crop rotations on soil microbial communities and soilborne diseases of potato. Soil Biol Biochem 40(6):1341–1351
Lazarovits G (2014) Discovering the value of compost. http://www.compost.org/English/PDF/free-webinars/CCC_Webinar_Feb_5_2014.pdf. Accessed 19 Oct 2014
Lazarovits G, Conn K, Abbasi P, Tenuta M (2005) Understanding the mode of action of organic soil amendments provides the way for improved management of soilborne plant pathogens. Acta Hortic 698:215–224
Litterick A, Wood M (2009) The use of composts and compost extracts in plant disease control. In: Walters D (ed) Disease control in crops: biological and environmentally friendly approaches. Wiley-Blackwell, Oxford, pp 93–121
Litterick AM, Harrier L, Wallace P, Watson CA, Wood M (2004) The role of uncomposted materials, composts, manures, and compost extracts in reducing pest and disease incidence and severity in sustainable temperate agricultural and horticultural crop production—a review. Crit Rev Plant Sci 23(6):453–479

Lockwood JL (1990) Relation of energy stress to behaviour of soil-borne plant pathogens and to disease development. In: Hornby D (ed) Biological control of soil-borne plant pathogens. CAB International, Wallingford, pp 197–214
Lodha S, Sharma SK, Aggarwal RK (2002) Inactivation of *Macrophomina phaseolina* propagules during composting and effect of composts on dry root rot severity and on seed yield of clusterbean. Eur J Plant Pathol 108(3):253–261
Lumsden R, Locke J, Adkins S, Walter J, Ridout C (1992) Isolation and localization of the antibiotics gliotoxin produced by *Gliocladium virens* from alginate prill in soil and soilless media. Phytopathology 82(2):230–235
MacDonald J (1982) Effect of salinity stress on the development of phytophthora root rot of *Chrysanthemum*. Phytopathology 72(2):214–219
Mahaffee W, Scheuerell S (2006) Compost teas: alternative approaches to the biological control of plant diseases. In: Bailey M, Lilley A, Timms-Wilson T, Spencer-Phillips P (eds) Microbial ecology of aerial plant surfaces. CAB International, London, pp 165–179
Manandhar T, Yami K (2008) Biological control of foot rot disease of rice using fermented products of compost and vermicompost. Sci World 6(6):52–57
St. Martin CCG (2014) Potential of compost tea for suppressing plant diseases. CAB Rev 9 (32):1–12
St. Martin CCG, Brathwaite RAI (2012) Compost and compost tea: principles and prospects as substrates and soil-borne disease management strategies in soil-less vegetable production. Biol Agric Hortic 28(1):1–33
St. Martin CCG, Dorinvil W, Brathwaite RAI, Ramsubhag A (2012) Effects and relationships of compost type, aeration and brewing time on compost tea properties, efficacy against Pythium ultimum, phytotoxicity and potential as a nutrient amendment for seedling production. Biol Agric Hortic 28(3):185–205
Matz C, Kjelleberg S, Givskov M (2007) Biofilms as refuge against predation. In: Kjelleberg S, Givskov M (eds) The biofilm mode of life: mechanisms and adaptations. Horizon Bioscience, Norwich, pp 195–213
Mazzola M (2004) Assessment and management of soil microbial community structure for disease suppression. Annu Rev Phytopathol 42:35–59. doi:10.1146/annurev.phyto.42.040803.140408
Mazzola M (2007) Manipulation of rhizosphere bacterial communities to induce suppressive soils. J Nematol 39(3):213–220
McKellar ME, Nelson EB (2003) Compost-induced suppression of pythium damping-off is mediated by fatty-acid-metabolizing seed-colonizing microbial communities. Appl Environ Microbiol 69(1):452–460
Mehta C, Palni U, Franke-Whittle I, Sharma A (2014) Compost: its role, mechanism and impact on reducing soil-borne plant diseases. Waste Manag 34(3):607–622
Merriman P (1976) Survival of sclerotia of *Sclerotinia sclerotiorum* in soil. Soil Biol Biochem 8 (5):385–389
Meyer JR, Linderman R (1986) Selective influence on populations of rhizosphere or rhizoplane bacteria and actinomycetes by mycorrhizas formed by *Glomus fasciculatum*. Soil Biol Biochem 18(2):191–196
Nelson E, Kuter G, Hoitink H (1983) Effects of fungal antagonists and compost age on suppression of *Rhizoctonia* damping-off in container media amended with composted hardwood bark. Phytopathology 73(10):1457–1462
Nitta T (1991) Diversity of root fungal floras: its implications for soil-borne diseases and crop growth. Jpn Agric Res Q 25(1):6–11
NOSB (2004) Compost tea task force report. National Organic Standards Board, Washington, DC
Ntougias S, Papadopoulou KK, Zervakis GI, Kavroulakis N, Ehaliotis C (2008) Suppression of soil-borne pathogens of tomato by composts derived from agro-industrial wastes abundant in Mediterranean regions. Biol Fertil Soils 44(8):1081–1090

Oritsejafor JJ (1986) Influence of moisture and pH on growth and survival of *Fusarium oxysporum* f.sp. elaeidis in soil. Trans Br Mycol Soc 87(4):511–517

Pal KK, Gardener BM (2006) Biological control of plant pathogens. Plant Health Instructor 2:1117–1142

Palmer A, Evans K, Metcalf D (2010) Characters of aerated compost tea from immature compost that limit colonization of bean leaflets by *Botrytis cinerea*. J Appl Microbiol 109(5):1619–1631

Pane C, Spaccini R, Piccolo A, Scala F, Bonanomi G (2011) Compost amendments enhance peat suppressiveness to *Pythium ultimum*, *Rhizoctonia solani* and *Sclerotinia minor*. Biol Control 56(2):115–124

Pane C, Celano G, Villecco D, Zaccardelli M (2012) Control of *Botrytis cinerea*, *Alternaria alternata* and *Pyrenochaeta lycopersici* on tomato with whey compost-tea applications. Crop Prot 38:80–86

Pang H, Zhang P, Duan C-J, Mo X-C, Tang J-L, Feng J-X (2009) Identification of cellulase genes from the metagenomes of compost soils and functional characterization of one novel endoglucanase. Curr Microbiol 58(4):404–408

Pantelides IS, Tjamos SE, Striglis IA, Chatzipavlidis I, Paplomatas EJ (2009) Mode of action of a non-pathogenic *Fusarium oxysporum* strain against *Verticillium dahliae* using Real Time QPCR analysis and biomarker transformation. Biol Control 50(1):30–36

Pascual JA, Garcia C, Hernandez T, Lerma S, Lynch JM (2002) Effectiveness of municipal waste compost and its humic fraction in suppressing *Pythium ultimum*. Microb Ecol 44(1):59–68

Pera A, Filippi C (1987) Controlling of fusarium wilt in carnation with bark compost. Biol Wastes 22(3):219–228

Phae CG, Shoda M, Kubota H (1990) Suppressive effect of B*acillus subtilis* and it's products on phytopathogenic microorganisms. J Ferment Bioeng 69(1):1–7

Pharand B, Carisse O, Benhamou N (2002) Cytological aspects of compost-mediated induced resistance against fusarium crown and root rot in tomato. Phytopathology 92(4):424–438

Postma J, Schilder MT, Bloem J, van Leeuwen-Haagsma WK (2008) Soil suppressiveness and functional diversity of the soil microflora in organic farming systems. Soil Biol Biochem 40 (9):2394–2406

Pratt JAR (2003) The use of green-waste compost in the biological control of *Fusarium* seedling blight caused by *Microdochium nivale*. Open University, Milton Keynes

Rangarajan A, Tuttle McGrath M, Bloemgren T (2001) Evaluation of two commercially available composts for managing phytophthora fruit rot of pumpkins. www.hort.cornell.edu/extension/commercial/vegetables/online/2001veg/pdfs/text/IPMfinalreportPumpkins.pdf. Accessed 15 Dec 2007

Ryckeboer J (2001) Biowaste and yard waste composts: microbiological and hygienic aspects: suppressiveness to plant diseases. Katholieke Universiteit Leuven, Faculteit Landbouwkundige en Toegepaste, Biologische Wetenschappen, Laboratorium voor Fytopathologie en Plantenbescherming

Ryu C-M, Hu C-H, Locy RD, Kloepper JW (2005) Study of mechanisms for plant growth promotion elicited by rhizobacteria in *Arabidopsis thaliana*. Plant Soil 268(1):285–292

Sang MK, Kim KD (2011) Biocontrol activity and primed systemic resistance by compost water extracts against anthracnoses of pepper and cucumber. Phytopathology 101(6):732–740

Sang MK, Kim JG, Kim KD (2010) Biocontrol activity and induction of systemic resistance in pepper by compost water extracts against *Phytophthora capsici*. Phytopathology 100 (8):774–783

Scheuerell S, Mahaffee W (2002) Compost tea: principles and prospects for plant disease control. Compost Sci Util 10(4):313–338

Scheuerell SJ, Mahaffee WF (2006) Variability associated with suppression of gray mold (*Botrytis cinerea*) on Geranium by foliar applications of nonaerated and aerated compost teas. Plant Dis 90(9):1201–1208

Serra-Wittling C, Houot S, Alabouvette C (1996) Increased soil suppressiveness to *Fusarium* wilt of flax after addition of municipal solid waste compost. Soil Biol Biochem 28(9):1207–1214

Siddiqui IA, Shaukat SS (2002) Mixtures of plant disease suppressive bacteria enhance biological control of multiple tomato pathogens. Biol Fertil Soils 36(4):260–268

Siddiqui Y, Meon S, Ismail R, Rahmani M, Ali A (2008) Bio-efficiency of compost extracts on the wet rot incidence, morphological and physiological growth of okra *Abelmoschus esculentus* [(L.) Moench]). Sci Hortic 117(1):9–14

Sivan A, Chet I (1989) The possible role of competition between *Trichoderma harzianum* and *Fusarium oxysporum* on rhizosphere colonization. Phytopathology 79(2):198–203

Spencer S, Benson D (1981) Root rot of *Aucuba japonica* caused by *Phytophthora cinnamomi* and *Phytophthora citricola* and suppressed with bark media. Plant Dis 65:918–921

Spencer S, Benson D (1982) Pine bark, hardwood bark compost, and peat amendment effects on development of *Phytophthora* spp. and Lupine root rot. Phytopathology 72(3):346–351

Spring D, Ellis M, Spotts R, Hoitink H, Schmitthenner A (1980) Suppression of the apple collar rot pathogen in composted hardwood bark. Phytopathology 70(12):1209–1212

St. Martin CCG (2013) Rotary drum compost and compost tea as substrates, amendments, and biocontrol agents for damping-off (Pythium ultimum) management in tomato (Solanum lycopersicum) The University of the West Indies, St. Augustine

St. Martin CCG, Ramsubhag A (2014) Potential of compost for controlling plant diseases. In: Sangeetha S, Kurucheve V, Jayaraman J (eds) Natural products for sustainable crop disease management. CAB International, Wallingford, pp 345–388

Stockwell C, Nelson E, Craft C (1994) Biological control of *Pythium graminicola* and other soilborne pathogens of turfgrass with actinomycetes from composts. Phytopathology 84:1113

Stone A, Vallad G, Cooperband L, Rotenberg D, Darby H, James R, Stevenson W, Goodman R (2003) Effect of organic amendments on soilborne and foliar diseases in field-grown snap bean and cucumber. Plant Dis 87(9):1037–1042

Stone A, Scheuerell S, Darby H, Magdoff F, Ray R (2004) Suppression of soilborne diseases in field agricultural systems: organic matter management, cover cropping, and other cultural practices. In: Magdoff F, Weil RR (eds) Soil organic matter in sustainable agriculture. CRC Press, Boca Raton, pp 131–177

Tenuta M, Conn KL, Lazarovits G (2002) Volatile fatty acids in liquid swine manure can kill microsclerotia of *Verticillium dahliae*. Phytopathology 92(5):548–552

Tilston E, Pitt D, Fuller M, Groenhof A (2005) Compost increases yield and decreases take-all severity in winter wheat. Field Crops Res 94(2):176–188

Timmusk S, Nicander B, Granhall U, Tillberg E (1999) Cytokinin production by *Paenibacillus polymyxa*. Soil Biol Biochem 31(13):1847–1852

Trankner A (1992) Use of agricultural and municipal organic wastes to develop suppressiveness to plant pathogens. In: Tjamos E, Papavizas G, Cook R (eds) Biological control of plant diseases: progress and challenges for the future. Plenum Press, New York, pp 35–42

Vallad GE, Goodman RM (2004) Systemic acquired resistance and induced systemic resistance in conventional agriculture. Crop Sci 44(6):1920–1934

Vallad GE, Cooperband L, Goodman RM (2003) Plant foliar disease suppression mediated by composted forms of paper mill residuals exhibits molecular features of induced resistance. Physiol Mol Plant Pathol 63(2):65–77

van Bruggen AH, Termorskuizen AJ (2003) Integrated approaches to root disease management in organic farming systems. Australas Plant Pathol 32(2):141–156

Van Schoor L, Denman S, Cook N (2009) Characterisation of apple replant disease under South African conditions and potential biological management strategies. Sci Hortic 119 (2):153–162

Vinale F, Flematti G, Sivasithamparam K, Lorito M, Marra R, Skelton BW, Ghisalberti EL (2009) Harzianic acid, an antifungal and plant growth promoting metabolite from *Trichoderma harzianum*. J Natl Prod 72(11):2032–2035

von der Weid I, Duarte GF, van Elsas JD, Seldin L (2002) *Paenibacillus brasilensis* sp. nov., a novel nitrogen-fixing species isolated from the maize rhizosphere in Brazil. Int J Syst Evol Microbiol 52(6):2147–2153

Wei G, Kloepper JW, Tuzun S (1991) Induction of systemic resistance of cucumber to *Colletotrichum orbiculare* by select strains of plant growth-promoting rhizobacteria. Phytopathology 81(11):1508–1512

Woltz SS, Jones JP (1981) Nutritional requirements of *Fusarium oxysporum*: basis for a disease control system. In: Nelson PE, Toussoun TA, Cook RJ (eds) Fusarium: diseases, biology and taxonomy. The Pennsylvania State University Press, University Park, pp 340–349

Yogev A, Raviv M, Hadar Y, Cohen R, Katan J (2006) Plant waste-based composts suppressive to diseases caused by pathogenic *Fusarium oxysporum*. Eur J Plant Pathol 116(4):267–278

Zaccardelli M, Perrone D, Pane C, Pucci N, Infantino A (2011) Control of corky root of tomato with compost and role of spore-forming bacteria to inhibit *Pyrenochaeta lycopersici*. Acta Hortic (ISHS) 914:393–396

Zhang W, Dick W, Hoitink H (1996) Compost-induced systemic acquired resistance in cucumber to *Pythium* root rot and anthracnose. Phytopathology 86(10):1066–1070

Zhang W, Han D, Dick W, Davis K, Hoitink H (1998) Compost and compost water extract-induced systemic acquired resistance in cucumber and Arabidopsis. Phytopathology 88 (5):450–455

Chapter 3
Soils and Crop Health in Rice–Wheat Cropping System Under Conservation Agriculture Scenario

D.P. Singh

3.1 Introduction

Wheat and rice are the major cereals besides maize for human consumption. Although both rice and wheat are grown in different cropping systems, wheat after rice is one of the world's principal agricultural production systems. Typically in the South Asian region, wheat is grown from November to April followed by rice during the monsoon from June–July to October–November. The rice–wheat system (RWS) has been practiced by farmers in Asia for past more than 1000 years and occupies 24–26 million ha (M ha) in Asia (Jing et al. 2010). Out of this, 13.5 M ha is in the Indo-Gangetic plains (IGP), amounting to 32 % of the total rice area and 42 % of the total wheat area in four countries. A negative yield trend and plateauing of productivity of rice and wheat have been experienced leading to excessive utilization of natural resource bases (Pathak et al. 2003). The main threatening factors for sustaining the productivity and production of RWS are the efficiency of current production practices, the scarcity of resources (water, labor, etc.), and climate and socioeconomic changes. Over the years, the soil organic matter content had reduced due to burning of crop residues after mechanical harvesting, and it is still a common practice under RWS. The soil and crop health is believed to be improved by incorporating crop residue into the soil using conservation agriculture (CA) practices. This review presents findings from recent research on resource conservation technologies involving tillage and crop establishment options that are enabling farmers to sustain productivity of intensive RWS through better soil and crop health. Much of this work has focused on understanding the effect of CA on soil and plant health aiming to produce rice and wheat cereals at a lower cost through reduced tillage and residue retention. Several kinds of these CA technologies are

D.P. Singh (✉)
Guyana Rice Development Board, Rice Research Station, Burma, Guyana
e-mail: dpkarnal@gmail.com

M.K. Meghvansi, A. Varma (eds.), *Organic Amendments and Soil Suppressiveness in Plant Disease Management*, Soil Biology 46,
DOI 10.1007/978-3-319-23075-7_3

being adopted by farmers according to their needs and conditions. In zero tillage, wheat seeds are drilled into unploughed fields which retain the residues from the rice crop. In reduced tillage, the seeds are surface sown onto rota-tilled soil. The surface seeding of wheat into standing rice or after rice harvest has been long used by farmers in parts of South Asia where soil moisture is generally too high after rice harvest and hampers conventional tillage. The CA is more sustainable and environmentally friendly and uses energy. Farmers may save on tillage costs, irrigation water, fossil fuel and can sow their wheat early at reduced or same cost by using CA (Chauhan et al. 2012; Koshal 2014). One ton of wheat grains remove about 24.5, 3.8, and 27.3 kg of N, P, and K, respectively, whereas similar production of rice grains removes 20.1 kg N, 4.9 kg P, and 25.0 kg K (Tandon and Sekhon 1988).

3.2 Rice and Wheat Ecosystem

Rice and wheat crops have different requirements of soil and water. Rice in most of the cases is transplanted in puddled soils and fields are generally kept in submergence condition. Puddling serves to break down soil aggregates. It reduces macroporosity and soil strength in the puddled layer and results in formation of dense zone of compaction (i.e., plow pan) in subsoil. The wheat is on the other hand grown in well-drained and in good tilth dry soil. Therefore, the RWS is an annual cycle of aerobic to anaerobic conditions for growing rice after wheat. The process results to changes in several physical, chemical, and biochemical conditions of soil, affecting availability of nutrients, root penetration, and moisture availability (Ponnamperuma 1985).

3.3 Conservation Agriculture and Soil Health

Crop production removes varying amounts of mineral nutrients depending on production and nutrient-supplying capacity of the soil. This process is influenced due to soil type, soil organic matter content, amount of nutrients applied, and removal or recycling of crop residues in the soil. Both rice and wheat are heavy feeders of nutrients. The long-term cultivation of RWS resulted in mining of major nutrients (N, P, K, and S) from the soil as well as created a nutrient imbalance, leading to deterioration in soil quality. Among nutrients, the deficiencies of N, P, and K are most extensive (Tandon and Sekhon 1988). Different types of soil aeration and tillage practices in RWS tend to influence soil health for crop growth as they will influence the number of detrimental and beneficial organisms in the rhizosphere. Conservation agriculture helps in maintaining a permanent or semipermanent organic soil cover. The growing of crop or use of dead mulch protects soil physically from sun, rain, and wind and feeds on soil biota. Mechanical tillage disturbs this process. Therefore, zero or minimum tillage and direct seeding are important elements of CA. The crop residues on the surface of soil under CA increase water infiltration and reduce erosion.

Among different tillages, the highest increase of porosity and field capacity was recorded in zero tillage in wheat–mung bean–rice cropping system in Bangladesh. Zero tillage also resulted in highest total N, P, K, and S in their available forms as compared to conventional, minimum, and deep tillages. The zero tillage with 20 % residue retention was therefore found most suitable for soil health and achieves optimum yield under the cropping system in Grey Terrace soil (Alam et al. 2014).

Soils in the IGP contain low organic matter due to RWS. Excessive nutrient mining of soils is one of the major causes of fatigue experienced in soils under the RW system. The RWS removes more quantities of nutrients than the amount added through fertilizers and recycled. Sulfur deficiency has also been observed in soils in NW region of India, particularly in soils that are coarse-textured, low in pH, and poor in organic matter (Sharma and Nayyar 2004). Rice requires more amount of micronutrient than that of wheat. Zn deficiency has become widespread in the IGP (Shukla and Behera 2011) and is more in rice and that of Mn is morein wheat. Deficiencies of other micronutrients such as Fe, Cu, and B are also on increase. Removal of all the straw from crop fields leads to K mining at alarming rates. Major K contents absorbed by plant (80–85 %) remain in rice and wheat crops. K is removed by crops than N and P, resulting in a negative K balance in the soil (Tandon and Sekhon 1988). The practice of RWS on long-term basis depletes soil of K in spite of the application of optimum doses of fertilizer K mainly due to non-incorporation of crop residues in soil (Tandon and Sekhon 1988). Fertilizer use in general is consistently increasing and so is the N–P_2O_5–K_2O ratio due to the imbalanced use of these nutrients. More and more N is being used with a very low rate of K application. The partial factor productivity of N, P, and K for food grain production has dropped from about 81 kg grain per kg of N, P, and K in 1966–1967 to 15 kg grain per kg N, P, and K in 2006–2007 (Benbi and Brar 2009). The efficiency of applied nutrients has been about 50 % for N, <25 % for P, and 40 % for K (Witt et al. 1999). The lower efficiency results due to leaching, runoff, gaseous emission, and fixation by soil. For improving the productivity of RWS, inclusion of short duration legumes between wheat and rice, balanced use of nutrients, incorporation of rice crop residues in soil after harvest, and application of a consortium of beneficial microbes for fast decomposition of residues may help to restore soil fertility.

3.4 Conservation Agriculture and Soil Microbial Population

The proper balance of beneficial microbes in soils in RWS is stressed to make the system more productive. In recent past, few studies have been conducted to show the effects of CA on population of microbes. The CA techniques are gaining popularity and therefore have led to the need for more research into their effects on soil and plant health. The soil inhibiting microbes (bacteria, fungi, and actinomycetes) play an important role in the soil ecosystem. Fungi are major decomposers of plant residues and release nutrients that sustain and stimulate plant growth in the process in the soil ecosystem. Some fungi possess properties antagonistic toward plant pathogens. Verhulst et al. (2009) studied the soil microbial population in RWS under zero tillage. Soil microbial biomass C and N were directly correlated with residue retained on the soil surface in both rainfed and the irrigated conditions. The soil microbial biomass is an important parameter to assess the ability of soil to store and cycle nutrients (C, N, P, and S) and organic matter. The suppression of crop pathogens is also related to total soil microbial biomass, which competes with pathogens for resources or acts as antagonist. Therefore, soil microbial biomass is an important indicator of soil quality. The rate of organic C input from plant biomass is a major factor in regulating the amount of microbial biomass in soil. The microbes get the energy from C of crop residue (Verhulst et al. 2009).

DNA-based methods to study complex fungal community structures have been used, and a combination of 18S-rDNA PCR amplification and TGGE community analysis helps in knowing the diversity, composition, and dynamics of the fungal community in bulk soil and in the rhizosphere. A well-developed and diverse rhizosphere community is thought to play a role in the suppression of pathogens (Jarosik et al. 1996). Knowledge of the structure and diversity of the fungal community in the rhizosphere will lead to a better understanding of pathogen–antagonist interactions. An array of molecular techniques, such as amplified ribosomal DNA (rDNA) sequencing, amplified rDNA restriction analysis, and temperature and denaturing gradient gel electrophoreses (TGGE and DGGE) of rDNA, has been applied to understand the microbial population structures in the environment. It resulted to tremendous increase in knowledge of microbial ecology and has revealed the existence of formerly unknown microorganisms (Smit et al. 1999).

Banerjee et al. (2006) conducted experiments to understand the effect of decrease in soil organic carbon on decreasing trends in productivity of RWS. The effect of soil organic carbon (SOC) on soil microbial biomass carbon (MBC) dynamics in the rice–wheat systems was also studied. The organic amendments and puddling of soil before rice transplanting increased SOC and MBC contents. The microbial biomass was low initially, reached its peak during the flowering stages in both rice and wheat, and declined thereafter. Microbial biomass carbon was linearly related to SOC in both rice and wheat. It was concluded that SOC could be used as a proxy for MBC. Sandeep et al. (2010) studied the effect of lantana (*Lantana camera* L.) residue incorporation on long-term basis (>12 years) on major soil microbes and on certain soil chemical properties in the rice–wheat cropping system at Palampur (Himachal Pradesh) with four levels of lantana incorporation and three tillage practices (no puddling, puddling, and soil compaction). After 12 crop cycles (2001–2002), *Lantana* residue application at 10, 20, and 30 Mg ha^{-1} increased soil organic carbon (7, 13, and 19 % over 1.29 g C kg^{-1} under no residue treatment) and pH (5.23–5.29 as against 5.12 in the control). *Lantana* incorporation at 10–30 kg ha^{-1} also recorded a significant increase in the bacterial (249–

369 × 10^4 CFU), fungal (148–220 × 10^4 CFU), actinomycete (79–144 × 10^4 CFU), and phosphorus-solubilizing microorganism (53–100 × 10^4 CFU) counts (0–0.15 m soil depth) compared to control. The most important variable contributing to rice and wheat yield was soil organic carbon ($R^2 = 86–95$ %), followed by bacteria and fungi.

Anju Rani (2012) studied the root parameters of wheat (analyzed region width, height, area, and diseased root area) in different treatments of CA under RWS (Table 3.1). Better root architect was recorded in plots where crop residues of both wheat and rice were left in field. The burning of crop residues on the other hand had negative effect on root health. The higher surface area of root invited a higher number of microbiota in soil whereas the higher root length helped plant to survive better in adverse water conditions (Anju Rani 2012). Likewise, highest CFU numbers per Petri dish were recorded in case of plots where residue of both rice

Table 3.1 Effect of different CA practices on microbial population in wheat rhizosphere soil

	Total colony counts/plate (9 cm diameter)					
	CFU/ Petri plate	Streptomycetes colonies	Bacteria colonies			*Aspergillus heteromorphus*
CA treatments (Main)		Chalky white colony	Dark yellow	Orange	Cream	Black colony
Removal of residue of rice and wheat	52.1	1.7	14.8	8.0	25.3	2.0
Incorporation of rice and wheat crop residue	34.0	2.2	14.0	1.3	16.4	0.0
Incorporation of rice residue and removal of wheat residue	32.1	1.4	10.4	1.7	18.4	0.0
Burning of residue of both rice and wheat	44.2	8.5	6.5	3.6	25.4	0.0
Burning of rice residue and removal of wheat residue	38.3	1.4	11.0	2.8	23.0	0.0
Retention of rice and wheat residue	54.1	1.3	13.4	1.5	37.4	0.0
Retention of rice residue and removal of wheat residue	25.9	1.3	2.0	1.5	20.3	0.0
Mean	40.1	2.5	10.3	2.9	23.8	0.3
CD (5 %)	3.3	1.2	2.5	1.1	2.1	0.3
N application treatments (Sub)						
N 100 kg/ha	39.0	4.2	5.9	2.2	26.0	0.5
N 150 kg/ha	50.3	1.7	16.6	3.2	28.4	0.3
N 200 kg/ha	31.0	2.0	8.4	3.5	16.9	0.0
Mean	40.1	2.6	10.3	3.0	23.8	0.3
CD (5 %)	2.2	0.8	1.7	0.7	1.4	0.2

and wheat was retained in field and allowed to decompose (Table 3.1). Significantly high (CFU/plate) were recorded in plots applied with N150 kg/ha as compared to N100 and N200 kg/ha. The residue incorporation in field in RWS also favored higher counts of bacteria and fungi. The counts of *Aspergillus heteromorphus* were only found in plots where residue of both crops was removed. The predominant fungal species found in the wheat rhizosphere under RWS and CA were *A. terreus*, *A. heteromorphus*, *Fusarium* spp., *Penicillium* spp., *Alternaria triticina*, and *Bipolaris sorokiniana*; bacteria and actinomycetes were also found. Bacterial counts were higher than fungal and actinomycete counts (Rani 2012). More studies are required to know what is going on in rhizosphere of rice and wheat in RWS under different soil tillage methods, residue incorporation, fertilizer doses, irrigation, soil types, and type of cultivars.

3.5 Conservation Agriculture and Plant Health

Conservation tillage affects the microbial biomass in the top 5–15 cm of soil by accumulating crop residue and organic matter and thus promotes survival of pathogens. Only a proportion of the rhizofloral population goes toward plant pathogens. Higher soil microbial activity may lead to competition effects that may ameliorate pathogen activity and survival and counteract a high pathogen inoculum pressure. Microbial antagonism in the root zone can lead to the formation of disease-suppressive soils (Sturz et al. 1997). Soil factors (soil water, aeration, compaction, porosity, and temperature) affect survival of soil-borne plant pathogens and their antagonists. Increased soil water can reduce disease through reducing plant water stress. Reduced soil aeration or temperature or increased soil water or compaction can predispose the host to infection and disease development. High soil water can also increase disease through increasing motility of the pathogen or diffusion of host exudates. Pore size of soil may limit activity or movement of the plant pathogens. Changes in soil physical factors also may limit diseases by affecting microbial antagonism (Rothrock 1992). The soil fertility and plant health are directly related. The poor land management and declining soil fertility often result in an increase in soil-borne pests. Therefore, management of soil health in ways that conserve and enhance a fully functional soil biota may improve crop yields and quality. A diverse soil community helps to reduce losses due to soil-borne pests, but also regulate decomposition of organic matter and toxic compounds, and thereby improve nutrient cycling and soil structure. Mezzalama et al. (2001) reported an increase in the incidence of wheat diseases like crown rot (*Fusarium pseudograminearum*) and tan spot (*Pyrenophora tritici-repentis*) under zero tillage.

Kohli and Fraschina (2009) reported higher incidence of foliar and spike diseases (species of *Alternaria*, *Fusarium*, *Helminthosporium*, and *Septoria*) in plots having wheat stubble on the soil surface and under zero tillage system. Of these, tan spot caused by *Pyrenophora tritici-repentis* (Died.) Drechs. teleomorph of

Drechslera tritici-repentis (Died.) Shoem. has become of worldwide importance and is associated with the increase in the conservation agriculture. Higher intensity of *Fusarium* head blight, FHB (*Fusarium graminearum*), was reported under zero tillage. The intensity of leaf rust disease caused by *Puccinia recondita* is not affected due to different tillage systems. Acevedo et al. (2009) reported an increase in the incidence of some diseases caused by *Fusarium*, *Septoria tritici*, *Helminthosporium tritici*, and *Erysiphe graminis* in wheat when grown under no tillage. The root length density, however, is higher in the surface soil of no-till (0–5 cm) when compared to conventional tillage (Martinez et al. 2008) which may be due to higher nutrients and microbial activity.

Trichoderma are free-living and fast-growing fungi in soil and root ecosystems of many plants and inhibit soil pathogens through antibiosis, antagonism, and competitive exclusion. Furthermore, *Pseudomonas* and *Trichoderma* species that function as biocontrol agents do not inhibit nitrogen fixers, arbuscular mycorrhizal fungi, and other beneficial microbes that positively impact plant growth. *Trichoderma* species have also been known to produce phytohormones and solubilize/mobilize phosphates. A number of *Trichoderma* species were isolated from different soil samples and were screened for their potential as biocontrol agents against known plant pathogenic fungi such as *Alternaria alternata* and *Curvularia* sp., *Bipolaris oryzae*, *Magnaporthe grisea*, and *Rhizoctonia solani*. The formulation increased rice yield by 40 % with increase in seedling vigor, plant height, number of tillers, and their carry-over effect on grain yield (Reddy and Lalithakumari 2009). Singh et al. (2002) also suggested that the adoption of RCT did not adversely affect soil biology. Nematodes which affect the grain yield in wheat were not found in significant numbers which may be due to regular flooding that takes place in the rice crops. The predominant fungal species found in the rhizosphere of wheat and rice were *Fusarium* species, *Drechslera rostrata*, and *Penicillium* species. Joshi et al. (2002) also did not notice any difference in severity of rice or wheat diseases in two tillage systems. However, in another study made by Sister et al. (2013), the disease severity of blast in rice was significantly lower in the no-tillage cropping system than in the conventional tillage system. The root-knot nematode (*Meloidogyne graminicola*) in RWS population density, root-knot index, and wide nematode to root biomass ration were reduced in case of zero till rice followed by zero till wheat and green manuring with *Sesbania* sp. as compared to other CA practices (Upadhyay et al. 2014). Ando et al. (2014) inoculated rice seedlings with *Burkholderia glumae* at germination and cultivated organic soils. The development of seedling rot symptoms was significantly suppressed on organic soils, but not on conventional soil. Such disease-suppressive activity of the organic soils was also observed on rice seedling damping-off disease caused by *B. plantarii*. The disease suppression was completely compromised once the organic soils were sterilized at 121 °C, indicating biological activity included in organic soils seems to be associated with the suppression of rice seedling diseases. Suppression of take-all of wheat, caused by *Gaeumannomyces graminis* var. *tritici*, is induced in soil after continuous wheat monoculture. It is attributed, in part, to selection of fluorescent pseudomonads capable to produce the antibiotic 2,4-diacetylphloroglucinol (Mazzola 2002). Such information on suppressive soils and their effects on the incidence of diseases of rice and wheat in cropping system need to be generated in the near future. The use of biological control agents in RWS will help in preventing and managing pests along with host resistance and cultural practices.

3.6 Conclusions

The sustainability of rice–wheat cropping system is dependent on both soil and plant health. The RWS will continue to be the most predominant among cropping systems in Indo-Gangetic region in spite of alternatives available mainly due to preference of farmers, minimum support prices of produce, as well as favorable policies of governments. It is also important to keep the South Asian region food secure since both wheat and rice are preferred cereal food. The studies conducted in the past indicated no major effect of RWS on plant health except decline in organic carbon over the years which may be a cause of concern. The conservation agriculture may play an important role in correcting the soil health and indirectly contribute to plant health positively. The incorporation of beneficial microbial population will go a long way in making positive soil and crop health in RWS besides optimum utilization of natural resources and fossil fuel in an eco-friendly manner.

References

Acevedo E, Martinez E, Silva P (2009) Constraints to zero tillage in Mediterranean environments. In: Lead papers, 4th world conference on conservation agriculture, New Delhi, 4–7 Feb 2009, pp 195–206

Alam MK, Islam MM, Salahin N, Hasanuzzaman M (2014) Effect of tillage practices on soil properties and crop productivity in wheat-mungbean-rice cropping system under subtropical climatic conditions. Sci World J 8:1–15

Ando S, Ito T, Kanno T, Kobayashi T, Morikawa T, Honda K, Tsushima S, Takahashi H (2014) Impact of organic crop management on suppression of bacterial seedling diseases in rice. Org Agric 4:187–196

Banerjee B, Aggarwal PK, Pathak H, Singh AK, Chaudhary A (2006) Dynamics of organic carbon and microbial biomass in alluvial soil with tillage and amendments in rice-wheat systems. Environ Monit Assess 119:173–189

Benbi DK, Brar JS (2009) A 25-year record of carbon sequestration and soil properties in intensive agriculture. Agron Sustain Dev 29:257–265

Chauhan BS, Gulshan M, Virender S, Jagadish T, Jat ML (2012) Productivity and sustainability of the rice wheat cropping system in the Indo-Gangetic plains of the Indian subcontinent: problems, opportunities, and strategies. Adv Agron 117:315–369

Jarosik V, Kovacikova E, Maslowska H (1996) The influence of planting location, plant growth stage and cultivars on microflora of winter wheat roots. Microbiol Res 151:177–182

Jing Q, van Herman K, Hengsdijk H (2010) Modelling biomass, nitrogen and water dynamics in rice-wheat rotations. Agric Syst 103:433–443

Joshi D, Duveiller E, Rutherford M, Kirk P, Singh Y, Singh US (2002) Effect of zero tillage on diseases of rice and wheat in uttaranchal state, India. In: Duveiller E, Bridge J, Rutherford M, Keeling S (eds) Rice-wheat consortium paper series 16, soil health and sustainability of the rice-wheat systems of the indo-gangetic plains. Proceedings of the end of project workshop, Kathmandu, Nepal, 7–10 May 2002, pp 19–23
Kohli MM, Fraschina J (2009) Adapting wheats to zero tillage in maize-wheat-soybean rotation system. In: Lead papers, 4th world conference on conservation agriculture, New Delhi, 4–7 Feb 2009, pp 212–222
Koshal AK (2014) Changing current scenario of rice-wheat system in indo-gangetic plain region of India. IJSRP 4:1–13
Martinez E, Juan-Pablo F, Silva P, Valle S, Edmundo A (2008) Soil physical properties and wheat root growth as affected by no-tillage and conventional tillage systems in a Mediterranean environment of Chile. Soil Tillage Res 99:232–244
Mazzola M (2002) Mechanisms of natural soil suppressiveness to soilborne diseases. Antonie Van Leeuwenhoek 81:557–564
Mezzalama M, Sayre KD, Nicol J (2001) Monitoring root rot diseases on irrigated, bed-planted wheat. In: Reeves J, McNab A, Rajaram S (eds) Proceedings of the Warren E. Kronstad symposium. DFCIMMYT, Mexico, pp 148–151
Pathak H, Aggarwal PK, Roetter R, Kalra N, Bandyopadhaya SK, Prasad S, Van Keulen H (2003) Modelling the quantitative evaluation of soil nutrient supply, nutrient use efficiency, and fertilizer requirements of wheat in India. Nutr Cycl Agroecosyst 65:105–113
Ponnamperuma FN (1985) Chemical kinetics of wetland rice soils relative to soil fertility. In: Wetland soils: characterization, classification and utilization. Proceedings of a workshop held 26 March–5 April 1984 under the joint sponsorship of the International Rice Research Institute; Soil Management Support Services (Agency for International development and United States department of Agriculture), and Bureau of Soils (Philippine Ministry of Agriculture). Published by the International Rice Research Institute, Los Banos, Laguna, pp 71–89
Rani A (2012) Effect of resource conservation agriculture practices on rhizosphere of wheat under wheat–rice cropping system. Project report, submitted for partial fulfillment of degree of M.Sc. in Microbiology and Biotechnology, Banasthali University, Rajasthan, pp 1–20
Reddy CA, Lalithakumari J (2009) Polymicrobial formulations for enhanced productivity of a broad spectrum of crops. In: Lead papers, 4th world conference on conservation agriculture, New Delhi, 4–7 Feb 2009, pp 94–101
Rothrock CS (1992) Tillage systems and plant diseases. Soil Sci 154:308–315
Sandeep S, Chander G, Varma TS (2010) Soil microbiological and chemical changes in rice-wheat cropping system at Palampur (Himachal Pradesh) after twelve years of *Lantana camara* L. residue incorporation. J Trop Agric 48:64–67
Sharma PK, Nayyar VK (2004) Diagnosing micronutrient related constraints to productivity in Muktsar, Patiala, Hoshiarpur and Ludhiana Districts. Information technology for sustainable agriculture in Punjab (IT-SAP) UNDP-TIFAC sponsored project. Punjab Remote Sensing Centre, pp 1–18
Shukla AK, Behera SK (2011) Zinc management in Indian agriculture: past, present and future. Indian J Fertil 7:14–33
Singh R, Malik RK, Singh S, Yadav A, Duveiller E (2002) Effect of tillage practice on rhizosphere fungi and diseases of rice and wheat in Haryana state, north India. In: Duveiller E, Bridge J, Rutherford M, Keeling S (eds) Rice-wheat consortium paper series 16, soil health and sustainability of the rice-wheat systems of the indo-gangetic plains. Proceedings of the end of project workshop, Kathmandu, Nepal, 7–10 May 2002, pp 1–7
Sister M, Raveloson H, Tharreau D, Dusserre J (2013) Conservation agriculture cropping system to limit blast disease in upland rainfed rice. Plant Pathol 63:373–381
Smit E, Leeflang P, Glandorf B, Van Elsas JD, Wernars K (1999) Analysis of fungal diversity in the wheat rhizosphere by sequencing of cloned PCR-amplified genes encoding 18s rRNA and temperature gradient gel electrophoresis. Appl Environ Microbiol 65:2614–2621

Sturz AV, Crater MR, Johnston HW (1997) A review of plant disease-pathogen interactions and microbial antagonism under conservation tillage in temperate, humid agriculture. Soil Tillage Res 41:169–189

Tandon HLS, Sekhon GS (1988) Potassium research and agricultural production in India. Fertilizer Development and Consultation Organization, New Delhi, pp 1–144

Upadhyay V, Vaish SS, Bhardwaj NR (2014) Effect of resource conservation practices and conventional practices on population dynamics of *Meloidogyne graminicola* under rice-wheat cropping system. Afr J Microbiol Res 8(32):3025–3028

Verhulst N, Govaerts B, Verachtert E, Kienle F, Limon-Ortega A, Deckers J, Dirk Raes D, Sayre KD (2009) The importance of crop residue management in maintaining soil quality in zero tillage systems; a comparison between long-term trials in rainfed and irrigated wheat systems. In: Lead papers, 4th world conference on conservation agriculture, New Delhi, 4–7 Feb 2009, pp 71–79

Witt C, Dobermann A, Abdulrachman S, Gines HC, Wang GH, Nagarajan R, Satawathananont S, Son TT, Tan PS, Tiem LV, Simbahan GC, Olk DC (1999) Internal nutrient efficiencies of irrigated lowland rice in tropical and subtropical Asia. Field Crop Res 63:113–138

Chapter 4
Developing Disease-Suppressive Soil Through Agronomic Management

R.S. Yadav, Jitendra Panwar, H.N. Meena, P.P. Thirumalaisamy, and R.L. Meena

4.1 Introduction

Plant diseases need to be controlled for maintaining the quality and quantity of food, feed, and fiber. Soilborne plant pathogens are one of the major limiting factors in most of the agroecosystems for the production of economical yields. Mostly they survive in bulk soil, but the parasitic relationship with crop plants is established in the rhizosphere. Soilborne pathogens caused numerous diseases like seed decay, pre- and postemergence damping off, wilting of roots, root rot, stem rot, crown rot, collar rot, decay of collar and fruits in trees, etc., and made serious losses to agricultural crops. These pathogens produce resting bodies in the soil which are long lasting and difficult to eliminate. The various diseases and symptoms are manifested by the plants which are difficult to diagnose and generally confused with the nondistinct symptoms caused by abiotic factors and/or due to lack of nutrients. Various approaches have been used to prevent, mitigate, or control the plant diseases. The practices for managing plant disease are largely based on genetic resistance in the host plants, management of the plant and its environment, and use of synthetic chemicals (Strange 1993). However, the use of agrochemicals needs to be ensured for safety of human health and environment (NRC 1996).

R.S. Yadav (✉) • H.N. Meena • P.P. Thirumalaisamy
ICAR-Directorate of groundnut Research, Junagadh 362001, Gujarat, India
e-mail: yadavrs2002@gmail.com; hariagro@gmail.com; thirumalaisamypp@yahoo.co.in

J. Panwar
Department of Biological Sciences, Centre for Biotechnology, Birla Institute of Technology and Science, Pilani 333031, India
e-mail: drjitendrapanwar@yahoo.co.in

R.L. Meena
Sardarkrushinagar Dantiwada Agricultural University, Sardarkrushinagar 385506, India
e-mail: ramji_meena@yahoo.com

M.K. Meghvansi, A. Varma (eds.), *Organic Amendments and Soil Suppressiveness in Plant Disease Management*, Soil Biology 46,
DOI 10.1007/978-3-319-23075-7_4

Moreover, issues like legal control on pesticide use (NRC 1996), their nontarget effects (Elmholt 1991), development of resistance in pathogens (Russell 1995), political pressure to ban the use of such hazardous chemicals, etc., are the concerns to think about other eco-friendly alternatives for stable agroecosystems. For this, developing disease-suppressive soils is one of the most eco-friendly viable options to reduce the plant disease pressure as well as to strengthen the agroecosystems for sustainable agriculture. Generally, competition, antibiosis, parasitism, enhancement of plant resistance, etc., are the major mechanisms employed in developing suppressive soils. Numerous factors like soil properties (Hoeper and Alabouvette 1996), soil microbial activity or the soil respiration (Van Os and Van Ginkel 2001), microbial diversity and composition (Garbeva et al. 2006), microbial population density (Tuitert et al. 1998), presence of antibiotic genes (Garbeva et al. 2006), agronomic management (Hoitink and Boehm 1999; Berg et al. 2002; Larkin and Honeycutt 2006), etc., are the key factors to determine the soil suppressiveness. Although, the mechanism behind soil suppressiveness is still not clear in many pathosystems. Both positive and negative correlations were reported between soil characteristics and suppressiveness, depending on the pathogens and the agroecosystems involved (Janvier et al. 2007).

The farm management practices used for crop cultivation not only promote the plant growth but also the soil suppressiveness having high efficacy for disease control without any additional cost and effect on the environment. Therefore, agronomic management practices are of multidimensional effects and have the high priority in contemporary agriculture (Martin 2003). The agricultural practices like crop rotation, tillage, fertilizers and organic amendments, use of microbes, etc., influence disease suppressiveness considerably. Nonetheless, many soil characteristics could interact, and hence, it will be very difficult to predict the precise effects of the agricultural practices on suppressiveness for specific disease and soil type (Janvier et al. 2007). Probably the soil suppressiveness is a combined effect of general and specific suppression, where the first relates to activity, biomass, and diversity and the second is the result of the presence of specific antagonistic groups. Therefore, knowledge about the process that results in increased soil suppressiveness is a prerequisite for its application under natural conditions. In this review, agronomic strategies for developing disease-suppressive soils for improved soil and plant health and productivity as well as for environmental benefits are discussed.

4.2 Suppressive Soils

The soils in which pathogens fail to establish or to produce disease are called disease-suppressive soils (Baker and Cook 1974). Two types of disease suppression have been described on biological basis, i.e., general and specific suppressions (Fig. 4.1). General suppression is the overall effect of the microbial community principally through resource competition which differs from specific suppression. The specific suppression relates with specific mode of action against pathogen populations (Weller et al. 2002). It is evident that most of the soils possess both

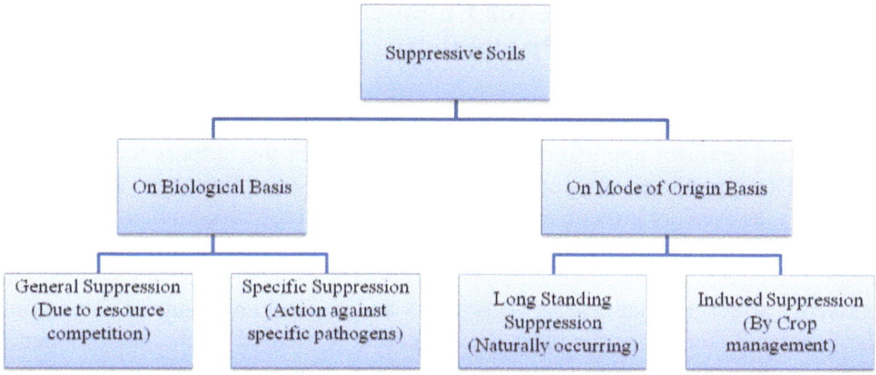

Fig. 4.1 Types of suppressiveness occurred in soils under different agroecosystems

general and specific suppressive activities at varying degrees which are greatly altered by management practices (Weller et al. 2002). The suppressive soils are also known for multiple soilborne pathogens and have been further categorized as long-standing or induced (Hornby 1983). Long-standing suppression is naturally associated with soil and is of unknown origin, whereas induced suppression develops as a result of crop management (Fig. 4.1). Some well-known examples of specific suppressive soils are Fusarium wilt of watermelon, take-all decline of wheat (Weller et al. 2002), *Rhizoctonia* damping off of cucumber, scab decline of potato, etc. (Menzies 1959).

Naturally occurring disease-suppressive soils have been well documented in a variety of cropping systems, and in many instances, the biological attributes contributing to suppressiveness have been identified. In spite of an understanding for mechanisms leading to the suppressive state, it is very difficult to realize the transfer of this knowledge into achieving effective field-level disease control. This might be due to the complex nature of biological control system and the inconsistent results for disease control in different agroecosystems under disease-suppressive soils (Pal and Gardener 2006). Therefore, greater emphasis is to be placed on manipulation of the cropping system to manage resident beneficial rhizosphere microorganisms as a means to suppress soilborne plant pathogens. Maintaining high levels of organic matter on the soil surface and incorporated into soil generally is associated with lower incidence and severity of root diseases (Bailey and Lazarovits 2003).

4.3 Factors Responsible for Enhancing Soil Quality, Plant Health, and Crop Productivity

The agronomic operations like plant species, land preparation, irrigation, and manure and fertilizer application are generally used by the farmers for crop cultivation. These practices considerably influenced the soil rhizosphere and biogeochemistry as well as growth and composition of microbial communities around plant roots. Plant roots and microorganisms are the vital component of the rhizosphere, and the total biomass and composition of rhizosphere microbial populations markedly affect interactions between plants and the soil environment. Therefore, the beneficial conditions for plant growth could be created by the use of amendments in the soil, breeding or engineering better plants, and manipulating plant/microorganism interactions. Plant root system, rhizosphere and rhizodepositions, soil properties, microbial diversity and microbiome, cultural practices, etc., are some of the major factors responsible for soil health and productivity of the crop plants (Fig. 4.2). These factors have positive influence on plant growth and development by facilitating plant establishment, enhanced nutrient availability, tolerance to stresses, improved plant protection, induced systemic plant disease resistance, etc. However, the benefits of root zone microbial biodiversity are still not certain in managed agroecosystems. Further management for disease control and yield maximization often minimized the community complexity and also disrupted the ecosystem stability. Therefore, the complexity of plant–soil–microbial interactions varied greatly, and the complete understanding of all the relationships involved is very difficult to be understood. Nevertheless, these beneficial biological

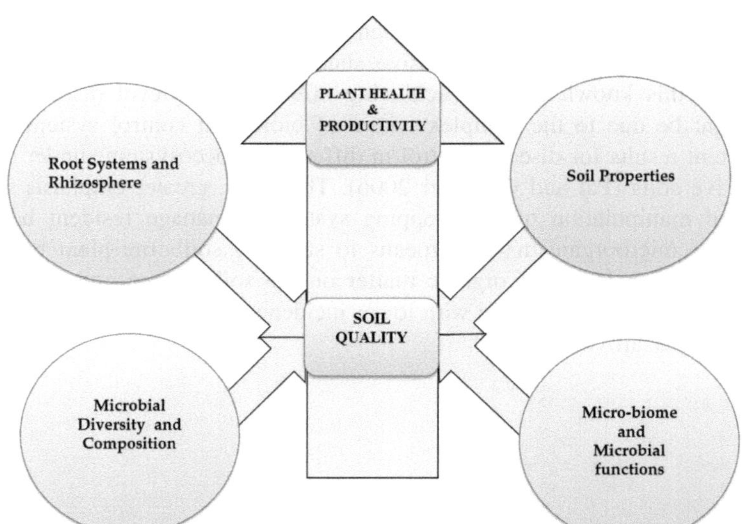

Fig. 4.2 Factors interacting with soil quality and affecting the plant health and productivity

interactions can be evaluated for better soil and plant health and also devised the management strategies accordingly.

4.3.1 Root System and Rhizosphere

Traditionally the root system was thought to provide anchorage and uptake of nutrients and water. However, it is the key element to a plant interacting with its surroundings by secreting various biochemical compounds as root exudates (Bais et al. 2006). Secretion of these compounds varies between different plant species (Rovira 1969), ecotypes (Micallef et al. 2009), and even distinct roots within a plant (Uren 2007). The diverse compounds released by plants as root exudates including sugars, proteins, fatty acids, flavonoids, amino acids, aliphatic acids, etc., create a unique environment in the rhizosphere (Badri and Vivanco 2009). All these different compounds are able to attract and initiate both symbiotic and pathogenic interactions within the rhizosphere (Bais et al. 2006).

Rhizodeposition that comprises of border cells, root debris, and root exudates are the major organic carbon source to the soil (Uren 2007) which could probably attract microorganisms that service the plant via biochemically active root system. Root exudates varied in composition and concentration depending on many factors like edaphic conditions, agronomic management (Bowen and Rovira 1999), age of the plant (De-la-Pena et al. 2010), soil type (Rovira 1969), biotic and abiotic factors (Flores et al. 1999), etc. All these factors also alter the microbial composition of the rhizosphere (Micallef et al. 2009) as these exudates are also used as growth substrates (Vandenkoornhuyse et al. 2007) by soil microbes for their population density and activities. Hence, the rhizosphere harbors many organisms having multiple effects on the plants like deleterious, beneficial, and neutral in action. The rhizosphere is also a battlefield where the complex rhizosphere community, both microflora and microfauna, interact with pathogens and influence the outcome of pathogen infection. Therefore, rhizosphere engineering may ultimately reduce our reliance on agrochemicals by replacing their functions with beneficial microbes, biodegradable biostimulants, or transgenic plants. For further details, see Ryan et al. (2009).

4.3.2 Soil Properties and Plant Health

Soils are highly diverse and dynamic in nature, allowing for habitation to diverse communities of microorganisms (Schloss and Handelsman 2006). The diverse communities of microbes have been associated with soils of varying texture (Girvan et al. 2003), nutrient content (Frey et al. 2004; Faoro et al. 2010), and soil pH (Fierer and Jackson 2006; Rousk et al. 2010). The bacterial community in soils was greatly influenced by soil pH (Fierer and Jackson 2006), and a strong correlation was observed between soil pH and the diversity and composition of bacterial communities across the biomes (Rousk et al. 2010). Thus, soil factors and plant root activities have been shown to strongly influence the soil microbial community.

Physicochemical properties of the soils like texture, structure, density, pH, EC, carbon content, nutrient content, C:N ratio, altitude, ratio of cations (Ca^{2+}, Mg^{2+}, and Al^{3+}), etc., were more or less correlated to suppressiveness (Faoro et al. 2010). These suppressive effects were found variable among different soil types like more effective in sandy organic matter poor soils (Tenuta and Lazarovits 2004) and reduced *Fusarium* diseases at high soil pH (Borrero et al. 2004). Soil biological properties like enzymes, respiration, microbial functions, etc., strongly influenced the soil suppressiveness against multiple soilborne pathogens. Fluorescein diacetate (FDA) hydrolysis is consistently related to suppressiveness of composts on *Pythium* (Chen et al. 1988). FDA hydrolysis was also proposed as a promising indicator for predicting organic matter suppressiveness (Hoitink and Boehm 1999). However, subsequent studies reported contrasting relationships for disease suppression in relation to both OM type and pathogen species (Yulianti et al. 2006). The substrate respiration was also considered as an important indicator for disease suppression as FDA hydrolysis which could be explained by the model of general suppression (Weller et al. 2002).

4.4 Microbial Diversity and Disease Suppression

The microorganisms in the rhizosphere are the key agents for changes in soil agroecosystems. The interactions between plant root systems and microorganisms have an intense effect on crop health, yield, and soil quality. Microorganisms like pathogenic fungi, oomycetes, bacteria, and nematodes adversely affect plant growth and health. In contrast, a wide range of microorganisms are also present which are beneficial to the plant and include nitrogen-fixing bacteria, endo- and ectomycorrhizal fungi, and plant growth-promoting bacteria and fungi (Pal and Gardener 2006). Several microorganisms have been suggested to be involved for general soil suppressiveness like *Trichoderma* spp. (Wiseman et al. 1996), *V. biguttatum* (Velvis et al. 1989), *Pseudomonas* population (Mazzola and Gu 2002), combination of *Pantoea*, *Exiguobacterium*, and *Microbacteria* (Barnett et al. 2006), etc., but their mode of action is still not clear. The nonpathogenic *fusaria* (the most common components of soil microbial communities) and deuteromycetes such as *Penicillium* species are strongly antagonistic to pathogenic *fusaria* (Fravel et al. 2003; Sabuquillo et al. 2005). Actinomycetes are also known to be a strong producer of antibiotics and have a direct influence on disease suppression (Mazzola et al. 2001). Recently fluorescent pseudomonads attained the highest percentage of positive correlation (73 %), followed by sporigenus bacteria (60 %) and *Trichoderma* spp. (56 %) with no cases of negative correlation with suppressiveness (Pal and Gardener 2006). These microbial groups are able to increase plant growth and development through production of phytohormones (Patten and Glick 1996), as biocontrol of phytopathogens in the root zone (Weller 1988), manipulation of ethylene levels (Glick et al. 1998), enhanced availability of minerals (Marschener and Römheld 1994), etc. Several species have been developed as biocontrol agents, with modes of action such as antibiotic production (Whipps 1997) and mycoparasitism (Harman et al. 2004).

Mycorrhizae are the dynamic symbionts between fungi and plants and occur on most terrestrial plant species. These fungi can prevent root infections by reducing the access sites and stimulating host defense. Various mechanisms employed by mycorrhizae to suppress plant pathogens includes intricate network of fungal hyphae around the roots, physical protection, chemical interactions, and other indirect effects like enhanced plant nutrition, increased root lignifications, biochemical changes of the plant tissues (Morris and Ward 1992), alleviation of abiotic stress, changes in mycorrhizosphere biology (Linderman 1994), etc. Specifically, disease protection by ectomycorrhizal fungi may involve multiple mechanisms including antibiosis, synthesis of fungistatic compounds by plant roots in response to mycorrhizal infection, and a physical barrier of the fungal mantle around the plant root (Duchesne 1994). Hence, the rich diversity of the soil microbes provides apparently the incessant resource for suppression of plant diseases (Elizabeth and Jo 1999).

4.5 Role of the Microbiome in Plant Health and Productivity

Soil microbiome provides an important role in disease-suppressive soils along with increased plant productivity (Mendes et al. 2011). Enhanced species richness and diversity resulted into quick recovery from the stresses which might be due to high functional redundancy within the soil microbiome (Nannipieri et al. 2003). The high functional redundancy in soil microbial diversity also confers protection against soilborne diseases (Brussaard et al. 2007; Mendes et al. 2011). This balanced microbiome due to enhanced microbial diversity does not allow pathogens to flourish (Mendes et al. 2011; Schnitzer et al. 2011). Many studies on disease-suppressive ability of particular taxons or group of microbes have been correlated with soil community as a whole (Garbeva et al. 2004; Mendes et al. 2011). For further details, see Chaparro et al. (2012).

Microbial community evenness has been also identified as one of the important factors for community functioning, soil health, and plant productivity (Crowder et al. 2010; Wittebolle et al. 2009). It ensures that no individual microbial taxum is able to flourish and/or upsetting the ecological balance (Elliott and Lynch 1994). Increased competition found in diverse and even microbial communities reduces the niche spaces available for potential invaders (Hillebrand et al. 2008; Naeem et al. 2000), and a lack of community microbial evenness has been associated with reduced plant productivity (Wilsey and Potvin 2000). It is suggested that when environmental fluctuations occur, even communities are quickly able to adapt to the new environment and sustain high productivity over time (Hillebrand et al. 2008; Wittebolle et al. 2009). These examples highlight the benefits of ensuring even and diverse microbial communities to produce healthy soil, high levels of nutrient cycling (Elliott and Lynch 1994) and to combat stress and disease (van Bruggen and Semenov 2000).

4.6 Mechanisms of Suppressive Soils

The ability of soil to suppress disease is of key importance in measuring soil productivity (Janvier et al. 2007). Many factors as discussed in the previous section determine the effectiveness of suppressive soils to combat the invading pathogens in soil–plant systems. The soil suppressiveness encompasses various mechanisms including competition, antibiosis, allelopathy, hyperparasitism, and induction of plant disease resistance (Fig. 4.3), which are being operative through different precursors like soil microbes, soil amendments, cropping systems, etc. (Haas and Défago 2005). Various soil bio-indicators like microbial biodiversity and composition (Garbeva et al. 2006), population density (Postma et al. 2008), the presence of specific antagonists (Postma et al. 2008), the presence of antibiotic genes (Garbeva et al. 2006), or combination of these have been related to soilborne disease suppressiveness. However, these mechanisms responsible for soil suppressiveness are not fully understood, and the effect may also differ depending on the host–pathogen systems (Janvier et al. 2007). The airborne diseases may also be reduced

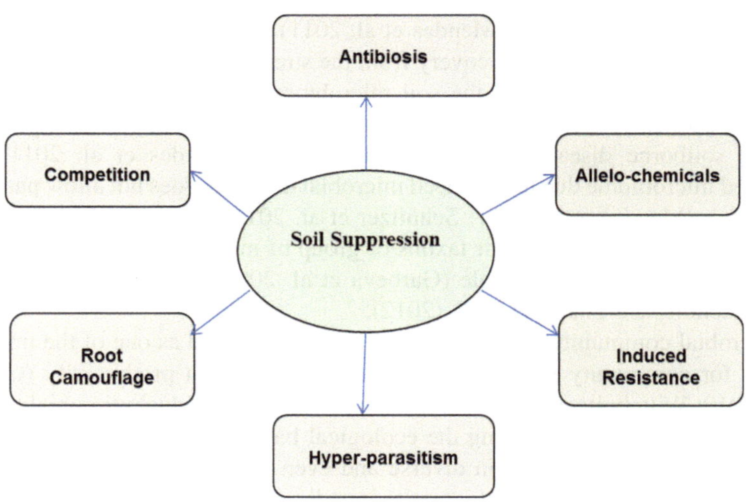

Fig. 4.3 Mechanisms of soil suppression in different agroecosystems

by soil microorganisms and induced systemic resistance (ISR) as reported in experiments under controlled conditions (Kloepper et al. 1999).

4.6.1 Allelochemicals

Allelochemicals are well known to influence a wide variety of soil and crop management-related processes (Sturz and Christie 2003). These processes include soil health, nutrient transfer, weed control, crop compatibility within rotations, residue management, plant growth and development, and disease-suppressive soils (Narwal 2000). Various plant-derived allelochemicals have been identified for weeds (Hoagland and Cutler 2000), fungal pathogens (Lovett and Hoult 1995), nematodes (Sukul 1992), and insects (Jacobsen 1989). Allelochemicals are secondary metabolites comprising lytic agents and enzymes (Glick et al. 1998), antibiotics (Bender et al. 1999), siderophores (Marschener and Crowley 1997), auxins (Patten and Glick 1996), volatile compounds (Claydon et al. 1987), and phytotoxic substances (Hoagland and Cutler 2000). However, the major source of allelochemicals in the rhizosphere is believed to be the plants. These allelochemicals are generated directly or indirectly, from precursor compounds released into the root zone and subsequently transformed through abiotic (i.e., oxidation) or biochemical reactions through the action of microbes or higher organisms (Tang et al. 1989). Suppression of *Meloidogyne incognita* by entomopathogenic nematodes has been proposed to be an allelopathic event mediated by symbiotic bacteria (Grewal et al. 1999).

4.6.2 Niche Competition and Microbiostasis

These are the mechanism that exists between pathogens and other microbial populations (Stephens et al. 1993). The siderophore-producing bacteria with high affinities for iron have been found to inhibit certain phytopathogens in iron-limited soils due to iron deficiency (Dowling et al. 1996). Similarly, by establishing partial sinks for nutrients, rhizobacteria can reduce the amount of carbon and nitrogen available for fungal spore germination and phytopathogen growth in the root zone (Elad and Baker 1985). The action of microbial population against pathogens was also proposed by altering the physical habitat rather than denial of the food source (Lockwood 1988) which were described as substrate antagonism (Lockwood 1986). In soil biostasis, microbial decomposers produce inhibitors during competitive interactions. The spectrum of inhibitors varies with microbial community composition. The inhibitors do not only affect the direct microbial competitors but have also negative effect on other soil-inhabiting organisms (Fig. 4.4). This ability of certain portions of a soil microbial population to impose fungistasis/biostasis appears to be relatively nonspecific. Thus, most isolates of actinomycetes, bacteria, and fungi were capable of initiating some degree of fungistasis (Lockwood 1964) or

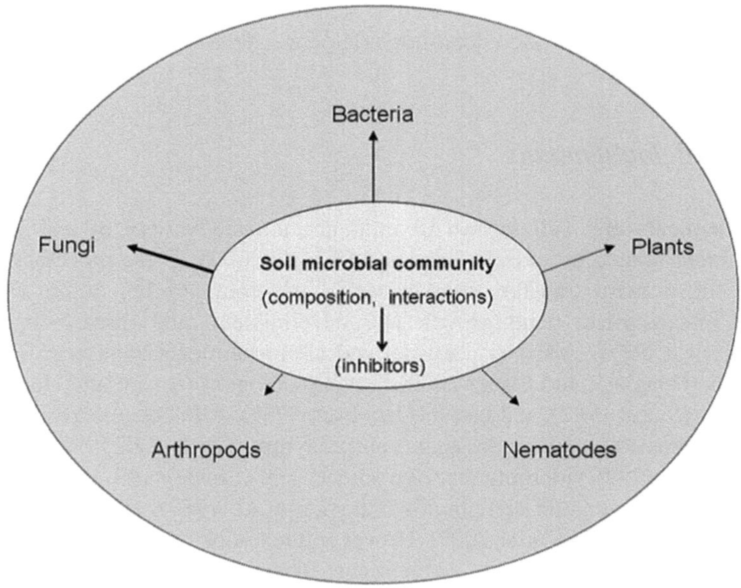

Fig. 4.4 Illustration of the soil biostasis concept. The length, weight, and pattern of the *arrows* illustrate the amount of supporting evidence for this concept (*Source*: Garbeva et al. 2011)

general microbiostasis (Ho and Ko 1986) when applied to sterilized soil or artificial soil, respectively. A more detailed discussion on biostasis can be found in Garbeva et al. (2011).

4.6.3 Antibiosis

Production of specific or nonspecific microbial metabolites, lytic agents, enzymes, volatile compounds, or other microbial toxins is often reported as agents of disease suppression (Fravel 1988; Lambert et al. 1987; Leyns et al. 1990). Antibiotics synthesized by rhizobacteria can contribute to microbial antagonism and persistence in the root zone soil (Kerry 2000). Antibacterial, antifungal, and anti-nematode activity has been identified in the antibiotic-producing strains of a wide range of bacterial genera, but most notably from *Agrobacterium* spp., *Pseudomonas* spp., *Bacillus* spp., *Trichoderma virens*, *Lysobacter* spp., *Pantoea agglomerans*, *Burkholderia cepacia*, etc. (Table 4.1). *P. fluorescens* bacteria that produce the antibiotic 2, 4-diacetylphloroglucinol (DAPG) are well known for their capacity to suppress diverse soilborne diseases (Weller et al. 2002), especially take-all disease of wheat (Raaijmakers and Weller 1998). Antibiotic production confers a competitive ecological advantage to the producer microbe; plants that stimulate root zone colonization by beneficial rhizobacteria will also benefit through the development

Table 4.1 Some of the antibiotics produced by biocontrol agents

Antibiotics	Source	Target pathogen	Disease	References
2, 4-diacetylphloroglucinol	*Pseudomonas fluorescens* F113	*Pythium spp.*	Damping off	Shanahan et al. (1992)
Agrocin 84	*Agrobacterium radiobacter*	*Agrobacterium tumefaciens*	Crown gall	Kerr (1980)
Bacillomycin D	*Bacillus subtilis* AU 195	*Aspergillus flavus*	Aflatoxin contamination	Moyne et al. (2001)
Bacillomycin fengycin	*Bacillus amyloliquefaciens* FZB 42	*Fusarium oxysporum*	Wilt	Koumoutsi et al. (2004)
Xanthobaccin A	*Lysobacter sp.* Strain SB-K88	*Aphanomyces cochlioides*	Damping off	Islam et al. (2005)
Gliotoxin	*Trichoderma virens*	*Rhizoctonia solani*	Root rots	Wilhite et al. (2001)
Herbicolin	*Pantoea agglomerans* C9-1	*Erwinia amylovora*	Fire blight	Sandra et al. (2001)
Iturin A	*B. subtilis* QST713	*Botrytis cinerea* and *R. solani*	Damping off	Paulitz and Belanger (2001), Kloepper et al. (2004)
Mycosubtilin	*B. subtilis* BBG100	*Pythium aphanidermatum*	Damping off	Leclere et al. (2005)
Phenazines	*P. fluorescens* 2–79 and 30-84	*Gaeumannomyces graminis* var. tritici	Take-all	Thomashow et al. (1990)
Pyoluteorin, pyrrolnitrin	*P. fluorescens* Pf-5	*Pythium ultimum* and *R. solani*	Damping off	Howell and Stipanovic (1980)
Pyrrolnitrin pseudane	*Burkholderia cepacia*	*R. solani* and *Pyricularia oryzae*	Damping off and rice blast	Homma et al. (1989)
Zwittermicin A	*Bacillus cereus* UW 85	*Phytophthora medicaginis* and *P. aphanidermatum*	Damping off	Smith et al. (1993)

Source: Pal and Gardener (2006)

of a protective root zone microflora. Similar antibiosis responses may also be delivered by a protective bacterial endophyte flora localized within specific tissues of the host (Sturz et al. 1999).

4.6.4 Induced Systemic Resistance in Plants

Induced systemic resistance (ISR) in plants occurs when root colonization by certain nonpathogenic rhizobacteria stimulates defense-related genes such as those encoding the production of jasmonate (van Wees et al. 1999), peroxidase (Jetiyanon et al. 1997), and enzymes involved in the synthesis of phytoalexins (van Peer et al. 1991). ISR is often described as a heightened state of defense-related preparedness, which may be expressed locally or systemically within the activated plant, and results in either delayed symptom development or reduced disease expression (Liu et al. 1995) but only after pathogen penetration. While bacterial strains can differ in their ability to induce resistance, multiple pathogens may be inhibited by individual strains of rhizobacteria, indicating a general defense mechanism being induced in the plant (Hoffland et al. 1996). Even so, no consistent structural alterations have been identified in plants subjected to ISR, and cultivar-specific variations in the level of the ISR response have been reported (see review by Van Loon et al. 1998; Sturz and Christie 2003). Some of the bacterial determinants and type of host resistance induced by biocontrol agents as described by Pal and Gardener (2006) are given in Table 4.2.

4.6.5 Root Camouflage

Root camouflage (Gilbert et al. 1994) is the concept to explain decreased microbial population densities in the rhizospheres of disease-resistant cultivars (Lochhead et al. 1940). This mechanism was postulated to attract soil pathogens on plant roots (i.e., rhizosphere) than in root-free soil so as to targeting the root system for pathogen attack. A reduction in the population densities of root zone microbial communities was observed to the levels of resistant donor parent in wheat cultivars and to that of the surrounding soil (Neal et al. 1970, 1973). Thus, the presence of the root system is believed to be masked. Further it refers to the mechanisms involved in regulating disease suppression and pathogen reduction as described by Sturz and Christie (2003).

Table 4.2 Bacterial determinants and types of host resistance induced by biocontrol agents

Bacterial strain	Plant species	Bacterial determinants	Type	References
Bacillus mycoides strain Bac J	Sugar beet	Peroxidase, chitinase, and β-1,3-glucanase	ISR	Bargabus et al. (2002)
Bacillus pumilus 203-6	Sugar beet	Peroxidase, chitinase, and β-1,3-glucanase	ISR	Bargabus et al. (2004)
Bacillus subtilis GB03 and IN937a	Arabidopsis	2,3-butanediol	ISR	Ryu et al. (2004)
Pseudomonas fluorescens strains CHA0	Tobacco	Siderophore	SAR	Maurhofer et al. (1994)
	Arabidopsis	Antibiotics (DAPG)	ISR	Iavicoli et al. (2003)
WCS374	Radish	Lipopolysaccharide	ISR	Leeman et al. (1995a, b)
		Siderophore		Leeman et al. (1995a, b)
		Iron-regulated factor		Leeman et al. (1995a, b)
WCS417	Carnation	Lipopolysaccharide	ISR	Van Peer and Schippers (1992)
	Radish	Lipopolysaccharide	ISR	Leeman et al. (1995a, b)
		Iron-regulated factor	ISR	Leeman et al. (1995a, b)
	Arabidopsis	Lipopolysaccharide	–	Van Wees et al. (1997)
	Tomato	Lipopolysaccharide	–	Duijff et al. (1997)
Pseudomonas putida strains	Arabidopsis	Lipopolysaccharide	–	Meziane et al. (2005)
WCS 358	Arabidopsis	Lipopolysaccharide	–	Meziane et al. (2005)
		Siderophore	–	Meziane et al. (2005)
BTP1	Bean	Z,3-hexenal	–	Ongena et al. (2004)
Serratia marcescens 90–166	Cucumber	Siderophore	ISR	Press et al. (2001)

Source: Pal and Gardener (2006)

4.7 Agronomic Management Practices to Develop Suppressive Soils

Agricultural practices like crop rotation (Cook et al. 2002), intercropping (Schneider et al. 2003), tillage and organic amendments (Tilston et al. 2002; Mazzola 2004; Stone et al. 2003), and their combinations (Garbeva et al. 2004)

spectacularly influenced the disease suppressiveness. Although the changes in soil suppressiveness due to altered agricultural management are often site-specific (Cook 2007). It is a well-known fact that agronomic practices have often lead to decreased soil fertility by loss in soil organic matter, soil microbial biomass, soil organisms, and soil structure (Bellamy et al. 2005; Khan et al. 2007); consequently, it affected the soil suppressiveness to control the diseases (Van Bruggen and Semenov 2000). Therefore, better understanding of the processes and the potential to maintain or reestablish disease suppressiveness is essential for developing sustainable agricultural practices.

It is uncertain up to what extent soil suppressiveness could be reestablished by agronomic management practices or by introduction of soil microbial communities in disturbed soils. Numerous strategies have been investigated, altering soil microbial communities to develop soil capacity for suppression of soilborne plant diseases. Crop management practices including crop rotation (Huber and Schneider 1982), input system like organic versus conventional (Workneh et al. 1993; van Bruggen 1995), and tillage and fertilization (Smiley 1978) will influence ecological processes that affect microbial communities involved in the suppression of soilborne plant pathogens. All these observations inferred that based on the knowledge of the operative biological mechanisms, the capacity exists to enhance or diminish the suppressive nature of the resident microbial community through timely application of the appropriate agronomic practices (Hoeper and Alabouvette 1996; Pankhurst et al. 2002). It is well evident that induction of soil suppressiveness is often mediated through transformations in soil microbial communities over time (Liu and Baker 1980; Larkin et al. 1993; Raaijmakers et al. 1997; Mazzola and Gu 2002). Hence, there may be a commendable opportunity to enhance the disease-suppressive state in the soils using various agronomic practices which would be the prerequisite for successful adoption of such disease control strategy. Some of the agronomic practices enhancing soil suppressiveness are summarized as mentioned below.

4.7.1 Organic Amendments

Various organic amendments like cover crops, animal and green manure, organic wastes, plant residues, composts and peats, etc., have been proposed to provide plant nutrition as well as control of diseases caused by soilborne pathogens (Steinberg et al. 2004; Widmer et al. 2002; Cotxarrera et al. 2002). These organic amendments have been successfully used to increase the soil suppressiveness to different diseases in agricultural and horticultural crops (Table 4.3). The effectiveness and the level of disease control obtained depend on many factors like chemical nature of the materials used, the composting process and degree of decomposition, type of microorganisms present, etc. These factors might be the probable reasons for contradictory reports for efficacy of disease control by organic amendments in the soil which seriously hinder the practical use of these amendments as

Table 4.3 Organic amendments and plant disease suppression

S. No.	Organic amendments	Disease suppression	References
1.	Vermicompost	*Phytophthora*	Szczech and Smolinska (2001)
2.	Vermicompost	Chickpea collar rot disease (*Sclerotium rolfsii*)	Sahni et al. (2008)
3.	Vermicompost	*Verticillium* wilt of eggplant	Elmer and Ferrandino (2009)
4.	Hairy vetch (*Vicia villosa*)	Fusarium wilt of watermelon	Zhou and Everts (2004)
5.	Swine manure	Microsclerotia of *Verticillium dahliae*	Tenuta et al. (2002)
6.	Compost tea	Damping off (*Pythium ultimum*)	Scheuerell and Mahaffee (2004)
7.	Composted hardwood bark	*Rhizoctonia* damping off	Nelson et al. (1983)
8.	*Brassica napus* seed meal amendment	Apple root pathogens	Mazzola et al. (2001)
9.	Broccoli residues	*Verticillium* wilt of cauliflower	Koike and Subbarao (2000)
10.	Composted swine Waste	*Rhizoctonia solani* on *Impatiens*	Diab et al. (2003)
11.	Composts	Damping off and root rot (*Pythium graminicola*) of creeping bent grass	Craft and Nelson (1996)
12.	*Brassica napus* seed meal	*Rhizoctonia* root rot	Cohen et al. (2005)
13.	Hardwood bark media	*Rhizoctonia solani*	Chung, et al. (1988)
14.	Synthetic and organic soil fertility amendments	Southern blight of tomato	Bulluck and Ristaino (2002)
15.	Composted municipal biowaste Composted cow manure	*Sclerotinia minor* (garden cress)	Pane et al. (2011)
16.	Vegetal composts	*Rosellinia necatrix* (avocado)	Bonilla et al. (2009)
17.	Fresh farmyard manure	*Rhizoctonia solani* (basil)	Tamm et al. (2010)
18.	Viticulture waste compost Composted cow manure	*Rhizoctonia solani* (garden cress)	Pane et al. (2011)
19.	Bark compost	*Pythium ultimum* (garden cress)	Erhart et al. (1999)

(continued)

Table 4.3 (continued)

S. No.	Organic amendments	Disease suppression	References
20.	Animal and vegetal composts	*Pythium ultimum* (garden cress)	Pane et al. (2011)
21.	Chipped euca-lyptus trimmings	*Phytophthora cinnamomi* (avocado)	Downer et al. (2001)
22.	Sludge vermicompost	*Phytophthora cinnamomi* (avocado)	Bender et al. (1992)
23.	Fresh and composted chicken manure	*Phytophthora cinnamomi* (white lupin)	Aryantha et al. (2000)
24.	Vegetal composts	*Fusarium spp.* on several hosts	Yogev et al. (2006)
25.	Vegetal com-post Poultry manure Green manure (legumes)	*Sclerotium rolfsii* (tomato)	Bulluck and Ristaino (2002)
26.	Horse manure Municipal green waste Wood shavings	*Verticillium dahliae* (eggplant)	Malandraki et al. (2008)
27.	Sewage sludge	*Laetisaria fuciformis*, *Pythium graminicola*, *R. solani*, *Sclerotinia homoeocarpa*, and *Typhula incarnate*	Nelson and Boehm (2002)
28.	*Brassica napus* seed meal	*Rhizoctonia* root rot (*Rhizoctonia solani* AG-5) in apple	Cohen et al. (2005)
29.	*Cruciferous* soil amendments	*Aphanomyces* root rot of peas	Papavizas (1966)
30.	Organic amendments	*Thielaviopsis basicola*	Papavizas (1968)
31.	Organic amendments	*Gaeumannomyces graminis* var. *Tritici* in wheat	Mazzola and Gu (2002), Weller et al. (2002), Tilston et al. (2002)
32.	Organic amendments	*Pythium splendens*	McKellar and Nelson (2003)
33.	Cotton-gin trash	*Sclerotium rolfsii*	Coventry et al. (2005)
34.	Organic amendments	*Macrophomina phaseolina*	Lodha (1995)

disease-suppressive materials (Termorshuizen et al. 2006). The effectivity and suppressive potential of organic amendments could be improved by inoculation of decomposed composts with specific strains of antagonistic microorganisms. Substantial effort has been made during the last decade for reliable indicators of organic matter-suppressive capability (Noble and Coventry 2005; Janvier et al. 2007). Testing of various organic matters on different pathosystems is the traditional approach for identification of characteristics responsible for disease suppression (Scheuerell et al. 2005; Termorshuizen et al. 2007). For instance, FDA hydrolysis assay has been correlated with organic matter decomposition (Schnurer and Rosswall 1982), peat (Boehm et al. 1997), and compost suppressiveness (Chen et al. 1988). The degree of decomposition of the amendments (Hoitink and Boehm 1999; Janvier et al. 2007) is also an important indicator for disease suppression in different plant species.

Significant changes in the correlation between suppressiveness and the level of decomposition have been reported for crop residues (Papavizas and Davey 1960), organic wastes (Kotsou et al. 2004), peats (Boehm et al. 1997), and composts (Diab et al. 2003). The biocontrol effect is sustained for as long as the parent organic matter remains constant for factors like particle size, salinity, pH, carbon-to-nitrogen ratio, lignin-to-cellulose ratio, and moisture (Hoitink et al. 1997). The microbial carrying capacity declines with decomposition of organic matter which ultimately declines the disease suppression. However, mostly biological control agents colonized naturally in the composts at indefinite extent which often leads to reduction in the efficacy or reproducibility effects between batches of composts. Therefore, an understanding of the influence of the degree of organic matter decomposition on the suppression of soilborne disease is essential to improve our predictive capability.

Plant residues left on or near the soil surface may contribute to an increase of disease suppressiveness through the promotion of the general microbial activity. When residues are buried, the pathogens are displaced from their niche to deeper layers in the soil and their ability to survive is severely decreased. Repeated incorporations of crop residues can affect a change in the activity of residue-borne microorganisms that in turn influence the decomposition of crop residues. Carbon released from this decomposition contributes to an increase of soil microbial activity and thereby enhances the level of general suppression. Developing disease-suppressive soils by introducing organic amendments and crop residue management takes time, but the benefits accumulate across successive years, thereby leading to an improvement of soil health and structure (Bailey and Lazarovits 2003).

4.7.2 Soil Solarization and Biofumigation/Biodisinfection

Solarization or solar heating is a method that uses the solar energy to enhance the soil temperature to levels at which many plant pathogens will be killed or sufficiently weakened to obtain significant control of the diseases. Solarization is a hydrothermal process, and its effectiveness is not only related to the temperature but also to the soil moisture. The efficiency of the process can be improved by combining soil solarization and organic amendments (Ndiaye et al. 2007; Oka et al. 2007). The duration of solarization is also an important factor determining the effectiveness of the treatment. Solarization does not destroy all soil microorganisms, but modifies the microbial balance in favor of the beneficial microorganisms. An important characteristic of soil solarization is its broad spectrum of activity including activity against fungi, nematodes, bacteria, weeds, arthropod pests, and some unidentified agents.

Disease control and yield increase have been reported after 2–3 years of solarization (Gallo et al. 2007). This long-term effect is probably due to both the reduction of the inoculums density and some induced level of disease suppressiveness of the soil. Many studies report that the efficacy of soil solarization is not only due to a decrease of pathogen populations but also to an increase of the density and activity of populations of antagonistic microorganisms such as *Bacillus spp.*, *Pseudomonas spp.*, and *Talaromyces flavus*. Several review papers are available that describe both the technology of solar heating and mechanisms involved in the control of pests, pathogens, and weeds by solarization (DeVay 1995; Katan 1996).

Biofumigation or biodisinfection is the strategy based on plastic mulching of the soil after incorporation of fresh organic matter which is suitable for cooler regions (Blok et al. 2000). Although the mechanisms involved are not fully understood, anaerobic fermentation of organic matter under plastic mulch and production of toxic metabolites are the two mechanisms considered to be contributed to the inactivation or destruction of pathogenic fungi. Therefore, two definitions have been proposed by Lamers et al. (2004), that is, biofumigation corresponds to the use of specific plant species containing identified toxic molecules, whereas biodisinfection refers to the use of high quantities of organic matter resulted in anaerobic conditions mainly responsible for the destruction of pathogens. For example, many species of *Brassicaceae* produce glucosinolates, a class of organic molecules that may represent a source of allelopathic control of various soilborne plant pathogens (Kirkegaard and Sarwar 1998).

4.7.3 Soil Tillage

However, it is very difficult to assess the role of tillage on disease suppression as its evaluation is often combined with the effects of other agricultural practices such as organic amendments and green manure burial, residue management, or crop rotations (Bailey and Lazarovits 2003). Therefore, tillage appears as giving conflicting effects on disease suppression. Conventional tillage results in considerable disturbance of the soil but removes residue from the surface. Tillage also disrupts hyphae, thereby affecting the ability of fungi such as *R. solani* to survive (Bailey and Lazarovits 2003). Reduced tillage can also favor pathogens by protecting the pathogen's refuge in the residue from microbial degradation, lowering soil temperature, increasing soil moisture, and leaving soil undisturbed (Bockus and Shroyer 1998).

A variable impact of conservation tillage practices on plant disease development has been reported depending on the specific regional crop–pathogen–environment interactions (Sturz et al. 1997; Bailey et al. 2001). Leaving plant debris on the surface or partially buried in the soil may facilitate the survival of some pathogens until the succeeding crop is planted, but conditions favorable for microbial antagonism of plant pathogens may also be increased (Baker and Cook 1974; Boosalis et al. 1981) under such systems. Soil physical and chemical properties, moisture and temperature, root growth, and pathogen vectors are all influenced by tillage practice, and consequently pathogen virulence, diversity, and host susceptibility are likewise influenced (Sumner et al. 1981). A list of the impacts of minimum tillage on specific crops and their associated pathogens can be found in Sturz et al. (1997). Plant residues left on or near the soil surface may contribute to the suppression of soilborne pathogens in minimum tillage systems.

4.7.4 Crop Rotation

Crop rotation is an agricultural management tool with ancient origins (Howard 1996). Besides the benefits like maintenance of soil health, soil organic matter, reduction in soil erosion, etc., crop rotation spectacularly declined the incidence of plant disease caused by soilborne pathogens (Pedersen and Hughes 1992). Monocropping generally led to the buildup of soil populations of specific plant pathogens resulting in the decline of crop yield and quality (Honeycutt et al. 1996). In contrary, crop rotation with resistant and/or less susceptible to specific pathogens enhanced the crop yield and quality because it declined the pathogen populations due to natural mortality and the antagonistic activities of root zone microorganisms (Fry 1982). Rotation is most successful in limiting the impact of biotrophic pathogens that require living host tissues or those pathogens with low saprophytic survival capability (Bailey and Duczek 1996). However, it is least successful in reducing disease caused by pathogens with a wide host range or that produce long-lived survival structures such as sclerotia or oospores (Umaerus et al. 1989). Crop choice in a rotation may also harvest microbial benefits beyond those normally associated with pathogen host range and saprophytic pathogen survival. For example, analysis of microbial populations in plant tissues and soils when clover preceded potato in a rotation revealed that 25 bacterial species were common to both clover and potatoes and represented 73 % of culturable bacteria recovered from clover roots and potato tubers (Sturz et al. 1998). Endophytic bacteria found inhibitorier to *R. solani* than the bacteria present in the root zone. Therefore, it emphasized that adaptation of bacteria to host plants can result in the expression of a mutually beneficial relationship (Sturz et al. 1998). Crop rotation also influences disease suppressiveness of the soil (Garbeva et al. 2006; Postma et al. 2008). The best examples are take-all disease (Weller et al. 2002), *Rhizoctonia solani* in wheat (Mazzola and Gu 2002), potato (Jager and Velvis 1995), sugar beet (Sayama et al. 2001), radish (Chet and Baker 1980), and cauliflower (Davik and Sundheim 1984). However, knowledge on the mode of action of *Rhizoctonia* disease decline is lacking. In most pathogen–crop combinations, it is unknown if the host crop or the pathogen itself are needed for the development of disease decline. In few cases, it was described that virulent *R. solani* was required to induce *Rhizoctonia* disease decline (Sayama et al. 2001).

4.7.5 Use of Beneficial Microbes

Agricultural management practices impact soil and rhizosphere microbial diversity and community structure. Management of soil properties is an important approach to promote the activities of beneficial microbes in the rhizosphere and thus limiting the densities and activities of soilborne pathogens to a tolerable level (Janvier et al. 2007). Furthermore, soil type is known to be a key determinant for soil microbial community structure (Garbeva et al. 2004). Adaptation of cultural practices has been proposed as a means to decrease the soil inoculum potential or increase the level of suppressiveness to diseases (Steinberg et al. 2007). Hence, it is evident that various cultural and management practices significantly influenced the microbial community structures and activities in the rhizosphere. Tillage (Feng et al. 2003), rotation (Lupwayi et al. 1998; Larkin 2003), use of mulches (Tiquia et al. 2002), cover crops (Schutter et al. 2001; Schutter and Dick 2002), and amendments (Parham et al. 2003; Pérez-Piqueres et al. 2006) are also known to influence the structure and activity of microbial communities.

Adding beneficial microorganisms to those already present in the soil can maximize plant nutrient uptake (Kirankumar et al. 2008), increase plant growth (Cummings 2009; Guiñazú et al. 2009), confer resistance to abiotic stress (Selvakumar et al. 2012), and suppress disease (De Vleesschauwer and Höfte 2009). These living microorganisms are dynamic and potentially self-sustaining, reducing the need for repeated applications, and can avoid the problem of pests and pathogens, evolving resistance to the treatments (Lucas 2011). A possible management technique is to apply plant growth-promoting rhizobacteria (PGPRs) as an agricultural treatment to minimize niche vacancy and effectively fill vacant niches. It has been shown that PGPRs colonize particularly and effectively in soils with low microbial biomass (Fliessbach et al. 2009) so inoculations are more likely to be successful. Beneficial microorganisms that thrive in this environment can more quickly take up space and nutrients made available for potential pathogen invaders and assist with achieving sustained niche occupancy (Kaymak 2011). In addition, PGPRs offer benefits of increased yields, nutrient acquisition, stress tolerance, and disease resistance to the plant host (Lugtenberg and Kamilova 2009). The application of PGPRs consortia has been shown to be even more effective than one treatment alone in suppressing disease (Ahemad and Khan 2011; Yang et al. 2011). This combination of beneficial microbes also had the added effect of stimulating plant N and P absorption (Hernandez and Chailloux 2004). Formulations of compost with beneficial bacteria have also shown the ability to suppress plant pathogens (Pugliese et al. 2011; Yang et al. 2011). The ability of formulations of multiple beneficial microbes to increase plant productivity and health hints at the potential of the entire microbiome and plants working together with mutually beneficial outcomes.

Sometimes, the same effect can be achieved by applying a microbial elicitor (compound produced by the microorganism and causes the desired effect). For example, exogenous application of the *Bacillus subtilis*-derived elicitor, acetoin (3-hydroxy-2-butanone), was found to trigger induced systemic resistance (ISR) and protect plants against *Pseudomonas syringae pv tomato* pathogenesis (Rudrappa et al. 2008). Similarly, adding low doses of *Chryseobacterium balustinum* AUR9 cell wall lipopolysaccharides, another bacterial elicitor, to A. thaliana reproduced systemic induction (Ramos Solano et al. 2008). Applications of living microbes or their elicitors have potential use for agricultural priming, the induction of ISR (Conrath and Loon 2009), which has been shown as an efficient way to increase pathogen resistance with little cost to the plant (De Vleesschauwer and Höfte 2009). An important addition to strategic management practices will be the development of crop species that are able to accomplish their own priming and ISR induction, which will reduce the use of microbial applications. Although, ideally, adding PGPRs as inoculants into the rhizosphere to exploit the immense benefits they provide is, potentially, an easy fix, there is still much inconsistency in their performance at the field scale (Morrissey et al. 2004; Mark et al. 2006). Research has begun to focus on how to cater the rhizosphere environment for PGPR rhizosphere colonization by means of rhizosphere engineering (Ryan et al. 2009), by understanding which PGPR traits are essential for rhizosphere competence (Barret et al. 2011), or by considering which indigenous soil microbial communities respond most favorably to inoculation (Bernard et al. 2012).

4.8 Conclusion

The evergrowing human population coupled with reduced natural resources and the need for more environmentally friendly agricultural practices have highlighted the need for sustainable farming. The intricacies of the plant–microbiome interaction and its impact on plant health and productivity need to be understood for obtaining healthier and more productive plants. Suppressive soils represent an underutilized resource for the control of soilborne pathogens of food, fiber, and ornamental crops. Early research identified the characteristics of soil suppressiveness and the major groups of microorganisms involved, but in recent past due to availability of molecular tools, it has been made possible to characterize and identify the factors and mechanisms responsible for genetic and functional determinants underlying the activity of some biologically suppressive soils. Adoption of different agronomic practices by the farmers spectacularly altered the soil microbiome and considerably enhanced soil suppressiveness to various soilborne diseases. The use of organic amendments or composts for the suppression of plant pathogens could be a promising and environmentally benign alternative to chemical pesticides. The deeper understanding of microbial ecology processes could also provide directions for possible manipulations of the community, leading to a reproducible suppressive amendment. Combining measures of microbial structural diversity with functional traits should be explored in relation to soil and root health in agricultural systems.

Manipulating soil quality to achieve an economic level of disease control via agronomic management has been deliberately reviewed with some skepticism. However, crop rotation, residue management practices, and various forms of organic amendments do contribute to the suppression of soilborne diseases. However, the level of understanding for the mechanisms involved in suppressive soils is still limited and not so clear. The benefits of applying organic amendments for disease control are incremental and long lasting depending upon soil ecosystems. The conventional agricultural systems need to be discouraged because of poor production efficiency due to reduced crop diversity, increased genetic uniformity, and shorter rotations. More attention is to be paid on conservation agriculture including maximum use of natural resources. Through the application of green and livestock manures, mulches, and composts, it is hoped that plant beneficial soil microbial populations will develop spontaneously. Selection of complementary rotation crops may also increase the buildup of beneficial microflora during successive field seasons. Plants can manage the development of beneficial microbial populations through the release of specific root exudates in the root zone. Recently, it has been proposed that plants may also be able to camouflage their presence to phytopathogens by blending into the soil microbial background through restricting the proliferation of root zone bacterial populations. Therefore, the future studies of biologically based soil suppressiveness will put new insights into the microbial ecology of agricultural soils and lay the foundation for the development of creative management strategies for the suppression of soilborne diseases.

References

Ahemad M, Khan MS (2011) Plant growth promoting rhizobacteria: recent advancements. Insight Microbiol 1:39–54

Aryantha IP, Cross R, Guest DI (2000) Suppression of *Phytophthora cinnamomi* in potting mixes amended with uncomposted and composted animal manures. Phytopathology 90:775–782

Badri DV, Vivanco JM (2009) Regulation and function of root exudates. Plant Cell Environ 32: 666–681

Bailey KL, Duczek LJ (1996) Managing cereal diseases under reduced tillage. Can J Plant Pathol 18:159–167

Bailey KL, Lazarovits G (2003) Suppressive soil borne diseases with residue management and organic amendments. Soil Till Res 72:169–180

Bailey KL, Gossen BD, Lafond GP, Watson PR, Derksen DA (2001) Effect of tillage and crop rotation on root and foliar diseases of wheat and pea in Saskatchewan from 1991 to 1998: univariate and multivariate analyses. Can J Plant Sci 81:789–803

Bais HP, Weir TL, Perry LG, Gilroy S, Vivanco JM (2006) The role of root exudates in rhizosphere interactions with plants and other organisms. Annu Rev Plant Biol 57:233–266

Baker KF, Cook RJ (1974) Biological control of plant pathogens. W.H. Freeman and Company, San Francisco, 433 pp

Bargabus RL, Zidack NK, Sherwood JW, Jacobsen BJ (2002) Characterization of systemic resistance in sugar beet elicited by a non-pathogenic, phyllosphere colonizing Bacillus mycoides, biological control agent. Physiol Mol Plant Pathol 61:289–298

Bargabus RL, Zidack NK, Sherwood JW, Jacobsen BJ (2004) Screening for the identification of potential biological control agents that induce systemic acquired resistance in sugar beet. Biol Control 30:342–350

Barnett SJ, Roget DK, Ryder MH (2006) Suppression of Rhizoctonia solani AG-8 induced disease on wheat by the interaction between *Pantoea*, *Exiguobacterium*, and *Microbacteria*. Austr J Soil Res 44:331–342

Barret M, Morrissey JP, O'Gara F (2011) Functional genomics analysis of plant growth-promoting rhizobacterial traits involved in rhizosphere competence. Biol Fertil Soils 47:729–743

Bellamy PH, Loveland PJ, Bradley RI, Lark RM, Kirk GJD (2005) Carbon losses from all soils across England and Wales 1978–2003. Nature 437:245–248

Bender GS, Casale WL, Rahimian M (1992) Use of worm-composted sludge as a soil amendment for avocados in Phytophthora-infested soil. In: Proceeding of second world avocado congress, Orange, CA, p 143

Bender CL, Rangaswamy V, Loper J (1999) Polyketide production by plant associated pseudomonads. Annu Rev Phytopathol 37:175–196

Berg G, Roskot N, Steidle A, Eberl L, Zock A, Smalla K (2002) Plant-dependent genotypic and phenotypic diversity of antagonistic rhizobacteria isolated from different *Verticillium* host plants. Appl Environ Microbiol 68:3328–3338

Bernard E, Larkin RP, Tavantzis S, Erich MS, Alyokhin A, Sewell G, Lannan A, Gross SD (2012) Compost, rapeseed rotation, and biocontrol agents significantly impact soil microbial communities in organic and conventional potato production systems. Appl Soil Ecol 52:2–9

Blok WJ, Lamers JG, Termorshuizen AJ, Bollen GJ (2000) Control of soilborne plant pathogens by incorporating fresh organic amendments followed by tarping. Phytopathology 90:253–259

Bockus WW, Shroyer JP (1998) The impact of reduced tillage on soilborne plant pathogens. Annu Rev Phytopathol 36:485–500

Boehm MJ, Wu T, Stone AG, Kraakman B, Iannotti DA, Wilson GE, Madden LV, Hoitink HAJ (1997) Cross polarized magic-angle spinning 13C nuclear magnetic resonance spectroscopic characterization of soil organic matter relative to culturable bacterial species composition and sustained biological control of pythium root rot. Appl Environ Microbiol 63:162–168

Bonilla N, Torés JA, Hermoso JM, González J, Cazorla FM, de Vicente A (2009) Biological control of avocado root rots by suppressive organic amendments. In: Elad et al. (eds) IOBC/WPRS Bulletin: biological control of fungal and bacterial plant pathogens, vol 43. Interlaken, Switzerland, pp 231–234

Boosalis MG, Doupnik B, Odvody GN (1981) Conservation tillage in relation to plant diseases. In: Piementel D (ed) Handbook of pest management in agriculture, vol 1. CRC, Boca Raton, FL, pp 445–474

Borrero C, Trillas MI, Ordovas J, Tello JC, Aviles M (2004) Predictive factors for the suppression of Fusarium wilt of tomato in plant growth media. Phytopathology 94:1094–1101

Bowen GD, Rovira AD (1999) The rhizosphere and its management to improve plant growth. Adv Agron 66:1–102

Brussaard L, de Ruiter PC, Brown GG (2007) Soil biodiversity for agricultural sustainability. Agric Ecosyst Environ 121:233–244

Bulluck LR, Ristaino JB (2002) Effect of synthetic and organic soil fertility amendments on southern blight, soil microbial communities, and yield of processing tomatoes. Phytopathology 92:181–189

Chaparro JM, Sheflin AM, Manter DK, Vivanco JM (2012) Manipulating the soil microbiome to increase soil health and plant fertility. Biol Fertil Soils 48:489–499

Chen W, Hoitink HAJ, Schmitthenner AF, Tuovinen OH (1988) The role of microbial activity in suppression of damping-off caused by *Pythium ultimum*. Phytopathology 78:314–322
Chet I, Baker R (1980) Induction of suppressiveness to *rhizoctonia solani* in soil. Phytopathology 70:994–998
Chung YR, Hoitink HAJ, Dick WA, Herr LJ (1988) Effects of organic matter decomposition level and cellulose amendment on the inoculum potential of *Rhizoctonia solani* in hardwood bark media. Phytopathology 78:836–840
Claydon N, Allan M, Hanson JR, Avent AG (1987) Antifungal alkyl pyrones of *Trichoderma harzianum*. Trans Br Mycol Soc 88:503–513
Cohen MF, Yamasaki H, Mazzola M (2005) Brassica napus seed meal soil amendment modifies microbial community structure, nitric oxide production and incidence of Rhizoctonia root rot. Soil Biol Biochem 37:1215–1227
Conrath U, Loon LCV (2009) Priming of induced plant defense responses. In: Van Loon LC (ed) Advances in botanical research, vol 51. Academic, London, pp 361–395
Cook RJ (2007) Management of resident plant growth-promoting rhizobacteria with the cropping system: a review of experience in the US Pacific Northwest. Eur J Plant Pathol 119:255–264
Cook RJ, Schillinger WF, Christensen NW (2002) Rhizoctonia root rot and take-all of wheat in diverse direct-seed spring cropping systems. Can J Plant Pathol 24:349–358
Cotxarrera L, Trillas-Gay MI, Steinberg C, Alabouvette C (2002) Use of sewage sludge compost and *Trichoderma asperellum* isolates to suppress fusarium wilt of tomato. Soil Biol Biochem 34:467–476
Coventry E, Noble R, Mead A, Whipps JM (2005) Suppression of allium white rot (*Sclerotium cepivorum*) in different soils using vegetable wastes. Eur J Plant Pathol 111:101–112
Craft CM, Nelson EB (1996) Microbial properties of composts that suppress damping-off and root rot of creeping bentgrass caused by *Pythium graminicola*. Appl Environ Microbiol 62: 1550–1557
Crowder DW, Northfield TD, Strand MR, Snyder WE (2010) Organic agriculture promotes evenness and natural pest control. Nature 466:109–112
Cummings SP (2009) The application of plant growth promoting rhizobacteria (PGPR) in low input and organic cultivation of graminaceous crops; potential and problems. Environ Biotechnol 5:43–50
Davik JE, Sundheim L (1984) Induced suppression of Rhizoctonia solani in Norwegian soils. Sci Rep Agric Univ Norway 63:1–12
De Vleesschauwer D, Höfte M (2009) Rhizobacteria-induced systemic resistance. In: Van Loon LC (ed) Advances in botanical research, vol 51. Elsevier, Burlington, pp 223–281
De-la-Pena C, Badri DV, Lei Z, Watson BS, Brandao MM, Silva-Filho MC, Sumner LW, Vivanco JM (2010) Root secretion of defense related proteins is development-dependent and correlated with flowering time. J Biol Chem 285:30654–30665
DeVay JE (1995) Solarization: an environmental-friendly technology for pest management. Arab J Plant Prot 13:56–61
Diab HG, Hu S, Benson DM (2003) Suppression of Rhizoctonia solani on impatiens by enhanced microbial activity in composted swine waste-amended potting mixes. Phytopathology 93:1115–1123
Dowling DN, Sexton R, Fenton A, Delany I, Fedi S, McHugh B, Callanan M, Moënne-Loccoz Y, O'Gara F (1996) Iron regulation in plant-associated *Pseudomonas fluorescens* M114: implications for biological control. In: Nakazawa K, Furukawa K, Haas D, Silver S (eds) Molecular biology of pseudomonads. ASM, Washington, DC, pp 502–511
Downer AJ, Menge JA, Pond E (2001) Association of cellulytic enzyme activities in eucalyptus mulches with biological control of *Phytophthora cinnamomi*. Phytopathology 91: 847–855
Duchesne LC (1994) Role of ectomycorrhizal fungi in biocontrol. In: Pfleger F, Linderman RG (eds) Mycorrhizae and plant health. APS, St. Paul, MN, pp 27–45

Duijff BJ, Gianinazzi-Pearson V, Lemanceau P (1997) Involvement of the outer- membrane lipopolysaccharides in the endophytic colonization of tomato roots by biocontrol *Pseudomonas fluorescens* WCS417r. New Phytol 135:325–334
Elad Y, Baker R (1985) The role of competition for iron and carbon in suppression of chlamydospore germination of *Fusarium* spp. by *Pseudomonas* spp. Phytopathology 75:1053–1059
Elizabeth ABE, Jo H (1999) Biocontrol of plant disease: a (Gram-) positive perspective-MiniReview. FEMS Microbiol Lett 171:1–9
Elliott LF, Lynch JM (1994) Biodiversity and soil resilience. In: Greenland DJ, Szabolcs I (eds) Soil resilience and sustainable land use. CAB International, Wallingford, pp 353–364
Elmer WH, Ferrandino FJ (2009) Suppression of verticillium wilt of eggplant by earthworms. Plant Dis 93(5):485–489
Elmholt S (1991) Side effects of propiconazole (Tilt 250 EC) on non-target soil fungi in a field trial compared with natural stress effects. Microb Ecol 22:99–108
Erhart E, Burian K, Hartl W, Stich K (1999) Suppression of *Pythium ultimum* by biowaste composts in relation to compost microbial biomass, activity and content of phenolic compounds. J Phytopathol 147:299–305
Faoro H, Alves AC, Souza EM, Rigo LU, Cruz LM, Al-Janabi SM, Monteiro RA, Baura VA, Pedrosa FO (2010) Influence of soil characteristics on the diversity of bacteria in the Southern Brazilian Atlantic Forest. Appl Environ Microbiol 76:4744–4749
Feng Y, Motta AC, Reeves DW, Burmester CH, van Santen E, Osborne JA (2003) Soil microbial communities under conventional till and no-till continuous cotton systems. Soil Biol Biochem 35:1693–1703
Fierer N, Jackson RB (2006) The diversity and biogeography of soil bacterial communities. Proc Natl Acad Sci U S A 103:626–631
Fliessbach A, Winkler M, Lutz MP, Oberholzer HR, Mäder P (2009) Soil amendment with *Pseudomonas fluorescens* CHA0: lasting effects on soil biological properties in soils low in microbial biomass and activity. Microb Ecol 57:611–623
Flores HE, Vivanco JM, Loyola-Vargas VM (1999) 'Radicle' biochemistry: the biology of root-specific metabolism. Trends Plant Sci 4:220–226
Fravel DR (1988) Role of antibiosis in the biocontrol of plant diseases. Annu Rev Phytopathol 26: 75–79
Fravel D, Olivain C, Alabouvette C (2003) *Fusarium oxysporum* and its biocontrol. New Phytol 157:493–502
Frey SD, Knorr M, Parrent JL, Simpson RT (2004) Chronic nitrogen enrichment affects the structure and function of the soil microbial community in temperate hardwood and pine forests. For Ecol Manag 196:159–171
Fry WE (1982) Principles of plant disease management. Academic, New York, 378 pp
Gallo L, Siverio F, Rodriguez-Perez AM (2007) Thermal sensitivity of *Phytophthora cinnamomi* and long-term effectiveness of soil solarisation to control avocado root rot. Ann Appl Biol 150: 65–73
Garbeva P, Van Veen JA, Van Elsas JD (2004) Microbial diversity in soil: selection of microbial populations by plant and soil type and implications for disease suppressiveness. Annu Rev Phytopathol 42:243–270
Garbeva P, Postma J, van Veen JA, van Elsas JD (2006) Effect of above-ground plant species on soil microbial community structure and its impact on suppression of *Rhizoctonia solani* AG3. Environ Microbiol 8:233–246
Garbeva P, Hol WHG, Termorshuizen AJ, Kowalchuk GA, de Boer W (2011) Fungistasis and general soil biostasis-A new synthesis: review. Soil Biol Biochem 43:469–477
Gilbert GS, Handelsman J, Parke JL (1994) Root camouflage and disease control. Phytopathology 84:222–225
Girvan MS, Bullimore J, Pretty JN, Osborn AM, Ball AS (2003) Soil type is the primary determinant of the composition of the total and active bacterial communities in arable soils. Appl Environ Microbiol 69:1800–1809

Glick BR, Shah S, Li J, Penrose DM, Moffat BA (1998) Isolation and characterization of ACC deaminase genes from two plant-growth promoting rhizobacteria. Can J Microbiol 44:833–843
Grewal PS, Lewis EE, Ventkatachari S (1999) Allelopathy: a possible mechanism of suppression of plant-parasitic nematodes by entomopathogenic nematodes. Nematology 1:735–743
Guiñazú LB, Andrés JA, Del Papa MF, Pistorio M, Rosas SB (2009) Response of alfalfa (Medicago sativa L.) to single and mixed inoculation with phosphate-solubilizing bacteria and *Sinorhizobium meliloti*. Biol Fertil Soils 46:185–190
Haas D, Défago G (2005) Biological control of soil-borne pathogens by fluorescent pseudomonads. Nat Rev Microbiol 3:307–319
Harman E, Howell R, Viterbo A, Chet I, Lorito M (2004) *Trichoderma* species-opportunistic, avirulent plant symbionts. Nat Rev Microbiol 2:43–56
Hernandez M, Chailloux M (2004) Las micorrizas arbusculares y las bacterias rizosfericas como alternativa a la nutricion mineral del tomate. Cultivos Tropicales 25:5–16
Hillebrand H, Bennett DM, Cadotte MW (2008) Consequences of dominance: a review of evenness effects on local and regional ecosystem processes. Ecology 89:1510–1520
Ho WC, Ko WH (1986) Microbiostasis by nutrient deficiency shown in natural and synthetic soils. J Gen Microbiol 132:2807–2815
Hoagland RE, Cutler SJ (2000) Plant microbial compounds as herbicides. In: Narwal SS, Hoagland RE, Dilday RH, Reigosa MJ (eds) Allelopathy in ecological agriculture and forestry, Proceedings of the III international congress on allelopathy in ecological agriculture and forestry, Dharwad, India, 18–21 Aug 1998. Kluwer Academic, London, pp 73–99
Hoeper H, Alabouvette C (1996) Importance of physical and chemical soil properties in the suppressiveness of soils to plant diseases. Eur J Soil Biol 32:41–58
Hoffland E, Hakulinen J, Van Pelt JA (1996) Comparison of systemic resistance induced by avirulent and non-pathogenic *Pseudomonas* species. Phytopathology 86:757–762
Hoitink H, Boehm M (1999) Biocontrol within the context of soil microbial communities: a substrate-dependent phenomenon. Annu Rev Phytopathol 37:427–446
Hoitink HAJ, Stone AG, Han DY (1997) Suppression of plant diseases by composts. HortSci 32: 184–187
Homma Y, Kato Z, Hirayama F, Konno K, Shirahama H, Suzui T (1989) Production of antibiotics by Pseudomonas cepacia as an agent for biological control of soil-borne plant pathogens. Soil Biol Biochem 21:723–728
Honeycutt CW, Clapham WM, Leach SS (1996) Crop rotation and N fertilization effects on growth, yield, and disease incidence in potato. Am Potato J 73:45–61
Hornby D (1983) Suppressive soils. Annu Rev Phytopathol 21:65–85
Howard RJ (1996) Cultural control of plant diseases: a historical perspective. Can J Plant Pathol 18:145–150
Howell CR, Stipanovic RD (1980) Suppression of Pythium ultimum induced damping - off of cotton seedlings by *Pseudomonas fluorescens* and its antibiotic, pyoluteorin. Phytopathology 70:712–715
Huber DM, Schneider RW (1982) The description and occurrence of suppressive soils. In: Schneider RW (ed) Suppressive soils and plant diseases. APS, St Paul, MN, pp 1–7
Iavicoli A, Boutet E, Buchala A, Métraux JP (2003) Induced systemic resistance in *Arabidopsis thaliana* in response to root inoculation with *Pseudomonas fluorescens* CHA0. Mol Plant-Microbe Interact 16:851–858
Islam TM, Hashidoko Y, Deora A, Ito T, Tahara S (2005) Suppression of damping-off disease in host plants by the rhizoplane bacterium *Lysobacter* sp. strain SB-K88 is linked to plant colonization and antibiosis against soilborne peronosporomycetes. Appl Environ Microbiol 71:3786–3796
Jacobsen M (1989) Botanical pesticides: past, present and future. In: Arnason JT, Philogene BJR, Morand P (eds) Insecticides of plant origin, vol 387, ACS symposium series. ACM, Washington, DC, pp 1–10

Jager G, Velvis H (1995) Dynamics of *Rhizoctonia solani* (black scurf) in successive potato crops. Eur J Plant Pathol 10:467–478

Janvier C, Villeneuve F, Alabouvette C, Edel-Hermann V, Mateille T, Steinberg C (2007) Soil health through soil disease suppression: which strategy from descriptors to indicators? Soil Biol Biochem 39:1–23

Jetiyanon K, Tuzun S, Kloepper J (1997) Lignification, peroxidase and superoxide dismutases as early plant defense reactions associated with PGPR-mediated induced systemic resistance. In: Kobayashi K, Homma Y, Kodama F, Kondo N, Akino S (eds) Plant growth-promoting rhizobacteria: present status and future prospects. OECD, Paris, France, pp 265–268

Katan J (1996) Cultural practices and soil-borne disease management. In: Utkhede RS, Gupta VK (eds) Management of soil-borne diseases. Kalyani, New Delhi, pp 100–122

Kaymak HC (2011) Plant growth and health promoting bacteria. In: Maheshwari DK (ed) Microbiology monographs, vol 18. Springer, Berlin, pp 45–79

Kerr A (1980) Biological control of crown gall through production of agrocin 84. Plant Dis 64: 25–30

Kerry BR (2000) Rhizosphere interactions and the exploitation of microbial agents for the biological control of plant-parasitic nematodes. Annu Rev Phytopathol 38:423–441

Khan SA, Mulvaney RL, Ellsworth TR, Boast CW (2007) The myth of nitrogen fertilization for soil carbon sequestration. J Environ Qual 36:1821–1832

Kirankumar R, Jagadeesh KS, Krishnaraj PU, Patil MS (2008) Enhanced growth promotion of tomato and nutrient uptake by plant growth promoting rhizobacterial isolates in presence of tobacco mosaic virus pathogen. Karnataka J Agric Sci 21:309–311

Kirkegaard JA, Sarwar M (1998) Biofumigation potential of brassicas. I. Variation in glucosinolate profiles of diverse field-grown brassicas. Plant Soil 201:71–89

Kloepper JE, Rodriguez-Ubana R, Zehnder GW, Murphy JF, Sikora E, Fernandez C (1999) Plant root–bacterial interactions in biological control of soilborne diseases and potential extension to systemic and foliar diseases. Austr Plant Pathol 28:21–26

Kloepper JW, Ryu CM, Zhang S (2004) Induce systemic resistance and promotion of plant growth by *Bacillus* spp. Phytopathology 94:1259–1266

Koike ST, Subbarao KV (2000) Broccoli residues can control verticillium wilt of cauliflower. Calif Agric 54:30–33

Kotsou M, Mari I, Lasaridi K, Chatzipavlidis I, Balis C, Kyriacou A (2004) The effect of olive oil mill wastewater (OMW) on soil microbial communities and suppressiveness against *Rhizoctonia solani*. Appl Soil Ecol 26:113–121

Koumoutsi A, Chen XH, Henne A, Liesegang H, Gabriele H, Franke P, Vater J, Borris R (2004) Structural and functional characterization of gene clusters directing nonribosomal synthesis of bioactive lipopeptides in *Bacillus amyloliquefaciens* strain FZB42. J Bacteriol 186:1084–1096

Lambert B, Leyns F, Van Rooyen L, Gosselé F, Papon Y, Swings J (1987) Rhizobacteria of maize and their fungal activities. Appl Environ Microbiol 53:1866–1871

Lamers J, Wanten P, Blok W (2004) Biological soil disinfestation: a safe and effective approach for controlling soilborne pests and diseases. Agroindustria 3:289–291

Larkin RP (2003) Characterization of soil microbial communities under different potato cropping systems by microbial population dynamics, substrate utilization, and fatty acid profiles. Soil Biol Biochem 35:1451–1466

Larkin RP, Honeycutt CW (2006) Effects of different 3-year cropping systems on soil microbial communities and rhizoctonia disease of potato. Phytopathology 96:68–79

Larkin RP, Hopkins DL, Martin FN (1993) Effect of successive watermelon plantings on *Fusarium oxysporum* and other microorganisms in soils suppressive and conducive to fusarium-wilt of watermelon. Phytopathology 83:1097–1105

Leclere V, Bechet M, Adam A, Guez JS, Wathelet B, Ongena M, Thonart P, Gancel F, Chollet-Imbert M, Jacques P (2005) Mycosubtilin overproduction by *Bacillus subtilis* BBG100 enhances the organism's antagonistic and biocontrol activities. Appl Environ Microbiol 71: 4577–4584

Leeman M, Van Pelt JA, Den Ouden FM, Heinbroek M, Bakker PAHM (1995a) Induction of systemic resistance by *Pseudomonas fluorescens* in radish cultivars differing in susceptibility to fusarium wilt, using novel bioassay. Eur J Plant Pathol 101:655–664

Leeman M, Van Pelt JA, Hendrickx MJ, Scheffer RJ, Bakker PAHM (1995b) Biocontrol of fusarium wilt of radish in commercial greenhouse trials by seed treatment with Pseudomonas fluorescens WCS374. Phytopathology 85:1301–1305

Leyns F, Lambert B, Joos H, Swings J (1990) Antifungal bacteria from different crops. In: Hornby D (ed) Biological control of soil-borne plant pathogens. CAB International, Wallingford, Oxfordshire, pp 437–444

Linderman RG (1994) Role of AM fungi in biocontrol. In: Pfleger FL, Linderman RG (eds) Mycorrhizae and plant health. APS Press, St. Paul, MN, pp 1–25

Liu S, Baker R (1980) Mechanism of biological control in soil suppressive to *Rhizoctonia solani*. Phytopathology 70:404–412

Liu D, Anderson NA, Kinkel LL (1995) Biological control of potato scab in the field with antagonistic *Streptomyces* scabies. Phytopathology 85:827–831

Lochhead AG, Timonin MI, West PM (1940) The microflora of the rhizosphere in relation to resistance of plants to soilborne plant pathogens. Sci Agric 20:414–418

Lockwood JL (1964) Soil fungistasis. Annu Rev Phytopathol 2:341–362

Lockwood JL (1986) Soilborne plant pathogens: concepts and connections. Phytopathology 76: 20–27

Lockwood JL (1988) Evolution of concepts associated with soilborne plant pathogens. Annu Rev Phytopathol 26:93–121

Lodha S (1995) Soil solarization, summer irrigation and amendments for the control of *fusarium oxysporum* f. sp. cumini and *macrophomina phaseolina* in arid soils. Crop Prot 14:215–219

Lovett JV, Hoult AHC (1995) Allelopathy and self defence in barley. In: Inderjit, Dakshini KMM, Einhellig FA (eds) Allelopathy: organisms, processes and applications. ACS symposium series, Washington, DC, pp 170–183

Lucas JA (2011) Advances in plant disease and pest management. J Agric Sci 149:91–114

Lugtenberg B, Kamilova F (2009) Plant-growth-promoting rhizobacteria. Annu Rev Microbiol 63: 541–556

Lupwayi NZ, Rice WA, Clayton GW (1998) Soil microbial diversity and community structure under wheat as influenced by tillage and crop rotation. Soil Biol Biochem 30:1733–1741

Malandraki I, Tjamos SE, Pantelides IS, Paplomatas EJ (2008) Thermal inactivation of compost suppressiveness implicates possible biological factors in disease management. Biol Control 44:180–187

Mark G, Morrissey JP, Higgins P, O'Gara F (2006) Molecular-based strategies to exploit Pseudomonas biocontrol strains for environmental biotechnology applications. FEMS Microbiol Ecol 56:167–177

Marschener P, Crowley DE (1997) Iron stress and pyoverdin production by a fluorescent pseudomonad in the rhizosphere of white lupin (*Lupinus albus* L.) and barley (*Hordeum vulgare* L.). Appl Environ Microbiol 63:277281

Marschener P, Römheld V (1994) Strategies of plants for the acquisition of iron. Plant Soil 165: 261–274

Martin FN (2003) Development of alternative strategies for management of soilborne pathogens currently controlled with methyl bromide. Annu Rev Phytopathol 41:325–350

Maurhofer M, Hase C, Meuwly P, Metraux JP, Defago G (1994) Induction of systemic resistance to tobacco necrosis virus by the root-colonizing *Pseudomonas fluorescens* strain CHA0: influence of the gacA gene and of pyoverdine production. Phytopathology 84:139–146

Mazzola M (2004) Assessment and management of soil microbial community structure for disease suppression. Annu Rev Phytopathol 42:35–59

Mazzola M, Gu YH (2002) Wheat genotype-specific induction of soil microbial communities suppressive to disease incited by *Rhizoctonia solani* anastomosis group (AG)-5 and AG-8. Phytopathology 92:1300–1307

Mazzola M, Granastein DM, Elfving DC, Mullinix K (2001) Suppression of specific apple root pathogens by *Brassica napus* seed meal amendment regardless of glucosinolate content. Phytopathology 91:673–679

Mckellar ME, Nelson EB (2003) Compost-induced suppression of Pythium damping-off is mediated by fatty-acid metabolizing seed-colonizing microbial communities. Appl Environ Microbiol 69:452–460

Mendes R, Kruijt M, de Bruijn I, Dekkers E, van der Voort M, Schneider JHM, Piceno YM, DeSantis TZ, Andersen GL, Bakker PAHM (2011) Deciphering the rhizosphere microbiome for disease-suppressive bacteria. Science 332:1097–1100

Menzies JG (1959) Occurrence and transfer of a biological factor in soil that suppresses potato scab. Phytopathology 49:648–652

Meziane H, Van der Sluis I, Van Loon LC, Hofte M, Bakker PAHM (2005) Determinants of *Pseudomonas putida* WCS358 involved in inducing systemic resistance in plants. Mol Plant Pathol 6:177–185

Micallef SA, Shiaris MP, Colon-Carmona A (2009) Influence of *Arabidopsis thaliana* accessions on rhizobacterial communities and natural variation in root exudates. J Exp Bot 60:1729–1742

Morris PF, Ward EWR (1992) Chemoattraction of zoospores of the plant soybean pathogen, *Phytophthora sojae*, by isoflavones. Physiol Mol Plant Pathol 40:17–22

Morrissey JP, Dow JM, Mark GL, O'Gara F (2004) Are microbes at the root of a solution to world food production? Rational exploitation of interactions between microbes and plants can help to transform agriculture. EMBO Rep 5:922–926

Moyne AL, Shelby R, Cleveland TE, Tuzun S (2001) Bacillomycin D: an iturin with antifungal activity against *Aspergillus flavus*. J Appl Microbiol 90:622–629

Naeem S, Knops J, Tilman D, Howe K, Kennedy T, Gale S (2000) Plant diversity increases resistance to invasion in the absence of covarying extrinsic factors. Oikos 91:97–108

Nannipieri P, Ascher J, Ceccherini MT, Landi L, Pietramellara G, Renella G (2003) Microbial diversity and soil functions. Eur J Soil Sci 54:655–670

Narwal SS (2000) Allelopathy in ecological agriculture. In: Narwal SS, Hoagland RE, Dilday RH, Reigosa MJ (eds) Allelopathy in ecological agriculture and forestry, Proceedings of the III international congress on allelopathy in ecological agriculture and forestry, Dharwad, India, 18–21 Aug 1998. Kluwer Academic, London, pp 11–32

Ndiaye M, Termorshuizen AJ, Van Bruggen AHC (2007) Combined effects of solarization and organic amendment on charcoal rot caused by *Macrophomina phaseolina* in the Sahel. Phytoparasitica 35:392–400

Neal JL Jr, Atkinson TG, Larson RI (1970) Changes in the rhizosphere microflora of spring wheat induced by disomic substitution of a chromosome. Can J Microbiol 16:153–158

Neal JL Jr, Larson RI, Atkinson TG (1973) Changes in rhizosphere populations of selected physiological groups of bacteria related to substitution of specific pairs of chromosomes in spring wheat. Plant Soil 39:209–212

Nelson EB, Boehm MJ (2002) Compost-induced suppression of turfgrass diseases. BioCycle 43: 51–55

Nelson EB, Kuter GA, Hoitink HAJ (1983) Effects of fungal antagonists and compost age on suppression of rhizoctonia damping-off in container media amended with composted hardwood bark. Phytopathology 73:1457–1462

Noble R, Coventry E (2005) Suppression of soil-borne plant diseases with composts: a review. Biocontrol Sci Technol 15(1):3–20

NRC Report (1996) Ecologically based pest management: new solutions for a new century. National Academy Science Press, Washington, DC, 144 pp

Oka Y, Shapira N, Fine P (2007) Control of root-knot nematodes in organic farming systems by organic amendments and soil solarization. Crop Prot 26:1556–1565

Ongena M, Duby F, Rossignol F, Fouconnier ML, Dommes J, Thonart P (2004) Stimulation of the lipoxygenase pathway is associated with systemic resistance induced in bean by a non-pathogenic Pseudomonas strain. Mol Plant-Microbe Interact 17:1009–1018

Pal KK, Gardener BMS (2006) Biological control of plant pathogens. Plant Health Instruct. doi:10.1094/PHI-A-2006-1117-02

Pane C, Spaccini R, Piccolo A, Scala F, Bonanomi G (2011) Compost amendments enhance peat suppressiveness to Pythium ultimum, Rhizoctonia solani and *Sclerotinia minor*. Biol Control 56:115–124

Pankhurst CE, McDonald HJ, Hawke BG, Kirkby CA (2002) Effect of tillage and stubble management on chemical and microbiological properties and the development of suppression towards cereal root disease in soils from two sites in NSW, Australia. Soil Biol Biochem 34:833–840

Papavizas GC (1966) Suppression of *aphanomyces* root rots of pea by *cruciferous* soil amendments. Phytopathology 56:1071–1075

Papavizas GC (1968) Survival of root-infecting fungi in soil. IV. Effect of amendments on bean root rot caused by Thielaviopsis basicola and on inoculum density of the causal organism. Phytopathology 58:421–428

Papavizas GC, Davey CB (1960) The rhizoctonia disease of bean as affected by decomposing green plant materials and associated microfloras. Phytopathology 51:516–522

Parham JA, Deng SP, Da HN, Sun HY, Raun WR (2003) Long-term cattle manure application in soil. II. Effect on soil microbial populations and community structure. Biol Fertil Soils 38:209–215

Patten CL, Glick BR (1996) Bacterial biosynthesis of indole-3-acetic acid. Can J Microbiol 42:207–220

Paulitz TC, Belanger RR (2001) Biological control in greenhouse systems. Annu Rev Phytopathol 39:103–133

Pedersen EA, Hughes GR (1992) The effect of crop rotation on development of the septoria disease complex on spring wheat in Saskatchewan. Can J Plant Pathol 14:152–158

Pérez-Piqueres A, Edel-Hermann V, Alabouvette C, Steinberg C (2006) Response of soil microbial communities to compost amendments. Soil Biol Biochem 38:460–470

Postma J, Schilder MT, Bloem J, Van Leeuwen-Haagsma WK (2008) Soil suppressiveness and functional diversity of the soil microflora in organic farming systems. Soil Biol Biochem 40:2394–2406

Press CM, Loper JE, Kloepper JW (2001) Role of iron in rhizobacteria mediated induced systemic resistance of cucumber. Phytopathology 91:593–598

Pugliese M, Liu BP, Gullino ML, Garibaldi A (2011) Microbial enrichment of compost with biological control agents to enhance suppressiveness to four soil-borne diseases in greenhouse. J Plant Dis Protect 118:45–50

Raaijmakers JM, Weller DM (1998) Natural plant protection by 2,4-diacetylphloroglucinol-producing Pseudomonas spp. in take-all decline soils. Mol Plant-Microbe Interact 11:144–152

Raaijmakers J, Weller DM, Thomashow LS (1997) Frequency of antibiotic-producing *Pseudomonas* spp. in natural environments. Appl Environ Microbiol 63:881–887

Ramos Solano B, Barriuso Maicas J, Pereyra de la Iglesia MT, Domenech J, Gutierrez Manero FJ (2008) Systemic disease protection elicited by plant growth promoting rhizobacteria strains: relationship between metabolic responses, systemic disease protection, and biotic elicitors. Phytopathology 98:451–457

Rousk J, Brookes PC, Baath E (2010) Investigating the mechanisms for the opposing pH relationships of fungal and bacterial growth in soil. Soil Biol Biochem 42:926–934

Rovira AD (1969) Plant root exudates. Bot Rev 35:35–57

Rudrappa T, Czymmek KJ, Pare PW, Bais HP (2008) Root-secreted malic acid recruits beneficial soil bacteria. Plant Physiol 148:1547–1556

Russell PE (1995) Fungicide resistance: occurrence and management. J Agric Sci 124:317–323

Ryan PR, Dessaux Y, Thomashow LS, Weller DM (2009) Rhizosphere engineering and management for sustainable agriculture. Plant Soil 321:363–383

Ryu CM, Farag MA, Hu CH, Reddy MS, Kloepper JW, Pare PW (2004) Bacterial volatiles induce systemic resistance in Arabidopsis. Plant Physiol 134:1017–1026

Sabuquillo P, De Cal A, Melgarejo P (2005) Dispersal improvement of a powder formulation of *Penicillium oxalicum*, a biocontrol agent of tomato wilt. Plant Dis 89:1317–1323

Sahni S, Sarma BK, Singh DP, Singh HB, Singh KP (2008) Vermicompost enhances performance of plant growth promoting rhizobacteria in *cicer arietinum* rhizosphere against *sclerotium rolfsii*. Crop Prot 27:369–376

Sandra AI, Wright CH, Zumoff LS, Steven VB (2001) *Pantoea agglomerans* strain EH318 produces two antibiotics that inhibit *Erwinia amylovora* in vitro. Appl Environ Microbiol 67:282–292

Sayama M, Homma Y, Furuya H, Takenaka S (2001) Some microbial properties of suppressive soil induced by successive inoculations of *rhizoctonia solani* anastomosis group 2–2. Soil Microorganisms 55:37–44

Scheuerell SJ, Mahaffee WF (2004) Compost tea as a container medium drench for suppressing seedling damping-off caused by *Pythium ultimum*. Phytopathology 94:1156–1163

Scheuerell SJ, Sullivan DM, Mahaffee WF (2005) Suppression of seedling damping-off caused by *Pythium ultimum*, *P. irregulare*, and *Rhizoctonia solani* in container media amended with a diverse range of Pacific Northwest compost sources. Phytopathology 95:306–315

Schloss PD, Handelsman J (2006) Toward a census of bacteria in soil. PLoS Comput Biol 2:92

Schneider O, Aubertot JN, Roger-Estrade J, Doré T (2003) Analysis and modelling of the amount of oilseed rape residues left at the soil surface after different soil tillage operations. In: 7th international conference on plant pathology, Tours France, 3–5 Dec 2003

Schnitzer SA, Klironomos JN, Hillerislambers J, Kinkel LL, Reich PB, Xiao K, Rillig MC, Sikes BA, Callaway RM, Mangan SA, van Nes EH, Scheffer M (2011) Soil microbes drive the classic plant diversity-productivity pattern. Ecology 92:296–303

Schnurer J, Rosswall T (1982) Fluorescein diacetate hydrolysis as a measure of total microbial activity in soil and litter. Appl Environ Microbiol 43:1256–1261

Schutter ME, Dick RP (2002) Microbial community profiles and activities among aggregates of winter fallow and cover-cropped soil. Soil Sci Soc Am J 66:142–153

Schutter ME, Sandeno JM, Dick RP (2001) Seasonal, soil type, and alternative management influences on microbial communities of vegetable cropping systems. Biol Fertil Soils 34: 397–410

Selvakumar G, Panneerselvam P, Ganeshamurthy AN, Maheshwari DK (2012) Bacterial mediated alleviation of abiotic stress in crops. In: Maheshwari DK (ed) Bacteria in agrobiology: stress management. Springer, New York, pp 205–224

Shanahan P, O'Sullivan DJ, Simpson P, Glennon JD, O'Gara F (1992) Isolation of 2,4-Diacetylphloroglucinol from a fluorescent pseudomonad and investigation of physiological parameters influencing its production. Appl Environ Microbiol 58:353–358

Smiley RW (1978) Colonization of wheat roots by *Gaeumannomyces graminis* inhibited by specific soils, micro-organisms and ammonium nitrogen. Soil Biol Biochem 10:175–179

Smith KP, Havey MJ, Handelsman J (1993) Suppression of cottony leak of cucumber with *Bacillus cereus* strain UW85. Plant Dis 77:139–142

Steinberg C, Edel-Hermann V, Guillemaut C, Pérez-Piqueres A, Singh P, Alabouvette C (2004) Impact of organic amendments on soil suppressiveness to diseases. In: Sikora RA, Gowen S, Hauschild R, Kiewnick S (eds) Multitrophic interactions in soil and integrated control. IOBC WPRS Bulletin/Bulletin OILB, pp 259–266

Steinberg C, Edel-Hermann V, Alabouvette C, Lemanceau P (2007) Soil suppressiveness to plant diseases. In: van Elsas JD, Jansson JC, Trevors JT (eds) Modern soil microbiology, 2nd edn. CRC, Boca Raton, pp 455–478

Stephens PM, Crowley JJ, O'Connel C (1993) Selection of pseudomonad strains inhibiting *Pythium ultimum* on sugarbeet seed in soil. Soil Biol Biochem 25:1283–1288

Stone AG, Vallad GE, Cooperband LR, Rotenberg D, Darby HM, James RV, Stevenson WR, Goodman RM (2003) Effect of organic amendments on soilborne and foliar diseases in field-grown snap bean and cucumber. Plant Dis 87:1037–1042

Strange RN (1993) Plant disease control: towards environmentally acceptable methods. Chapman and Hall, New York, 354 pp

Sturz AV, Christie BR (2003) Beneficial microbial allelopathies in the root zone: the management of soil quality and plant disease with rhizobacteria. Soil Till Res 72:107–123

Sturz AV, Carter MR, Johnston HW (1997) A review of plant disease, pathogen interactions and microbial antagonism under conservation tillage in temperate humid agriculture. Soil Till Res 41:169–189

Sturz AV, Christie BR, Matheson BG (1998) Associations of bacterial endophyte populations from red clover and potato crops with potential for beneficial allelopathy. Can J Microbiol 44:162–167

Sturz AV, Christie BR, Matheson BG, Arsenault WJ, Buchanan N (1999) Endophytic bacterial communities in the periderm of potato tubers and their potential to improve resistance to soil-borne plant pathogens. Plant Pathol 48:360–370

Sukul NC (1992) Plants antagonistic to plant-parasitic nematodes. Indian Rev Life Sci 12:23–52

Sumner DR, Doupnik B Jr, Boosalis MG (1981) Effects of reduced tillage and multiple cropping on plant diseases. Annu Rev Phytopathol 19:167–187

Szczech M, Smolinska U (2001) Comparison of suppressiveness of vermicomposts produced from animal manures and sewage sludge against *Phytophthora nicotianae* Breda de Haan var. *nicotianae*. J Phytopathol 149:77–82

Tamm L, Thürig B, Bruns C, Fuchs J, Köpke U, Laustela M, Leifert C, Mahlberg N, Nietlispach B, Schmidt C (2010) Soil type, management history, and soil amendments influence the development of soil-borne (*Rhizoctonia solani*, *Pythium ultimum*) and air-borne (*Phytophthora infestans*, *Hyaloperonospora parasitica*) diseases. Eur J Plant Pathol 2010(127):465–481

Tang CS, Komai K, Huang RS (1989) Allelopathy and the chemistry of the rhizosphere. In: Chou CH, Waller GR (eds) Phytochemical ecology: allelochemicals, mycotoxins and allomones, vol 9, Monograph series. Institute of Botany, Academia Sinica, Taipei

Tenuta M, Lazarovits G (2004) Soil properties associated with the variable effectiveness of meat and bone meal to kill microsclerotia of *Verticillium dahliae*. Appl Soil Ecol 25:219–236

Tenuta M, Conn KL, Lazarovits G (2002) Volatile fatty acids in liquid swine manure can kill microsclerotia of *Verticillium dahliae*. Phytopathology 92:548–552

Termorshuizen AJ, van Rijn E, van der Gaag DJ, Alabouvette C, Chen Y, Lagerlof J, Malandrakis AA, Paplomatas EJ, Ramert B, Ryckeboer J, Steinberg C, Zmora-Nahum S (2006) Suppressiveness of 18 composts against 7 pathosystems: variability in pathogen response. Soil Biol Biochem 38:2461–2477

Termorshuizen AJ, van Rijn E, van der Gaag DJ, Alabouvette C, Chen Y, Lagerlof J, Malandrakis AA, Paplomatas EJ, Ramert B, Ryckeboer J, Steinberg C, Zmora-Nahum S (2007) Suppressiveness of 18 composts against 7 pathosystems: variability in pathogen response. Soil Biol Biochem 38:2461–2477

Thomashow LS, Weller DM, Bonsall RF, Pierson LS (1990) Production of the antibiotic phenazine-1-carboxylic acid by fluorescent pseudomonas in the rhizosphere of wheat. Appl Environ Microbiol 56:908–912

Tilston EL, Pitt D, Groenhof AC (2002) Composted recycled organic matter suppresses soil-borne diseases of field crops. New Phytol 154:731–740

Tiquia SM, Lloyd J, Herms DA, Hoitink HAJ, Michel FC (2002) Effects of mulching and fertilization on soil nutrients, microbial activity and rhizosphere bacterial community structure determined by analysis of T-RFLPs of PCR-amplified 16S rRNA genes. Appl Soil Ecol 21:31–48

Tuitert G, Szczech M, Bollen GJ (1998) Suppression of *Rhizoctonia solani* in potting mixtures amended with compost made from organic household waste. Phytopathology 88:764–773

Umaerus VR, Scholte K, Turkensteen LJ (1989) Crop rotation and the occurrence of fungal diseases in potatoes. In: Vos J, Van Loon CD, Bollen GJ (eds) Effects of crop rotation on potato production in the temperate zones. Kluwer Academic, Dordrecht, pp 171–189

Uren NC (2007) Types, amounts, and possible functions of compounds released into the rhizosphere by soil-grown plants. In: Pinton R, Varanini Z, Nannipieri P (eds) The rhizosphere biochemistry and organic substances at the soil-plant interface, 2nd edn. CRC, Boca Raton, pp 1–21

van Bruggen AHC (1995) Plant disease severity in high-input compared to reduced-input and organic farming systems. Plant Dis 79:976–983

van Bruggen AH, Semenov AV (2000) In search of biological indicators for soil health and disease suppression. Appl Soil Ecol 15:13–24

Van Loon LC, Bakker PAHM, Pieterse CMJ (1998) Systemic resistance induced by rhizosphere bacteria. Annu Rev Phytopathol 36:453–483

van Os GJ, van Ginkel JH (2001) Suppression of pythium root rot in bulbous Iris in relation to biomass and activity of the soil microflora. Soil Biol Biochem 33:1447–1454

Van Peer R, Schippers B (1992) Lipopolysaccharides of plant-growth promoting *Pseudomonas* sp. strain WCS417r induce resistance in carnation to fusarium wilt. Neth J Plant Pathol 98: 129–139

van Peer R, Niemann GJ, Schippers B (1991) Induced resistance and phytoalexin accumulation in biological control of fusarium wilt of carnation by *Pseudomonas* sp. Strain WCS417r. Phytopathology 81:728–734

Van Wees SC M, Pieterse CMJ, Trijssenaar A, Van't Westende Y, Hartog F (1997) Differential induction of systemic resistance in *Arabidopsis* by biocontrol bacteria. Mol Plant-Microbe Interact 10:716–724

van Wees SCM, Luijendijk M, Smoorenburg I, van Loon LC, Pieterse CMJ (1999) Rhizobacteria-mediated induced systemic resistance (ISR) in *Arabidopsis* is not associated with a direct effect on expression of known defense-related genes but stimulates the expression of the jasmonate-inducible gene *Atvsp* upon challenge. Plant Mol Biol 41:537–549

Vandenkoornhuyse P, Mahe S, Ineson P, Staddon P, Ostle N, Cliquet JB, Francez AJ, Fitter AH, Young JP (2007) Active root- inhabiting microbes identified by rapid incorporation of plant derived carbon into RNA. Proc Natl Acad Sci U S A 104:16970–16975

Velvis H, van den Boogert PHJF, Jager G (1989) The role of antagonism in decline of Rhizoctonia solani inoculum in soil. Soil Biol Biochem 21:125–129

Weller DM (1988) Biological control of soilborne plant pathogens in the rhizosphere with bacteria. Annu Rev Phytopathol 26:379–407

Weller DM, Raaijmakers JM, McSpadden Gardener BB, Thomashow LS (2002) Microbial populations responsible for specific suppression to plant pathogens. Annu Rev Phytopathol 40:309–348

Whipps JM (1997) Development in the biological control of soil-borne plant pathogens. Adv Bot Res 26:1–84

Widmer TL, Mitkowski NA, Abawi GS (2002) Soil organic matter and management of plant-parasitic nematodes. J Nematol 34:289–295

Wilhite SE, Lunsden RD, Strancy DC (2001) Peptide synthetase gene *in Trichoderma virens*. Appl Environ Microbiol 67:5055–5062

Wilsey BJ, Potvin C (2000) Biodiversity and ecosystem functioning: importance of species evenness in an old field. Ecology 81:887–892

Wiseman BM, Neate SM, Keller KO, Smith SE (1996) Suppression of *rhizoctonia solani* anastomosis group 8 in Australia and its biological nature. Soil Biol Biochem 28:727–732

Wittebolle L, Marzorati M, Clement L, Balloi A, Daffonchio D, Heylen K et al (2009) Initial community evenness favours functionality under selective stress. Nature 458:623–626

Workneh F, van Bruggen AHC, Drinkwater LE, Shennan C (1993) Variables associated with corky root and phytophthora root rot of tomatoes in organic and conventional farms. Phytopathology 83:581–589

Yang X, Chen L, Yong X, Shen Q (2011) Formulations can affect rhizosphere colonization and biocontrol efficiency of *Trichoderma harzianum* SQR-T037 against fusarium wilt of cucumbers. Biol Fertil Soils 47:239–248

Yogev A, Raviv M, Hadar Y, Cohen R, Katan J (2006) Plant waste-based composts suppressive to diseases caused by pathogenic *Fusarium oxysporum*. Eur J Plant Pathol 116:267–278

Yulianti T, Sivasithamparam K, Turner DW (2006) Saprophytic growth of *Rhizoctonia solani* Kuhn AG2-1 (ZG5) in soil amended with fresh green manures affects the severity of damping-off in canola. Soil Biol Biochem 38:923–930

Zhou XG, Everts KL (2004) Suppression of fusarium wilt of watermelon by soil amendment with hairy vetch. Plant Dis 88:1357–1365

Chapter 5
Natural Mechanisms of Soil Suppressiveness Against Diseases Caused by *Fusarium*, *Rhizoctonia*, *Pythium*, and *Phytophthora*

Prashant P. Jambhulkar, Mahaveer Sharma, Dilip Lakshman, and Pratibha Sharma

5.1 Introduction

Soilborne fungal and oomycete plant pathogens are among the major factors limiting the productivity of agroecosystems and are often difficult to control with conventional strategies such as the use of resistant host cultivars and synthetic fungicides. Due to limitations in the effectiveness of fungicides and a lack of successful plant-based resistance, enhancement of soil-based natural disease suppression could be an effective option to control disease (Weller et al. 2002). This suppressive effect has been attributed to diverse microbial communities of bacteria, fungi, and protozoa and is reported to affect pathogen survival, growth in bulk and rhizosphere soil, and root infection (Barnett et al. 2006). Maintaining a high level of organic matter (OM) on the soil surface or incorporation of OM into the soil is generally associated with lower incidence and severity of root diseases. Natural disease-suppressive soils probably are the best examples in which the indigenous microflora effectively protects plants against soilborne pathogens. Soil microbes

P.P. Jambhulkar (✉)
Agricultural Research Station (MPUAT), Banswara 327001, RJ, India
e-mail: ppjambhulkar@gmail.com

M. Sharma
ICAR-Directorate of Soybean Research, Khandwa Road, Indore 452001, India
e-mail: mahaveer620@gmail.com

D. Lakshman
Floral and Nursery Plant Research Unit and Sustainable Agricultural System Laboratory, USDA-, Beltsville, MD 20705, USA
e-mail: Dilip.Lakshman@ars.usda.gov

P. Sharma
Division of Plant Pathology, Indian Agricultural Research Institute, New Delhi 110012, India
e-mail: psharma032003@yahoo.co.in

M.K. Meghvansi, A. Varma (eds.), *Organic Amendments and Soil Suppressiveness in Plant Disease Management*, Soil Biology 46,
DOI 10.1007/978-3-319-23075-7_5

and disease-causing phytopathogens share the common rhizosphere and their interaction prior to crop sowing and/or in the rhizosphere, subsequently influencing both plant growth and productivity (Penton et al. 2014).

Suppressive soils to soilborne plant diseases have been described as "those in which disease development is minimal even in the presence of a virulent pathogen and susceptible plant host" (Mazzola 2007). Though some experts (Bruehl 1987) argue for limiting the use of the term disease suppressiveness to situations involving a clear biological component, there is a plethora of evidence for the role of both biotic and abiotic elements of the soil having roles in disease suppression. Physical and chemical characteristics of the soil, including pH, OM, and clay content, can operate in the suppression of plant diseases directly or indirectly through their impact on soil microbial activity. Although these abiotic characteristics of the soil can contribute to disease suppression, soil suppressiveness (SS) is often directly or indirectly a function of the activity of soil microorganisms or microbial metabolites. Another frequently quoted definition of "suppressive soil" from Baker and Cook (1974) is "soils in which the pathogen does not establish or persist, establishes but causes little or no damage, or establishes and causes disease for a while but thereafter the disease is less important, although the pathogen may persist in the soil." However, it is difficult to precisely define the term "suppressive soil" (Hornby 1983) simply because there are many types of suppressiveness acting in the rhizosphere. The terms "pathogen suppressive" and "disease suppressive" have often been used interchangeably (Weller et al. 2002), but the former refers to the suppression of the pathogen growing saprophytically on decaying OM in the soil or surviving in the soil and is suppressed when the pathogen is growing parasitically in the host, while in latter case the term usually refers to suppression of the pathogen growing as parasite in the host (Hornby 1983). On the basis of speed with which soils become suppressive, soil suppressiveness (SS) is distinguished as induced suppression which does not show maximum effectiveness even immediately after application of inductive treatment, and in some cases monoculture (MC) is required to achieve maximum effectiveness. On the other hand, "introduced suppression" (IS) was found to be effective immediately after treatment. Hoper and Alabouvette (1996) distinguished pathogen suppression (the ability of the soil to limit the inoculum density of the pathogen and its saprophytic activity) with disease suppression (the capacity of the soil to restrict disease development) under ideal host–pathogen environmental conditions. The disease suppressiveness can be designated in soils in which disease development is minimal even in the presence of a virulent pathogen and susceptible plant host (Mazzola 2002). Suppressive soils have been described for many soilborne pathogens, including *Gaeumannomyces graminis* var. *tritici* (Andrade et al. 2011), *Fusarium oxysporum* (Alabouvette 1999), *Aphanomyces euteiches* (Persson et al. 1999), *Heterodera avenae* (Kerry 1988), *Phytophthora cinnamomi* (Keen and Vancov 2010), *P. infestans* (Andrivon 1994), *Rhizoctonia solani* (Wiseman et al. 1996), and *Plasmodiophora brassicae* (Murakami et al. 2000). Natural disease-suppressive soils are the best examples in which the activities of specific soil and rhizosphere microorganisms keep susceptible plants mostly free from infection in spite of ample exposure to/load of virulent inoculum of soilborne pathogens. For most of the disease-suppressive soils, however, the consortia of microorganisms and the mechanisms involved in pathogen control have not yet been understood. In this chapter we focus on recent progresses made toward unraveling the mechanisms of natural soil suppressiveness against four specific soilborne pathogens, viz., *Fusarium, Rhizoctonia, Pythium,* and *Phytophthora*. However, there is still a need to understand mechanisms underlying for the occurrence of disease suppression and strategies to enhance the suppressiveness through manipulating agricultural management practices eventually to create consistently suppressive soils for the management of soilborne diseases and phytopathogens.

5.2 Categories of Soil Suppressiveness

There are two mechanisms of soil suppressiveness (general and specific) according to the spectrum of microorganisms involved in the process. The "general soil-suppressive potential" is linked to abiotic and biotic substrate characteristics that are not related to the microorganism or group of antagonistic microorganisms in particular, while in "specific soil-suppressive potential," the suppression is related to the action of one or few organisms in the substrate (Termorshuizen and Jeger 2008).

5.2.1 General Soil Suppressiveness

General suppression of the crop diseases occurs when a high microbial activity is created in the soil environment/rhizosphere which inhibits the propagation of pathogen propagules. It occurs when a large number of different microorganisms compete with pathogens for nutrients and/or produce general antibiotics that reduce pathogen survival and growth. In compost there is a slow release of nutrients which supports beneficial activity of the microflora. General suppression is often enhanced by the addition of OM, certain agronomic practices, or the buildup of soil fertility (Stone et al. 2004) which consequently can increase soil microbial activity. The general suppressive potential of suppressive soil is explained by the ability of practices/materials to sustain sufficient microbial activity over time, fed by slow degradation of complex carbon compounds, particularly the polymeric carbohydrates (Hoitink et al. 1996). No one microorganism is responsible for general suppression (Alabouvette 1986; Cook and Baker 1983) and the suppressiveness is not transferable between soils. Thus, the entire soil microbial community increases nutrient withdrawal, resulting in fungistasis of fungal pathogen propagules or competition for colonization of rhizosphere zones which are rich in radical exudates. When an inoculum of a pathogen is added to pairs of raw and sterilized soil samples, the effect of general suppression becomes apparent by the greater severity of disease on a host grown in the sterilized soil as compared to the raw soil (Weller et al. 2002). The extent of production of antifungal microbial metabolites varied with the species (de Boer et al. 2003) and showed positive relationship between microbial diversity and general disease suppression of different pathogens (Garbeva et al. 2006; Postma et al. 2008; Benitez and McSpadden Gardner 2009). It may be due to synergistic interaction between microbial populations producing secondary metabolites or to greater collective efficiency in the removal of nutrients (Garbeva et al. 2011). Each pathogen is usually preferentially associated with one type of suppressive potential. *Pythium* spp. and *Phytophthora* spp. have propagules with small amounts of nutrients and depend on exogenous carbon sources for germination to affect host plants. They are described as highly sensitive to microbial nutrient competition and antibiosis and related to general suppression (Aryantha et al. 2000). The control of pathogens such as *Pythium*, *Fusarium*, and *Phytophthora* has often been related to general suppression due to OM amendments (Weller et al. 2002). Under such conditions, a broad variety of microbial species creates a competitive environment suppressive to pathogens (Serra-Wittling et al. 1996; Stone et al. 2001).

5.2.2 *Specific Soil Suppressiveness*

In contrast to general soil suppressiveness, "specific suppression occurs when the individual or selected groups of microorganisms compete with "pathogens for nutrient and produce specific antibiotics during a certain stage in the life cycle of a pathogen to reduce its survival (Weller et al. 2002). Specific suppression is considered to be generated through the activities of one or several populations of organisms. Specific suppression is more qualitative, owing to more specific effects of individual or select group of microorganisms antagonistic to the pathogen during some stage in its life cycle (Cook and Baker 1983). Transferability is the key factor of specific suppression (Andrade et al. 1994; Westphal and Becker 1999) and the term "transferable suppression" has been used synonymously with specific suppression. Activity in suppressive soils is because of their ability to combine general and specific suppression. This combination acts as a continuum in the soil, although they may be affected differently by edaphic, climatic, and agronomic conditions. Weller et al. (2002) observed that most suppressive soils maintain their activity when brought into the greenhouse or laboratory while assessing the mechanisms of suppression under more controlled and reproducible conditions. Biotic and abiotic variables affect the structure and activity of microbial populations including pathogens and their antagonists which eventually help in disease suppression. Specific soil suppressiveness depends on microorganisms that operate as biological control agents emerged after the thermophilic phase. Many conducive soils possess properties with regard to microorganisms involved in disease suppression, while other attributes are unique to specific pathogen-suppressive soil systems. Modes of action of biocontrol agents (BCAs) include inhibition of the pathogen by antimicrobial compounds (antibiosis), competition for iron through production of siderophores, competition for sites of colonization and nutrients supplied by seeds and roots, induction of plant resistance mechanisms, inactivation of pathogen germination factors present in seed or root exudates through allelopathy, degradation of pathogenicity factors of the pathogens such as toxins, and parasitism that may involve production of extracellular cell wall-degrading enzymes such as chitinase and β-1,3-glucanase that can lyse pathogen cell walls (Keel and Défago 1997; Whipps 1997). None of the mechanisms are necessarily mutually exclusive, and frequently, several modes of action are exhibited by a single BCA. Indeed, for some BCAs, different mechanisms or combinations of mechanisms may be involved in the suppression of different plant diseases (Whipps 2001). So the organisms operative in pathogen suppression do so via diverse mechanisms including competition for nutrients, antibiosis, and induction of host resistance. Nonpathogenic *Fusarium* spp. and fluorescent *Pseudomonas* spp. also play critical roles in naturally occurring soils that are suppressive to Fusarium wilt. The suppression of take-all of wheat (*Triticum aestivum*), caused by *G. graminis* var. *tritici*, is induced in the soil after continuous wheat monoculture and is attributed in part to selection of fluorescent *Pseudomonas* spp. with capacity to produce the antibiotic 2,4-diacetylphloroglucinol (DAPG). The cultivation of orchard soils with specific wheat varieties induces suppressiveness to *Rhizoctonia* root rot of apple (*Malus domestica*) caused by *R. solani* AG 5 (Mazzola and Gu 2002). Long-standing suppression is a biological condition naturally associated with the soil and is called natural disease suppression. Its origin is not known and it appears to survive in the absence of crops in the field (Weller et al. 2002). In contrast, long-term adoption of crop management practices that supply higher levels of biologically available carbon inputs either through crop residues or addition of composts and organic manures can support higher levels of suppression. This occurs through changes to the composition and activity of the soil microbial community (Gupta et al. 2011; Postma et al. 2003). Induced suppressiveness is initiated and sustained by practice of monoculture in the presence of pathogens (Weller et al. 2002). Soils suppressive to take-all disease of wheat and barley, caused by the fungal pathogen *G. graminis* var. *tritici*, are referred to as take-all decline soils and are well-known examples of induced suppressiveness. The Fusarium wilt-suppressive soils from Châteaurenard (France) and Salinas Valley (CA, USA) are among the best examples of long-standing suppressive soils. Wheat cultivars that stimulate disease suppression enhance populations of specific fluorescent pseudomonad strains with antagonistic activity toward this pathogen.

Sterilization by autoclaving and gamma radiation can eliminate both general and specific suppression. General suppression is reduced but not eliminated by soil fumigation and usually remains after treatment at up to 70 °C moist heat (Weller et al. 2002). Pasteurization can eliminate specific suppression but this characteristic is not a prerequisite for specific suppression. Another strategy which allows confirmation of the biological basis of suppression involves transfer of suppressiveness to raw, conducive, fumigated, or sterilized soil by addition of 0.1–10 % or less (wt/wt) of the suppressive soil into the conducive soil (Weller et al. 2002).

5.3 Mechanism of Soil Suppressiveness

Mechanisms of the majority of cases of soil suppressiveness are unknown or unproven, but most explanations for suppressiveness involve microbial antagonism like antibiosis, competition, predation, and parasitism (Mazzola 2007). Investigations into the causes of disease-resistant and disease-tolerant soils sometimes revealed correlations with certain chemical and physical properties which are not an actual determinant of suppressiveness (Weller et al. 2002). Abiotic factors most frequently cited are the proportion and type of clay, acidity, and moisture: moist acid soils and dry alkaline soils tend to be unfavorable to the growth of pathogens.

While reviewing the work done on soil suppressiveness which suggests that most of the mechanisms reported to be responsible for reduction of diseases in plants involve microbiological changes in the bulk soil, the rhizosphere soil, and/or the rhizoplane, resulting in antagonism of the pathogen. However, globally, there are various schools of thoughts and opinions on the mechanism involved which largely state that different microbial antagonists are responsible for the proliferation of disease-causing pathogens. Some of the mechanisms of disease-suppressive soils are described below.

5.3.1 Organic Matter-Mediated Mechanism of Soil Suppression

The application of organic matter (OM) such as animal manure, green manure, and peat has been proposed for conventional agriculture to improve soil structure and fertility (Conklin et al. 2002; Cavigelli and Thien 2003) and to reduce disease incidence caused by soilborne pathogens (Litterick et al. 2004; Noble and Coventry 2005). Studies revealed that OM can be very effective in reducing pathogens such as species of *Fusarium* (Szczech 1999), *Phytophthora* (Szczech and Smolińska 2001), *Pythium* (McKellar and Nelson 2003; Veeken et al. 2005), *R. solani* (Diab et al. 2003), and *Sclerotinia* (Coventry et al. 2005). There are different mechanisms to explain the suppressive capacity of organic amendments: enhanced activities of antagonistic microbes (Hoitink and Boehm 1999), increased competition against pathogens for resources that cause fungistasis (Lockwood 1990), release of fungitoxic compounds during OM decomposition (Tenuta and Lazarovits 2002), or induction of systemic resistance in host plants (Pharand et al. 2002). The inconsistent disease control results obtained with OM amendments with both suppressive (disease reduction) and conducive (disease increase) effects produced skepticism in farmers about the use of these materials.

The suppressive capacity of all OM types against soilborne pathogens was evaluated by Bonanomi et al. (2007) which suggested that the OM was suppressive in 45 % and nonsignificant in 35 % of the cases, while in 20 % of the cases, a significant increase of disease incidence was found. OM amendment resulted in

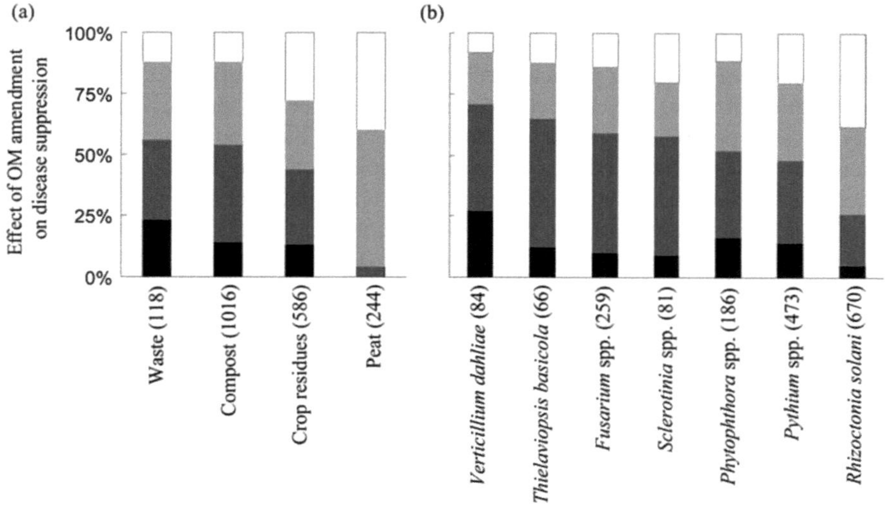

Fig. 5.1 Effect of OM amendments on disease suppression (*black* highly suppressive, *dark gray* suppressive, *gray* null, *white* conducive) in relation to different OM types (**a**) and soilborne fungal pathogens (**b**). Total percentage of suppressive cases is the sum of highly suppressive and suppressive. Only pathogens with more than 50 study cases (*numbers in brackets*) are shown (Bonanomi et al. 2007; reproduced with permission)

highly suppressive conditions (disease reduction >80 %) only in 12 % of the cases. Considering all OM types together, the suppressive capacity of the amendments varied largely with respect to different pathogens (Fig. 5.1a). Suppression was very high for both *V. dahliae* and *T. basicola* (>65 %), above 50 % of cases for *Fusarium* spp., *Sclerotinia* spp., and *Phytophthora* spp., and slightly below 50 % for *Pythium* spp. In contrast, effective control of *R. solani* was achieved only in 26 % of cases (Fig. 5.1b).

In the following paragraphs, we discuss specific mechanisms involved in OM-mediated disease suppression. Though these mechanisms are discussed individually, they act in consortia to carry out disease suppression.

5.3.1.1 Microbiostasis

Nutrient stress to soil microbial community results in repression of microbial spore germination and growth; this phenomenon is called microbiostasis or fungistasis for repression of fungal spores. Microbiostasis is an adaptive feature, as it protects the propagule from the energy losses or even death that might occur if germination occurred in the absence of a host. Microbiostasis can be overcome by inputs of external energy-rich nutrients such as root and seed exudates or organic amendments such as plant residues or manures (Lockwood 1990). Soil microbiostasis could be beneficial to microorganisms because it would be advantageous to their successful colonization in suitable habitats (Ko 2003). Germination of fungal conidia and the chlamydospores of *Fusarium* spp. is restricted because of insufficiency of energy-yielding nutrients as they require an external source of energy for germination in vitro. The competition for energy sources by the microbial community is a strong energy sink; exudation from ^{14}C-labeled fungal propagules increases in response to energy stress in the soil. However, propagules also lose energy and viability because of respiration (Hyakumachi et al. 1989). Losses in propagule energy can lead to a reduction in biological function. Addition of new energy sources to the soil system can initially destroy fungistasis, but fungistasis resumes, typically at a higher fungistatic level, after the sources have been slightly degraded (Lockwood 1990). Addition of sucrose and asparagine, or seed exudates, to compost-amended suppressive soil reduces the level of suppressiveness in a dose-dependent, linear relationship (Chen et al. 1988). In addition, compost harvested from the center, i.e., the thermophilic region, of a hardwood bark compost pile was conducive and of lower microbial activity and biomass and higher reducing sugars than the suppressive, lower-temperature outer region of the same pile. However, within days, the conducive material (incubated at room temperature) became suppressive; during the same period, the microbial activity increased and the reducing sugar content declined to levels comparable to those in the suppressive, outer-region compost (Stone et al. 2004).

Preemptive metabolism of exudate from a seed that initiates germination of pathogen propagules can induce microbiostasis and thus prevent disease; this is an indirect form of biological control because the pathogen is not directly antagonized. McKellar and Nelson (2003) elegantly described this phenomenon for BCA and compost-mediated suppression of damping-off of cotton caused by *Pythium ultimum*. The BCA *Enterobacter cloacae* metabolizes plant exudates required for germination and infection. *P. ultimum* oospores and sporangia germinate, grow, and infect cotton seeds in response to long-chain fatty acids (e.g., linoleic acid) released by the seeds as they germinate. *E. cloacae* inoculated onto cotton seeds competitively metabolizes the fatty acids and prevents *P. ultimum* germination, thereby suppressing the disease. Fatty acid uptake and oxidation mutants of *E. cloacae* do not prevent germination. In addition, there is no evidence to suggest that *E. cloacae* produces compounds inhibitory to the *Pythium* propagules (e.g., antibiotics) or is directly engaged in parasitism (van Dijk and Nelson 2000). In addition, populations of linoleic acid-metabolizing bacteria and actinobacteria were higher in the seed-colonizing microbial consortium from the suppressive compost than from the consortium isolated from the conducive compost. Individual isolates were not as suppressive as the suppressive microbial consortium, and linoleic acid metabolism varied greatly among isolates. This suggests that competition for linoleic acid was a strong determinant of damping-off suppression and that suppression was generated not by single isolates but by the combined activities of the linoleic acid-degrading microbial consortium supported by the suppressive compost substrate (McKellar and Nelson 2003).

5.3.1.2 Microbial Colonization of Pathogen Propagules

Pathogen propagules incubated in compost-amended potting mixes and organic residue-amended field soils are typically colonized by higher densities of bacterial and fungal propagules and, in some cases, protozoa, than in conducive or non-amended soils (Toyota and Kimura 1993). Colonized fungal spores germinate less readily and lyse and die more rapidly than noncolonized spores (Lockwood 1990). Bacterial colonization increased the rate of lysis, reduced the germination potential, and decreased the virulence of spores of various *Cochliobolus* spp.—the causal agents of root rots of grasses (Fradkin and Patrick 1985). Adherence might be an important component of biological control in and of itself; bacterial–fungal, fungal–fungal, and fungal–nematode interactions might be mediated by specific adherence mechanisms.

5.3.1.3 Destruction of Pathogen Propagules

Microbial antagonists generate hyphal lysis and degradation of chlamydospores, oospores, conidia, sporangia, and zoospores. Sporangia of *Phytophthora* spp. were destroyed after bacterial colonization of the sporangial surface. Sporangia nearing maturity release substances attractive to both microorganisms and microfauna. *Trichoderma* spp. can stimulate oospore formation, hyphal lysis, and chlamydospore formation in *Phytophthora* spp. (Costa et al. 2000). *Pseudomonas stutzeri* and *Pimelobacter* spp. isolated from chlamydospores of *F. oxysporum* f. sp. *raphani* (incubated in a manure-amended field soil) prevented chlamydospore formation or reduced chlamydospore germination (Toyota and Kimura 1993).

5.3.1.4 Antibiosis

Antibiosis is "antagonism mediated by specific or nonspecific metabolites of microbial origin, by lytic agents, volatile compounds, or other toxic substances" (Fravel 1988). The evidence for the role of antibiotics in the biocontrol of plant diseases has been extensively reviewed by Fravel (1988). *Pseudomonas* spp. that produce the antibiotic DAPG have been implicated in suppression of take-all of wheat, Fusarium wilt of pea, cyst nematode and soft rot of potato, and *Thielaviopsis* root rot of tobacco (Weller et al. 2002). Antibiotic production has also been implicated in the suppression of damping-off (causal agent *P. ultimum*) by *Gliocladium virens* (Howell and Stipanovic 1983).

5.3.1.5 Competition for Substrate Colonization

Most plant pathogens are weak saprophytes, and competition in the soil environment for organic substrates is strong. Pathogens that grow saprophytically on plant residues can be managed by pre-colonizing plant residues with nonpathogens, termed as the *possession principle* by Leach (1938) (Cook and Baker 1983). In studies of competitive interactions in soil aggregate colonization, closely related fungal species (other *F. oxysporum* formae speciales) strongly inhibited colonization by *F. oxysporum* f. sp. *raphani*. Other fungal genera moderately inhibited colonization, and bacterial species mildly inhibited colonization. *Burkholderia cepacia*, an antibiotic-producing bacterial species, also strongly inhibited colonization (Toyota et al. 1996). *P. nunn*, a saprophytic species of *Pythium*, outcompetes *P. ultimum* for colonization of added organic substrates, resulting in nutrient deprivation and production of survival structures by *P. ultimum*. In many cases, these structures are of lower inoculum potential, resulting in a reduction in the disease potential of *P. ultimum* (Paulitz and Baker 1988).

5.3.1.6 Competition for Root Infection Sites

Potato root colonization by the nonpathogenic fungal species *F. equiseti* was found effective in suppression of *Verticillium* wilt. Root colonization by *V. dahliae* was positively related to wilt incidence and negatively related to root colonization by *F. equiseti*. Sudangrass-cropped fields had the highest soil and root inoculum of *F. equiseti* and had the lowest wilt incidence. However, it is not clear if the increased *F. equiseti* colonization directly impacts *V. dahliae* colonization and disease incidence (Davis et al. 1996). Nonpathogenic strains of *F. oxysporum* compete with pathogenic strains for colonization of the root (Benhamou and Garand 2001) and other plant tissues (Postma and Luttikholt 1996) and might thereby contribute to suppression of Fusarium wilt.

5.3.1.7 Induced Systemic Resistance

Induced resistance has recently been implicated in some suppressive soil systems. Nonpathogenic *F. oxysporum* soil isolates induced systemic resistance in watermelon to Fusarium wilt (Larkin et al. 1996). Paper mill residual compost induced resistance to Fusarium wilt of tomato, resulting in a reduction in fungal colonization of root tissues. Suppression was associated with reduced fungal colonization of the tomato roots due to an increase in physical barriers (callose-enriched, multilayered wall appositions and osmiophilic deposits) to fungal penetration (Pharand et al. 2002). Tomato plants grown in compost-amended peat without inoculation with *F. oxysporum* did not exhibit increased physical barriers. An increased level of suppression and physical protection occurred when suppressive compost was inoculated with *P. oligandrum*, a species of *Pythium* known to induce resistance in tomato crop (Pharand et al. 2002). Composted pine bark container media was suppressive to *Pythium* root rot and foliar anthracnose of cucumber (Zhang et al. 1996), whereas dark peat container media was not suppressive to either disease. Cucumber and *Arabidopsis* plants grown in the composted pine bark expressed higher levels of β-1,3-glucanase (Zhang et al. 1998) and peroxidase (Zhang et al. 1996) than those grown in peat. Split-root experiments suggested that the resistance mechanism in cucumber was systemic (Zhang et al. 1996).

5.3.2 Compost-Mediated Mechanism of Soil Suppression

Compost is an organic material subjected to aerobic biological decomposition, during which temperatures of 40–70 °C are reached as a result of microbial activity. This process allows both the sanitization of the material (from human and plant pathogens and weed seeds) and its stabilization. Composts prepared from a variety of organic wastes are naturally suppressive against diseases caused by *Fusarium, Rhizoctonia, Pythium,* and *Phytophthora.* Only 20 % of all composts are suppressive against damping-off caused by *Rhizoctonia* and less than 10 % of all composts induced systemic resistance in plants (Hoitink and Boehm 1999). Furthermore, mechanisms that confer suppressive potential to composts depend on various factors as discussed below.

5.3.2.1 Hydraulic Conductivity and Free Air and Water Accessibility

The free air capacity of composts compared with some soils and peats is higher, which not only helps to improve plant growth but also has positive effect on the severity of rotting diseases of plant roots. Tree bark composts usually have an air capacity of over 25 % and a percolation rate of more than 2.5 cm/min and they suppress root rots. This suggests the importance of air capacity in those diseases where free water is important in the asexual multiplication of fungi (Aviles et al. 2011). It is well known that the manipulation of water potential as a control strategy is significant in diseases caused by oomycetes, particularly the possibility of producing adverse conditions for as long as possible during zoospore formation (Hardy and Sivasithamparam 1991). A negative water potential inhibits zoospore release from the sporangia of several *Phytophthora* spp. (Wilcox and Mircetich 1985). Thus, in order to reduce the incidence of disease due to these root rot pathogens, the necessary components of the growth media should be chosen in the proper amounts together with the correct irrigation system and watering strategy (Ownley et al. 1990).

5.3.2.2 Effect of pH and Electrical Conductivity in Interfering Nutrient Availability to the Pathogens

The majority of *Phytophthora* root rot diseases are inhibited by a low pH. The low pH reduced sporangium formation, zoospore release, and motility. For this reason the use of *Sphagnum* moss with low pH is beneficial in reducing *Phytophthora* and *Pythium* spp. High pH values of certain composts made from agricultural and industrial wastes were found suppressive against Fusarium wilt severity in various crops. The pH of the plant growth medium as a determinant of Fusarium wilt severity is associated with the availability of macro- and micronutrients and is important for growth, sporulation, and the virulence of *F. oxysporum* (Jones et al. 1991). A high pH reduces the availability of nutrients such as phosphorus (P), magnesium, manganese, copper (Cu), zinc (Zn), and iron (Fe) in organic growth. Borrero et al. (2004) showed a significant positive correlation between Fusarium wilt severity and final availability in the growth media of Cu on the one hand and the final nutrient status in the plants of Fe, Cu, and P on the other.

The lignin/cellulose ratio of wastes affects the duration of the composting process. Substrates with high lignin and low cellulose content do not immobilize a large amount of nitrogen, but this can be amended with essential micronutrients such as calcium and magnesium in order to improve the potential for growth of the majority of crops (Aviles et al. 2011). Hardwood bark and sewage sludge decompose well and do not require the addition of micronutrients. However, a high level of chloride, in the form of ions or as salt, can neutralize the suppressive effect of compost against *Phytophthora* root rot. There are contrasting reports presented by Pane et al. (2011) which show negative correlation between the damping-off induced by *Sclerotinia minor* and the salinity of compost-amended plant growth media. It is also important to note that phytotoxicity due to manganese available in certain bark composts must be amended with calcium carbonate before use.

5.3.2.3 Source of Nitrogen and C/N Ratio in Disease Suppression

High nitrogen levels and high ammonium to nitrate ratios increase Fusarium wilt incidence and severity. Thus, nitrate-amended composts may help to reduce Fusarium wilt diseases in ornamental (carnation, chrysanthemum) and horticultural crops (cucumber, tomato, asparagus, pea, radish, etc.) (Huber and Thompson 2007). Plants grown in bark compost immobilize mainly ammonium nitrogen and the nitrate nitrogen remains available for plant growth. However, sewage sludge compost (with a low C/N ratio) releases ammonium and consequently increases Fusarium wilt, even under colonization by BCAs capable of suppressing this wilt under other conditions (Hoitink et al. 1993). Cotxarreraa et al. (2002) used compost from vegetables and animal wastes, sewage sludge, and yard wastes and found it to reduce Fusarium wilt in tomato to a high degree. Low availability of ammonia in this compost may cause the direct effect of a high C/N ratio of other materials included in the compost, in addition to the negative effects of high pH and the reduced availability of Fe, Cu, and Zn on the pathogen.

5.3.2.4 Degree of Decomposed Compost

The degree of decomposition of compost has a strong effect on the rate of disease suppression. Immature compost could not suppress damping-off of cucumber seedlings caused by *P. aphanidermatum*. Fresh undecomposed OM mixed with *Trichoderma* does not exert biological control of *R. solani*. The synthesis of lytic enzymes involved in the parasitism of pathogens by *Trichoderma* is repressed in fresh OM due to high glucose concentrations. In mature composts, where concentration of nutrients such as glucose is low, the sclerotia of *R. solani* are killed by parasites and biological control prevails (Hoitink et al. 2001). On the other hand, the disease suppression potential of excessively stabilized compost is lost as it does not support microbial activity.

5.3.2.5 Role of Microbial Communities in Suppressive Potential of Compost

The environment around the compost plant, the system of composting used, and the composition of the raw material all affect the species richness and therefore the degree and spectrum of suppressive effect (Castano et al. 2011). The high temperature reached during the thermophilic phase of composting kills or inactivates all pathogens as well as beneficial microorganisms; thus, the composts are generally free of plant pathogens (Noble and Roberts 2004). As the temperature falls below 40 °C, mesophilic microorganisms colonize the semipasteurized compost; this is reinforced during the curing phase when there is also recolonization by surrounding antagonists, which develops the disease suppression capacity of the compost (Hoitink and Boehm 1999). Composts with high lignocellulosic substances are mostly colonized by *Trichoderma* spp. The microbial community that induced suppression of *Pythium* damping-off in cotton were populations of bacteria and actinobacteria capable of metabolizing fatty acids (linoleic acid) and thereby reducing the sporangium germination of *P. ultimum* (McKellar and Nelson 2003). Bonanomi et al. (2010) concluded that fluorescein diacetate hydrolysis, basal respiration, microbial biomass, total culturable bacteria, fluorescent pseudomonad counts, and *Trichoderma* populations gave the best predictions of disease suppression. Mechanisms involved in the phenomenon of disease suppression included competition, antibiosis, or hyperparasitism (Hoitink et al. 1993). According to Hoitink and Boehm (1999), the majority of composts suppress *Pythium* and *Phytophthora* root rot, while only 20 % of composts naturally suppress *Rhizoctonia* damping-off and very few (<10 %) induce resistance in plants. The type of organic amendment in compost has a clear positive effect on bacterial density and in particular, on the number of spore-forming bacteria, with an increase directly correlated with the dose of compost. The majority of the spore-forming bacteria isolated from the compost used in this study and selected during the composting process showed in vitro antibiotic activity against soilborne phytopathogenic fungi such as *F. oxysporum, F. solani*, and *R. solani*. Moreover, a greater decrease in damage by *Pyrenochaeta lycopersici* to tomato roots has been found in the same soil amended with compost (Zaccardelli et al. 2006, 2010). These results confirm the assumption that compost obtained from the organic fraction of municipal solid wastes produced an increase of suppressiveness against phytopathogenic fungi due to a change i,n the composition of the soil microbial community and a modification of the relationships among microorganisms—both competitive and/or antagonistic—producing a decrease in the activity of plant pathogens (Zaccardelli et al. 2013).

5.3.2.6 Arbuscular Mycorrhizal Fungi (AM Fungi) and Disease Suppression

Among beneficial soil microorganisms, the mycorrhizal fungi, particularly arbuscular mycorrhizae (AM), the most common fungal association formed almost with more than 90 % of cultivated plants, are gaining importance due to their varied benefits to plants. AM fungi offer many benefits to plants through a multiple action via absorption of nutrients, particularly P, water absorption, disease resistance, heavy metal toxicity, resistance to salt stress, etc. (Azcon-Aguilar and Barea 1996). AM fungi exert profound effects on other rhizosphere microorganisms either directly or indirectly via the host through a phenomenon termed the mycorrhizosphere effect by Linderman (1988) where most beneficial bacteria do inhabit and interact synergistically to stimulate plant growth. These interactions play a role in the suppression of fungal and nematode pathogens. Significant yield enhancement through field application of AM fungal inoculum has been recorded in a variety of crop plants (Sharma et al. 2010). Augmentation of these mycorrhizal fungi either through inoculation or through managing soil and crop management systems such as adopting conservation tillage and crop rotations (Sharma et al. 2012) can promote plants to cope up with many biotic and abiotic stresses and eventually sustain plant productivity. AM fungi protect plant roots from disease infection through several mechanisms as given below:

- One mechanism is via the changes in microbial communities that are produced as the mycorrhizosphere develops. There is strong evidence that shifts in microbial community structure and the resulting microbial changes can influence the growth and health of plants (Linderman 2000). Secilia and Bagyaraj (1987) isolated more pathogen-antagonistic actinomycetes from the rhizosphere of AM plants than from nonmycorrhizal controls, an effect that also depended on the AM fungus involved. AM fungi and other plant growth-promoting rhizobacteria (PGPR) share a common rhizosphere. AM fungi and PGPR may interact and cooperate in several ways, including their mutual establishment in the rhizosphere, improvement in plant rooting, enhancement of plant growth and nutrition, biological control of root pathogens, and improved nodulation in the case of legumes (Barea et al. 1996).

- Many authors suggested that the ability of AM-colonized plants to protect from root pathogens can be ascribed to an increased nutritional status in the host plant due to presence of the AMF. However, the effectiveness of AM fungi to suppress the) disease is dependent on the AM fungus involved and the substrate and the host plant (Whipps 2004). AM-mediated P-nutritional plants are more tolerant because these plants with a high phosphorus status are less sensitive to pathogen damage. Recently, Li et al. (2007) also found that AM fungi-associated bacteria (AMB) from the genus *Paenibacillus* have biocontrol ability against *Pythium*, which causes damping-off of cucumber. The possible antagonistic mechanisms of AMB against plant pathogens have been suggested to be the same as those of PGPR, i.e., competition for nutrients such as Fe, production of antibiotics, or production of fungal cell wall-degrading enzymes (Compant et al. 2005). Bharadwaj et al. (2008) suggested that some AMB could contribute to the often described ability of AM fungi to inhibit pathogens, acquire mineral nutrients, and modify plant root growth.
- Non-nutritional mechanisms are also important because mycorrhizal and nonmycorrhizal plants with the same internal phosphorus concentration may still be differentially affected by pathogens (Cardoso and Kuyper 2006). Such non-nutritional mechanisms include activation of plant defense systems, changes in exudation patterns and concomitant changes in mycorrhizosphere populations, increased lignifications of cell walls, and competition for space for colonization and infection sites. The mycorrhizal fungi protect plant roots from diseases by providing a physical barrier to the invading pathogen. A few examples of physical exclusion have been reported (Ingham 1991). However, some studies have shown that nematodes can penetrate the fungal mat (Maronek 1981), but still, disease development was affected adversely. Activation of plant defense mechanisms, including the development of systemic resistance, has also been proposed by Pozo et al. (2002). Among the compounds involved in plant defense (Bowles 1990) studied in relationship to AM formation are phytoalexins, enzymes of the phenylpropanoid pathway, chitinases, β-1,3-glucanases, peroxidases, pathogenesis-related (PR) proteins, callose, hydroxyproline- rich glycoproteins (HRGP), and phenolics (Gianinazzi-Pearson et al. 1994).
- By providing antagonistic chemicals and plant root exudates, AM fungi can produce a variety of antibiotics and other toxins that act against pathogenic organisms. Meyer and Linderman (1986) found that the number of sporangia and zoospores formed by cultures of *Phytophthora cinnamomi* was reduced by the application of extracts of rhizosphere soil from AM plants. Furthermore, Caron (1989) reported a reduction) in *Fusarium* populations in the soil surrounding mycorrhizal tomato roots as compared with the soil of nonmycorrhizal controls.

5.4 Soil Suppression of Soilborne Pathogens

5.4.1 Mechanism of Soil Suppressiveness Against Fusarium spp.

Fusarium wilt is caused by pathogenic *F. oxysporum*, a soilborne fungus, and it attacks many plant species. *Fusarium* spp. have good competitive saprophytic abilities and populations can increase after organic amendments. However, similar to *Pythium* spp., many *Fusarium* spp. are poor competitors and cannot colonize organic substrates previously colonized by other organisms. Natural suppressiveness of soils to Fusarium wilt was first recognized in the nineteenth century by Atkinson et al. (1975) and was later described for other soils around the globe (Peng et al. 1999; Dominguez et al. 2001). The suppressiveness is specific only to Fusarium wilts and not effective against diseases caused by nonvascular *Fusarium* species including *F. roseum* and *F. solani*, *F. subglutinans*, or other soilborne pathogens (Deacon and Berry 1993; Fravel et al. 2003). Such soils share many of the same biological and physical properties and several abiotic features including soil pH, OM content, and clay content, which play roles in disease suppression (Amir and Alabouvette 1993; Hoper and Alabouvette 1996). As early as 1970, Smith observed and reported that entities responsible for suppressiveness may be pleomorphic bacteria closely adhering to the stunted and lysed germ tube of chlamydospores of Fusarium wilt pathogen. But they were absent or few in number in steamed (54 °C) conducive soil. Soil pH also plays a significant role in soil suppressiveness and host susceptibility to Fusarium wilt pathogens (Barea et al. 1998). In clay loam soil at pH 8.0, the soil was suppressive; at 7.0, disease incidence significantly increased; and at pH 6.0, disease incidence was significantly higher than at pH 8.0 and 7.0. These factors pointed toward the presence of bacteria in sandy loam soil which is suppressive against Fusarium wilt pathogen as bacteria prefer alkaline soils.

Long-standing suppression operates in most Fusarium wilt-suppressive soils, but there are only a few examples of induced suppression. For example, suppressiveness to *F. oxysporum* f. sp. *niveum* (Larkin et al. 1993) was induced following continuous cropping of melon and watermelon, respectively. Interestingly, the induced suppressiveness in these cases was associated with continuous cropping of partially resistant cultivars, whereas induction of suppressiveness against other soilborne pathogens normally involves monoculture of susceptible cultivars (Whipps 1997). Incorporation of certain organic amendments into the soil may induce soil suppressiveness against soilborne and foliar pathogens. The soil microfauna/soil microbiome plays a significant role in natural and induced disease suppression. The possible mechanisms of induced soil suppressiveness include pathogen suppression, induced systemic resistance within host, and microbial interaction which takes place in the rhizosphere and which involves competition for nutrients and antibiosis (Andrews and Harris 2000). An example of induced soil suppressiveness with wild rocket (WR) found that sustainable disease suppression was maintained for 21 days after *F. oxysporum* f. sp. *radicis-cucumerium* inoculation of cucumber seedlings and transplantation into the WR-amended soil. It was observed that there was delayed onset of disease with reduced incidence, which demonstrates that the impact of soil suppressiveness on root diseases begins after inoculation of the pathogen. Among the bacterial and fungal genera responsible for Fusarium wilt suppressiveness are *Alcaligenes* sp. (Sayyed and Patel 2011), *Bacillus, Trichoderma* (Sivan and Chet 1989; Jambhulkar et al. 2011), *Pseudomonas* spp. (Mazurier et al. 2009), actinomycetes (Larkin et al. 1996), and nonpathogenic *F. oxysporum* (Olivain et al. 2006). Although the introduction of representative strains of each of these genera increased the level of soil suppressiveness in most cases, the introduction of large populations is unlikely to reproduce the microbial community structure and interactions that occur naturally in suppressive soils. In these soils, natural suppressiveness is associated with a reduction in the saprophytic growth and inhibition of chlamydospore germination of pathogenic *F. oxysporum* (Weller et al. 2002). This suppressiveness has been attributed mainly to the activity of nonpathogenic *F. oxysporum* and fluorescent *Pseudomonas* spp., and for both microbial groups, similar mechanisms including competition and induced systemic resistance were shown to be active (Fravel et al. 2003). Particularly interesting from the work of Lemanceau et al. (1993) is the intimate and complementary association between these two groups of microorganisms; in combination they provided enhanced disease suppression mediated by competition for iron via siderophores produced by the pseudomonads and for carbon by nonpathogenic *F. oxysporum* strain Fo47 (Lemanceau et al. 1993). The work by Duijff et al. (1998), using a glucuronidase GUS-marked strain of pathogenic *F. oxysporum* f. sp. *lini* and a pvdinaZ-marked derivative of *P. putida*, WCS358, supported and extended earlier observations that suppression by the nonpathogenic *Fusarium* strain is related to reductions in both population density and metabolic activity of the pathogen on the root surface; it also showed that competition for iron both contributes to the suppression by *Pseudomonas* and enhances the biological activity of the nonpathogenic *F. oxysporum* strain. Among a large collection of bacteria, fungi, and actinomycetes isolated from this suppressive soil, only nonpathogenic *F. oxysporum* isolates consistently suppressed the disease in both microwave-treated and natural soil. Induced systemic resistance was the primary mode of action for several of these isolates, but it is not yet clear if the mechanism is similar to that described for induced systemic resistance by rhizobacteria. Strains of nonpathogenic *F. oxysporum* differ considerably in their efficacy against Fusarium wilt. For example, strain Fo20 was the least effective of eight strains tested, whereas Fo47 proved to be the most effective against Fusarium wilt (Alabouvette et al. 1993). From this observation, we can infer that the composition of nonpathogenic *F. oxysporum* populations remained relatively stable over a considerable period of time, consistent with the long-standing nature of the suppressiveness of these soils.

To date, soil suppressiveness to Fusarium wilt disease has been ascribed to carbon and iron competition between pathogenic *F. oxysporum* and nonpathogenic *F. oxysporum* and fluorescent pseudomonads. Mazurier et al. (2009) studied the role of bacterial antibiosis in Fusarium wilt suppressiveness by comparing the densities, diversity, and activity of fluorescent species producing DAPG (*phl*D+) or phenazine (*phz*C+) antibiotics (Fig. 5.2). The frequencies of *phl*D+ populations were similar in the suppressive and conducive soils but their genotypic diversity differed significantly. However, *phl*D+ genotypes from two soils were equally effective in suppressing Fusarium wilt, either alone or in combination with nonpathogenic *F. oxysporum* strain Fo47. A mutant deficient in DAPG production provided a similar level of control as its parental strain, suggesting that this antibiotic does not play a major role. In contrast, phzC+ pseudomonads were only detected in suppressive soil. Representative phzC+ isolates of five distinct genotypes did not suppress Fusarium wilt on their own but acted synergistically in combination with strain Fo47. This increased level of disease suppression was attributed to phenazine production as the phenazine-deficient mutant was not effective. These results suggest, for the first time, that redox-active phenazines produced by fluorescent pseudomonads contribute to the natural soil suppressiveness to Fusarium wilt disease and may act synergistically with carbon competition by resident nonpathogenic *F. oxysporum*.

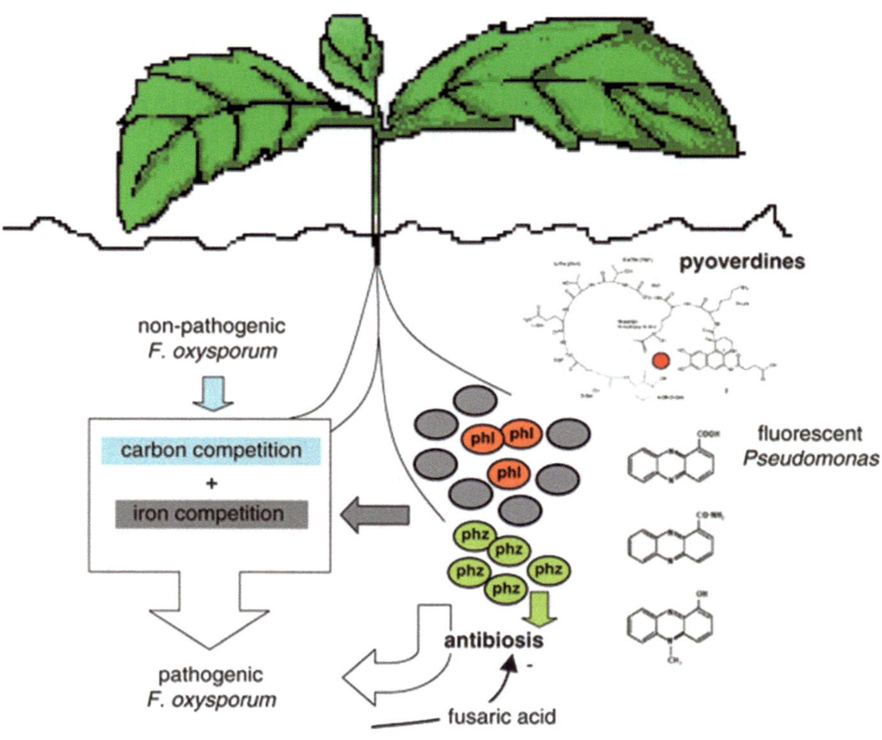

Fig. 5.2 Schematic model presenting the proposed mechanisms that contribute to the natural soil suppressiveness to Fusarium wilt. *phl* 2,4-diacetylphloroglucinol (DAPG), *phz* phenazine (Mazurier et al. 2009; reproduced with permission)

A higher level of carbon competition in the suppressive soil is generated due to the significantly higher microbial biomass in the suppressive soil as compared with the conducive soil. On this background of general competition, the higher density of nonpathogenic *F. oxysporum* in the suppressive soil further increases the carbon competition. The suppressive soil also differs from the conducive soil in its lower concentration of extractable iron, due to a high pH and $CaCO_3$ content, making pyoverdine-mediated iron competition between the pathogen and the fluorescent pseudomonads stronger in the suppressive than in the conducive soil. Carbon and iron competition act in synergy to suppress the saprophytic growth of pathogenic *F. oxysporum*, leading to a reduced activity and rate of root infection (Mazurier et al. 2009).

5.4.2 Mechanism of Disease Suppressiveness Against **Rhizoctonia** *spp.*

R. solani Kuhn is a soilborne fungus that causes disease in many economically important crop plants worldwide. Strains of the fungus are traditionally grouped into genetically isolated anastomosis groups (AGs) based primarily on hyphal anastomosis reactions and are further subdivided into intraspecific groups (ISGs) (Bolton et al. 2010). *Rhizoctonia*-suppressive soils reduced the severity of diseases caused by *R. solani* due to successive growing of a given plant host, which in general has been attributed to increased antagonism by *Trichoderma* spp. (Liu and Baker 1980). Ghini et al. (2007) evaluated the contribution and relationship of abiotic factors (pH, electrical conductivity, OM content, N total, P, Ca, Mg, S, Na, Fe, Mn, Cu, Zn, B, cation exchange capacity) and biotic factors (total microbial activity evaluated by CO_2 evolution and fluorescein diacetate hydrolysis; culturable bacteria, fungi, actinomycetes, protozoa, fluorescent pseudomonads, and *Fusarium* spp.) to the suppressiveness of soils to *R. solani*. Studies have reported that in highly suppressive soils of forest and pasture/fallow ground areas, several abiotic variables and fluorescein diacetate hydrolysis correlated with the suppression of *R. solani*; and this set of variables have explained more than 98 % of suppressiveness (Ghini and Morandi 2006). However, suppressive soils possessed higher populations of *Trichoderma* spp. than the corresponding conducive soil.

The soil suppressive to *Rhizoctonia* root rot of apple, caused by *R. solani* AG 5, was identified in Washington State (Mazzola and Gu 2002). However, the relative *Rhizoctonia*-suppressive capacity of the indigenous soil microbial community diminished with increasing age of the orchard block. The change in soil suppressiveness corresponded with a decrease in apple root colonization by actinomycetes and *Burkholderia cepacia* and a transformation in species composition of the fluorescent pseudomonad populations. While *P. putida* dominated the fluorescent pseudomonad community in non-planted orchard soil, a precipitous decline in its population was observed with increasing age of the orchard. *P. putida* was supplanted by *P. fluorescens* bv. III and *P. syringae* in the soil in response to planting apple. Likewise, isolates of *P. putida* from these soils provided biological control of *R. solani* AG 5. As observed in other systems, the *Rhizoctonia*-suppressive nature of the non-planted orchard soil was abolished by steam pasteurization (Mazzola 2007).

The mechanism for suppressing the pathogenic activity of *R. solani* differs from that of reducing its saprophytic activity in the case of damping-off. This aspect contrasts with the process during the events of damping-off caused by *Pythium* spp. For the latter, the frequency of seed colonization is directly related to the number of propagules until the colonization frequency reaches its maximum and is also correlated to the incidence of damping-off. Hence, different approaches to biological control need to be employed for *R. solani* and *Pythium* spp. (Kasuya et al. 2006). Microorganisms capable of suppressing these two kinds of pathogens also are known to be different. It was demonstrated that, although >70 different commercial composted pine bark amended potting mixes were effective in controlling damping-off of radish by *Pythium* spp., only one-fifth of those provided adequate control of *R. solani* damping-off because the latter was controlled by a much narrower spectrum of antagonistic microorganisms (Abbasi et al. 1999).

The feasibility of using organic amendments such as compost, animal manures, and organic industrial by-products in order to suppress soilborne plant pathogens has been well documented (Hoitink and Boehm 1999; Cheuk et al. 2005; Noble and Coventry 2005). Composts prepared from agricultural waste and used in container media or as soil amendments may have highly suppressive effects against diseases caused by a variety of soilborne plant pathogens. Barakat and Al-Masri (2009) amended sheep manure with *T. harzianum* and investigated its suppressiveness against damping-off of bean (*Phaseolus vulgaris*) for a 24-month period. Disease reduction was 50 % after 24 months with the highest concentration of organic amendment (10 %). Disease reduction increased with increasing concentration of organic amendment and with the duration of the incubation time. A combination of *T. harzianum* and sheep manure reduced both the total fungal population and the *R. solani* population after 12 and 24 months.

5.4.3 Mechanism of Disease Suppressiveness Against **Pythium** *and* **Phytophthora**

Damping-off and root rot caused by *Pythium* are considered to be the most devastating diseases of greenhouse crops. Biological control of *Pythium* is a promising environmentally friendly approach. Many factors affect the suppression of diseases in compost-amended soil affected with *Pythium* spp. These factors include compost type, OM quality and quantity, and associated level of microbial activity. Lightly decomposed OM colonized by a diverse microflora is very suppressive to diseases caused by *Pythium* spp. in container systems (Stone et al. 2004). This mechanism is being exploited by many nursery growers using tree barks in container system to suppress root rots in woody perennials. Apart from this, much of the evidence suggests damping-off of cucumber is suppressed with composts prepared from cattle manure, licorice roots, municipal biosolids, and sugarcane residues (Jenana et al. 2009). *Pythium* species are poor microbial competitors that strictly depend on the production of effective survival structures. They have the ability to germinate rapidly and grow in response to plant-derived seed or root exudate molecules to initiate plant infections. Carbohydrates and amino acids are the primary exudate components responsible for stimulating sporangium and oospore germination and initiating *Pythium*-seed interaction in the soil. Suppressive soil has greater mean concentrations of sodium, sulfate, and chloride than conducive soils; only chloride is inhibitory to *P. ultimum*. When conducive soils were amended with chloride at concentrations found in suppressive soil, colonizations of leaf debris by *P. ultimum* were partially suppressed. In suppressive soils, *P. oligandrum* was the most commonly isolated primary colonizing fungus and tended to be found at higher propagule densities than observed in conducive soils. When propagule densities of *P. oligandrum* were increased artificially in conducive soils, colonization and subsequent inoculum increases of *P. ultimum* were reduced. Suppressiveness was overcome by successive soil amendments with dried leaf debris, which resulted in progressive reductions in the frequencies of colonization by *P. oligandrum*. Apparently, soils with elevated chloride concentrations allow *P. oligandrum* to successfully compete with *P. ultimum*, and thus, the former increases its propagule density and further suppresses the saprophytic activity of *P. ultimum* (Martin and Hancock 1986).

The sphagnum peat system has been used as a model system to investigate the impact of OM quality on *Pythium* damping-off suppression (Boehm and Hoitink 1992; Boehm et al. 1997). Peats harvested from the top layers of a bog (very slightly decomposed sphagnum moss or light peat) are suppressive to *Pythium* damping-off. As a light peat decomposes, it loses the ability to suppress *Pythium* damping-off. Suppression is supported for 1–7 weeks. The loss of suppressiveness is related to (1) a decline in microbial activity as measured by the rate of hydrolysis of fluorescein diacetate (FDA) activity, (2) a shift in the culturable bacterial community composition from one in which 10 % of the isolates have the potential to suppress *Pythium* damping-off to one in which less than 1 % have this potential, and (3) a decline in carbohydrate content as determined by ^{13}C NMR spectroscopy (Boehm et al. 1997).

The following characteristics of the container system are responsible for suppression of *Pythium* damping-off:

1. Many types and sources of organic amendments consistently generate suppression.
2. Suppression is generated immediately after high-rate organic amendment (unless the organic substrate is raw).
3. Suppression is for a short duration (ranges from 1 week to 1 year).
4. Suppression is positively related to microbial activity.

Soil suppressiveness of diseases caused by *Phytophthora* spp. is considered to be the result of general suppression. Many types of organic materials suppress diseases caused by *Phytophthora* spp. The duration of suppression is similar to that of diseases caused by *Pythium* diseases, and suppression occurs soon after organic amendment. However, in contrast to suppression of *Pythium* spp., in which pathogen populations typically do not decline, in most documented systems of *Phytophthora* spp., propagules undergo microbial colonization, germination, and lysis. Bioassays determining the suppressiveness of soils have been used widely for various diseases with a variety of approaches and indicator plants. Such techniques may be used to determine the relative potential of the antagonistic population of a soil. Thus, blue lupin seedlings are used as indicator plant hosts to measure the suppressiveness of soils that are infested with *P. cinnamomi* (Duvenhage et al. 1991).

5.5 Conclusion

Soil suppressiveness research has clearly demonstrated that the phenomenon exists and is microbiologically mediated. However, there is considerably more uncertainty surrounding the identity of causal microbial agents and ecological processes that result in disease-suppressive soils. Many studies appear to have commenced with an assumption that suppression is specific. While it is likely that the principal mode of suppression will vary with each incidence of pathogen-suppressive soil, each study should commence by attempting to ascertain whether suppression is specific or general. We believe that this approach is justified as the outcomes provide a sound rationale for allocating resources toward future research efforts. The past dominance of culture-based studies has imposed limitations on our ability to test a specific suppression hypothesis. While not without their limitations, microbiomic methods currently provide the best tool for examining this question. Suppression cannot be achieved for all pathogens in question as the factors predicted to suppress different diseases are different for each pathogen. Suppressive soils are an asset to mankind as suppressive OM or compost can be produced but suppressive soil is not a renewable resource.

Acknowledgment The authors thank Dr. John Hammond (USDA-ARS, FNPRU, Beltsville, MD) for his critical reading and suggestions for this manuscript.

References

Abbasi PA, Miller SA, Meulia T, Hoitink HAJ, Kim JM (1999) Precise detection and tracing of *trichoderma hamatum* 382 in compost-amended potting mixes by using molecular markers. Appl Environ Microbiol 65(12):5421–5426

Alabouvette C (1986) *Fusarium* wilt suppressive soils from the Chateaurenard region: review of a 10-year study. Agronomie 6:273–284
Alabouvette C (1999) Fusarium wilt suppressive soils: an example of disease-suppressive soils. Austr Plant Pathol 28:47–64
Alabouvette C, Lemanceau P, Steinberg C (1993) Recent advances in the biological control of fusarium wilts. Pestic Sci 37:365–373
Amir H, Alabouvette C (1993) Involvement of soil abiotic factors in the mechanisms of soil suppressiveness to *Fusarium* wilts. Soil Biol Biochem 25:157–164
Andrade OA, Mathre DE, Sands DC (1994) Natural suppression of take-all disease of wheat in Montana soils. Plant Soil 164:9–18
Andrade O, Campillo R, Peyrelongue A, Barrientos L (2011) Soils suppressive against *Gaeumannomyces graminis* var. *tritici* identified under wheat crop monoculture in southern Chile. Ciencia Investigation Agraria 38(3):345–356
Andrews JH, Harris RF (2000) The ecology and biogeography of microorganisms on plant surfaces. Annu Rev Phytopathol 38:145–180
Andrivon D (1994) Dynamics of the survival and infectivity to potato tubers of sporangia of *Phytophthora infestans* in three different soils. Soil Biol Biochem 26:945–952
Aryantha IP, Cross R, Guest DI (2000) Suppression of *Phytophthora cinnamomi* in potting mixes amended with uncomposted and composted animal manures. Phytopathology 90:775–782
Atkinson TG, Neal JL Jr, Larson RI (1975) Genetic control of the rhizosphere microflora of wheat. In: Bruehl GW (ed) Biology and control of soil-borne plant pathogens. The American Phytopathological Society, St. Paul, pp 116–122
Aviles M, Borrero C, Trillas MI (2011) Review on compost as an inducer of disease suppression in plants grown in soilless culture. Dyn Soil Dyn Plants 5 (special issue 2):1–11
Azcon-Aguilar C, Barea JM (1996) Arbuscular mycorrhizas and biological control of soil-borne plant pathogens. An overview of the mechanisms involved. Mycorrhiza 6:457–464
Baker KF, Cook RJ (1974) Biological control of plant pathogens. Freeman, San Francisco, 433 pp
Barakat RM, Al-Masri MI (2009) *Trichoderma harzianum* in combination with sheep manure amendment enhances soil suppressiveness of fusarium wilt of tomato. Phytopathol Mediterr 48:385–395
Barea JM, Tobar RM, Azcón-Aguilar C (1996) Effect of a genetically-modified *Rhizobium meliloti* inoculant on the development of arbuscular mycorrhizas, root morphology, nutrient uptake and biomass accumulation in *Medicago sativa* L. New Phytol 134:361–369
Barea JM, Andrade G, Bianciotto V, Dowling D, Lohrke S, Bonfantr P, O'Gara F, Azcon-Aguilar C (1998) Impact on arbuscular mycorrhiza formation of *Pseudomonas* strains used as inoculants for biocontrol of soil-borne fungal plant pathogens. Appl Environ Microbiol 64:2304–2307
Barnett SJ, Roget DK, Ryder MH (2006) Suppression of *Rhizoctonia solani* AG-8 induced disease on wheat by the interaction between *Pantoea*, *Exiguobacterium* and *Microbacteria*. Aust J Soil Res 44:331–342
Benhamou N, Garand C (2001) Cytological analysis of defense-related mechanisms induced in pea root tissues in response to colonization by nonpathogenic *Fusarium oxysporum* Fo47. Phytopathology 91:730–740
Benitez M, McSpadden Gardner BB (2009) Linking sequence to function in soil bacteria: sequence-derived isolation of novel bacteria contributing to soilborne plant disease suppression. Appl Environ Microbiol 75:915–924
Bharadwaj DP, Lundquist P, Alstrom S (2008) Arbuscular mycorrhizal fungal spore-associated bacteria affect mycorrhizal colonization, plant growth and potato pathogens. Soil Biol Biochem 40:2494–2501
Boehm MJ, Hoitink HAJ (1992) Sustenance of microbial activity in potting mixes and its impact on severity of *Pythium* root rot of Poinsettia. Phytopathology 82:259–264
Boehm MJ, Wu T, Stone AG, Kraakman B, Lannotti DA, Wilson EG (1997) Cross polarized magic-angle spinning 13C nuclear magnetic resonance spectroscopic characterization of soil

organic matter relative to culturable bacterial species composition and sustained biological control of *Pythium* root rot. Appl Environ Microbiol 63:162–168

Bolton MD, Panella L, Campbell L, Khan MFR (2010) Temperature, moisture and fungicide effects in managing *Rhizoctonia* root and crown rot of sugar beet. Phytopathology 100:689–697

Bonanomi G, Antignani C, Scala F (2007) Suppression of soil borne fungal diseases with organic amendments. J Plant Pathol 89(3):311–324

Bonanomi G, Antignani V, Capodilupo M, Scala F (2010) Identifying the characteristics of organic soil amendments that suppress soilborne plant diseases. Soil Biol Biochem 42:136–144

Borrero C, Trillas MI, Ordovas J, Tello JC, Aviles M (2004) Predictive factors for the suppression of *Fusarium* wilt of tomato in plant growth media. Phytopathology 94:1094–1101

Bowles DJ (1990) Defence related proteins in higher plants. Annu Rev Biochem 59:873–907

Bruehl GW (1987) Soilborne plant pathogens. Macmillan, New York

Cardoso IM, Kuyper TW (2006) Mycorrhizas and tropical soil fertility. Agric Ecosyst Environ 116:72–84

Caron M (1989) Potential use of mycorrhizae in control of soilborne diseases. Can J Plant Pathol 11:177–179

Castano R, Borrero C, Aviles M (2011) Organic matter fractions by SP-MAS 13 C NMR and microbial communities involved in the suppression of fusarium wilt in organic growth media. Biol Control 58:286–293

Cavigelli MA, Thien SJ (2003) Phosphorus bioavailability following incorporation of green manure crops. Soil Sci Soc Am J 67:1186–1194

Chen W, Hoitink HAJ, Schmitthenner AF, Tuovinen OH (1988) The role of microbial activity in suppression of damping-off caused by *Pythium ultimum*. Phytopathology 78:314–322

Cheuk W, Lo KV, Coperman R, Joliffee P, Fraser BS (2005) Disease suppression on greenhouse tomatoes using plant waste compost. J Environ Sci Health 40(30):449–461

Compant S, Reiter B, Sessitsch A, Nowak J, Clément C, Barka EA (2005) Endophytic colonization of *Vitis vinifera* L. by plant growth-promoting bacterium *Burkholderia* sp. strain PsJN. Environ Microbiol 71:1685–1693

Conklin AE, Erich MS, Liebman M, Lambert D, Gallandt ER, Halteman WA (2002) Effects of red clover (*Trifolium pratense*) green manure and compost soil amendments on wild mustard (*Brassica kaber*) growth and incidence of disease. Plant Soil 238:245–256

Cook RJ, Baker KF (1983) The nature and practice of biological control of plant pathogens. The American Phytopathological Society, St. Paul, p 539 pp

Costa JL, Menge JA, Casal WL (2000) Biological control of *Phytophthora* root rot of avocado with microorganisms grown in organic mulches. Br J Microbiol 31:239–246

Cotxarreraa L, Trillas Gay MI, Steinberg C, Alabouvette C (2002) Use of sewage sludge compost and *Trichoderma asperellum* isolates to suppress fusarium wilt of tomato. Soil Biol Biochem 34:467–476

Coventry E, Noble R, Mead A, Whipps JM (2005) Suppression of *Allium* white rot (*Sclerotium cepivorum*) in different soils using vegetable wastes. Eur J Plant Pathol 111:101–112

Davis JR, Huisman OC, Westermann DT, Hafez SL, Everson DO, Sorenson LH, Schneider AT (1996) Effects of green manures on verticillium wilt of potato. Phytopathology 86:444–453

de Boer W, Verheggen P, Gunnewiek P, Kowalchuk GA, van Veen JA (2003) Microbial community composition affects soil fungistasis. Appl Environ Microbiol 69:835–844

Deacon JW, Berry LA (1993) Biocontrol of soil-borne plant-pathogens—concepts and their application. Pestic Sci 37:417–426

Diab H, Hu S, Benson DM (2003) Suppression of *Rhizoctonia solani* on impatiens by enhanced microbial activity in composted swine waste amended potting mixes. Phytopathology 93:1115–1123

Dominguez J, Negrin MA, Rodriguez CM (2001) Aggregate water-stability, particle size and soil solution properties in conducive and suppressive soils to fusarium wilt of banana from Canary islands (Spain). Soil Biol Biochem 33:449–455

Duijff BJ, Pouhair D, Olivain C, Alabouvette C, Lemanceau P (1998) Implication of systemic induced resistance in the suppression of fusarium wilt of tomato by *Pseudomonas fluorescens* WCS417r and by nonpathogenic *Fusarium oxysporum* Fo47. Eur J Plant Pathol 104:903–910

Duvenhage JA, Kotze JM, Maas EMC (1991) Suppressive soils and biological control of *Phytophthora* root rot. In: South African avocado growers' association yearbook, vol 14, pp 6–11

Fradkin A, Patrick ZA (1985) Effect of matric potential, pH, temperature, and clay minerals on bacterial colonization of conidia of *Cochliobolus sativus* and on their survival in soils. Can J Plant Pathol 7:19–27

Fravel DR (1988) Role of antibiosis in the biocontrol of plant diseases. Annu Rev Phytopathol 26:75–91

Fravel D, Olivain C, Alabouvette C (2003) *Fusarium oxysporum* and its biocontrol. New Phytol 157(3):493–502

Garbeva P, Postma J, van Veen JA, van Elsas JD (2006) Effect of above-ground plant species on soil microbial community structure and its impact on suppression of *Rhizoctonia solani* AG3. Environ Microbiol 8:233–246

Garbeva P, Silby MW, Raaijmakers JM, Levy SB, Boer WD (2011) Transcriptional and antagonistic responses of *Pseudomonas fluorescens* Pf0-1 to phylogenetically different bacterial competitors. ISME J 5(6):973–985

Ghini R, Morandi MAB (2006) Biotic and abiotic factors associated with soil suppressiveness to *Rhizoctonia solani*. Sci Agric 63(2):153–160

Ghini R, Patrício FRA, Bettiol W, Almeida IMG, Maia AHN (2007) Effect of sewage sludge on suppressiveness to soil-borne plant pathogens. Soil Biol Biochem 39:2797–2805

Gianinazzi-Pearson V, Gollotte A, Dumas-Gaudot E, Franken P, Gianinazzi S (1994) Gene expression and molecular modifications associated with plant responses to infection by arbuscular mycorrhizal fungi. In: Daniels M, Downic JA, Osbourn AE (eds) Advances in molecular genetics of plant-microbe interactions. Kluwer, Dordrecht, pp 179–186

Gupta VVSR, Rovira AD, Roget DK (2011) Principles and management of soil biological factors for sustainable rainfed farming systems. In: Tow P, Cooper I, Partridge I, Birch C (eds) Rainfed farming systems'. Springer, New York, pp 149–184

Hardy GESJ, Sivasithamparam K (1991) Sporangial responses do not reflect microbial suppression of *Phytophthora dreschsleri* in composted Eucalyptus bark mix. Soil Biol Biochem 23:757–765

Hoitink HAJ, Boehm MJ (1999) Importance of physical and chemical soil properties in the suppressiveness of soils to plant disease. Eur J Soil Biol 32:41–58

Hoitink HAJ, Boehm MJ, Hadar Y (1993) Mechanisms of suppression of soilborne plant pathogens in compost-amended substrates. In: Hoitink HAJ, Keener HM (eds) Science and engineering of composting: design, environmental, microbiological and utilization aspects. Renaissance, Worthington, pp 601–621

Hoitink HAJ, Stone AG, Grebus ME (1996) Suppression of plant diseases by composts. In: De Bertoldi M, Sequi P, Lemmes B, Papi T (eds) The science of composting, vol 1. Blackie Academic & Professional, Glasgow, pp 373–381

Hoitink HAJ, Krause MS, Han DY (2001) Spectrum and mechanisms of plant disease control with composts. In: Stoffela Kahn BA (ed) Compost utilization in horticultural cropping systems. Lewis, Boca Raton, pp 263–273

Hoper H, Alabouvette C (1996) Importance of physical and chemical soil properties in the suppressiveness of soils to plant diseases. Eur J Soil Biol 32:41–58

Hornby D (1983) Suppressive soils. Annu Rev Phytopathol 21:65–85

Howell CR, Stipanovic RD (1983) Gliovirin, a new antibiotic from *Gliocladium virens*, and its role in the biological control of *Pythium ultimum*. Can J Microbiol 29:321–324

Huber DM, Thompson IA (2007) Nitrogen and plant disease. In: Datnoff LE, Elmer WH, Huber DM (eds) Mineral nutrition and plant disease. The American Phytopathological Society, St Paul, pp 31–44

Hyakumachi M, Loffler HJM, Lockwood JL (1989) Methods for the determination of carbon loss from fungal propagules incubated in soil. Soil Biol Biochem 21:567–571

Ingham ER (1991) Interactions among mycorrhizal fungi, rhizosphere organisms, and plants. In: Barbosa P, Krischik VA, Jones CG (eds) Microbial mediation of plant-herbivore interactions. Wiley, New York, pp 169–197

Jambhulkar PP, Babu SR, Ameta GS (2011) Comparative efficacy of fungicides and antagonists against fusarium wilt of chickpea. J Mycol Plant Pathol 41(3):399–401

Jenana RKB, Haouala R, Triki MA, Godon JJ, Hibar K, Khedher MB, Henchi B (2009) Compost, compost extract and bacterial suppressive action on *Pythium aphanidermatum* in tomato. Pakistan J Bot 41(1):315–327

Jones JB, Jones JP, Stall RE, Zitter TA (1991) Compendium of tomato diseases. The American Phytopathological Society, St Paul, p 100 pp

Kasuya M, Olivier AR, Ota Y, Tojo M, Honjo H, Fukui R (2006) Induction of soil suppressiveness against *Rhizoctonia solani* by incorporation of dried plant residues into soil. Phytopathology 96:1372–1379

Keel C, Défago G (1997) Interactions between beneficial soil bacteria and root pathogens: mechanisms and ecological impact. In: Gange AC, Brown VK (eds) Multitrophic interactions in terrestrial system. Blackwell Science, Oxford, pp 27–47

Keen B, Vancov T (2010) *Phytophthora cinnamomi* suppressive soils. In: Mendez-Vilas A (ed) Current research, technology and education topics in applied microbial biotechnology. Formatex Research Center, Badajoz, Spain, pp 239–250

Kerry BR (1988) Fungal parasites of cyst nematodes. Agric Ecosyst Environ 24:293–295

Ko WH (2003) Germination of fern spores in natural soils. Am Fern J 93(2):70–75

Larkin RP, Hopkins DL, Martin FN (1993) Effect of successive watermelon plantings on *Fusarium oxysporum* and other microorganisms in soils suppressive and conducive to fusarium-wilt of watermelon. Phytopathology 83:1105–1116

Larkin RP, Hopkins DL, Martin FN (1996) Suppression of fusarium wilt of watermelon by non-pathogenic *Fusarium oxysporum* and other microorganisms recovered from a disease-suppressive soil. Phytopathology 86:812–819

Lemanceau P, Bakker PAHM, Dekogel WJ, Alabouvette C, Schippers B (1993) Antagonistic effect of non-pathogenic *Fusarium oxysporum* Fo47 and pseudobactin 358 upon pathogenic *Fusarium oxysporum* f. sp. *dianthi*. Appl Environ Microbiol 59:74–82

Li B, Ravnskov S, Xie G, Larsen J (2007) Biocontrol of pythium damping-off in cucumber by arbuscular mycorrhiza-associated bacteria from the genus *Paenibacillus*. BioControl 52:863–875

Linderman RG (1988) Mycorrhizal interactions with the rhizosphere effect. Phytopathology 78:366–371

Linderman RG (2000) Effects of mycorrhizas on plant tolerance to diseases. In: Kapulnik Y, Douds DD Jr (eds) Arbuscular mycorrhizas: physiology and function. Kluwer, Dordrecht, pp 345–365

Litterick AM, Harrier L, Wallace P, Watson CA, Wood M (2004) The role of uncomposted materials, composts, manures, and compost extracts in reducing pest and disease incidence and severity in sustainable temperate agricultural and horticultural crop production: a review. Crit Rev Plant Sci 23:453–479

Liu S, Baker R (1980) Mechanism of biological control in soil suppressive to *Rhizoctonia solani*. Phytopathology 70:404–412

Lockwood JL (1990) Relation of energy stress to behaviour of soilborne plant pathogens and to disease development. In: Hornby D (ed) Biological control of soilborne plant pathogens. CAB International, Wallingford, UK, pp 197–214

Maronek DM (1981) Mycorrhizal fungi and their importance in horticultural crop production. In: Jaurch R. Hortic Rev 3:172–213

Martin FN, Hancock JG (1986) Association of chemical and biological factors in soils suppressive to *Pythium ultimum*. Phytopathology 76:1221–1231

Mazzola M, Gu YH (2002) Wheat genotype-specific induction of soil microbial communities suppressive to disease incited by *Rhizoctonia solani* anastomosis group (AG)-5 and AG-8. Phytopathology 92:1300–1307

Mazurier S, Corberand T, Lemanceau P, Raaijmakers JM (2009) Phenazine antibiotics produced by fluorescent pseudomonads contribute to natural soil suppressiveness to fusarium wilt. ISME J 3(8):977–991

Mazzola M (2002) Mechanisms of natural soil suppressiveness to soilborne diseases. Antonie Van Leeuwenhoek 81(1-4):557–564

Mazzola M (2007) Manipulation of rhizosphere bacterial communities to induce suppressive soils. J Nematol 39(3):213–220

McKellar ME, Nelson EB (2003) Compost-induced suppression of Pythium damping-off is mediated by fatty-acid metabolizing seed-colonizing microbial communities. Appl Environ Microbiol 69:452–460

Meyer JR, Linderman RG (1986) Selective influence on populations of rhizosphere or rhizoplane bacteria and actinomycetes by mycorrhizas formed by *Glomus fasciculatum*. Soil Biol Biochem 18:191–196

Murakami H, Tsushima S, Shishido Y (2000) Soil suppressiveness to clubroot disease of Chinese cabbage caused by *Plasmodiophora brassicae*. Soil Biol Biochem 32:1637–1642

Noble R, Coventry E (2005) Suppression of soil-borne plant diseases with composts: a review. Biocontrol Sci Technol 15:3–20

Noble R, Roberts SJ (2004) Eradication of plant pathogens and nematodes during composting: a review. Plant Pathol 5:548–568

Olivain C, Humbert C, Nahalkova J, Fatehi J, L'Haridon F, Alabouvette C (2006) Colonization of tomato root by pathogenic and non-pathogenic *Fusarium oxysporum* strains inoculated together and separately into the soil. Appl Environ Microbiol 72(2):1523–1531

Ownley BH, Benson DM, Bilderback TE (1990) Physical properties of container media and relation to severity of phytophthora root rot of rhododendron. J Am Soc Hortic Sci 115:564–570

Pane C, Spaccini R, Piccolo A, Scala F, Bonanomi G (2011) Compost amendments enhance peat suppressiveness to *Pythium ultimum, Rhizoctonia solani* and *Sclerotinia minor*. Biol Control 56:115–124

Paulitz T, Baker R (1988) The formation of secondary sporangia by *Pythium ultimum*: the influence of organic amendments and *Pythium nunn*. Soil Biol Biochem 20:151–156

Peng HX, Sivasithamparam K, Turner DW (1999) Chlamydospore germination and fusarium wilt of banana plantlets in suppressive and conducive soils are affected by physical and chemical factors. Soil Biol Biochem 31:1363–1374

Penton CR, Gupta VVSR, Tiedje JM, Neate SM, Ophel-Keller K et al (2014) Fungal community structure in disease suppressive soils assessed by 28S LSU gene sequencing. PLoS ONE 9(4), e93893

Persson L, Larsson-Wikström M, Gerhardson B (1999) Assessment of soil suppressiveness to *Aphanomyces* root rot of pea. Plant Dis 83:1108–1112

Pharand B, Carisse O, Benhamou N (2002) Cytological aspects of compost-mediated induced resistance against fusarium crown and root rot in tomato. Phytopathology 92:424–438

Postma J, Luttikholt AJG (1996) Colonization of carnation stems by a nonpathogenic isolate of *Fusarium oxysporum* and its effect on *F. oxysporum* f. sp. *dianthi*. Can J Bot 74:1199–1205

Postma J, Montanari M, van den Boogert PHJF (2003) Microbial enrichment to enhance the disease suppressive activity of compost. Eur J Soil Biol 39:157–163

Postma J, Schilder MT, Bloem J, van Leeuwen-Haagsma WK (2008) Soil suppressiveness and functional diversity of the soil microflora in organic farming systems. Soil Biol Biochem 40:2394–2406

Pozo MJ, Cordier C, Dumas-Gaudot E, Gianinazzi S, Barea JM, Azcon-Aguilar C (2002) Localized versus systemic effect of arbuscular mycorrhizal fungi on defence responses to Phytophthora infection in tomato plants. J Exp Bot 53:525–534

Sayyed RZ, Patel PR (2011) Biocontrol potential of siderophore producing heavy metal resistant *Alcaligenes* sp. and *Pseudomonas aeruginosa* RZS3 vis-à-vis organophosphorus fungicide. Indian J Microbiol 51(3):266–272

Secilia J, Bagyaraj DJ (1987) Bacteria and actinomycetes associated with pot cultures of vesicular-arbuscular mycorrhizas. Can J Microbiol 33:1069–1073

Serra-Wittling C, Houot S, Alabouvette C (1996) Increased soil suppressiveness to fusarium wilt of flax after addition of municipal solid waste compost. Soil Biol Biochem 28:1207–1214

Sharma MP, Sharma SK, Gupta GK (2010) Arbuscular mycorrhizal fungi and plant growth-promoting rhizobacteria in the control of soil-borne diseases of selected oil-seed crops. In: Khachatourians GG, Arora DK, Rajendran TP, Srivastava A (eds) Agriculturally important microorganisms, an international multi-volume series, vol II. Academic World International, Bhopal, pp 45–75

Sharma MP, Gupta S, Sharma SK, Vyas AK (2012) Effect of tillage and crop sequences on arbuscular mycorrhizal symbiosis and soil enzyme activities in soybean (*Glycine max* L. Merril) rhizosphere. Indian J Agric Sci 82:25–30

Sivan A, Chet I (1989) The possible role of competition between *Trichoderma harzianum* and *Fusarium oxysporum* on rhizosphere colonization. Phytopathology 79:198–203

Stone AG, Traina SJ, Hoitink HAJ (2001) Particulate organic matter composition and pythium damping-off of cucumber. Soil Sci Soc Am J 65:761–770

Stone AG, Scheuerell SJ, Darby HM (2004) Suppression of soil borne diseases in field agricultural systems: organic matter management, cover cropping and other cultural practices. In: Magdoff F and Weil R (eds) Soil organic matter in sustainable agriculture. CRC, Boca Raton, pp 132–164, Ch 5

Szczech MM (1999) Suppressiveness of vermicompost against fusarium wilt of tomato. J Phytopathol 147:155–161

Szczech M, Smolinśka U (2001) Comparison of suppressiveness of vermicomposts produced from animal manures and sewage sludge against *Phytophthora nicotianae* Breda de Haan var. *nicotianae*. J Phytopathol 149:77–82

Tenuta M, Lazarovits G (2002) Ammonia and nitrous acid from nitrogenous amendments kill the microsclerotia of *Verticillium dahliae*. Phytopathology 92:255–264

Termorshuizen AJ, Jeger MJ (2008) Strategies of soilborne plant pathogenic fungi in relation to disease suppression. Fungal Ecol 1:108–114

Toyota K, Kimura M (1993) Colonization of chlamydospores of *Fusarium oxysporum* f. sp. *raphani* by soil bacteria and their effects on germination. Soil Biol Biochem 25(193):197

Toyota K, Ritz K, Young IM (1996) Microbiological factors affecting the colonisation of soil aggregates by *Fusarium oxysporum* f. sp. *raphani*. Soil Biol Biochem 28:1513–1521

van Dijk K, Nelson EB (2000) Fatty acid competition as a mechanism by which *Enterobacter cloacae* suppresses *Pythium ultimum* sporangium germination and damping-off. Appl Environ Microbiol 66:5340–5347

Veeken AHM, Blok WJ, Curci F, Coenen GCM, Temorshuizen AJ, Hamelers HVM (2005) Improving quality of composted biowaste to enhance disease suppressiveness of compost-amended, peat based potting mixes. Soil Biol Biochem 37:2131–2140

Weller DM, Raaaijmakers JM, MacSpadden Gardener BB, Thomashow LS (2002) Microbial populations responsible for specific soil suppressiveness to plant pathogens. Annu Rev Phytopathol 40:309–348

Westphal A, Becker JO (1999) Biological suppression and natural population decline of *Heterodera schachtii* in a California field. Phytopathology 89:434–440

Whipps JM (1997) Developments in the biological control of soil-borne plant pathogens. Adv Bot Res 26:1–134

Whipps JM (2001) Microbial interactions and biocontrol in the rhizosphere. J Exp Bot 52(suppl 1):487–511. doi:10.1093/jexbot/52.suppl_1.487

Whipps JM (2004) Prospects and limitations for mycorrhizas in biocontrol of root pathogens. Can J Bot 82:1198–1227

Wilcox TL, Mircetich SM (1985) Influence of soil water matric potential on the development of phytophthora root and crown rots of mahaleb cherry. Phytopathology 75:648–653
Wiseman BM, Neate SM, Keller KO, Smith SE (1996) Suppression of *Rhizoctonia solani* anastomosis group 8 in Australia and its biological nature. Soil Biol Biochem 28:727–732
Zaccardelli M, Perrone D, Del Galdo A, Giordano I, Villari G, Bianco M (2006) Multidisciplinary approach to validate compost use in vegetable crop systems in Campania Region (Italy): effect of compost fertilization on processing tomato in field cultivation. Acta Hortic 700:285–288
Zaccardelli M, Pane C, Perrone D, Pucci N, Infantino A (2010) Control of corky root of tomato with compost and role of spore-forming bacteria to inhibit *Pyrenochaeta lycopersici*. In: III international symposium on tomato diseases, Ischia, NA Italy, 25–30 Jul 2010
Zaccardelli M, DeNicola F, Villecco D, Scotti R (2013) The development and suppressive activity of soil microbial communities under compost amendment. J Soil Sci Plant Nutr 13(3):730–742
Zhang W, Hoitink HAJ, Dick WA (1996) Compost-induced systemic acquired resistance in cucumber to pythium root rot and anthracnose. Phytopathology 86:1066–1070
Zhang W, Han DY, Dick WA, Davis KR, Hoitink HAJ (1998) Compost and compost water extract induced systemic acquired resistance in cucumber and *Arabidopsis*. Phytopathology 88:450–455

Chapter 6
Agricultural Soil Health and Pea Footrot Disease Suppressiveness

Ebimieowei Etebu

6.1 Introduction

Peas are self-pollinating annual herb of 30–150 cm long. They are propagated from seed at a recommended density of between 70 and 100 plants/m^2 (Davies et al. 1985; Knott 1999). Their growing seasons vary from 80 to 150 days depending on geographical region (Davies et al. 1985). They are grown in over 87 countries all over the world providing food for humans and feed for domestic animals (Hulse 1994; McPhee 2003). Major pea-producing countries are France, Russia, Ukraine, Denmark and the UK in Europe, China and India in Asia, Canada and the USA in North America, Chile in South America, Ethiopia in Africa and Australia (FAO 1994). Pea ranks fourth next to soybean, groundnut and beans in global legume production (Hulse 1994); global production amounts to about 10.5 million tonnes of dry pea and 7 million tonnes of fresh peas (Duke 1981; FAO 2001). Peas are a good source of proteins, fat, carbohydrate, crude fibre, ash, calcium, phosphorus, sodium, potassium, iron, thiamine, riboflavin, niacin and ascorbic acid (Duke 1981; Hulse 1994). Notwithstanding their enormous nutritional qualities and position in the total worldwide trade of pulses, the cultivation and production of peas are challenged by an array of constraints. Top among the constraints affecting pea production are diseases and pests, especially root and footrot diseases (Graham and Vance 2003). Root and footrot diseases occur wherever peas are grown in the world (Hagedorn 1976; Persson et al. 1997). Several soil fungi associate with pea roots and are responsible for the diseases. Some of these fungi include *Aphanomyces euteiches*, *Pythium ultimum*, *Rhizoctonia solani*, *Thielaviopsis basicola*, *Fusarium oxysporum*, *Ascochyta pinodella*, *Sclerotinia*

E. Etebu (✉)
Department of Biological Sciences, Niger Delta University, Amassoma, Wilberforce Island, Bayelsa State, Nigeria
e-mail: eetebu@ndu.edu.ng

M.K. Meghvansi, A. Varma (eds.), *Organic Amendments and Soil Suppressiveness in Plant Disease Management*, Soil Biology 46,
DOI 10.1007/978-3-319-23075-7_6

sclerotiorum and *Nectria haematococca* (anamorph *Fusarium solani* f. sp. *pisi*) (Hagedorn 1991; Hwang and Chang 1989; Oyarzun et al. 1993a).

Nectria haematococca (anamorph *F. solani* f. sp. *pisi* is the most important fungus (Hwang and Chang 1989) among the suite of fungi responsible for root rot disease complex in peas. It is a soilborne pathogen responsible for pea footrot disease in particular. *N. haematococca* is pathogenic on all commercial processing pea cultivars (Hagedorn 1991; Grünwald et al. 2003), accounting for as much as 57 % of pea yield losses (Kraft 1984; Oyarzun 1993). Neither genetic resistance nor chemical control is effective in the control of pea footrot disease; as such the disease is controlled or managed only through avoidance of fields with high disease potential. Identifying agricultural fields with a high disease potential has therefore been paramount in the implementation of preventive measures (Oyarzun 1993). Like several other soilborne plant diseases, the ability of *N. haematococca* to cause disease in peas is, in part, dependent on the health of the soil on which peas are grown. Some soils, referred to as 'suppressive soils', either completely inhibit disease initiation or truncate its progression in susceptible plants, in spite of favourable conditions for disease incidence and development (Cook and Baker 1983; Schippers 1992).

This chapter is therefore intended to discuss the factors that make for agricultural soil health with respect to pea footrot disease suppression or otherwise. The chapter is divided into four sections under the following subheadings: the causal pathogen, pea footrot disease symptoms and assessment, molecular basis of pea footrot disease and the potential role of agricultural soil health indices in pea footrot disease suppressiveness.

6.2 The Causal Pathogen: *Nectria haematococca*

Footrot of peas is caused by the soilborne fungus *Nectria haematococca*. The fungus is a member of a heterogeneous group of ascomycetous fungi composed of both homothallic and heterothallic groups (Booth 1971). Members of mating population VI (MPVI) infect and cause disease on nine plant species and one animal species; occur as secondary/tertiary pathogens, in 14 species of plants; and can exist as saprophytes in soil (Van Etten and Kistler 1988; Funnell and Van Etten 2002). They are best known and studied as pathogens of the garden pea (*Pisum sativum*), where they are often referred to as *Fusarium solani* f. sp. *pisi*, reflecting the anamorphic stage of the fungus (Funnell et al. 2001). The fungus *F. solani* survives in soils and plant debris and infects a wide variety of crops where they cause diseases with symptoms such as wilting, rotting of seeds, damping off and root and tuber rots, among others. Usually, asexual spores are produced, but under certain conditions perithecial stages identified as *Nectria haematococca* are found (Booth 1971; Matuo and Snyder 1972). Colonies of *F. solani* f. sp. *pisi* grown on freshly prepared potato dextrose agar are characterised by typical blue-green to buff-coloured sporodochia. Macroconidia are hyaline, measuring between 27 and

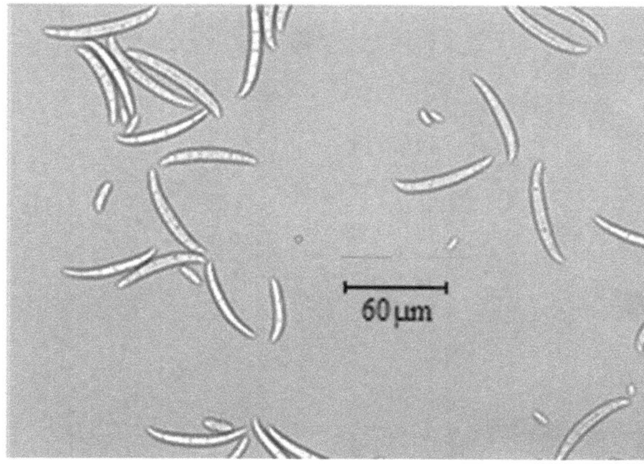

Fig. 6.1 Conidia of *N. haematococca* isolated from agricultural soil with pea footrot disease history in the UK (*Courtesy*: E. Etebu)

60 µm long, 'canoe-shaped' conidia in transverse view, and a distinct 'foot cell' at the lower end, and divided by several cross-walls (Fig. 6.1) (Kraft 2001; Etebu 2008).

Morphological traits are not sufficient to differentiate pea pathogenic strains of *N. haematococca*, from nonpathogenic forms, both of which are known to exist in agricultural soils. Hence recent studies of this fungus aimed at understanding its pea pathogenicity potentials have been hinged on DNA-based molecular techniques (Oyarzun 1993; Etebu and Osborn 2009).

6.3 Pea Footrot Disease Symptoms and Assessment

As earlier mentioned, pea footrot disease is caused by the soilborne fungus *N. Haematococca*. Some workers prefer to call the disease *Fusarium* root rot of peas to reflect the anamorphic stage of the causal pathogen (Hagedorn 1991). The disease was first reported in the USA and Europe at about 1918 (Kraft 2001). The growth of *N. haematococca* and subsequent infection of the pea plant are facilitated by chemical exudates formed by the root system. The fungus penetrates the plant through the tap root just above the point of cotyledon attachment (Short and Lacy 1976; Integrated Pest Management 2002) (see Fig. 6.2 showing pea plant and infection locus of *N. haematococca*).

Disease symptoms are produced on infected peas as early as 3 days of contact with pathogenic forms of the fungus (Funnell et al. 2001); early symptoms appear as reddish-brown streaks at the primary and secondary roots and later coalesce to form a dark reddish-brown lesion (see Fig. 6.3) on the primary root up to the soil line. Externally, symptoms are characterised by stunted growth, yellowing and necrosis at the base of the stem (Fig. 6.4) (Kraft and Kaiser 1993; Kraft 2001; Etebu 2008). Poor crop rotations, high soil temperatures (22–30 °C) and moist,

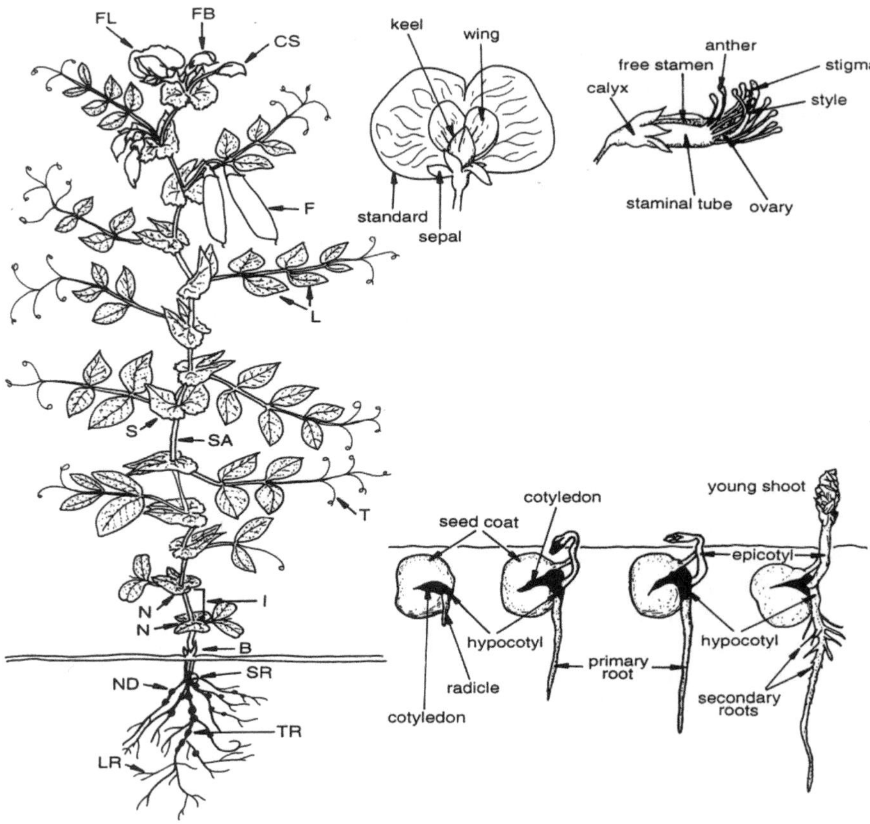

Fig. 6.2 The pea plant. FL = open flower; FB = flower bud; CS = clam shell; F = fruit; L = leaflets; S = stipule; SA = stem axis; T = tendril; N = node; I = internode; B = bracts; ND = nodule; SR = seed remnant; TR = taproot; and LR = lateral root. Flower parts: sepal, keel, wing, standard, calyx, staminal tube, anther, free stamen, stigma, style and ovary. Germinating seed parts: cotyledon, radicle, hypocotyl, seed coat, primary root, epicotyl, secondary roots and young shoot (*Courtesy*: F. Muehlbauer) (*Source*: Kraft 2001)

acidic (pH 5.1–6.2), low fertility and compact soils would generally facilitate pea footrot disease following infection by the causal pathogen (Kraft 1984; Tu 1994).

Assessment and grading disease severity is very vital in plant pathological experiments. Disease assessment and grading are usually aimed at evaluating disease resistance or susceptibility among different varieties of a given species of plants or of a given variety of crop under different agricultural practices. Disease is usually graded through the use of scales often ranging from 0 to 5 or 1 to 9 such that low figures on the scale depicts a corresponding low degree of damage in the infected plant and vice versa (Infantino et al. 2006). Research on the assessment of footrot disease on peas has focused on laboratory and greenhouse experiments (Han et al. 2001). Studies conducted to assess the response of peas to footrot disease

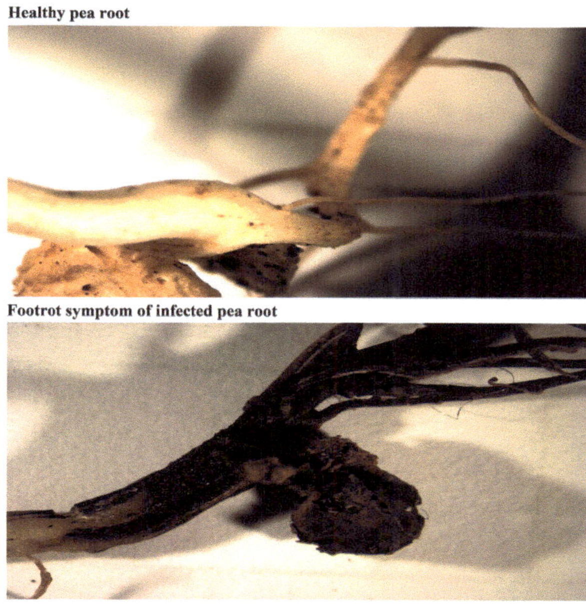

Fig. 6.3 Root symptom of pea footrot disease (*Source*: Etebu 2008). (**a**) Healthy pea root. (**b**) Footrot symptom of infected pea root

Fig. 6.4 Early field symptoms of pea footrot disease (*Source*: Processors and Growers Research Organisation 1997)

Fig. 6.5 Greenshaft peas showing various degrees of footrot disease (*Source*: Etebu and Osborn 2011a)

have often centred on a disease index (DI) scale ranging from 0 to 5 to show the differential symptomatic effect on pea roots infected with *N. haematococca*: 0 = no discolouration; 5 = totally discoloured roots (Biddle 1984; Oyarzun et al. 1997; Grünwald et al. 2003). Using the same scale, Grünwald and associates (2003) described the various levels of footrot disease severity of the scale as follows: 0 = no symptoms; 1 = slight hypocotyl lesions; 2 = lesions coalescing around epicotyls and hypocotyls; 3 = lesions starting to spread into the root system with the root tip starting to be infected; 4 = epicotyl, hypocotyl and root system almost completely infected and only slight amount of white, uninfected tissue left; 5 = completely infected root. This scale is largely subjective and requires a great deal of technical expertise. A simpler and yet objective and accurate assessment scale has recently been developed and adopted in recent studies (Etebu and Osborn 2009, 2010, 2011a, b, c). These authors, whilst maintaining the disease index (DI) scale of 0–5, defined the different stages of disease severity to be 0 = no root discolouration; 1 = 1–20 % discolouration; 2 = 21–40 % discolouration; 3 = 41–60 % discolouration; 4 = 61–80 % discolouration; and 5 $\geq$ 81 % discolouration (see Fig. 6.5).

6.4 Molecular Basis of Pea Footrot Disease

The interaction between *P. sativum* and *N. haematococca*, which leads to footrot disease of the former, is in many ways similar to what is known with other plant-soil-fungal interaction systems. Fungal pathogens are able to cause disease in plants after they get established within the tissues. Plants generally resist microbial infection by producing antimicrobial secondary metabolites, either during their normal course of development or in response to pathogen attack or stress (Kotchoni and Gachomo 2006). Antimicrobial metabolites, formed constitutively within plants in course of their normal development and growth, are generally termed phytoanticipin. These metabolites potentially protect plants, wherein they are formed, from attack against a wide range of pathogens (Mansfield 1983; Osbourn 1996). Conversely, phytoalexins are produced in response to pathogenic attack or stress. As such, they are usually restricted to the infection locus and the surrounding cells (Paxton 1980, 1981; Grayer and Harborne 1994; Smith 1996). Interestingly, a number of studies have also shown that pathogenic attack on plants elicits phytoalexinic response in both disease-resistant and disease-susceptible plants, but the rate and amount of phytoalexin produced in resistant plants are significantly higher than in susceptible ones (Morrissey and Osbourn 1999; Van Etten et al. 2001).

Resistance genes play a vital role in conferring resistance on plants when induced by pathogenic invasion, primarily through signal transduction which leads to the activation of defence genes (Dangl and Jones 2001; Kotchoni and Gachomo 2006). Numerous defence genes have been identified, most of which occur within plants as multigene families (Douglas et al. 1987). Genes associated with inducible defence responses encode, among others, hydrolytic enzymes such as chitinases and glucanases and a number of other 'pathogenesis-related (PR) proteins' whose functions are yet not properly understood (Bowels 1990; van Loon et al. 1994). In addition, these genes also encode enzymes involved in the synthesis of antimicrobial phytoalexins (Dixon and Paiva 1995). The garden pea produces an isoflavonoid phytoalexin (+) pisatin, and many of the fungi that are pathogenic on the pea plant are able to detoxify pisatin via demethylation (Van Etten et al. 1989). The high virulence of pathogenic forms of *N. haematococca* MPVI population on peas has been linked with the capacity of the fungus to detoxify the pea phytoalexin, pisatin (Kistler and Van Etten 1984). All field isolates, pathogenic on peas, are known to produce a microsomal cytochrome P450 monooxygenase enzyme called pisatin demethylase (pdm). Pisatin demethylase is encoded by pisatin demethylase activity (*PDA*) genes (Van Etten et al. 1995; Funnell et al. 2002; Liu et al. 2003), and these catalyse the detoxification of pisatin, via demethylation (Fig. 6.6). All naturally occurring isolates of *N. haematococca* that lack the ability to demethylate pisatin (PDA^{-}) normally lack *PDA* genes and are not pathogenic on peas (Ciuffetti and Van Etten 1996; Wu and Van Etten 2004).

However, natural isolates of *N. haematococca* vary quantitatively in pisatin demethylating ability, and as a result, three whole cell phenotypic groups have been classified (Matthews and Van Etten 1983). These include PDA^{-}, PDA^{L} and PDA^{H}. The first group (PDA^{-}) lacks the ability to detoxify pisatin. The second group (PDA^{L}) produces low levels of pisatin demethylase enzyme after long exposure to pisatin, whilst the third group (PDA^{H}) rapidly produces moderate to high levels of pisatin demethylase enzyme on exposure to pisatin. In some earlier publications, PDA^{L} phenotype has been referred to as PDA^{n} or PDA^{LL} and the PDA^{H} phenotype as PDA^{i}, PDA^{SH} or PDA^{SM} (Mackintosh et al. 1989). Deductions

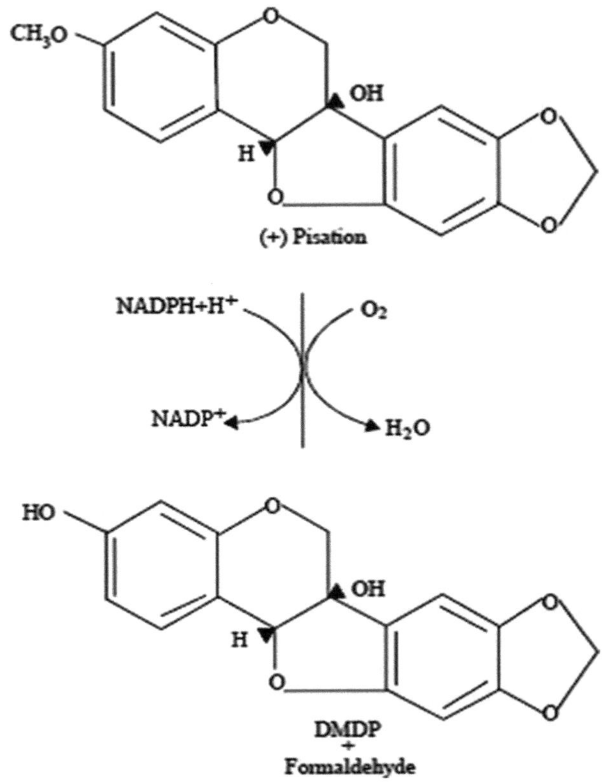

Fig. 6.6 Detoxification of pisatin through demethylation. Pisatin demethylated to 3,6a-dihydroxy-8, 9-methylenedioxypterocarpan (DMDP) (*Source*: Matthews and Van Etten 1983)

from earlier conventional genetic studies inferred that *PDA* was inherited as a single gene, and in addition, the *PDA* gene was considered to be an absolute requirement for any microorganism to infect and cause disease on peas (Kistler and Van Etten 1984). This position was later debunked as studies with site-directed disruption experiments of *PDA* genes proved otherwise. Disrupting the *PDA* gene in a pea pathogenic fungus was expected to render it nonpathogenic on peas, but this was not the case. Although site-directed disruption of the *PDA* gene reduced the capacity of the fungus to cause disease, it did not render the gene-disrupted strains completely nonpathogenic. The reduction in pathogenicity in *PDA* site-disrupted strains on peas rather than becoming nonpathogenic raised questions on previous conventional genetic studies and begged for answers. This apparent inconsistency was explained and synchronised by two findings. Firstly, *PDA* gene is located on a 1.6 million base pair (Mb) conditional dispensable chromosome (Wasmann and Van Etten 1996), and secondly, some additional gene(s) located on the same dispensable chromosome were observed as additional requirement for high virulence on peas (Wasmann and Van Etten 1996; Etebu and Osborn 2009, 2011b).

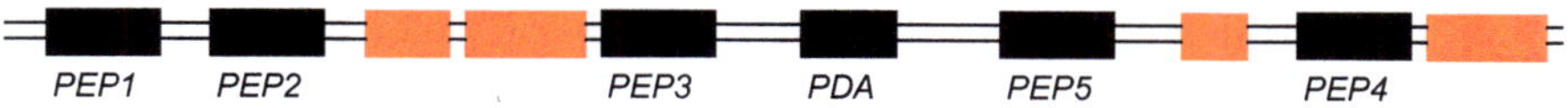

Fig. 6.7 Schematic representation of the PEP gene cluster. The cluster represents the PEP gene cluster found in strain 77-13-7 of *N. haematococca* MPVI and contains six genes that are expressed during infection of pea (black rectangles) and four ORFs with homology to different class II fungal transposable elements (orange rectangles) [Adopted from Han et al. (2001), Temporini and Van Etten (2002)]

These additional genes which are also expressed in pathogenic forms of *N. haematococca* during pea infection are *PEP1*, *PEP2*, *PEP3*, *PEP4* and *PEP5* (Fig. 6.7), all clustered together with the *PDA* gene (Temporini and Van Etten 2002).

All highly virulent isolates possess at least one homologue of each of the six genes except *PEP4* (Temporini and Van Etten 2002). Of the six genes located on the dispensable chromosome of *N. haematococca*, *PDA1*, *PEP1*, *PEP2* and *PEP5* are generally termed the 'pea pathogenicity' (*PEP*) cluster essentially because each of these genes is able to independently confer pathogenic properties to nonpathogenic isolates of *N. haematococca* that lack the conditional dispensable chromosome (Ciuffetti and Van Etten 1996; Han et al. 2001). The role of *PEP1*, *PEP2* and *PEP3* in pea pathogenicity is yet unknown (Idnurm and Howlett 2001; Temporini and Van Etten 2002). Whilst *PDA* is responsible for detoxification of pisatin, the *PEP5* gene is suggested to be involved in the efflux of pisatin (Han et al. 2001). An excellent review by Van Etten and associates (2001) observed that *N. haematococca* tolerates pisatin through degradative (*PDA*) and non-degradative (*PEP5*) means, suggesting that both types of tolerance mechanisms may operate in synergism during infection. They further suggested that elimination of both mechanisms in *N. haematococca* may be required to make it nonpathogenic. Although *PEP3* is usually not included among the pea pathogenicity genes (Han et al. 2001; Liu et al. 2003), apparently because of its inability to independently confer pathogenic attributes when used to transform nonpathogenic strains, several workers have shown that it apparently plays a significant role in pea pathogenicity. This position stems from the fact that whereas homologues of pea pathogenicity genes (*PDA1*, *PEP1*, *PEP2* and *PEP5*) could sometimes occur in isolates with low virulence, *PEP3* homologue is the only gene that is present exclusively in highly virulent isolates, pathogenic on peas (Temporini and Van Etten 2002; Han et al. 2001). The importance of the *PEP3* gene in pea pathogenicity culminating to pea footrot disease is further elucidated by the fact that the *PDA* and *PEP3* genes are physically located between *PEP2* and *PEP5* (Fig. 6.6) and isolates lacking *PDA* and/or *PEP3* are not highly virulent even when other pea pathogenicity genes are present in the cluster of genes (Temporini and Van Etten 2002). Similarly, recent molecular studies aimed at targeting pea pathogenicity genes of *N. haematococca* as a means to quantify population of pathogenic forms of the pathogen in soil identified *PDA*, *PEP3* and *PEP5* as major determinants of pea footrot disease (Etebu and Osborn 2010, 2011a).

6.5 Potential Role of Soil Health Indices on Pea Footrot Disease Suppressiveness

Disease incidence and severity among different plant-pathogen interactions are, in part, directly related to pathogen inoculum density (Bhatti and Kraft 1992; Sugha et al. 1994; Navas-Cortés et al. 2000). Although this phenomenon has also been demonstrated between peas and *N. haematococca* (Etebu and Osborn 2011a), literature is awash with reports of several other studies involving plant-pathogen interactions (including the interaction between peas and *N. haematococca*) where this relationship is not always the norm (Ristaino 1991; Oyarzun et al. 1994; Etebu and Osborn 2011c). Different soils differentially affect the inoculum potential of *F. solani* f. sp. *pisi* in peas even when the inoculum density of the pathogen is the same in all soils. This differential effect of virulent spores of *N. haematococca* in peas clearly indicates that the interactions between a susceptible pea plant and its specific pathogen, leading to pea footrot disease, are largely dependent on the soil environment.

Soil is a complex and dynamic biological system that houses numerous organisms involved in recycling organic matter and associated nutrients and in the process modulates the outcome of many plant-pathogen interactions. Soils have the capacity to either facilitate or suppress the incidence, severity and/or progression of plant disease. Depending on the side of the divide it tilts, a soil would be considered healthy or otherwise. A healthy soil from an agricultural view point lies in its ability to suppress the activity of plant pathogens, such that disease incidence, progression and/or severity on susceptible host plants would be significantly delayed or completely obliterated, in spite of the presence of a pathogen and climatic conditions favourable for disease (Schippers 1992; Abawi and Widmer 2000; Van Bruggen and Semenov 2000). Soils with this capability are referred to as suppressive soils (Alabouvette 1990) as opposed to conducive or receptive soils. The capability of suppressive soils to control the pathogenic activity of pathogens is dependent on inherent biotic and abiotic soil properties (Alabouvette et al. 1982).

6.5.1 Biotic Factors Affecting Agricultural Soil Health and Pea Footrot Disease Suppressiveness

Biotic factors include all aspects of association between plants and other organisms, particularly microorganisms in soils. The composition of microbial communities plays very crucial roles in the fertility/health status of agricultural soils, and these have been exploited in agricultural practice for decades (van Veen et al. 1997; Girvan et al. 2003; Nannipieri et al. 2003). All natural soils are able to suppress the activity of plant pathogens, in some way, by reason of the presence and activity of its resident soil microorganisms. So the concept of disease suppressive soils is often described in terms of both general suppression and specific suppression (Cook and

Baker 1983). General suppression is a component of disease suppressive soils, so manipulating the biotic components of soils has always been directed at achieving specific suppression against specific plant disease(s). Among the varied practices adopted to achieve specific suppression is the practice of amending crop soils with organic matter to restore and improve soil quality. Hence, the application of organic amendments aimed at improving agricultural soil health through enhancement of soil suppressiveness, especially for soilborne diseases, has received a considerable resurgence in recent times.

Biotic components of soil quality commonly measured during research/experimentation include soil organic matter, respiration, microbial biomass (total bacteria and fungi) and mineralisable nitrogen (Stevenson 1994). Although soil organic matter is generally considered to be a biological factor, a recent review by Etebu and Osborn (2012) treated it as a chemical component of abiotic factors, because plants and animals living in soil usually account for less than 5 % of the soil organic carbon (Stevenson 1994). Biotic factors affecting pea footrot disease include initial pathogen (*N. haematococca*) density, microbial biomass and soil microbial richness and diversity.

6.5.1.1 Initial Density of *N. haematococca*

Although the inoculum potential of a soil, defined as the pathogenic energy present to cause infection (Bouhot 1979), is dependent on many factors, the pathogen inoculum load present in soil at the outset of cultivation is generally known to significantly dictate the incidence and severity of soilborne diseases among plants (Cullen et al. 2001; Goud and Termorshuizen 2003). As a result, it is a common practice to allow fallow periods between susceptible crops to repress build-up of high inoculum load in fields to avoid disease outbreaks in such fields. A 6-year rotation period is thought to deter the build-up of *N. haematococca* and is therefore practised with respect to pea cultivation in some European countries (Oyarzun et al. 1993b). This practice may not be very effective in the management and control of the disease because it does not guarantee a significant reduction of the pathogen inoculum load (Etebu and Osborn 2010). Employing other reliable and effective means to identify agricultural fields with high disease potentials, prior to crop planting, is pivotal in the management of pea footrot disease. Since soils with high disease potential are generally characterised by high initial inoculum density of plant pathogen(s) (Rush and Kraft 1986; Bhatti and Kraft 1992; Navas-Cortés et al. 2000), there has to be a reliable means of quantifying pathogenic forms of *N. haematococca* in agricultural fields, prior to pea cultivation. Until very recently, isolation and quantification of *N. haematococca* in soil had relied on the use of peptone-pentachloronitrobenzene agar (PPA) (Oyarzun et al. 1994). Although this medium is considered to be a *Fusarium*-selective medium, it is not specific to *Fusarium solani* (Dhingra and Sinclair 1995), neither does it discriminate between pathogenic and nonpathogenic forms of the pea footrot pathogen (Etebu and Osborn 2009, 2010).

Goud and Termorshuizen (2003) attempted to quantify *N. haematococca* in soil, using molecular approaches targeting ITS regions. Unfortunately, like culture-dependent assays, molecular assays targeting the ITS region were equally unsuitable because it also does not discriminate between pathogenic and nonpathogenic forms of the pathogen (Suga et al. 2000). The discovery of a cluster of six pea pathogenicity genes (*PDA*, *PEP1*, *PEP2*, *PEP3*, *PEP4 and PEP5*) which pathogenic strains of the fungus are known to possess has helped in no little ways to the development of molecular assays that differentially detects and quantifies pathogenic forms of the fungus in soil without recourse to culture. In particular, molecular assays targeting three of the pathogenicity genes (*PDA*, *PEP3* and *PEP5*) have been developed and validated. The assays showed that gene copy numbers of each of the three genes (*PDA*, *PEP3* and *PEP5*) quantified from soil DNA were comparable to the number of pea pathogenic forms of *N. haematococca* in soil (Etebu and Osborn 2010, 2011a). In a related review article, Etebu and Osborn (2011d) opined that the *PEP3* gene would be the most ideal indicator gene to target in the molecular quantification of pea pathogenic forms of *N. haematococca* in soil, because of all the six genes linked with pea pathogenicity, the *PEP3* homologue is the only gene that is present exclusively in highly virulent pea pathogenic isolates (Han et al. 2001; Temporini and Van Etten 2002). *PEP3* gene copy numbers of up to 100 g^{-1} soil would constitute a threshold number for infection, potentially capable of causing economically significant pea footrot disease. So agricultural fields having this density of *PEP3* gene copies at the outset of pea cultivation could be considered pea footrot disease suppressive soils if viable peas planted thereon do not show appreciable degree of pea footrot disease.

6.5.1.2 Microbial Biomass

Soil microbial biomass represents the fraction of the soil responsible for the energy, nutrient cycling and regulation of organic matter transformation (Gregorich et al. 1994; Turco et al. 1994). The biological activities of nutrient cycling and organic matter decomposition are facilitated by soil organisms particularly microorganisms, and these are largely concentrated in the topsoil ($\leq$30 cm deep). Microbial communities constitute the first line of soil inhabitants that change, both in structure and diversity, in the event of any change in soil conditions. Changes in microbial populations and activities therefore indicate a real change in soil health (Pankhurst et al. 1995). Although soil microorganisms constitute a very small fraction of total soil organic matter, the rate of organic matter decomposition and nitrogen mineralization is directly related to the microbial biomass of the soil, and the rate at which organic matter decomposes is a measure of soil health (Jenkinson 1988; Singh 1995; Carter et al. 1999). The works aimed at studying the effect of soil microbial biomass and pea footrot disease suppressiveness in agricultural soils are grossly limited, and some of the few existing studies have often focused on the relationship between microbial biomass (measuring biomass indices such as carbon and nitrogen) and yield without reference to pea footrot disease. A very recent study with peas showed that soil microbial biomass positively correlates with pea dry matter yields (Jannouraa et al. 2014). The relationship between microbial biomass and pea footrot disease was not specifically reported in this recent work, but some relatively earlier works had shown that soil organic matter, which includes microbial biomass among others, influences the health of agricultural soils with respect to pea footrot disease suppression (Etebu and Osborn 2011c, 2012).

6.5.1.3 Microbial Diversity

Microbial communities play vital roles in the acquisition and recycling of nutrients required for maintenance of soil structure, degradation of pollutants and the biological control of plant and animal pests, as well as the sustenance of agricultural soil health and plant growth and productivity (Bossio and Scow 1995; Hill et al. 2000). The significance of biological diversity, often simply termed 'biodiversity', in ecological studies has been in limelight and appreciated since as early as the 1950s. Biodiversity is an index of community stability and could be defined as a measure of variability among living organisms. This includes diversity within species, between species and of ecosystems (Swift 1974; Harper and Hawksworth 1995; Nielsen and Winding 2002). Biodiversity studies began with plant and animal communities up till the 1960s. Microbiologists began to investigate the impact of biodiversity on the function and structure of microbial communities from about the 1960s, and from then up till now, the subject of microbial diversity has continually been accorded due recognition and significance in ecological studies (Swift 1974). A case in point is the formulation of the 'Diversities International Research Program' in 1991 and the Biodiversity Treaty that was issued from the United Nations Conference on Environment and Development in 1992 in Rio de Janeiro, Brazil. These institutional drives were intended to promote scientific investigations into the origins and conservation of biodiversity and the impact of biodiversity on ecological functions (Colwell 1996). Closely linked to the subject of biodiversity is the concept of 'resilience'. This concept was first introduced into ecological parlance by Holling (1973) to explain the non-linear dynamics observed in ecosystems. Ecological resilience was defined as the amount of disturbance an ecosystem could withstand without altering self-organised processes and structures.

Biodiversity as an index of resilience enhances the efficiency and stability of some functions of the ecosystems (Tilman and Downing 1994; Tilman et al. 1996), and these are dependent on the diversity of functional groups of soil organisms in the ecosystem, as well as the species diversity within these groups (Walker 1992). A resilient agricultural soil would therefore be a soil with diverse species of microorganisms, as components of its microbial community, with none enjoying an exclusive dominance status in terms of abundance (Pankhurst et al. 1996). A biologically diverse soil would be resilient and be able to suppress plant disease. Soilborne pathogens are suppressed by soils through a variety of ways; these include induced resistance, direct parasitism, nutrient competition and direct inhibition through antibiotics secreted by beneficial organisms (Sullivan 2001). Changes in soil microbial diversity result from ecosystem management, global change (Bossio and Scow 1995) and agricultural practices such as organic amendments (Pankhurst et al. 1996; Girvan et al. 2003, 2004). These practices impact agricultural soil health either negatively or positively resulting to plant disease receptive (conducive) or suppressive soils, respectively. There has been considerable development of techniques for characterising and measuring diversity, in particular at the molecular level for both culturable and non-culturable microorganisms (Rondon et al. 2000; Theron and Cloete 2000). Biodiversity is studied at three levels of complexities, and these include genetic (intraspecies diversity), species (numbers of species) and ecological (community diversity) (Harper and Hawksworth 1995). Species richness or abundance is considered to be the fundamental measures of biodiversity (Magurran 1988). Recent microbial diversity studies have often relied on indices derived from formulae put forward by different workers. Two commonly used indices are the Simpson's diversity index and Shannon-Weaver diversity index. In particular, the Simpson's diversity index has been used to measure the fungal diversity of agricultural soils with pea footrot disease histories in the UK, and the interrelationship between fungal diversity and pea footrot disease in those soils was studied some years ago (Etebu 2008). Specifically, fungal richness/biodiversity within the soils was investigated through the generation of terminal restriction fragments using labelled FAM-ITS4 and ITS1 primers in a molecular assay, and Simpson's diversity index (SDI) used as measure of the diversity of fungi (TRFs) was calculated from the formula

$$SDI = 1 - \frac{\Sigma n(n-1)}{N(N-1)}$$

where SDI = Simpson's diversity index, n = relative abundance of the different terminal restriction fragments (TRF) and N = total number of TRF (Fowler et al. 2005).

Results of the study herein referred to showed that whilst fungal richness was significantly different in different agricultural fields, a perceived inverse correlation to footrot disease was not significant ($P = 0.05$), but it was nonetheless significantly, positively correlated to shoot length and to total plant dry weight. This indicates that fungal richness promotes pea plant growth, and this by extension further supports the widely accepted view that agricultural soils endowed with numerous fungal species boosts the yield of food crops.

6.5.2 Abiotic Factors Affecting Agricultural Soil Health and Pea Footrot Disease Suppressiveness

The biotic environment could be considered to be the sole determinate factor responsible for receptivity or suppressiveness of agricultural soils, whilst the prevailing abiotic factors simply play a modulating role (Oyarzun et al. 1998). This seems to be the case with pea footrot disease suppressive soils. Soil chemical factors such as pH, total oxidised nitrogen, soluble ammonium nitrogen, carbon/total nitrogen ratio (C/N), phosphate and potassium have been shown to be positively related to pea footrot disease (significant at $P \leq 0.05$), indicating that a decrease of these factors in soil would render such soils suppressive to pea footrot disease whether or not the threshold density of pathogenic forms of *N. haematococca* (100 g^{-1}soil) required for disease is present. Plant growth and microbial growth are both limited by nitrogen availability in many ecosystems (Kaye and Hart 1997). Although fertilisers are often applied to soil in the majority of agricultural management practices, peas are relatively unresponsive to fertilisers, particularly nitrogen, except when nodulation is poor or fails completely (Muehlbauer et al. 1983). This is because peas, in association with *Rhizobium*, are capable of fixing atmospheric nitrogen which meets their requirement for high yield (Crozat et al. 1994). The form of nitrogen (NO_3 or NH_4) has been noted as an important factor when it comes to its role in disease suppression in soil (Janvier et al. 2007). Studies have shown that pea footrot disease is not influenced by total ammonium nitrogen (NH_4-N) but total oxidised nitrogen (TON) does, as the latter has been observed to have a significantly positive correlation to the disease. Of further interest is the fact that TON is also positively related to inoculum density of *N. haematococca* in soil quantified via pathogenicity genes (Oyarzun et al. 1998; Etebu and Osborn 2010, 2011c). Since peas are capable of meeting their nitrogen requirements through atmospheric nitrogen fixation, excess TON not utilised by pea plants in soil infested with pathogenic *F. solani* f. sp. *pisi* may be utilised by pea footrot disease pathogen for growth and reproduction, thereby increasing the chances of inoculum proliferation in soil.

A predictive disease model for pea footrot disease was recently identified and proposed as $DI = 1.97 + [(3.48 \times \text{phosphate}) + (-0.66 \times C/N)]$, where DI represents disease index (0 = no disease; 5 = maximum disease); phosphate is measured in mg/g soil; N represents total ammonium nitrogen, also measured in mg/g soil; and C represents soil organic carbon measured as percentage loss of ignition (LOI). Both potential predictors contribute significantly ($P < 0.05$) to the variability of pea footrot disease which is often observed on different agricultural soils. Whilst phosphate contributed 31 % of the variation in pea footrot disease, C/N ratio accounted for an additional 11 %. The model showed that the relative abundance of three soil chemical factors, phosphate, carbon and nitrogen, in part, determines whether or not a soil would be suppressive to pea footrot disease. Whilst phosphate positively correlated to pea footrot disease, as part of the same model, C/N ratio was found to be negatively correlated to the same disease (Etebu and Osborn 2011c).

What this portends is that a combination of a relative decrease in phosphate and a high C/N ratio would render a soil suppressive to pea footrot disease. Expression of the *PDA* gene, responsible for pea footrot pathogenicity in *N .haematococca*, is known to be suppressed in culture by glucose and amino acids (Straney and Van Etten 1994; Khan and Straney 1999). It could therefore mean that carbon existing as sugars and carbohydrates in soil could, depending on the relative amount of total ammonium nitrogen, suppress the expression of the *PDA* gene in *N. haematococca* required to initiate footrot disease in peas.

6.6 Conclusion

Although pea footrot disease is largely dependent on the interaction between the pea plant itself and the causal pathogen *N. haematococca*, the incidence, progression and severity of the disease are often modulated by soil factors such as microbial biomass, microbial richness and diversity, pH, total oxidised nitrogen, phosphate and potassium. Agricultural soils with inputs that seek to decrease phosphate and total nitrogen, depending on the relative amount of total carbon, would generally be suppressive to pea footrot disease whether or not the threshold density of pathogenic forms of *N. haematococca* (100 g^{-1} soil) required for disease is present. In contrast soils with high carbon/total nitrogen (C/N) ratio value would potentially also render a soil suppressive to pea footrot disease whether or not the threshold density of pathogenic forms of *N. haematococca* (100 g^{-1} soil) required for disease is present.

References

Abawi GS, Widmer TL (2000) Impact of soil health management practices on soil borne pathogens, nematodes and root diseases of vegetable crops. Appl Soil Ecol 15:37–47

Alabouvette C (1990) Biological control of Fusarium wilt pathogens in suppressive soils. In: Hornby D (ed) Biological control of soil-borne plant pathogens. CAB International, Wallinford, pp 27–43

Alabouvette C, Couteaudier Y, Louvet J (1982) Comparison de la réceptivité de différents sols et substrats de culture aux fusarioses vasculaires. Agronomie 2:1–6

Bhatti MA, Kraft JM (1992) Effects of inoculum density and temperature on root rot and wilt of chickpea. Plant Dis 76:50–54

Biddle AJ (1984) A prediction test for pea footrot and the effect of the previous legumes. In: British crop protection conference: pests and diseases, pp 773–777

Booth C (1971) The genus *Fusarium*. Commonwealth Mycological Institute, Kew

Bossio DA, Scow KM (1995) Impact of carbon and flooding on the metabolic diversity of microbial communities in soils. Appl Environ Microbiol 61:4043–4050

Bouhot D (1979) Estimation of inoculum density and inoculum potential: techniques and their values for disease prediction. In: Schippers B, Gams W (eds) Soil-borne plant pathogens. Academic, London, pp 250–278

Bowels DJ (1990) Defense-related proteins in higher plants. Annu Rev Biochem 59:873–907

Carter MR, Grogorich EG, Angers DA, Beare MH, Sparling GP, Wardle DA, Voroney RP (1999) Interpretation of microbial biomass measurements for soil quality assessment in humid temperate regions. Can J Soil Sci 79:507–520

Ciuffetti LM, Van Etten HD (1996) Virulence of a pisatin demethylase-deficient *Nectria haematococca* MPVI isolate is increased by transformation with a pisatin demethylase gene. Mol Plant Microbe Interact 9:787–792

Colwell RR (1996) Microbial diversity and biotechnology. In: Reaka-Kudla ML, Wilson DE, Wilson EO (eds) Biodiversity II: understanding and protecting our biological resources. Joseph Henry Press, Washington, DC, pp 279–288

Cook RJ, Baker KF (1983) The nature and practice of biological control of plant pathogens. The American Phytopathological Society, St Paul

Crozat Y, Aveline A, Coste F, Gillet JP, Domenach AM (1994) Yield performance and seed production pattern of field-grown pea and soybean in relation to N nutrition. Eur J Agron 3:135–144

Cullen DW, Lees AK, Toth IK, Duncan JM (2001) Conventional PCR and real-time quantitative PCR detection of *Helminthosporium solani* in soil and potato tubers. Eur J Plant Pathol 107:387–398

Dangl JL, Jones JDG (2001) Plant pathogens and integrated defense responses to infection. Nature 411:826–833

Davies DR, Berry GJ, Heath MC, Dawkins TCK (1985) Pea (*Pisum sativum* L.). In: Summerfield RJ, Roberts EH (eds) Grain legume crops. Williams Collins Sons, London, pp 147–198

Dhingra OD, Sinclair JB (1995) Basic plant pathology methods, 2nd edn. CRC, Boca Raton

Dixon RA, Paiva NL (1995) Stress-induced phenylpropanoid metabolism. Plant Cell 7:1085–1097

Douglas C, Hoffman H, Schulz W, Hahlbrock K (1987) Structure and elicitor or u.v.-light-stimulated expression of two 4-coumarate: CoA ligase genes in parsley. EMBO J 6:1189–1195

Duke JA (1981) Handbook of legumes of world economic importance. Plenum, New York

Etebu E (2008) Molecular detection and quantification of the pea footrot disease pathogen (*Nectria haematococca*) in agricultural soils: a potential model for disease prediction. PhD thesis, The University of Sheffield, Sheffield, 275 p

Etebu E, Osborn AM (2009) Molecular assays reveal the presence and diversity of genes encoding pea footrot pathogenicity determinants in *Nectria haematococca* and in agricultural soils. J Appl Microbiol 106(5):1629–1639

Etebu E, Osborn AM (2010) Molecular quantification of the pea footrot disease pathogen (*Nectria haematococca*) in agricultural soils. Phytoparasitca 38:447–454

Etebu E, Osborn AM (2011a) Molecular prediction of pea footrot disease in agricultural soils. Asian J Agric Sci 3(6):417–426

Etebu E, Osborn AM (2011b) Pea footrot disease depends on the combination of pathogenicity genes in *Nectria haematococca*. Asian J Agric Sci 3(3):156–161

Etebu E, Osborn AM (2011c) A potential model for pea footrot disease prediction. Asian J Agric Sci 3(3):177–186

Etebu E, Osborn AM (2011d) In search of target gene(s) to quantify pea pathogenic *Nectria haematococca* in agricultural soils. Curr Res J Biol Sci 3(3):195–208

Etebu E, Osborn AM (2012) A review of indicators of healthy agricultural soils with pea footrot disease suppression potentials. Sust Agric Res 2(1):235–250

Food and Agriculture Organisation of the United Nations (2001) 2001 production year book, Rome, Italy

Food and Agriculture Organization of the United Nations (1994) 1994 production year book, Rome, Italy

Fowler J, Cohen L, Jarvis P (2005) Practical statistics for field biology, 2nd edn. Wiley, West Sussex

Funnell DL, Van Etten HD (2002) Pisatin demethylase genes are on dispensable chromosomes while genes for pathogenicity on carrot and ripe tomato are on other chromosomes in *Nectria haematococca*. Mol Plant Microbe Interact 15:840–846

Funnell DL, Matthews PS, Van Etten HD (2001) Breeding for highly fertile isolates of *Nectria haematococca* MPVI that are highly virulent on pea and in planta selection for virulent recombinants. Phytopathology 91:92–101
Funnell DL, Matthews PS, VanEtten HD (2002) Identification of new pisatin demethylase genes (PDA5 and PDA7) in Nectria haematococca and non-mendelian segregation of pisatin demethylating ability and virulence on pea due to loss of chromosomal elements. Fungal Genet Biol 37:121–133
Girvan MS, Bullimore J, Pretty JN, Osborn AM, Ball AS (2003) Soil type is the primary determinant of the composition of the total and active bacterial communities in arable soils. Appl Environ Microbiol 69:1800–1809
Girvan MS, Bullimore J, Ball AS, Pretty JN, Osborn AM (2004) Response of active bacterial and fungal communities in soils under winter wheat to different fertilizer and pesticide regimes. Appl Environ Microbiol 70:2692–2701
Goud JC, Termorshuizen AJ (2003) Quality of methods to quantify microsclerotia of *Verticillium dahliae* in soil. Eur J Plant Pathol 109:523–534
Graham PH, Vance CP (2003) Legumes: importance and constraints to greater use. Plant Physiol 131:872–877
Grayer RJ, Harborne JJ (1994) A survey of antifungal compounds from higher plants 1982-1993. Phytochemistry 37:19–42
Gregorich EG, Carter MR, Angers DA, Monreall CM, Ellert BH (1994) Towards a minimum data set to assess soil organic-matter quality in agricultural soils. Can J Soil Sci 74:367–385
Grünwald NJ, Coffman VA, Kraft JM (2003) Sources of partial resistance to *Fusarium* root rot in the *Pisum* core collection. Plant Dis 87:1197–2001
Hagedorn DJ (1976) Handbook of pea diseases. Cooperative Extension Publications, University of Wisconsin, Madison, A 1167
Hagedorn DJ (1991) Handbook of pea diseases. Cooperative Extension, University of Wisconsin, Madison
Han YN, Liu XG, Benny U, Kistler HC, Van Etten HD (2001) Genes determining pathogenicity to pea are clustered on a supernumerary chromosome in the fungal plant pathogen *Nectria haematococca*. Plant J 25:305–314
Harper JL, Hawksworth DL (1995) Preface. In: Hawksworth DL (ed) Biodiversity, measurement and estimation. Chapman & Hall, London, pp 5–12
Hill GT, Mitkowski NA, Aldrich-Wolfe L, Emele LR, Jurkonie DD, Ficke A, Maldonado-Ramire-S, Lync ST, Nelson EB (2000) Methods for assessing the composition and diversity of soil microbial communities. Appl Soil Ecol 15:25–36
Holling CS (1973) Resilience and stability of ecological systems. Annu Rev Ecol Syst 4:1–23
Hulse JH (1994) Nature, composition and utilization of food legumes. In: Muehlbauer FJ, Kaiser WJ (eds) Expanding the production and use of cool season food legumes. Kluwer, Dordrecht, pp 77–97
Hwang SF, Chang KF (1989) Incidence and severity of root rot disease complex of field peas in northeastern Alberta in 1988. Can Plant Dis Surv 69:139–141
Idnurm A, Howlett BJ (2001) Pathogenicity genes of phytopathogenic fungi. Mol Plant Pathol 2:241–255
Infantino A, Kharrat M, Riccioni L, Coyne CJ, McPhee KE, Grünwald NJ (2006) Screening techniques and sources of resistance to root diseases in cool season food legumes. Euphytica 147:201–221
Integrated Pest Management (2002) Reports on plant diseases: root rots of pea. RPD No. 911, http://www.ipm.uiuc.edu/diseases/series900/rpd911/. Accessed on 24 Feb 2015
Jannouraa R, Joergensena RG, Brunsb C (2014) Organic fertilizer effects on growth, crop yield, and soil microbial biomass indices in sole and intercropped peas and oats under organic farming conditions. Eur J Agron 52:259–270

Janvier C, Villeneuve F, Alabouvette C, Edel-Hermann V, Mateille T, Steinberg C (2007) Soil health through soil disease suppression: which strategy from descriptors to indicators. Soil Biol Biochem 39:1–23

Jenkinson DS (1988) Soil organic matter. In: Alan W (ed) Russell's soil conditions and plant growth, 11th edn. Longman, Essex, pp 564–607

Kaye JP, Hart SC (1997) Competition for nitrogen between plants and soil microorganisms. Trends Ecol Evol 12:139–143

Khan R, Straney DC (1999) Regulatory signals influencing expression of the *PDA1* gene of *Nectria haematococca* MPVI in culture and during pathogenesis of pea. Mol Plant-Microbe Interact 12:733–742

Kistler HC, Van Etten HD (1984) Regulation of pisatin demethylation in *Nectria haematococca* and its influence on pisatin tolerance and virulence. J Gen Microbiol 130:2605–2613

Knott CM (1999) Irrigation of field peas (*Pisum sativum*): response to timing at different crop growth stages. J Agric Sci 132:417–424

Kotchoni SO, Gachomo EW (2006) The reactive oxygen species network pathways: an essential prerequisite for perception of pathogen attack and the acquired disease resistance in plants. J Biosci 31:389–404

Kraft JM (1984) Fusarium root rot. In: Hagedorn DJ (ed) Compendium of pea diseases. The American Phytopathological Society, St. Paul, pp 30–31

Kraft JM (2001) Fusarium root rot. In: Kraft JM, Pfleger FL (eds) Compendium of pea diseases. The American Phytopathological Society, St. Paul, MN, pp 13–14

Kraft JM, Kaiser WJ (1993) Screening for disease resistance in pea. In: Singh KB, Saxena MC (eds) Breeding for stress tolerance in cool-season legumes. Wiley, Chichester, pp 123–144

Liu XG, Inlow M, Van Etten HD (2003) Expression profiles of pea pathogenicity (PEP) genes *in vivo* and *in vitro*, characterization of the flanking regions of the pep cluster and evidence that the pep cluster region resulted from horizontal gene transfer in the fungal pathogen *Nectria Haematococca*. Curr Genet 44:95–103

Mackintosh SF, Matthews DE, Van Etten HD (1989) Two additional genes for pisatin demethylation and their relationship to the pathogenicity of *Nectria haematococca* on pea. Mol Plant-Microbe Interact 2:354–362

Magurran AE (1988) Ecological diversity and its measurement. Princeton University Press, Princeton

Mansfield JW (1983) Antimicrobial Compounds. In: Callow JA (ed) Biochemical plant pathology. Wiley, Chichester, pp 237–265

Matthews DE, Van Etten HD (1983) Detoxification of the phytoalexin pisatin by a fungal cytochrome P450. Arch Biochem Biophys 224:494–505

Matuo T, Snyder WC (1972) Host virulence and the hypomyces stage of *Fusarium solani* f. sp. *pisi*. Phytopathology 62:731–735

McPhee K (2003) Dry pea production and breeding – a mini-review. Food Agric Environ 1:64–69

Morrissey JP, Osbourn AE (1999) Fungal resistance to plant antibiotics as a mechanism of pathogenesis. Microbiol Mol Biol Rev 63:708–724

Muehlbauer FJ, Short RW, Kraft JM (1983) Description and culture of dry peas, USAD-ARS agricultural reviews and manuals. Western series no. 37. Western Region, Oakland

Nannipieri P, Ascher J, Ceccherini MT, Landi L, Pietramellara G, Renella G (2003) Microbial diversity and soil functions. Eur J Soil Sci 54:655–670

Navas-Cortés JA, Alcalá-Jiménez AR, Hau B, Jiménez-Díaz RM (2000) Influence of inoculum density of races 0 and 5 of *Fusarium oxysporum* f. sp. *ciceris* on development of Fusarium wilt in chickpea cultivars. Eur J Plant Pathol 106:135–146

Nielsen MN, Winding A (2002) Microorganisms as indicators of soil health, Ministry of Environment. Technical report no. 388. National Environmental Research Institute, Denmark

Osbourn AE (1996) Preformed antimicrobial compounds and plants defense against fungal attack. Plant Cell 8:1821–1831

Oyarzun PJ (1993) Bioassay to assess root rot in pea and effect of root rot on yield. Neth J Plant Pathol 99:61–75

Oyarzun P, Gerlagh M, Hoogland AE (1993a) Pathogenic fungi involved in root rot of peas in the Netherlands and their physiological specialization. Neth J Plant Pathol 99:23–33

Oyarzun P, Gerlagh M, Hoogland AE (1993b) Relation between cropping frequency of peas and other legumes and foot and root rot in peas. Neth J Plant Pathol 99:35–44

Oyarzun PJ, Dijst G, Maas PWT (1994) Determination and analysis of soil receptivity to *Fusarium solani* f. sp. *pisi* causing dry root rot of peas. Phytopathology 82:834–842

Oyarzun PJ, Dijst G, Zoon FC, Maas PWT (1997) Comparison of soil receptivity to *Thielaviopsis basicola, Aphanomyces euteiches*, and *Fusarium solani* f. sp. *pisi* causing root rot in pea. Phytopathology 87:534–541

Oyarzun P, Gerlagh M, Zadoks JC (1998) Factors associated with soil receptivity to some fungal root rot pathogens of peas. Appl Soil Ecol 10:151–169

Pankhurst CE, Hawke BG, McDonald HJ, Kirkby CA, Buckerfield JC, Michelsen P, O'Brien KA, Gupta VVSR, Doube BM (1995) Evaluation of soil biological properties as potential bioindicators of soil health. Aust J Expt Agric 35:1015–1028

Pankhurst CE, Ophel-Keller K, Doube BM, Gupta VVSR (1996) Biodiversity of soil microbial communities in agricultural systems. Biodiv Conserv 5:197–209

Paxton J (1980) A new working definition of the term phytoalexin. Plant Dis 64:734

Paxton JD (1981) Phytoalexins-A working redefinition. Phytopathol Z 101:106–109

Persson L, Bødker L, Larsson WM (1997) Prevalence and pathogenicity of foot and root rot pathogens of pea in southern Scandinavia. Plant Dis 81:171–174

Processors and Growers Research Organisation (1997) Pea and bean pests, diseases and disorders. Processors and Growers Research Organisation, Thornhaugh

Ristaino JB (1991) Influence of rainfall, drip irrigation and inoculum density on the development of phytophthora root and crown rot epidemics and yield in bell pepper. Phytopathology 81:922–929

Rondon MR, August PR, Betterman AD, Brady SF, Grossman TH, Liles MR, Loiacono KA, Lynch BA, MacNeil IA, Minor C, Tiong CL, Gilman M, Osburne MS, Clardy J, Handelsman J, Goodman RM (2000) Cloning and soil metagenome: a strategy for assessing the genetic and functional diversity of uncultured microorganisms. Appl Environ Microbiol 66:2541–2547

Rush CM, Kraft JM (1986) Effects of inoculum density and placement on Fusarium root rot of peas. Phytopathology 76:1325–1329

Schippers B (1992) Prospects for management of natural suppressiveness to control soilborne pathogens. In: Tjamos EC, Papavizas GC, Cook RJ (eds) Biological control of plant diseases, progress and challenges for the future, vol 230, NATO ASI, life sciences. Plenum, New York, pp 21–34

Short GE, Lacy ML (1976) Carbohydrate exudation from pea seeds: effect of cultivar, seed age, seed color and temperature (and relation to fungal rots). Phytopathology 66:182–187

Singh H (1995) Nitrogen mineralization, microbial biomass and crop yield as affected by wheat residue placement and fertilizer in a semi-arid tropical soil with minimum tillage. J Appl Ecol 32:588–595

Smith CJ (1996) Accumulation of phytoalexins: defence mechanism and stimulus response system. New Phytol 132:1–45

Stevenson FJ (1994) Humus chemistry: genesis, composition, reactions. Wiley, New York

Straney DC, Van Etten HD (1994) Characterization of the *PDA1* promoter of *Nectria haematococca* and identification of a region that binds a pisatin-responsive DNA binding factor. Mol Plant-Microbe Interact 7:713–725

Suga H, Hasgawa T, Mitsui H, Kageyama K, Hyakumachi M (2000) Phylogenetic analysis of the phytopathogenic fungus Fusarium solani based on the rDNA-ITS region. Mycol Res 104:1175–1183

Sugha SK, Kapoor SK, Singh BM (1994) Factors influencing Fusarium wilt of chickpea (*Cicer arietinum* L.). Indian J Mycol Plant Pathol 24:97–102

Sullivan P (2001) Sustainable management of soil-borne plant diseases, soil systems guide. Appropriate Technology Transfer for Rural Areas, Fayetteville, p 19

Swift MJ (1974) Species diversity and the structure of microbial communities in terrestrial habitats. In: Anderson JM, McFadyen A (eds) The role of terrestrial and aquatic organisms in decomposition processes. Blackwell, Oxford, UK, pp 185–221

Temporini ED, Van Etten HD (2002) Distribution of the pea pathogenicity (PEP) genes in the fungus *Nectria haematococca* mating population VI. Curr Genet 41:107–114

Theron J, Cloete TE (2000) Molecular techniques for determining microbial diversity and community structure in natural environments. Crit Rev Microbiol 26:37–57

Tilman D, Downing JA (1994) Biodiversity and stability in grasslands. Nature 367:363–365

Tilman D, Wedin D, Knops J (1996) Productivity and sustainability influenced by biodiversity in grassland ecosystems. Nature 379:718–720

Tu JC (1994) Effects of soil compaction, temperature, and moisture on the development of the Fusarium root complex of pea in southwestern Ontario. Phytoprotection 75:125–131

Turco RF, Kennedy AC, Jawson MD (1994) Microbial indicators of soil quality. In: Doran JW, Coleman DC, Bezdicek DF, Stewart BA (eds) Defining soil quality for a sustainable environment. Soil Science Society of America, Madison, pp 73–90

van Bruggen AHC, Semenov AM (2000) In search of biological indicators for soil health and disease suppression. Appl Soil Ecol 15:13–24

Van Etten HD, Kistler HC (1988) *Nectria haematococca*, mating population I and VI. In: Sidhu GS (ed) Advances in plant pathology, vol 6, Genetics of plant pathogenic fungi. Academic, New York, pp 190–206

Van Etten HD, Matthews DE, Matthews PS (1989) Phytoalexin detoxification: importance for pathogenicity and practical implications. Annu Rev Phytopathol 27:143–164

Van Etten HD, Sandrock RW, Wasmann CC, Soby SD, Mccluskey K, Wang P (1995) Detoxification of phytoanticipins and phytoalexins by phytopathogenic fungi. Can J Bot 73:518–525

Van Etten H, Temporini E, Wasmann C (2001) Phytoalexin (and phytoanticipin) tolerance as a virulence trait: why is it not required by all pathogens? Physiol Mol Plant Pathol 59:83–93

van Loon LC, Pierpoint WS, Boller T, Conejero V (1994) Recommendations for naming plant pathogenesis-related proteins. Plant Mol Biol Rep 12:245–264

van Veen JA, van Overbeek LS, van Elsas JD (1997) Fate and activity of microorganisms introduced into soil. Microbiol Mol Biol Rev 61:121–135

Walker BH (1992) Biological diversity and ecological redundancy. Conserv Biol 6:18–23

Wasmann CC, Van Etten HD (1996) Transformation-mediated chromosome loss and disruption of a gene for pisatin demethylase decrease the virulence of *Nectria haematococca* on pea. Mol Plant-Microbe Interact 9:793–803

Wu QD, Van Etten HD (2004) Introduction of plant and fungal genes into pea (*Pisum sativum* L.) hairy roots reduces their ability to produce pisatin and affects their response to a fungal pathogen. Mol Plant-Microbe Interact 17:798–804

Chapter 7
Managing the Phytotoxicity and Inconsistent Nematode Suppression in Soil Amended with Phytonematicides

Phatu W. Mashela, Zakheleni P. Dube, and Kgabo M. Pofu

7.1 Introduction

The global withdrawal of environment-unfriendly synthetic nematicides from agrochemical markets resulted in the emergence of various alternatives for managing plant-parasitic nematodes (Chedekal 2013; Stirling 2014). However, the introduced alternatives had inherent drawbacks. For instance, most crude extracts from plants with acceptable efficacies on suppression of nematodes were highly phytotoxic and could therefore not be sanctioned for use in crop husbandry. The European and Mediterranean Plant Protection Organization (EPPO 2010) and other such legal entities in various countries have zero tolerance on products that induce phytotoxicity on crops which are being protected against pests. Invariably, some non-phytotoxic products have had inconsistent results on target pests, which raised credibility issues for their registration. Incidentally, most products to be used in agricultural pests such as plant-parasitic nematodes have to undergo registration after intensive efficacy and non-phytotoxic trials.

Plant-parasitic nematodes are among the most injurious soilborne pests in cropping systems, with yield losses ranging from 5 % to 15 % (Stirling 2014) and translating to billions of US dollars (Chitwood 2002; Khan et al. 2008). Following the withdrawal of highly effective nematicides, the use of nematode-resistant genotypes had been in the forefront as a management strategy of choice in reducing

P.W. Mashela (✉) • Z.P. Dube
School of Agricultural and Environmental Sciences, University of Limpopo, Private Bag X1106, Sovenga 0727, Republic of South Africa
e-mail: phatu.mashela@ul.ac.za; zakheleni_dube@yahoo.com

K.M. Pofu
Agricultural Research Council – Vegetable and Ornamental Plants Institute, Private Bag X293, Pretoria 0001, Republic of South Africa
e-mail: MarthaP@arc.agric.za

M.K. Meghvansi, A. Varma (eds.), *Organic Amendments and Soil Suppressiveness in Plant Disease Management*, Soil Biology 46,
DOI 10.1007/978-3-319-23075-7_7

nematode densities to below injurious levels. However, in plant genotypes without nematode resistance such as watermelon (*Citrullus lanatus*), peppers (*Capsicum annuum*) and potatoes (*Solanum tuberosum*), the yield losses escalate to as high as 50 % and at times to complete crop failure due to infection by the root-knot (*Meloidogyne* species) nematodes (Pofu et al. 2012). Reliance on nematode resistance was not sustainable due to the existence of nematode races and sensitivity of nematode-resistant genotypes to environmental factors such as high soil temperature (Dropkin 1969), salinity (Mashela et al. 1992) and honeydew-inducing foliar insects (Pofu et al. 2012). Lack of nematode-resistant genotypes in certain economically important crops and incompatibility of intergeneric nematode-resistant rootstocks and scions also negated the widespread adoption of nematode resistance technology (Pofu et al. 2012). Notwithstanding the listed drawbacks and the degree of nematode resistance in a given cultivar, the extent of crop losses is also depended upon the aggressiveness of the target nematode. For example, *Meloidogyne* species, lesion nematode (*Pratylenchus* species), sting nematode (*Belonolaimus longicaudatus*) and burrowing nematode (*Radopholus similis*) are highly aggressive, and, therefore, each may induce excessive damage to the host plants. In contrast, the citrus nematode (*Tylenchulus semipenetrans*) is not aggressive, but could be highly damaging in soils with salinity problems (Mashela and Nthangeni 2002; Duncan 2009).

Management of plant-parasitic nematodes in cropping systems is indispensable if crop enterprises are to be profitable and thereby improve food security on a global scale. Due to various setbacks on nematode resistance, organic amendments and/or other biological agents were tested on a grand scale for the suppression of population nematode densities. Notably, higher plants, biocontrol agents and fungi have since provided a broad spectrum of active compounds for use in nematode management (Chitwood 2002; Okwute 2012; Chedekal 2013). Phytonematicides as an alternative management strategy in nematode suppression comprise a class of plant-based nematicides, which are available as aqueous plant extracts (Egunjobi and Afolami 1976; Rossner and Zebitz 1987; Chedekal 2013), methanol plant extracts (Usman 2013), ethanol plant extracts (Khan et al. 2008), oilcakes (Muller and Gooch 1982), essential oils (Meyer et al. 2008), fermented crude plant extracts (Kyan et al. 1999; Ncube 2008; Pelinganga and Mashela 2012; Pelinganga et al. 2013a), powders (Ahmad et al. 2013) or granules (Mashela et al. 2008, 2011, 2012).

Phytonematicides differ from conventional organic amendments, which may include crop residues, manures, compost, organic manures, agro-industrial wastes and sewage sludge (Castagnone-Sereno and Kermarrec 1991; D'Addabbo 1995; Thoden et al. 2011; Stirling 2014). Generally, phytonematicides were introduced to mitigate the drawbacks of conventional organic amendments in suppression of plant-parasitic nematodes (Mashela 2002), which include (1) inconsistent results in nematode suppression; (2) large quantities (10–500 t/ha) which were required to achieve nematode suppression; (3) unavailability of the materials; (4) high transport costs to haul the materials from the production site to that of use; (5) negative period, with the subsequent time-lag to allow for microbial decomposition in order to avoid negative periods; and/or (6) decrease in soil pH, which inherently imbalances the availability of essential nutrient elements in the soil (Jafee et al. 1994; Belair and Tremblay 1995; McSorley and Gallaher 1995; Mashela 2002; Kimpinski et al. 2003; Thoden et al. 2011; Stirling 2014). Inputs for most phytonematicides are locally collected from indigenous plants (Muller and Gooch 1982; Akhtar and Malik 2000; Oka 2010; Mashela et al. 2011; Ahmad et al. 2013), which possess complex allelochemical compounds (Chitwood 2002; Okwute 2012). In purified formulation, most phytonematicides lose their nematode suppression capabilities (Wuyts et al. 2006; Oka 2010; Ntuli and Caboni 2012; Okwute 2012) and are accompanied by unacceptable high phytotoxicity levels on crops being protected against nematodes (Mian and Rodriguez-Kabana 1982a, b; Meyer et al. 2008). Generally, phytonematicides rely on allelochemicals as their active ingredients and are used in vivo for defence against invading pathogens (Rice 1984; Inderjit et al. 1999). Roots of certain allelochemical-producing plants exude copious quantities of allelochemicals to provide competitive edge against competitors during interference (Inderjit et al. 1999; Rice 1984). The objective of this overview was to provide the dosage model as an alternative strategy in managing plant-parasitic nematodes with specific reference to addressing efficacy, phytotoxicity and inconsistent result issues of phytonematicides.

7.2 Distinction Between Phytonematicides and Organic Amendments

In the original overview on organic amendments, Muller and Gooch (1982) noted that between 1971 and 1981, out of 33 organic amendment trials, those with at least 91 % success frequency on nematode suppression were in the form of powders and oilcakes from neem (*Azadirachta indica*), peanuts (*Arachis hypogaea*) and castor (*Ricinus communis*). Later, other reviews (Alam 1993; Ferraz and de Freitas 2004; Oka 2010) confirmed that neem extracts, particularly those from seed kernels, had high bioactivities on nematode populations. Mashela et al. (2011) introduced a classical model on the ground leaching technology (GLT) system, with the research focus being on powdered plant products from selected plant organs with the view of ameliorating the numerous drawbacks of conventional organic amendments in smallholder tomato (*Solanum lycopersicum*) farming systems in South Africa. In the GLT system, powdered materials were derived from unshelled dried castor bean, fever tea (*Lippia javanica*) leaves, wild cucumber (*Cucumis myriocarpus*) fruit and wild watermelon (*Cucumis africanus*) fruit. In all cases, the four products each consistently reduced population densities of *Meloidogyne* species and *T. semipenetrans*. Overall, the GLT system uses 0.2–0.7 powdered materials/ha for 4000 tomato plants when compared with 10–500 t organic amendments/ha required to effect consistent results in nematode suppression (Mashela 2002). In order to distinguish the powdered materials with their small quantities required in suppression of nematodes relative to large quantities required in conventional organic amendments, the former were referred to as phytonematicides (Mashela et al. 2011). Phytonematicides at the concentration used are intended to consistently suppress population densities of the target nematodes, while stimulating growth of the protected crops instead of inhibiting plant growth and productivity (Pelinganga 2013). Certain phytonematicides can be highly effective in nematode suppression. Incidentally, the efficacy of powdered materials from *C. myriocarpus* fruit in nematode suppression was similar to those of aldicarb and fenamiphos nematicides (Mashela et al. 2008).

The major distinctions between phytonematicides and conventional organic amendments could be:

1. The empirically based small quantities applied to achieve consistent nematode suppression under diverse conditions as opposed to large quantities.
2. In GLT system there is gradual release of active ingredients from crude extracts into the rhizosphere which is achieved through irrigation water or rainfall as opposed to microbial degradation in conventional organic amendments.
3. Phytonematicide products mimic synthetic chemical nematicides since they could be commercially packaged in relatively small containers with label information which includes active ingredients, along with efficacy features.
4. Phytonematicides like most non-fumigant nematicides do not have negative periods and could therefore be applied as post-planting products.
5. These products are required to comply with relevant legislation in terms of avoiding health risks to end users, nontarget organisms and the environment.

The drawback of the GLT system was its high labour costs since products were manually applied, which rendered the system less appealing to large commercial tomato producers (Mashela et al. 2011). An alternative technology, referred to as botinemagation (Mashela et al. 2011), was developed for use in large-scale tomato farming systems, where crude extracts from fermented plant organs were used through drip irrigation systems. Using dried fruits of *C. myriocarpus* and *C. africanus* fruits, fermented crude extracts as liquid formulations consistently reduced population densities of *Meloidogyne* species in tomato production (Pelinganga et al. 2013a, b).

Not all plant organs contain allelochemicals with nematicidal properties. In South Africa, Van Wyk et al. (2002) listed 372 plant species on the basis of their toxicity to humans and animals, which were for the purpose of this discussion classified into six using their degree of toxicity (Table 7.1). Approximately 22.6, 18.3 and 6.7 % of the listed plants were described as being poisonous, very poisonous and deadly, respectively, to humans and livestock. The degree of toxicity to humans and animals does not confer a plant a better status to be a candidate for serving as source of phytonematicides. *Cucumis myriocarpus* and *R. communis*, from which two phytonematicides were developed for the GLT system (Mashela 2002; Mashela and Nthangeni 2002), were regarded as being poisonous and very poisonous, respectively (Van Wyk et al. 2002). However, the deadly oleander (*Nerium oleander*) and tamboti (*Spirostachys africana*) did not have phytonematicidal properties against *Meloidogyne* species (Mashela et al. 2011).

Table 7.1 Count and percent count of plants clustered according to the degree of toxicity to humans and livestock in South Africa

Classification	Count	%
Not really poisonous	14	3.8
Poisonous	84	22.6
Very poisonous	68	18.3
Deadly	25	6.7
Causes skin allergies or contact dermatitis	18	4.8
Poisonous to animals	163	13.8
Total	**372**	**100**

Statistics developed from Van Wyk et al. (2002)

In contrast, certain plants listed as 'not really poisonous', namely, fever tea (*Lippia javanica*) and *Brassica* species, produced potent phytonematicides (Mashela et al. 2010). Among the listed plant species, only 0.8 % plant species were tested against nematodes in South Africa, with only 0.55 % having some nematicidal properties. At a global level, among the 45 papers of biological control agents of nematodes discussed at the 2014 International Congress of Nematology in Cape Town, South Africa, 42, 36, 18 and 4 % were on botanicals, fungi, bacteria and enzymes, respectively. The highest percentage of phytonematicide papers clearly illustrated the potential importance and interest in this group of biological control agents in plant nematology.

7.3 Efficacy of Phytonematicides

The majority of in vitro trials have had in excess of 90 % suppression of nematode numbers from phytonematicides (Okwute 2012). However, due to their high phytotoxicities and restricted measures (EPPO 2010), a large number of botanicals with potent nematicidal properties do not make it beyond in vitro tests. Notwithstanding the high rejection of most products, detailed assessments on mode of action for certain phytonematicides had been undertaken.

7.4 Mode of Action of Phytonematicides

The distinguishing feature of synthetic pesticides is their single active ingredients, with clearly defined bioactivities. In synthetic insecticides, such single active ingredients had high incidents of insect resistance, particularly in insects with high reproductive capabilities (Nzanza and Mashela 2012). However, although certain nematode species have high reproductive capabilities, resistance to synthetic nematicides in plant-parasitic nematodes had not been observed (Van Gundy and McKenry 1975). In contrast to synthetic pesticides, phytonematicides have multiple active ingredients, with complementary modes of action. For instance, in wild garlic (*Tulbaghia violacea*), the plant bulb contains sacrid volatile oils and sulpho-oxides—each being a derivative of allicin, which has insecticidal and nematicidal properties (Vijayalakshmi et al. 1996; Nzanza and Mashela 2012; Mashela et al. 2012). In insects the mode of action for the allicin derivatives had been identified as antifeedant, repellent and insecticidal (Vijayalakshmi et al. 1996; Dhanalakshmi 2006). Similarly, in insects, azadirachtin in neem had been shown to have antifeedant, repellent and anti-ovipositor properties, with capabilities for delaying or preventing moulting in insects. Apparently, using phytopesticides confers a broad spectrum of active ingredients, with multiple modes of action. In phytonematicides, observations on mode of action had been limited to chemotaxis, juvenile motility, egg hatch, juvenile mortality or juvenile paralysis, with limited information on behavioural responses of adult nematodes.

7.4.1 Chemotaxis

Chemotaxis is a phenomenon where nematodes direct their movement according to the gradient of selected chemical cues in the environments (Bargmann and Mori 1997). Positive chemotaxis occurs when movement is towards the increasing gradient of chemical cues. Conversely, movement towards the opposite direction of the increasing gradient is described as negative chemotaxis (Bargmann and Mori 1997). The nematode is literally exposed to both liquid- and airborne volatilised chemicals in the air-water interface of the soil, which could either be water-soluble and/or volatile chemoattractants or chemorepellents. According to Bargmann and Mori (1997), water-soluble chemoattractants are detected by chemoreceptors in nematodes at micromolar concentrations, while the volatile chemoattractants are detected at picomolar concentrations. Water-soluble and volatile chemoattractants are used for short- and long-distance chemotaxes, respectively (Prot 1980). In contrast, water-soluble and volatile chemorepellents are toxic and could cause either paralysis or death of the nematode. In phytonematicides, both chemoattractants and chemorepellents are important. Chemoattractant phytonematicides may disorientate the nematode from being guided by chemoattractant cues produced by potential host plants, thereby deferring penetration and attack of host by nematodes (Wuyts et al. 2006). In contrast, chemorepellents may induce various behavioural changes in the nematode, including paralysis and death (Bargmann and Mori 1997).

The body of a nematode is 'wired' with chemoreceptors, particularly on the frontal and cervical regions (Ferraz and Brown 2002), suggesting that chemoattractants and chemorepellents play important roles in behavioural activities of nematodes. Plants release numerous chemicals through exudation, leaching, volatilisation and microbial degradation for different reasons (Stirling 2014). Similarly, phytonematicides release potent chemicals either through leaching, volatilisation or microbial degradation (Mashela et al. 2011, 2012). Generally, increasing concentrations of phytochemicals could interfere with chemotaxis in one of three ways: no effect (neutral chemotaxis), attract (positive chemotaxis) and repel (negative chemotaxis). Responses characterised by these three phases in the environment subscribe to density-dependent growth (DDG) curves (Salisbury and Ross 1992; Liu et al. 2003), which constitute an important part of this review.

Using purified phytochemical compounds, Wuyts et al. (2006) demonstrated that certain chemical compounds from *Philenoptera violacea* in the Fabaceae family had similar and/or different effects on chemotaxis—which is dependent much on the nematode species. In their work (Wuyts et al. 2006), among the tested chemical compounds produced through the shikimic acid pathway, 26 % repelled *R. similis*, 2.6 % attracted this nematode, while 45 % were neutral. In contrast, of the 37 % tested chemical compounds on *P. penetrans*, even those that were chemorepellent to *R. similis* had no effect on this nematode. In contrast, some chemorepellents to *R. similis* were also repellent to *M. incognita*. Although the approach used by Wuyts et al. (2006) did not provide information on physiological activities of the target chemicals in nematode bodies, it provided broad clues in terms of what we want to convey using DDG patterns later on in this overview. The three nematode species used depicted neutrality to the largest number of chemical compounds produced through the mevalonate pathway, followed by inhibition of motility and then repellence as depicted in chemotaxis (Wuyts et al. 2006). A remarkable feature in the work of Wuyts et al. (2006) was, therefore, the agreement of their observations with the concept of DDG patterns, particularly with the observed repeated neutral responses.

7.4.2 *Motility*

Juveniles from unhatched eggmasses which were previously exposed to crude extracts from leaves of Borelin remained motile, while those exposed to crude extracts of garlic bulb or neem seed kernels had impact on juvenile motility (Agbenin et al. 2005). According to DDG principles, different concentrations of phytonematicides might have no effect (neutral) on, stimulate and/or inhibit motility of nematodes (Salisbury and Ross 1992; Liu et al. 2003). Wuyts et al. (2006) observed that a chemical compound which was neutral in one nematode species could inhibit juvenile motility in another nematode species, vice versa. Similarly, those that were chemoattractants in chemotaxis for one nematode species might be neutral and/or inhibitive in juvenile motility for another nematode species. Oka et al. (2000) showed that essential oils from 12 of 27 plants immobilised more than 80 % *M. javanica* J2s after a 2-day exposure, with immobilisation being amenable to DDG patterns.

7.4.3 Egress

Egress in *M. incognita* was inversely proportional to concentrations of crude extracts from garlic and neem (Agbenin et al. 2005; Chedekal 2013). Although egress is a physical process, in most plant-parasitic nematodes, it is stimulated by external chemical cues from roots (Prot 1980). According to Wuyts et al. (2006), some phytochemical compounds were neutral towards egg hatch, while others were inhibitive. In contrast, one flavanone, which is a hesperetin chemical compound, was both stimulatory and inhibitive to egress in *R. similis* (Wuyts et al. 2006). Most active ingredients from phytonematicides have the capability to penetrate eggmasses, where J1s become exposed to aqueous solutions (Hirschmann 1985; Parmar 1987; Agbenin et al. 2005). Incidentally, the materials interfered with stylet development, rendering it incapable of piercing through the eggshell and, therefore, resulting in complete failure of egress (Hirschmann 1985; Parmar 1987).

Using in vitro trials, essential oils from 27 different plant species, at 1000 μL/L only 30 % of plants inhibited egress, while at 600 μL/L only 15 % of plants had oils with inhibitive properties (Oka et al. 2000). Ojo and Umar (2013) demonstrated that crude extracts from testa of cocoa bean (*Theobroma cacao*) plants had significantly higher effects on egress of *M. javanica* than oil palm fibre, with differences attributed to different chemical constituents. Cocoa bean testa contains alkaloids and flavonoids, with egress inhibition being directly proportional to the concentration of the listed chemical compounds (Ojo and Umar 2013). However, in the same study, Ojo and Umar (2013) observed that oil palm fibre, which was devoid of alkaloids and flavonoids, had negligent effects on egress. Okeniyi et al. (2013) demonstrated increasing concentrations (0, 10, 25, 50 and 100 %) of leaf crude extracts from the coastal golden leaf (*Bridelia micrantha*), euphorbia (*Mallotus oppositifolius*), abeere (*Hunteria umbellate*) and citron (*Citrus medica*)—each increased inhibition of egress in *M. incognita*. Removal of eggs from the chemical compounds resulted in reversal of the extract effects.

7.4.4 Mortality

In vitro exposure of *Meloidogyne* J2s to crude extracts from hen's nettle (*Fleurya interrupta*), panicled peristrophe (*Peristrophe bicalyculata*) and king of bitters (*Andrographis paniculata*) resulted in 100 % mortalities (Mukherjee and Sukul 1978). Similarly, high *Meloidogyne* J2 mortalities were observed in crude extracts from leaves of marigold (*Tagetes* species), Indian gooseberry (*Emblica officinalis*) and Christ's thorn (*Carissa carandas*) during in vitro exposure (Toida and Moriyama 1978; Haseeb et al. 1980). Also, in vitro exposure of *M. incognita* J2s to crude extracts or aqueous extracts from fresh leaves of various plants resulted in high mortalities (Agbenin et al. 2005; Chedekal 2013). Similarly, crude extracts of either cocoa bean testa or oil palm fibre resulted in high mortalities of *M. javanica* juveniles (Ojo and Umar 2013). Juvenile mortalities were directly proportional to increasing concentrations of phytonematicides and exposure time (Agbenin et al. 2005). In some instances, 'mortalities' were reversible when J2s were removed from the chemicals (Wuyts et al. 2006).

7.4.5 *Paralysis*

Paralysis involves irreversible interference of nematicides with the nervous systems of J2s. Generally, affected J2s can still wiggle, but have complete loss of coordinated mobility. Phytonematicide-induced paralysis reports on plant-parasitic nematodes are uncommon. An exceptional case is that in Ntalli et al. (2011), where paralysis of *Meloidogyne* J2s was regularly observed when exposed to aliphatic ketones from rue (*Ruta chalepensis*).

7.5 Variation in Efficacy of Phytonematicides

Incidentally, biological entities respond to various abiotic and/or biotic factors through a myriad of complex processes and mechanisms. For instance, when various plant-parasitic nematodes infect plants at population densities below the tolerance limit, plant growth is invariably stimulated (Wallace 1973), while at high population densities, growth is reduced (Seinhorst 1967). Similarly, infection by different nematode species on various legumes either stimulated, had no effect on or inhibited nodulation and/or nitrogen fixation (Huang 1987). Vesicular-arbuscular mycorrhizal (VAM) fungi on various host plants also resulted in positive, neutral or negative growth responses (Smith 1987). Different fertilisers and/or salinity levels can also induce such growth responses in plants. In soil allelochemical residue (SAR) trials, it was shown that while SAR effects from one phytonematicide stimulated growth of the successor crop, SAR effects consistently reduced population densities of *Meloidogyne* species (Mashela and Dube 2014), with reduced population densities subscribing to similar inconsistent growth patterns (Zasada and Ferris 2003). Mashela (2014) showed that SAR effects had inhibitive effects on nodulation by *Bradyrhizobium japonicum* in cowpea (*Vigna unguiculata*). Others (Mashela and Dube 2014) argued that for phytonematicides to be successful, their inhibition concentration range to nematodes should overlap the stimulation range to the crop being protected against nematodes.

Sites of action in organisms by allelochemicals are not yet established. However, cucurbitacins from fruits of wild *Cucumis* species were shown to have the potential to inhibit cell division in cancer at high concentrations, while the materials were highly cytotoxic to healthy cells (Lee et al. 2010). In contrast, when used at low concentrations, cytotoxicity was avoided, but division of healthy cells was stimulated, thereby rendering the materials cancerous (Lee et al. 2010). These observations in cancer trials provided clues on the site of action of cucurbitacins—the cellular level.

Reports which demonstrated that conventional organic amendments increased population densities of nematodes in Europe (Belair and Tremblay 1995; Kimpinski et al. 2003), had no effect on nematode numbers in Florida, USA (Jafee et al. 1994; McSorley and Gallaher 1995) and reduced nematode numbers (Stirling 2014) raised credibility issues on organic amendments due to the 'perceived' inconsistent results (McSorley 2011). The efficacy of phytonematicides is dependent upon the concentration of allelochemicals in the organ used for processing the intended products. Generally, the accumulation of secondary metabolites in organs varies from season to season (Mudau et al. 2008), with high inconsistent results in nematode suppression and high phytotoxicities during certain seasons. However, the variability that leads to inconsistent results should not be confounded with DDG patterns in allelochemical-containing products. Although the variability of concentrations of allelochemicals in a particular organ could be associated with DDG patterns in certain cases, DDG principles are primarily related to responses of living entities in response to increasing concentrations of allelochemicals ex vitro. In organs such as fruits or bulbs where the accumulation of secondary metabolites appears to level off with maturity, variability in efficacy of phytonematicides on nematode suppression had mostly been due to different concentrations in the processed product (Meyer et al. 2008). Generally, sources that result in the final product being of high variability are undesirable, particularly when commercial products are envisaged. On the basis of the three phases (stimulation, neutral and inhibition) being characterised by different concentration ranges, one could argue that the various materials of plant origin did not have 'inconsistent' results on nematode suppression, but what was being observed in a particular time was a reflection of differences in concentrations with respect to the allelochemicals involved.

7.5.1 Density-Dependent Response Patterns in Phytonematicides

At low concentrations, crude extracts of neem leaf were shown to stimulate growth of maize (*Zea mays*) and tomato seedlings, while at high concentrations, the opposite occurred (Egunjobi and Afolami 1976; Rossner and Zebitz 1987). Similarly, Inderjit et al. (1999) noted that at low concentrations root leachate from golden crownbeard (*Verbesina encelioides*) consistently stimulated plant growth of various plant species. Also, at low concentrations, nemarioc-B phytonematicide stimulated growth of tomato seedlings, where the product was viewed as having a 'fertiliser effect' (Mashela 2002). However, detailed analysis of essential nutrient elements in leaves did not support the 'fertiliser effect' view since the product had negligible effect on accumulation of essential nutrient elements. In subsequent studies (Mafeo et al. 2011a, b; Pelinganga et al. 2012, 2013a, b), it was shown that various plant variables (y-axis) when subjected to lines of the best fit on increasing concentrations of nemarioc-A (x-axis) invariably resulted in quadratic relationships, which is a strong indicator for the existence of DDG patterns (Salisbury and Ross 1992; Liu et al. 2003). A myriad of complex models regarding DDG patterns exist in biological entities, including plant-parasitic nematodes (Ferris and Wilson 1987; Duncan and McSorley 1987). The DDG tenets are closely related to the original conceptual framework of the carrying capacity (Nicholson 1933), which had since been used in a wide range of disciplines. DDG patterns have three distinct growth responses: stimulated, saturated (neutral) and inhibited growth (Salisbury and Ross 1992; Liu et al. 2003), with biological indices which had been used to unravel diverse biological responses to increasing pressures from their environments. DDG principles have the ultimate aim of improving decision-making systems in sustainable management of natural resources. Generally, plants, nematodes and microbes respond to increasing concentrations of allelochemicals through DDG patterns (Rice 1984; Ferris and Wilson 1987; Zasada and Ferris 2003; Liu et al. 2003), with attempts to investigate the mechanisms involved still being at conceptual stages, except that the site of action is at the cellular level (Lee et al. 2010).

Biological entities respond to increasing concentrations of allelochemicals in phytonematicides through DDG patterns, which comprise three phases, namely, stimulation, neutral and inhibition phases (Salisbury and Ross 1992; Liu et al. 2003; Pelinganga et al. 2012, 2013a, b). DDG patterns are an advanced modification of the 1933 Nicholson's carrying capacity model, which had been adapted and used in various disciplines. Liu et al. (2003) quantified concentrations of allelochemicals which lead to three stages that characterise DDG patterns for various organisms using the curve-fitting allelochemical response dosage (CARD) computer-based model. The CARD model quantifies the three phases through seven biological indices: (1) threshold stimulation (D_m) = the allelochemical concentration that initiates the stimulation phase, (2) saturation point (R_h) = the concentration that terminates stimulation or starts the neutral phase, (3) 0 % inhibition (D_0) = the concentration that terminates the neutral phase, (4) 50 % inhibition (D_{50}) = the concentration at half the distance of the inhibition phase, (5) 100 % inhibition (D_{100}) = the concentration at the end of the inhibition phase), (6) the sensitivity index (k) = provides the degree of sensitivity of an organism to the test product and (7) the coefficient of determination (R^2) = provides the degree of the strength of the CARD model. Generally, stimulated (D_m–R_h) and inhibited (D_0–D_{100}) growth concentrations are ideal representatives for phytonematicides and herbicides, respectively. The CARD model had since been empirically adapted to generate phytonematicide concentrations which stimulate plant growth while reducing population densities of nematodes using fruits as organs of preference in order to avoid confounding variability of allelochemical concentrations in the source and the actual concentration of allelochemicals in the processed product (Mafeo and Mashela 2010; Pelinganga and Mashela 2012). Using the three phases of the CARD model, we are currently in a position to argue that observations that nematode populations were not consistently suppressed by application of conventional organic amendments, which were dubbed ‘inconsistent’ since the materials sometimes stimulated (Belair and Tremblay 1995; Kimpinski et al. 2003), had no effect on (Jafee et al. 1994; McSorley and Gallaher 1995; Thoden et al. 2011) or inhibited population densities of nematodes (Mashela et al. 2011), were biologically incorrect. Incidentally, it should also be noted that not all plant organs or species have allelochemicals which have potent nematicidal properties (Mashela et al. 2011). In most plants with nematicidal allelochemicals, due to their auto-allelopathy, the chemical compounds in vivo are compartmentalised in organs not always preferred for use in conventional organic amendments. For instance, in *C. myriocarpus* fruit, cucurbitacin A is compartmentalised in seeds (Jeffrey 1978), which are hardly used in conventional organic amendments for fear of spreading the ‘weed’ through seed dispersal. Similarly, in neem the active ingredient, azadirachtin, is primarily concentrated in seed kernels (Parmar 1996).

Another important feature of DDG patterns in the CARD model is that the variable (y-axis) and the concentration of allelochemicals (x-axis) invariably have quadratic relationships (Salisbury and Ross 1992; Liu et al. 2003; Pelinganga et al. 2013a, b; Pelinganga and Mashela 2012). On this basis, should there be a positive linear instead of quadratic relationship between dependent and independent variables, results could be suggesting that the concentrations of the phytonematicide used were within the stimulation range—as observed in various trials. Incidentally, no effective response of dependent variables over increasing independent variable levels could suggest that phytonematicide concentrations were either within the neutral range (R_h–D_0) or below D_m. In contrast, negative linear relationship invariably suggested that the concentration of allelochemicals tested was within the inhibition range (D_0–D_{100}). In biology, literature is replete with responses to abiotic and/or biotic factors that can be described as relationships that have positive (D_m–R_h), neutral (R_h–D0) or negative linear responses (D_0–D_{100}) (Salisbury and Ross 1992). Interestingly, such responses had not attracted attention as those in conventional organic amendments, where the responses were broadly viewed as evidence that phytonematicides were unsuitable for use in management of plant-parasitic nematodes since they were unpredictable.

7.5.2 Fluctuations in Concentrations of Allelochemicals In Vivo

Allelochemicals in plants are produced for ‘unknown’ physiological roles through various pathways, with the major ones being the (a) shikimic acid pathway, (b) malonic acid pathway and (c) mevalonic acid pathway (Lai 2008). Concentrations of any allelochemical within the pathways are in continuous state of fluctuation as depicted by a large number of precursors and reversible chemical reactions which are linked to the end of glycolysis just prior to the Krebs cycle of respiration

(Lai 2008). Primarily, the formation of secondary metabolites helps to remove excess end products along the respiration pathway, particularly the acetyl co-A at the end of glycolysis (Campbell 1990). Responses to a phytonematicide in the plant being protected against nematodes are primarily a reflection of the phytonematicide concentration level at the time that particular organ was harvested for the development of the phytonematicide in question. For instance, phytonematicides derived from leaves such as those of fermented crude extracts from *L. camara* plants have the tendency of being highly inconsistent in nematode suppression due to seasonal variation of active ingredients in leaves. Also, the drying conditions of the organ after harvest might have deleterious effects on concentrations of the allelochemicals (Makkar 1991). For instance, shade-, sun- and oven-dried plant materials from the same plant organ may eventually contain different chemical concentrations due to differential chemical losses through volatilisation. Also, exposure of the harvested materials to rainfall as is common in maturation of conventional organic amendments may result in leaching out of allelochemicals since they are primarily nonstructural.

Timing of nematode sampling with reference to the initial application time of the phytonematicide could also be an important factor to consider in the perceived 'inconsistent' effect of organic materials in nematode management. In GLT systems, nematode sampling for *Meloidogyne* species was empirically established at 56 days after inoculation of plants (Mashela et al. 2011). When Maila and Mashela (2013) increased the sampling time from 56 to 150 days in citrus seedlings treated with nemarioc-AG and nemafric-BG, the highest population densities of *T. semipenetrans* were in phytonematicide-treated plants than in untreated controls. The unexpected observation was explained on the basis of cyclic growth patterns of population nematode densities, which subscribe to DDG patterns due to inherent competition for infection sites (Fig. 7.1). Generally, soon after the application of a phytonematicide, the product reduced population nematode densities, while those in untreated control increased, resulting in a situation where growth in the two populations was unsynchronised in a way that when the treated reached the trough the control was reaching the peak (Maila and Mashela 2013). By 150 days, nematode numbers under the untreated control were approaching the trough, while those from the treated seedlings were approaching the peak after reaching the trough within approximately 56-day application interval. Pofu and Mashela (2014) quantified the cyclic growth of population nematode densities of *Meloidogyne* species in four hemp (*Cannabis sativa*) cultivars and concluded that from inoculation to the peak of the nematode densities approximately 56 days were required, which was in agreement with the 56-day application interval for phytonematicides in GLT systems.

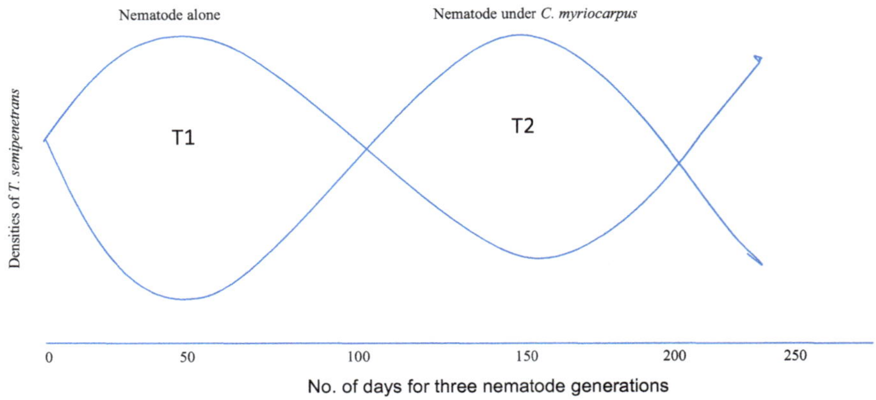

Fig. 7.1 Relative cyclic population densities of *Tylenchulus semipenetrans* on rough lemon under untreated control and nemarioc-AG-treated pots at 150 days after inoculation with 25,000 nematodes

7.5.3 Confounding Survival Adaptations with Phytonematicidal Effects

Nematodes have evolved unique survival strategies, which rendered them the status of 'undefeatable enemies' after attempts to annihilate them failed. The survival strategies had been classified into (1) intrinsic adaptations in the life cycle of the nematode and (2) extrinsic rapid responses to environmental stresses (McSorley 2003). Intrinsic adaptations occur at three levels: (a) diapause in the egg (J1), (b) developmental dormancy prior to egress in various nematodes and (c) sex reversal, mainly in *Meloidogyne* species (Triantaphyllou 1973; Papadopoulou and Triantaphyllou 1982). Extrinsic rapid responses to environmental stresses (cryptobiosis = anabiosis) involve modifications in nematode cuticles which eventually decrease their permeability to water and related gases during J2, J3 and J4 stages, depending on the nematode species (Bird and Bird 1991; McSorley 2003). Cryptobiotic responses to drought, low temperature, osmotic stress, low oxygen and high concentrations of toxic chemical compounds had been referred to as anhydrobiosis, cryobiosis, osmobiosis, anoxybiosis and chemiobiosis, respectively (McSorley 2003). Both intrinsic and extrinsic adaptations might in many respects be confounded to nemastatic responses observed in non-fumigant synthetic nematicides (Van Gundy and McKenry 1975). For example, when eggs used in hatching in vitro trials are allowed gradual permeation of chemicals to J1s, juveniles may enter the diapause stage, with the resultant failure of egress. Similarly, when cryptobiosis coincides with the application of any nematicide, the product might be rendered unfit for the intended purpose. Notwithstanding, conditions should be improved and specified during in vitro trials to establish efficacy of phytonematicides on nematodes in order to avoid confounding survival strategies induced by gradual adverse effects on various stages of nematodes with the effects of phytonematicides.

7.6 Magnitude of Phytotoxicity in Phytonematicides

Allelochemicals as active ingredients in phytonematicides are naturally phytotoxic to other plant species during interference (Wuyts et al. 2006; Okwute 2012; Ntuli and Caboni 2012). In banana (*Musa acuminata*) trial, application of 200–400 g powdered neem seed kernels per mat at 6-month application interval induced phytotoxicity—characterised by complete wilting prior to fruiting (Musabyimana et al. 2000). Additionally, in survivor plants, the inflorescence failed to emerge (Musabyimana et al. 2000), resulting in a condition called choking, where the inflorescence could not emerge through the whorl of the pseudostem. Wild garlic (*Tulbaghia violacea*) bulbs contain sacrid volatile oils and sulpho-oxides—both being derivatives of allicin (Vijayalakshmi et al. 1996). Crude extracts of garlic bulb at 50 % concentration reduced population densities of plant-parasitic nematodes, but was highly phytotoxic to tomato seedlings (Sukul et al. 1974; Egunjobi and Afolami 1976). However, at 20 % concentration, there were no noticeable effects on tomato plant growth, while the product suppressed population densities of *M. incognita* (Agbenin et al. 2005). Oil from clove (*Eugenia caryophyllata*), when drenched using 0.1, 0.2 and 0.3 % concentrations at 0, 2, 5 and 7 days prior to transplanting cucumber (*Cucumis sativus*), muskmelon (*C. melo*), pepper and tomato seedlings, all concentrations were highly phytotoxic to all crops while reducing nematode populations (Meyer et al. 2008). Sensitivities of seedlings to clove oil from *E. caryophyllata* varied with plant species, with tomato seedlings being the most sensitive among all the test plants (Meyer et al. 2008). Generally, at transplanting, seedlings from various crops were all affected by oil at 0.2 and 0.3 % concentrations. The product contains eugenol as an active ingredient, which is naturally herbicidal at low concentrations (Walter et al. 2001; Tworkoski 2002; Waliwitiya et al. 2005; Bainard et al. 2006; Boyd and Brennan 2006; Boyd et al. 2006). Incide ntally, oilcakes from different plant species have high levels of phytotoxicity to various crops at various concentrations (Haseeb et al. 1980; Mian and Rodriguez-Kabana 1982a, b; Muller and Gooch 1982; Parmar 1996). Ahmad et al. (2013) demonstrated that ground leaves of adulsa (*Justicia adhatoda*) at 3 % (w/w) concentration were highly phytotoxic to tomato seedlings. Similar phytotoxic effects were observed from high concentrations of *L. camara* root extracts on various plant species (Shaukat et al. 2003).

Two phytonematicides from fruits of indigenous *Cucumis* species in South Africa are available in granular formulation, nemarioc-AG and nemafric-BG (Mashela et al. 2011), and liquid formulation, nemarioc-AL and nemafric-BL (Pelinganga et al. 2013a). Nemarioc-AG phytonematicide was shown to be highly phytotoxic to eight monocotyledonous and ten dicotyledonous crops, with most crops failing to emerge when 5 g crude extracts were applied as pre-emergent drenches (Mafeo and Mashela 2009a, b, 2010). Similarly, both nemafric-BL and nemarioc-AL were highly phytotoxic to tomato seedlings when applied at above 10 % concentration after transplanting (Pelinganga and Mashela 2012; Pelinganga et al. 2013a, b). Nemafric-BL has cucurbitacin B ($C_{32}H_{48}O_8$), while nemarioc-AL contains two active ingredients, namely, cucumin ($C_{27}H_{40}O_9$) and leptodermin ($C_{27}H_{38}O_8$) (Rimington 1938; Jeffrey 1978). Except in rare cases such as pyrethrins that account for 80 % global uses of botanical pesticides, in purified form most active ingredients of phytonematicides, including azadirachtin-containing products, are not effective on nematode suppression, while they are highly phytotoxic to crops (Wuyts et al. 2006; Okwute 2012). Subsequently, most active ingredients in phytonematicides are applied as crude extracts.

7.7 Management of Phytotoxicity in Phytonematicides

Due to phytotoxicity and its zero tolerance in m ost legislative frameworks on products used in agriculture, literature is replete with efficacy trials which do not go beyond in vitro status. Using the concept of DDG patterns, there are basically three concentration ranges, namely, stimulation, neutral and inhibition concentration ranges (Fig. 7.2). Using the latter, we developed the concept of mean concentration stimulation range in an attempt to answer the farmers' question 'How much concentration of nemarioc-AL or nemafric-BL to apply?' which was followed by 'What is the application interval for the recommended concentration?'. The two questions were empirically answered, with avoidance of phytotoxicity and the efficacy of the products on nematode suppression in mind.

7.7.1 Establishing the Mean Stimulation Concentration Range

The potential uses of the CARD model rely on the availability of empirically generated data (Mafeo et al. 2011a, b, c; Pelinganga et al. 2012, 2013a, b). As an illustration, an experiment was conducted on tomato plants inoculated with 5000 eggs and second-stage juveniles (J2s) of *M. incognita*/plant and subjected to 0, 2, 4, 8, 16, 32 and 64 % concentrations of nemafric-BL (Fig. 7.3). At 56 days after initiating the treatments, plant variables were subjected to analysis of variance, with significant ($P \leq 0.05$) treatment means (Table 7.2) being further subjected to the CARD model to generate the quadratic relationships.

From the CARD-generated biological indices (Table 7.3), the actual values of R_h for the variables measured were computed (Table 7.4). The mean actual D_m and actual R_h values were used to establish the concentration stimulation range (CSR), which is representative of the stimulated growth in the test plant (Table 7.5).

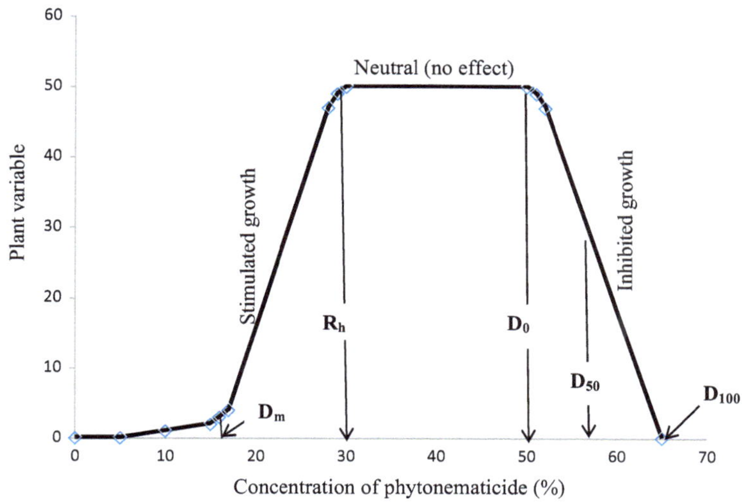

Fig. 7.2 Three distinct growth responses in density-dependent growth patterns

Fig. 7.3 Tomato seedlings for generating mean stimulation concentration range

Half the distance of the integrated CSR is referred to as the mean concentration stimulation range (MCSR). By definition, MCSR is the concentration of a phytonematicide which stimulates plant growth, while suppressing population densities of the target nematode (Pelinganga et al. 2013a; Mashela et al. 2014) and is quantified as:

Table 7.2 Means of plant growth in tomato seedlings in responses to increasing concentration of nemarioc-AL phytonematicide

	Dry shoot (g)	Dry root (g)	Plant height (cm)	Stem diameter (mm)
0	12.030	3.474	97.390	7.556
2	12.180	3.159	100.330	6.916
4	12.180	3.891	97.900	7.218
8	12.630	3.331	95.640	7.509
16	12.160	3.264	102.880	7.007
32	9.140	2.338	94.030	6.725
64	8.620	1.650	89.330	6.543
ns	ns	ns	ns	ns

ns not significant at $P \leq 0.05$

Table 7.3 Curve-fitting allelochemical response dosage-generated biological indices on tomato seedlings over six concentrations of nemarioc-AL phytonematicide

Biological index	Dry shoot mass	Dry root mass	Plant height	Stem diameter	Mean
Threshold stimulation (D_m)	2.533	2.195	2.734	1.534	2.224
Saturation point (R_m)	0.713	0.321	1.98	0.078	0.773
0 % inhibition (D_0)	11.482	9.209	19.942	5.420	10.961
50 % inhibition (D_{50})	164.897	59.003	2899.739	1603.200	957.165
100 % inhibition (D_{100})	703.5	170.3	2902.473	1604.735	1110.842
k	4	1	2	4	–
Sensitivity ranking: $\sum k = 11$					
$P \leq$	0.01	0.01	0.01	0.05	

Table 7.4 Demonstration of how MCSR is computed from threshold stimulation and saturation point biological indices

	Concentration of nemarioc phytonematicide				
Biological index	DSM[x]	DRM	PHT	SDR	Mean
Threshold stimulation (D_m)	2.533	2.195	2.734	1.534	2.249
Saturation point (R_m)	0.713	0.321	1.980	1.612	0.773
Actual R_h value	3.246	2.516	4.714	1.612	3.022
Mean concentration stimulation range (MCSR)					2.63 %

MCSR = (D_m + adjusted R_h)/2 = ($D_m + D_m + R_h$)/2 = ($2D_m + R_h$)/2 = D_m + (R_h/2) = 2.244 + (0.773)/2 = 2.244 + 0.3865 = 2.6305 = 2.63 % concentration would be non-phytotoxic to tomato plants

$$\text{MCSR} = (D_m + \text{adjusted } R_h)/2 = (D_m + D_m + R_h)/2 = (2D_m + R_h)/2 = D_m + (R_h/2)$$

Using actual mean D_m and R_h, values in the MCSR formula provided the values of 2.63 and 2.99 % for nemafric-BL and nemarioc-AL, respectively, in tomato plants (Pelinganga 2013). The MCSR value, which is empirically based on a series of phytonematicide concentrations, should be interpreted alongside the overall k-value of the plant to the test phytonematicide. The usefulness of a given product for use as a phytonematicide is entirely dependent on the overall sensitivity ($\sum$k) of the plant being protected to the product used (Liu et al. 2003). The k-values, which are plant and product specific, are generated using the CARD model and are defined as the number of *In* $(D+1)$ transformations that serves as a biological indicator of the degree of sensitivity of an organism to increasing concentrations of an allelochemical (Liu et al. 2003). The lower is the mean k-value, the higher is the sensitivity of the plant to the test allelochemicals and vice versa (Liu et al. 2003; Mafeo and Mashela 2010; Pelinganga and Mashela 2012). In CARD model, as the mean sensitivity ($\sum$k/n) values increase, coefficients of determination (R^2) also increase to a peak, where $k=$ i, followed by decreases from $i+1$ transformations until the model stops running (Liu et al. 2003). The three DDG patterns and the selected biological indices for nemarioc-AL phytonematicide on tomato plants were illustrated for various potential purposes (Fig. 7.4).

In both nemafric-BL and nemarioc-AL phytonematicides, MCSR values were established as being equivalent to 3 % concentration (Pelinganga et al. 2013b). In other words, for every 3 L stock solution of nemafric-BL or nemarioc-AL phytonematicides, 100 L chlorine-free water is used for application through drip irrigation. After empirically determining the amount to be applied per irrigation, the next step is to determine the application interval, which allows the computation of the application frequency—a factor required in the computation of dosage (D) = MCSR × application frequency.

7.7.2 Determining Phytonematicide Application Interval

The application interval (T) in days for the derived MCSR cannot be established using the CARD model since the latter is exclusively used when the x-axis represents increasing concentration of allelochemicals (Liu et al. 2003). The concept 'weeks-of-30-day-month' for the x-axis was developed for *Meloidogyne* species, where the x-axis was equivalent to 0, 1, 2, 3 and 4 'weeks-of-30-day-month' (Pelinganga and Mashela 2012; Pelinganga et al. 2013b). The unit 'weeks-of-30-day-month' was developed to enhance the capability of a phytonematicide to break the life cycle of *Meloidogyne* species since their life cycles under optimum conditions in tropical and subtropical areas is approximately 30 days. In *T. semipenetrans* with the life cycle of approximately 42 days, the unit would be

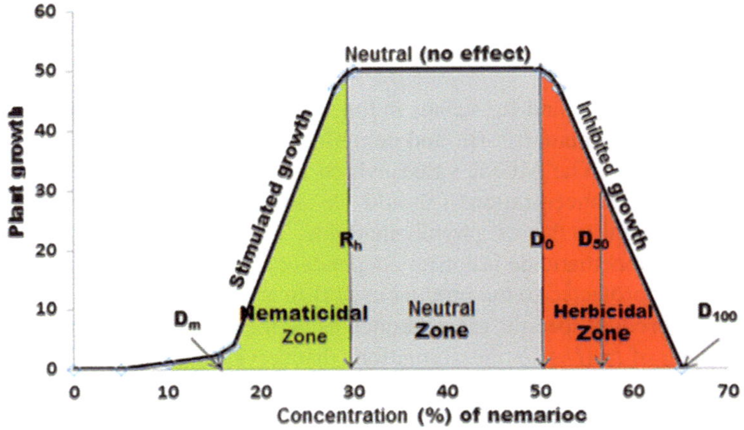

Fig. 7.4 Application of concentration model

'weeks-of-42-day-month'. Since empirical information is required to establish the application interval, experiments are usually established with each tomato seedling inoculated with 5000 eggs and J2s of *M. incognita* under greenhouse conditions (Pelinganga and Mashela 2012). Nematode population densities were managed using the empirically established MCSR value of 3 % for nemafric-BL at 0-day (untreated control), 7.5-day (1 week × 30 days/4 weeks), 15-day (2 weeks × 30 days/4 weeks), 22.5-day (3 weeks × 30 days/4 weeks) and 30-day (4 weeks × 30 days/4 weeks) application interval. At 56 days after the treatment, plant variables (y-axis) are then subjected to ANOVA, with significantly ($P \leq 0.05$) different treatment means being subjected to lines of the best fit to generate the quadratic relationships ($Y = b_2x^2 + b_1x + a$), where the optimum application interval was determined using $x = -b_2/2b_1$ in weeks (Table 7.5). In nemafric-BL 3 % and nemarioc-AL 3 %, the application intervals were 18 and 16 days, respectively (Pelinganga et al. 2012, 2013a). Doubling the concentration from 3 % to 6 % concentration had negligent effect on application interval of nemarioc-AL phytonematicide, but increased that of nemafric-BL from 18 days (2.40 weeks × 30 days/4 weeks) to 20 days (Pelinganga et al. 2013a).

In the use of nematicides applied into the soil, the concept of dosage is important and should be distinguished from dose and concentration (Van Gundy and McKenry 1975). Dose is an amount of chemical taken up by the target pest to effect detrimental behavioural changes, which may include disruption of juvenile development in eggs, egress, disoriented motility and/or mortality in nematodes (Van Gundy and McKenry 1975). In contrast, dosage (D) is the product of concentration (C) and the application frequency (T_{ca}), which could be summarised as:

$$D(\%) = C\ (\%) \times T_{ca}$$

The T_{ca} is the proportion of the crop cycle (days) to the application interval (days),

Table 7.5 Optimum application interval of nemarioc-AL phytonematicide at 3 % concentration on tomato seedlings

Variable	Quadratic relation	R^2	x
Dry root mass (g)	$Y = -0.3838x^2 + 1.5878x + 6.901$	0.92	2.07
Dry shoot mass (g)	$Y = -1.3405x^2 + 5.0903x + 44.374$	0.64	2.82
Dry fruit mass (g)	$Y = -0.8833x^2 + 4.9781x + 17.914$	0.65	1.90
Plant height (cm)	$Y = -0.7202x^2 + 3.6208x + 57.275$	0.88	2.51
Stem diameter (mm)	$Y = -0.3427x^2 + 1.1775x + 12.945$	0.65	1.72
			2.20

with the factor being unit-less. For instance, at 56 days under greenhouse or microplot conditions, T_{ca} values for nemafric-BL 3 % and nemafric-AL 3 % were 3.11 and 3.50, respectively. The model is primarily for seasonal crops, but can also be adapted for perennial crops since nematode population dynamics for various crops, particularly in citriculture, are well established (Duncan 2009).

7.8 Soil Allelochemical Residual Effects

The soil allelochemical residue (SAR) effects investigate post-application effects of phytonematicides on various successor crops and nematode population densities. Increasing the concentration and shortening the application intervals inherently increase the dosage in the soil and, therefore, might defeat the purpose of establishing the MCSR and T_{ca} values which are intended to ameliorate phytotoxicity. Doubling the concentration of phytonematicides may have negligent effects on the application interval, but serious consequences on dosage (Pelinganga et al. 2013a) and, thereby, SAR effects. The SAR effects of phytonematicides from *Cucumis* species were shown to have inhibitive effects of nodulation in *B. japonicum* (Mashela and Dube 2014) while having stimulation effects on growth of sweet-stem sorghum (Mashela 2014). In both cases, SAR effects reduced population nematode densities of *Meloidogyne* species. Additional work is still being under way to understand the chemistry of the SAR effects from phytonematicides.

7.9 Conclusion

Higher plants provide a broad spectrum of active ingredients for use in the management of plant-parasitic nematodes, with their principal drawback being phytotoxicity since their active ingredients comprise allelochemicals. The development of a phytonematicide where phytotoxicity is to be avoided consists of a series of steps. Firstly, there is need to establish whether the plant organ intended for use as a phytonematicide has the potential to reduce population nematode densities under in vitro and/or ex vitro conditions. Secondly, a series of concentrations with known bioactivity effects on the target nematode under greenhouse conditions are used to establish the MCSR value, which is a non-phytotoxic concentration to the crop which is to be protected against nematodes. Thirdly, the MCSR is used to establish the application interval (days), which should be based on the unit that would allow the product to interrupt the life cycle of the target nematode. Using the proposed procedures, commercial phytonematicides could be a reality in the management of plant-parasitic nematodes.

Acknowledgements The authors are grateful to the Land Bank Chair of Agriculture—University of Limpopo, the Flemish Interuniversity Council (VLIR) and the Agriculture Research Council of South Africa for funding the Green Technologies Research Programme at the University of Limpopo.

References

Agbenin NO, Emechebe AM, Marley PS, Akpa AD (2005) Evaluation of nematicidal action of some botanicals on *Meloidogyne incognita in vivo* and *in vitro*. J Agric Rural Dev Trop Subtrop 106:29–39

Alam MM (1993) Bioactivity against phytonematodes. In: Parmar BS, Randhawa NS (eds) Neem research and development. Society of Pesticide Science, New Delhi

Ahmad F, Siddiqui MA, Babalola OO (2013) Characterization of nematicidal activity of plant residues and their application with moisture approach against *Meloidogyne incognita* in tomato. Afr J Agric Res 8:93–101

Akhtar M, Malik A (2000) Role of organic soil amendments and soil organisms in the biological control of plant-parasitic nematodes: a review. Bioresour Technol 74:35–47

Bainard LD, Isman MB, Upadhyaya MK (2006) Phytotoxicity of clove oil and its primary constituent eugenol and the role of leaf epicuticular wax in the susceptibility to these essential oils. Weed Sci 54:833–837

Bargmann CI, Mori I (1997) Chemotaxis and thermotaxis. In: Blumenthal T, Meyer BJ, Priess JR, Riddle DL (eds) C. elegans II. Cold Spring Harbour Laboratory, New York, pp 717–737

Bird AF, Bird J (1991) The structure of nematodes, 1st edn. Academic, San Diego

Boyd NS, Brennan EB (2006) Burning nettle, common purslane, and rye response to a clove oil herbicide. Weed Technol 20:646–650

Boyd NS, Brennan EB, Fennimore SA (2006) Stale seedbed techniques for organic vegetable production. Weed Technol 20:1052–1057

Belair M, Tremblay N (1995) The influence of chitin-urea amendments applied to an organic soil on a *Meloidogyne hapla* population and on the growth of greenhouse tomato. Phytoprotection 76:75–80

Campbell NA (1990) Biology, 1st edn. Cummings, Redwood City

Castagnone-Sereno P, Kermarrec A (1991) Invasion of tomato roots and reproduction of *Meloidogyne incognita* as affected by raw sewage sludge. Suppl J Nematol 23:724–728

Chedekal AN (2013) Effect of four leaf extracts on egg hatching and juvenile mortality of root-knot nematode *Meloidogyne incognita*. Int J Adv Life Sci 6:68–74

Chitwood DJ (2002) Phytochemical based strategies for nematode control. Annu Rev Phytopathol 40:221–249

D'Addabbo T (1995) The nematicidal effect of organic amendments: review of the literature 1982–1994. Nematol Mediterr 23:299–305
Dhanalakshmi DN (2006) Studies on storability of indigenous materials and their utilization on okra sucking pests. M.Sc. (Agri.) dissertation, University of Agricultural Sciences, Dharwad
Dropkin VH (1969) The necrotic reaction of tomato and other hosts resistant to *Meloidogyne:* reversal by temperature. Phytopathology 59:1632–1637
Duncan LW (2009) Managing nematodes in citrus orchards. In: Ciancio A, Mukerji KG (eds) Integrated management of fruit crops and forest nematodes. Springer, Milton Keynes, pp 135–173
Duncan LW, McSorley R (1987) Modeling nematode population. In: Veetch JA, Dickson DW (eds) Vistas on nematology: a commemoration of the twenty- fifth anniversary of the society of nematologist. Society of Nematologists, Hyattsville, pp 377–389
Egunjobi OA, Afolami SO (1976) Effects of neem (*Azadirachta indica*) leaf extracts on population of *Pratylenchus brachyurus* and on the growth and yield of maize. Nematologica 22:125–132
EPPO (2010) Efficacy of evaluation of plant protection products: phytotoxicity assessment. European and mediterranean plant protection organisation. EPPO PP1/135(3)
Ferraz ICCB, Brown DJF (2002) An introduction to nematodes: plant nematology, 1st edn. Pensoft, Sofia-Moscow
Ferraz S, de Freitas LG (2004) Use of antagonistic plants and natural products. In: Chen ZX, Chen SY, Dickson DW (eds) Nematology advances and perspectives. Tsinghua University Press, Beijing
Ferris H, Wilson LT (1987) Concepts and principles of population dynamics. In: Veetch JA, Dickson WD (eds) Vistas on nematology: a commemoration of the twenty-fifth anniversary of the society of nematologist. Society of Nematologists, Hyattsville, pp 372–376
Haseeb A, Singh B, Khan AM, Saxena SK (1980) Effect of watering and mode of application of oil cakes and nematodes on their efficiency in controlling root- knot nematode on tomato. Acta Bot Indica 8:193–195
Huang JS (1987) Interactions of nematodes with *Rhizobia*. In: Veetch JA, Dickson WD, Soc N (eds) Vistas on nematology: a commemoration of the twenty-fifth anniversary of the society of nematologist. Society of Nematologist, Hyattsville, pp 301–306
Hirschmann H (1985) The genus *Meloidogyne* and morphological characters differentiating its species. In: Sasser JN, Carter CC (eds) An advanced treatise on *Meloidogyne*, vol I, Biology and control. North Carolina State University Graphics, Raleigh
Inderjit K, Asakawa C, Dakshini KMM (1999) Allelopathic potential of *Verbesina encelioides* root leachate in soil. Can J Bot 77:1419–1424
Jafee BA, Ferris H, Stapleton JJ, Norton MVK, Muldoon AE (1994) Parasitism of nematodes by the fungus *Hirsutella rhossiliensis* as affected by certain organic amendments. J Nematol 26:152–161
Jeffrey C (1978) Cucurbitaceae. In: Launert E (ed), Flora Zambezi, 1st edn. London
Kimpinski J, Gallant CF, Henry R, Macleod JA, Sanderson JB, Sturz AV (2003) Effect of compost and manure soil amendments on nematodes and yields of potato and barley: a 7-year study. J Nematol 35:289–293
Khan IA, Sayed M, Shaukat SS, Handoo ZA (2008) Efficacy of four plant extracts on nematodes associated with papaya in Sindh, Pakistan. Nematol Mediterr 36:93–98
Kyan T, Shintani M, Kanda S, Sakurai M, Ohashi H, Fujisawa A, Pongdit S (1999) Kyusei nature farming and the technology of effective microorganisms. Asia Pacific Natural Agriculture Network, Bangkok
Lai E (2008) Secondary metabolites and plant defense. In: Taiz L, Zeiger E (eds) A companion to plant physiology, 1st edn. I.K. International, New Delhi
Lee DH, Lwanski GB, Thoennissen NH (2010) Cucurbitacin: ancient compound shedding new light on cancer treatment. Sci World J 10:413–418

Liu DL, An M, Johnson IR, Lovett JV (2003) Mathematical modeling of allelopathy. III. A model for curve-fitting allelochemical dose responses. Nonlinear Biol Toxicol Medic 1:37–50

Mafeo TP, Mashela PW, Mphosi MS (2011a) Allelopathic responses of various seeds to crude extracts of *Cucumis myriocarpus* fruits when used as a pre- emergent bio-nematicide. Acta Hortic (ISHS) 938:409–414

Mafeo TP, Mashela PW, Mphosi MS (2011b) Sensitivity of selected Alliaceae seedlings to crude extracts of *Cucumis myriocarpus* fruits. Afr J Agric Res 6:158–164

Mafeo TP, Mashela PW, Mphosi MS, Pofu KM (2011c) Modelling responses of maize, millet and sorghum seedlings to crude extracts of *Cucumis myriocarpus* fruit as pre-emergent bio-nematicide. Afr J Agric Res 6:3678–3684

Mafeo TP, Mashela PW (2010) Allelopathic inhibition of seedling emergence in dicotyledonous crops by *Cucumis* bio-nematicide. Afr J Biotechnol 9:8349–8354

Mafeo TP, Mashela PW (2009a) Responses of germination in tomato, watermelon and butternut squash to a *Cucumis* bio-nematicide. J Agric Environ Sci 6:215–219

Mafeo TP, Mashela PW (2009b) Responses of four monocotyledonous crops to crude extracts of *Cucumis myriocarpus* fruit as a pre-emergent bio- nematicide. Afr Crop Sci Proc 9:631–634

Maila KD, Mashela PW (2013) Responses of the citrus nematode (*Tylenchulus semipenetrans)* to nemarioc-A phytonematicide with and without effective micro-organisms in citrus production. Afr Crop Sci Soc Conf 11:177 (Abst.)

Makkar HPS (1991) Quantification of tannins in tree foliage. IAEA working document. IAEA, Viena

Mashela PW, Duncan LW, McSorley R (1992) Salinity reduces resistance to *Tylenchulus semipenetrans* in citrus rootstocks. Nematropica 22:7–12

Mashela PW (2002) Ground wild cucumber fruits suppress numbers of *Meloidogyne incognita* on tomato in micro-plots. Nematropica 32:13–19

Mashela PW, Nthangeni ME (2002) Efficacy of *Ricinus communis* fruit meal with and without *Bacillus* species on suppression of *Meloidogyne incognita* and growth of tomato. J Phytopathol 150:399–402

Mashela PW, Shimelis HA, Mudau FN (2008) Comparison of the efficacy of ground wild cucumber fruits, aldicarb and fenamiphos on suppression of the root-knot nematode in tomato. J Phytopathol 156:264–267

Mashela PW, Shimelis HA, De Waele D, Mokgalong MN, Mudau FN, Ngobeni LG (2010) Fevertea (Lippia javanica) as a root-knot nematode suppressant in tomato production. Afr Plant Prot 6:1–6

Mashela PW, Pofu KM, Nzanza B (2012) Suitability of *Brassica oleracea* leaves in managing *Meloidogyne incognita* through the ground leaching technology system under microplot conditions. Acta Agric Scand 63:19–24

Mashela PW, De Waele D, Pofu KM (2011) Use of indigenous *Cucumis* technologies as alternative to synthetic nematicides in management of root- knot nematodes in low-input agricultural farming systems: a review. Sci Res Essay 6:6762–6768

Mashela PW (2014) Soil allelochemical residue effects in a tomato cowpea rotation – nodulation and productivity of cowpea and nematode suppression. Acta Agric Scand 64:372–375

Mashela PW, Dube ZP (2014) Soil allelochemical residue effects of nemafric-BL and nemarioc-AL phytonematicides on soil health, growth of sweet sorghum and *Meloidogyne* species. Acta Agric Scand 64:79–84

Mashela PW, Pofu KM, Dube ZP (2014) Managing phytotoxicities in phytonematicides: the dosage model. In: 6th international conference of nematology, pp 69–70

McSorley R (2003) Adaptations of nematodes to environmental extremes. Florida Entomol 86:138–142

McSorley R (2011) Overview of organic amendments for management of plant- parasitic nematodes, with case studies from Florida. J Nematol 43:69–81

McSorley R, Gallaher RN (1995) Cultural practices improve crop tolerance to nematodes. Nematropica 25:53–60

Meyer SLF, Lakshman DK, Zasada IA, Vinyard BT, Chitwood DJ (2008) Phytotoxicity of clove oil to vegetable crop seedings and nematotoxicity to root-knot nematodes. Horti Technol 18:631–638
Mian IH, Rodriguez-Kabana R (1982a) Soil amendments with oilcakes and chicken litter for control of *Meloidogyne arenaria*. Nematropica 12:205–220
Mian IH, Rodriguez-Kabana R (1982b) Survey of the nematicidal properties of some organic materials available in Alabama as amendments to soil for control of *Meloidogyne arenaria*. Nematropica 12:235–246
Mudau FN, Mogotlane ID, Mashela PW, Soundy P (2008) Seasonal variation of total antioxidant contents of wild bush tea. Acta Hortic (ISHS) 802:273–275
Mukherjee SN, Sukul NC (1978) Nematicidal action of three species of wild herb. Indian J Res 2:12
Muller R, Gooch PS (1982) Organic amendments in nematode control: an examination of the literature. Nematropica 12:319–326
Musabyimana T, Saxena RC, Kairu EW, Ogol KZR (2000) Powdered neem seed and cake for management of banana weevil and parasitic nematodes. Phytoparasitica 28:321–330
Ncube L (2008) Evaluation of effective micro-organisms (EM) on soil chemical properties and yield of selected vegetables in the Eastern Cape, South Africa. MSc dissertation, University of Fort Hare, Alice, South Africa
Nicholson AJ (1933) The balance of animal populations. J Anim Ecol 2:132–178
Ntalli NG, Manconi F, Leonti M, Maxia A, Caboni P (2011) Aliphatic ketones from *Ruta chalepensis* (Rutaceae) induce paralysis on root-knot nematodes. J Agric Food Chem 59:7098–7103
Ntuli NG, Caboni P (2012) Botanical nematicides: a review. Agric Food Chem 60:9929–9940
Nzanza B, Mashela PW (2012) Control of whiteflies and aphids in tomato by fermented plant extracts of neem leaf and wild garlic. Afr J Biotechnol 11:16077–16082
Ojo GT, Umar I (2013) Evaluation of some botanicals on root-knot nematode (*Meloidogyne javanica*) in tomato (*Lycopersicon esculentum* Mill) in Yola Adamawa State, Nigeria. Biol Forum 5:31–36
Oka Y (2010) Mechanisms of nematode suppression by organic soil amendments: a review. Appl Soil Ecol 44:101–115
Oka Y, Nacar S, Putievsky E, Ravid U, Yaniv Z, Spiegel Y (2000) Nematicidal activity of essential oils and their components against the root-knot nematode. Phytopathology 90:710–715
Okeniyi MO, Afolami SO, Fademi OA, Oduwaye OF (2013) Effect of botanical extracts on root-knot nematode (*Meloidogyne incognita*) infection and growth of cashew (*Anacardium occidentale*) seedlings. Acad J Biotechnol 1:81–86
Okwute SK (2012) Plants as potential sources of pesticidal agents: a review. In: Soundararajan RP (ed) Pesticides: advances in chemical and botanical pesticides. In Tech, pp 208–232
Papadopoulou J, Triantaphyllou AC (1982) Sex differentiation in *Meloidogyne incognita* and anatomical evidence of sex reversal. J Nematol 14:549–566
Parmar BS (1987) An overview of neem research and use in India during the years 1983–1986. In: Proceedings of the 3rd international neem conference abstract, Nigeria
Parmar BS (1996) An overview of neem research and use in India during the years 1983–1986. In: Schmutter H, Ascher KRS (eds) Natural pesticides from the neem tree and other tropical plants, vol 3, Proceedings of the international neem conference. GTZ, Eschborn, pp 50–80
Pelinganga OM (2013) Developing bio-pesticides using indigenous *Cucumis africanus* and *Cucumis myriocarpus* fruits for tomato production systems. PhD thesis, University of Limpopo, Sovenga
Pelinganga OM, Mashela PW (2012) Mean dosage stimulation range of allelochemicals from crude extracts of *Cucumis africanus* fruit for improving growth of tomato plant and suppressing *Meloidogyne incognita* numbers. J Agric Sci 4:8–12
Pelinganga OM, Mashela PW, Mphosi MS, Mafeo TP (2013a) Using computer- based model to determine phytotoxicity concentration of nemarioc-A phytonematicide in tomato production. Afr Crop Sci Soc Conf (Abst) 11:168

Pelinganga OM, Mashela PW, Mphosi MS, Nzanza B (2013b) Optimizing application frequency of diluted (3%) fermented *Cucumis africanus* fruit in tomato production and nematode management. Acta Agric Scand 63(3):278–282

Pelinganga OM, Mashela PW, Nzanza B, Mphosi MS (2012) Baseline information on using fermented crude extracts from *Cucumis africanus* fruit for suppression of *Meloidogyne incognita* and improving growth of tomato plant. Afr J Biotechnol 11:11407–11413

Pofu KM, Mashela PW, Shimelis H (2012) Intergeneric grafting in watermelon for managing *Meloidogyne* species: a review. Sci Res Essay 7:107–113

Pofu KM, Mashela PW (2014) Density-dependent growth patterns of *Meloidogyne javanica* on hemp cultivars: establishing nematode-sampling timeframes in host-status trials. Am J Exp Agric 4:639–650

Prot JC (1980) Migration of plant-parasitic nematodes towards plant roots. Rev Nematol 3:305–318

Rossner J, Zebitz CPW (1987) Effect of neem products on nematodes and growth of tomato (*Lycopersicon esculentum*) plants. In: Schmutterer H, Ascher KRS (eds) Naturalpesticides from the neem tree (*Azadirachta indica* A. Juss) andothertropical plants, vol 3, Proceedings of the international neem conference. GTZ, Eschborn, pp 611–621

Rice EL (1984) Allelopathy, 1st edn. Academic, New York

Rimington P (1938) Medicinal and poisonous plants of South and East Africa. University of Natal Press, Pietermaritzburg

Salisbury FB, Ross CW (1992) Plant physiology, 2nd edn. Wadsworth, Belmont

Seinhorst JW (1967) The relationship between population increase and population density in plant-parasitic nematodes. 3. Definition of the terms host, host- status and resistance. 4. The influence of external conditions on the regulation of population density. Nematologica 13:429–442

Shaukat SS, Siddiqui IA, Ali NI (2003) Nematicidal and allelopathic responses of *Lantana camara* root extract. Phytopathol Mediterr 42:71–78

Smith GS (1987) Interactions of nematodes with mycorrhizal fungi. In: Veetch JA, Dickson DW (eds) Vistas on nematology: a commemoration of the twenty- fifth anniversary of the society of nematologist. Society of Nematologist, Hyattsville, pp 292–300

Stirling GR (2014) Biological control of plant parasitic nematodes: soil ecosystem management in sustainable agriculture, 2nd edn. CAB International, Wallingford

Sukul NC, Das PK, De GC (1974) Nematicidal action of edible crops. Nematologica 20:187–191

Thoden TC, Korthals GW, Termorshuizen AJ (2011) Organic amendments and their influences on plant-parasitic and free-living nematodes: a promising method for nematode management? Nematology 13:133–153

Toida Y, Moriyama H (1978) Effects of the Marigold on the control of nematodes in Mulberry. 2. Nematicidal effects of the Mexican marigold by application. Helminthol B 49:696, Abstr

Triantaphyllou AC (1973) Environmental sex differentiation of nematodes in relation to pest management. Annu Rev Phytopathol 11:441–462

Tworkoski T (2002) Herbicide effects of essential oils. Weed Sci 50:42–431

Usman A (2013) Studies on the integrated management of phytonematodes attacking some vegetable crops. PhD thesis, Aligarh Muslim University, Aligarh

Van Gundy SD, McKenry MV (1975) Action of nematicides. In: Horsefall JG, Cowling EB (eds) Plant disease: an advanced treatise, Ith edn, How diseaseis managed. Academic, New York

Van Wyk B, Heerden F, Oudshoorn B (2002) Poisonous plants of South Africa. Briza, Pretoria

Vijayalakshmi K, Subhashini B, Shivani VK (1996) Plant in pest control using garlic and onion. Centre for Indian Knowledge System, Chennai

Wallace HR (1973) Nematode ecology and plant disease, 1st edn. Edward Arnold, London

Waliwitiya R, Isman MB, Vernon RS, Riseman A (2005) Insecticidal activity of selected monoterpenoids and rosemary oil to *Agriotes obscurus* (Coleoptera: Elateridae). J Econ Entomol 98:1560–1565

Walter M, Jaspers MV, Eade K, Frampton CM, Stewart A (2001) Control of *Botrytis cinerea* in grape using thyme oil. Austr Plant Pathol 30:21–25

Wuyts N, Swennen R, De Waele D (2006) Effects of plant phenylpropanoid pathway products and selected terpenoids and alkaloids on the behaviour of the plant parasitic nematodes *Radopholus similis, Pratylenchus penetrans* and *Meloidogyne incognita*. Nematology 8:89–101

Zasada IA, Ferris H (2003) Sensitivity of *Meloidogyne javanica* and *Tylenchulus semipenetrans* to isothiocyanates in laboratory assays. Phytopathology 93:747–750

Chapter 8
Suppressiveness in Different Soils for *Rhizoctonia solani*

Silvana Pompeia Val-Moraes

8.1 Introduction

Often a limiting factor in conventional crop production management, suppressiveness may change soil characteristics or alter the incidence of diseases caused by soilborne pathogens. Generally, while these diseases are rare in undisturbed natural ecosystems, they can be severe in conventional production systems and often become a limiting factor. Agricultural soils suppressive to soilborne plant pathogens occur worldwide, and this characteristic results from both biotic and abiotic factors, in a variety of intricate mechanisms. Since soil becomes suppressive to target pathogens, determination of its main physical, chemical, and biological attributes can be useful for comprehension of the mechanisms of suppressiveness and to reveal information in other areas where the same pathogen is a problem. Despite the importance of soil microbial communities in regulating soil ecosystem-level processes, such as the nutrient cycle and organic matter decomposition, little is known about the structure of these microbial communities and the factors that influence them in soils (Val-Moraes et al. 2013). Soil quality is considered an integrative indicator of environmental quality, food security, and economic viability. Thus, soil itself serves as a potential indicator for monitoring sustainable land management; a healthy soil supports high levels of activity, internal nutrient cycling, resilience to disturbance, and biological diversity (Sharma et al. 2011). Handling soil microbial communities using soil and crop management practices is a basic strategy in developing sustainable agricultural systems (Van Bruggen 1995). It is known that a range of specific soil microorganisms play an important role in

S.P. Val-Moraes (✉)
UNESP, Univ. Estadual Paulista, Department Technology – Biochemistry of Microorganisms and plants Laboratory (LBMP), Via de Acesso Prof. Paulo Donato Castellane, s/n, 14884-900 Jaboticabal, São Paulo, Brazil
e-mail: valmoraes.silvana@gmail.com

M.K. Meghvansi, A. Varma (eds.), *Organic Amendments and Soil Suppressiveness in Plant Disease Management*, Soil Biology 46,
DOI 10.1007/978-3-319-23075-7_8

suppressing soilborne plant diseases, as well as in plant growth promotion (Kennedy and Smith 1995). Soil microbial diversity is reduced, and eventually an increase in crop diseases occurs in conventional agricultural systems with low crop diversity, increased genetic uniformity, and high mineral nutrient inputs, which results in the need for chemicals to control plant diseases and pests, which may cause environmental pollution (Sturz and Christie 2003). However, in natural ecosystems with diverse vegetation types, soilborne diseases are rarely observed. Soil suppressiveness is regularly described for *Rhizoctonia solani*, which is of high importance among soilborne, plant pathogens, and damages a large number of hosts worldwide. The overarching aim was the focus on recent progress toward unraveling the microbial basis of suppressive soils.

8.2 Characteristics of Suppressive Soils

Few soils with experimentally demonstrated natural suppressiveness to soilborne plant pathogenic fungi were found in nature. Suppressive soils are common in ecologically balanced environments with ecosystems in climax, where the physicochemical and microbiological constituents of the soil are stabilized (Schneider 1982). Soil microbe and phytopathogen interactions can occur before crop sowing and/or in the rhizosphere, later influencing both plant growth and productivity (Penton et al. 2014). Dobbs and Hinson (1953) first described this phenomenon, referred to as widespread soil fungistasis. Healthy soils are essential to an ecosystem's ability to remain intact or to recover from disturbances such as drought, climate change, pest infestation, pollution, and human exploitation including agriculture (Ellert et al. 1997). Research on suppressive microbial communities has concentrated on bacteria, although fungi can also influence soilborne disease (Penton et al. 2014). The development of biological disease suppression in soil supporting monocultures or continuous cropping is a widespread natural phenomenon, yet the microbial mechanisms are often poorly understood (Berendsen et al. 2012). However, general suppression is often enhanced by addition of organic matter, certain agronomic practices, or the buildup of soil fertility (Rovira and Wildermuth 1981), all of which can increase soil microbial activity. No one microorganism is responsible for general suppression (Alabouvette 1986; Cook and Baker 1983), and the suppressiveness is not transferable between soils (Cook and Baker 1983; Rovira and Wildermuth 1981). Suppressive soils undoubtedly owe their activity to a combination of general and specific suppression. The two function as a continuum in the soil, although they can cause different effects in edaphic, climatic, and agronomic conditions (Rovira and Wildermuth 1981). In most of the cases, adding mature compost to a soil induces disease resistance (Sullivan 2004). Suppressive soils also have been differentiated according to their longevity (Kumar et al. 2012). Hornby (1983) divided suppressive soils into long-standing suppression and induced suppression. The former type of suppression is a biological condition naturally associated with the soil; its origin is not known and appears to survive in the absence of plants. Most suppressive soils maintain their activity when brought into the greenhouse or laboratory, which facilitates the assessment of their properties and mechanisms of suppression under more controlled and reproducible conditions (Kumar et al. 2012). The first step is to determine whether suppressiveness can be destroyed by pasteurization (moist heat, 60 °C for 30 min) (Shipton et al. 1973), by using selective biocides (e.g., novobiocin or chloropicrin), or by harsher treatments (e.g., steam, methyl bromide, autoclaving, or gamma radiation) (Wiseman et al. 1996; Weller et al. 2002). Both general and specific suppressions are eliminated by autoclaving and gamma radiation. General suppression is reduced but not eliminated by soil fumigation and usually survives at 70 °C moist heat (Cook and Rovira 1976). A second step, which allows confirmation of the biological basis of suppression, involves transfer of suppressiveness to a raw conducive, fumigated, or sterilized soil by addition of 0.1–10 % (w/w) or less of the suppressive soil. The impact of soil edaphic factors on disease development in soil transfer studies is minimized when suppressive and conducive soils are diluted into a common background soil, allowing a direct comparison of the introduced microbiological components. Composts have been used for centuries to maintain soil fertility and plant health. Hoitink (2004) reported the control of phytopathogens with composts, which indicates their disease suppressive nature. Bent et al. (2008) reported 5- to 16- fold reductions in population of root-knot nematode as compared to identical but pasteurized soil 2 months after infestation. Since the earliest observations of antagonistic disease suppressing soil microorganisms more than 70 years ago, plant pathologists have been fascinated by the idea that such microorganisms could be used as environmental friendly biocontrol agents, both in the field and in greenhouses (Kumar et al. 2012). Penton et al. (2014), in studies of the agricultural fields located in the wheat-cropping region in South Australia, which have been under continuous cropping for more than 10 years, observed that differences in pathogen inoculum levels between suppressive and non-suppressive soils were small, indicating similar pathogen pressure.

8.3 *Rhizoctonia*

In 1858, Julius Kühn, the founder of agricultural phyto-medicine, called a fungus *Rhizoctonia solani*, which he had isolated from infected potatoes, the root killer.

Indeed, soilborne plant pathogenic fungi seem to be more difficult to control, probably because their epidemiological statuses and ecological requirements are yet to be characterized. Among the soilborne fungi, some directly produce damage, such as *R. solani*, *R. violacea*, *Phytophthora drechsleri*, *Pythiumaphani dermatum*, and *Rhizopus arrhizus*, these being responsible at various extents for severe root rotting of the tubers or damping-off of seedlings. Others like *Polymyxa betae* have an indirect role by transmitting the beet necrotic yellow vein virus (BNYVV) responsible for rhizomania disease (Liu and Lewellen 2007). However, the soilborne microorganism, *R. solani*, is the potential threat to the farmers cultivating sugar beet. *R. solani* is a phytopathogenic fungus which is present in the soil in very low densities, but is able to cause disease in different plant species because in soil it is saprotrophic and a facultative parasite. This species is complex and includes 14 anastomosis groups (AGs) which in turn include many different subgroups. The former are characterized by the ability of their members to anastomose within a group, while the latter through various biochemical, nutritional, molecular, or phenotypic traits. However, apart from AG B which includes fungi that establish a pseudo-symbiotic association with orchids, the other AG includes plant pathogenic members which all have very broad host spectra. AG 3 appears to have the narrowest host spectrum and it is responsible for diseases mainly in plants of the *solanaceous* family. The other AG can attack plants from various botanic families, and similarly, one plant can be attacked by various AGs of *R. solani*. In the case of sugar beet, AG 4 of *R. solani* and to a lesser extent AG 2-2 can cause damping-off of seedlings, while necroses and root rotting of adult tubers are due exclusively to AG 2-2. This subgroup can also cause damages in carrots (Janvier et al. 2006), tomato (de Gurfinkel et al. 1994), or pine (Guillemaut 2003). Moreover, within this subgroup AG 2-2, three populations of *R. solani* have been identified: the population AG 2-2IIIB occurs mainly in northern European countries (the Netherlands, Germany), while AG 2-2IV concerns southern European countries (Spain, France). The third one, AG 2-2LP, concerns mainly bulbs (Guillemaut 2003). The first one causes more severe and noticeable damage in corn in the Netherlands than the second one in France. However, corn in France can harbor and allow the development of *R. solani*, causing a strong primary inoculum toward the forthcoming susceptible crop of sugar beet. The damage caused by this disease is variable and may lead to complete loss of the yield. The control of this disease is the main problem because the chemical control is not environmentally friendly and there is only partial genetic resistance available against the disease, which is not sufficient for the control. Even long rotation without sugar beet does not guarantee the complete control of the disease because of two main reasons: firstly, *R. solani* has a broad host spectrum and can survive on the intermediary cultures and weeds, and secondly this fungus has the ability to survive by making sclerotia. The biological control has not been consistently successful against this pathogen, but the potential antagonistic activity of soilborne microflora has recently been assessed (Zachow et al. 2008). Therefore, there is a need for a pioneering research approach to find new methods to control this devastating pathogen. A high between-season mobility of patches was observed when sugar beet was monocropped (Hyakumachi 1996). The patches never occurred at the same place where they were observed in the previous season. In the preliminary studies, higher suppression toward the disease caused by *R. solani* AG 2-2 was observed in the soil from within the disease patches than in the soil from healthy areas in the same laboratory (Guillemaut 2003). The resulting hypothesis is that the increased suppressiveness inside the diseased patches may be due to the accumulation of the antagonistic microflora against *R. solani* AG 2-2. This accumulation of antagonistic microorganisms and higher suppression may explain the patch mobility between seasons.

8.4 Hosts and Geographic Distribution

The pathogen has a great number of host species and has been found in seeds of *Brassica* spp. (broccoli, Brussels sprouts, cabbage, Chinese cabbage, kale, kohlrabi, mustard, rutabaga, turnip), *Capsicum* spp. (peppers), *Citrus* spp. (lemon, sweet lemon, pomelo, key lime, and more), *Gossypium* spp. (cotton), *Lycopersicon esculentum* (tomato), *Phaseolus* spp. (bean, string bean, field bean, flageolet bean, French bean, garden bean, haricot bean, pop bean, or snap bean), *Spinacia oleracea* (spinach), *Vigna unguiculata* (yard-long bean, bora, bodi, long-podded cowpea, asparagus bean, pea bean, snake bean, or Chinese long bean), *Zea mays* (maize), and *Zinnia elegans* (common zinnia, youth-and-old-age) (Neergaard 1977).

In Latin America, *R. solani* occurs in Mexico, all countries of Central America, and the Caribbean and in South America in the Amazon region of Peru and Brazil, the coffee zone of Colombia, and the northwestern region of Argentina. Other countries that have described the disease are the USA, Japan, the Philippines, Burma, and Sri Lanka and as a minor pathogen in Kenya and Malawi (Gálvez et al. 1989).

8.5 Biology and Transmission

In nature, *R. solani* exists as many strains, differing in cultural appearance, physiology, and pathogenicity. Naturally occurring strains or isolates differ in mycelium, growth rate, saprophytic deportment, and enzyme production (Abawi 1989). The teleomorph of *Thanatephorus cucumeris* may occur and form a hymenial layer at the base of plants and/or the underside of soil aggregates during periods of high humidity and rainfall. Basidia are short and barrel shaped with stout straight sterigmata, while basidiospores are smooth, thin walled, and hyaline (Abawi 1989). *R. solani* is a very common soilborne pathogen (Sneh et al. 1991) with a great diversity of host plants (Table 8.1 modified).

8.6 Treatment Versus Control

Because *R. solani* has a mondial distribution, including in uncultivated soils, execution and eradication are usually not effectual field control measures. The fungus can be eradicated from infected greenhouse soil by steaming at 60 °C for 30 min. *R. solani* infection may be reduced by various cultural practices. In Colombia the infection is less severe during the wet rainy season if the beans are planted on raised beds that promote good drainage. Shallow planting minimizes seedling damage so that less seedling tissue is exposed to the inoculum. Seeds planted 7.5 cm deep developed more root rot and hypocotyl injury than seed planted only 2.5 cm deep (Abawi 1989). Continuous planting of beans in the same field

Table 8.1 Diseases arranged by anastomosis groups and host range of *Rhizoctonia solani* (According to Sneh et al. (1991))

Anastomosis group	Diseases	Host
AG 1-IA	"Sheath blight"/"sheath spot"	Rice
	"Sclerotial disease"/"leaf blight"/"banded leaf"	Corn
	"Leaf blight"/"banded leaf"	Sorghum
	"Leaf blight"	Bean, soybean
	"Summer blight"	Crimson clover
	"Southern blight"	Camphor seedlings
	"Brown patch"	Turfgrass
	"Web blight"	Bean, rice, soybean, figs, leguminous woody plants, hortensia
AG 1-IB	"Rot"	Cabbage
	"Bottom rot"	Lettuce
	"Damping-off"	Buckwheat, soybean, flax, pine
	"Damping-off"/"crown root rot"	Carrot
AG 2-1	"Damping-off"	Crucifers
	"Bud rot"	Strawberry
	"Leaf blight"	Tulip
	"Root rot"	Japanese radish, subterranean clover
AG 2-2IIIB	"False sheath blight"	Rice
	"Sheath blight"	Mat rush, ginger, gladiolus
	"Black scurf"	Edible burdock
	"Brown patch"	Turfgrass
	"Crown/brace rot"	Corn
	"Damping-off"	Sugar beet, tree seedlings, chrysanthemum
	"Root rot"	Konjak, Chinese yam
AG 2-2IV	"Root rot"/"leaf blight"	Sugar beet
	"Large patch"	Turfgrass
AG 3	"Black scurf"/"stem/stolon cankers"	Potatoes
	"Target spot"	Tobacco
	"Leaf blight"	Tomato
	"Brown spot"	Eggplant
AG 4 (HG I/HGII/HGIII)	"Fruit rot"	Tomato
	"Stem rot"	Pea
	"Damping-off"/"stem canker"	Potato
	"Damping-off"/"root rots"	Soybean, loblolly pine seedlings, onion, stevias, pea, snap bean, cotton, peanuts, slash
	"Pod rot"	Snap bean

(continued)

Table 8.1 (continued)

Anastomosis group	Diseases	Host
AG 5	"Black scurf"	Potato
	"Brown patch"	Turfgrass
	"Root rot"	Beans, soybeans, adzuki beans
AG 8	"Bare patches"	Cereals
AG 9	"Weak pathogen"	Crucifers, potatoes

increases the inoculum density of *R. solani*. However, crop rotation with non-host crops reduces the incidence of bean root rot even though it does not completely eradicate the pathogen. Fungus populations rapidly decline in soil planted with wheat, oats, barley, or maize. Population levels remain relatively high in soil planted with susceptible bean, pea, or potato plants. An alternative to crop rotation would be the incorporation of selected residues or decomposable material. In addition, many antagonists or mycoparasites such as *Trichoderma* species have effectively reduced activities of *R. solani* when incorporated with organic amendments or applied directly on the seed. Deep plowing is another cultural practice that is effective in reducing surface inoculum of *R. solani* and thus disease incidence. Turning under soil and crop residue to a deep 20–25 cm was found to reduce *Rhizoctonia* root rot on beans for 3 years (Abawi 1989).

Fungicides that are effective against *R. solani* include PCNB (the most commonly used fungicide to control *R. solani*), benomyl, carboxin, Busan 30A, thiram, zineb, chloroneb, and others. These fungicides are commonly applied as seed treatments (1–3 g i.a./kg seed) before or during planting (Abawi 1989). Biological control with *Trichoderma harzianum*, *Bacillus subtilis*, and *B. licheniformis* reduced *Rhizoctonia* root rot. Seed treatment and root drenching with bacterial suspensions with 0.5 % chitin were more effective against *R. solani* in *Capsicum annuum* (Sid et al. 2003) than addition of the organisms without chitin.

8.7 Induction of Suppressiveness

Disease control relies largely on the treatment of preplant soils with broad-spectrum pesticides, such as methyl bromide, that are being phased out of agricultural production (Weller et al. 2002). Soils that have not undergone *Malus domestica* (apple) cultivation are suppressive to replant disease. But, in contrast to the take-all and *Solanum tuberosum* (potato) scab-suppressive soils that are induced by monoculture, orchard soils become progressively more conducive to replant disease the longer the orchard is in production. Mazzola (1999) demonstrated this phenomenon by introducing an inoculum of *R. solani* AG 5 (a member of the replant pathogen complex) (Mazzola 1997) into soils collected from orchard blocks in their first to fifth years of growth and from nearby noncultivated areas. Apple seedling growth was significantly reduced in soils from the third-, fourth-, and fifth-year blocks as compared to growth in noncultivated soil or in soil from first- to second-year blocks.

Healthy soils are suppressive soils; thus, disease suppressiveness can be considered as an indicator of soil health. However, suppressiveness is a complex process that depends on several factors. Moreover, its measure, through pathogen-specific bioassays, if possible, is time and labor intensive. That is why it would be very interesting, and useful, to find other soil characteristics highly related to soil suppressiveness, but easier to measure. This need for indicators of soil health is a real concern, from the field scale to the global level. Therefore, it is necessary to define an exact strategy, from sampling to validation, which would allow for the proposal of indicators (Janvier et al. 2007).

8.8 Conclusion

Although soil quality involves physical and chemical characteristics in addition to biological ones, soil health is primarily an ecological characteristic. Ecosystem health has been defined in terms of ecosystem stability and resilience in response to a disturbance or stress. Accordingly, it is suggested that indicators for soil health could be found by monitoring responses of the soil microbial community to the application of different stress factors at various intensities. Therefore, indicators for soil health could also function as indicators for disease suppressiveness. Disease suppression can be viewed as manifest ecosystem stability and health.

References

Abawi GS (1989) Root rots. In: Schwartz HF, Pastor-Corrales MA (eds) Bean production problems in the tropics, 2nd edn. Centro Intern Agric Tropical (CIAT), Cali, pp 105–157

Alabouvette C (1986) Fusarium wilt suppressive soils from the Chateaurenard region: review of a 10-year study. Agronomie 6:273–284

Bent E, Loffredo A, McKenry MV, Becker JO, Borneman J (2008) Detection and investigation of soil biological activity against Meloidogyne incognita. J Nematol 40(2):109–118

Berendsen RL, Pieterse CMJ, Bakker PAHM (2012) The rhizosphere microbiome and plant health. Trends Plant Sci 17:478–486

Cook RJ, Baker KF (1983) The nature and practice of biological control of plant pathogens. The American Phytopathology Society, St. Paul, MN, 539p

Cook RJ, Rovira AD (1976) The role of bacteria in the biological control of *Gaeumannomyces graminis* by suppressive soils. Soil Biol Biochem 8:269–273

DeGurfinkel S, Gasoni L, Fortugno C (1994) Pathogenicity of some sterile fungi isolated from plants and soils in Argentina. Rev Investig Agropec 25:143–149

Dobbs CG, Hinson WH (1953) A widespread fungistasis in soil. Nature (London) 172:197–199

Ellert BH, Clapperton MJ, Anderson DW (1997) An ecosystem perspective of soil quality. In: Gregorich EG, Carter MR (eds) Soil quality for crop production and ecosystem health, vol 25, Developments in soil science. Elsevier, Amsterdam, pp 115–141

Gálvez EGE, Mora B, Pastor-Corrales MA (1989) Web blight. In: Schwartz HF, Pastor-Corrales MA (eds) Bean production problems in the tropics, 2nd edn. Centro Intern Agric Tropical (CIAT), Cali, pp 195–209

Guillemaut C (2003) Identification et etude de l'ecologie de *Rhizoctonia solani*, responsablede lamaladie de pourriture brune de labettravesucrière. Ecologie Microbienne. Université de Bourgogne, Dijon

Hoitink HAJ (2004) Disease suppression with compost: history, principles and future. In: International conference on soil and compost eco-biology, 15th –17th Sept 2004, León, Spain

Hornby D (1983) Suppressive soils. Annu Rev Phytopathol 21:65–68

Hyakumachi M (1996) Rhizoctonia disease decline. In: Sneh B, Jabaji-Hare S, Neate S, Dijst G (eds) Rhizoctonia species: taxonomy, molecular biology, ecology, pathology and disease control. Kluwer, The Netherlands, pp 227–235

Janvier C, Steinberg C, Villeneuve F, Mateille T, Alabouvette C (2006) Towards indicators of soil biological quality: use of microbial characteristics. In: Raaijmakers JM, Sikora RA (eds) Multitrophic interactions in soil. Bulletin OILB-WPRS, 29, pp 37–40

Janvier C, Villeneuvea F, Alabouvetteb C, Edel-Hermannb V, Mateillec T, Steinberg C (2007) Soil health through soil disease suppression: which strategy from descriptors to indicators? Soil Biol Biochem 39(1):1–23

Kennedy AC, Smith KL (1995) Soil microbial diversity and sustainability of agricultural soil. Plant Soil 170:75–86

Kumar P, Khare S, Dubey RC (2012) Diversity of bacilli from disease suppressive soil and their role in plant growth promotion and yield enhancement. New York Sci J 5(1)

Liu HY, Lewellen RT (2007) Distribution and molecular characterization of resistance breaking isolates of Beet necrotic yellow vein virus in the United States. Plant Dis 91:847–851

Mazzola M (1997) Identification and pathogenicity of *Rhizoctonia* spp. isolated from apple roots and orchard soils. Phytopathology 87:582–587

Mazzola M (1999) Transformation of soil microbial community structure and *rhizoctonia*-suppressive potential in response to apple roots. Phytopathology 89:920–927

Neergaard P (1977) Seed pathology. Wiley, New York

Penton CR, Gupta VVSR, Tiedje JM, Neate SM, Ophel-Keller K, Gillings M, Harvey P, Pham A, Roget DK (2014) Fungal community structure in disease suppressive soils in the Mediterranean climatic region assessed by 28S LSU gene sequencing. PLoS ONE 9(4):e 93893

Rovira AD, Wildermuth GB (1981) The nature and mechanisms of suppression. In: Asher MJC, Shipton P (eds) Biology and control of take-all. Academic, London, pp 385–415, 538 p

Schneider RW (1982) Suppressive soils and plant disease. APS, St. Paul, 88p

Sharma SK, Ramesh A, Sharma MP, Joshi OP, Govaerts B, Steenwerth KL, Karlen DL (2011) Microbial community structure and diversity as indicators for evaluating soil quality. In: Lichtfouse E (ed) Biodiversity, biofuels, agroforestry and conservation agriculture, vol 5. Springer, New York, pp 317–358

Shipton PJ, Cook RJ, Sitton JW (1973) Occurrence and transfer of a biological factor in soil that suppressive take-all of wheat in Eastern Washington. Phytopathology 63:511–517

Sid AA, Ezziyyani M, Sánchez CP, Candela ME (2003) Effect of chitin on biological control activity of *Bacillus spp.* and *Trichoderma harzianum* against root rot disease in pepper (*Capsicum annuum*) plants. Eur J Plant Pathol 109(6):633

Sneh B, Burpee L, Ogoshi A (1991) Identification of Rhizoctonia species. APS Press, St Paul

Sturz AV, Christie BR (2003) Beneficial microbial allelopathies in the root zone: the management of soil quality and plant disease with rhizobacteria. Soil Till Res 72:107–123

Sullivan P (2004) Sustainable management of soil-borne plant diseases. https://attra.ncat.org/attra-pub/summaries/summary.php?pub=283

Val-Moraes SP, Pedrinho EAN, Eliana Lemos EGM, Carareto-Alves LM (2013) Molecular identification of fungal communities in a soil cultivated with vegetables and soil suppressiveness to *Rhizoctonia solani*. Appl Environ Soil Sci 2013:1–7

Van Bruggen AHC (1995) Plant disease in high-input compared to reduced input and organic farming systems. Plant Dis 79:976–983
Weller DM, Raaijmakers JM, Gardener Brian BM, Thomashow LS (2002) Microbial populations responsible for specific soil suppressiveness to plant pathogens. Annu Rev Phytopathol 40:309–348
Wiseman BM, Neate SM, Keller KO, Smith SE (1996) Suppression of *Rhizoctonia solani* anastomosis group 8 in Australia and its biological nature. Soil Biol Biochem 28:727–732
Zachow C, Tilcher R, Berg G (2008) Sugar beet associated bacterial and fungal communities show a high indigenous antagonistic potential against plant pathogens. Microbial Ecol 55:119–129

Part II
Concepts in Plant Disease Management Involving Microbial Soil Suppressiveness

Chapter 9
Microbial Suppressiveness of *Pythium* Damping-Off Diseases

Mona Kilany, Essam H. Ibrahim, Saad Al Amry, Sulaiman Al Roman, and Sazada Siddiqi

9.1 Introduction

Soilborne plant pathogens causing wilts, root and crown rots, and damping-off are major yield-limiting factors in the production of fiber, food, and ornamental crops. Most soilborne pathogens are difficult to control by conventional strategies such as the use of synthetic fungicides. The lack of reliable chemical controls, the occurrence of fungicide resistance in pathogens, and the breakdown or circumvention of host resistance by pathogen populations are among the key factors underlying potentials to develop other control measures. The search for alternative strategies has also been stimulated by public concerns about the adverse effects of soil fumigants such as methyl bromide on the environment and human health. Cook and Long (1995) postulated that many plant species have developed a defense strategy against soilborne pathogens that involves the selective stimulation and support of populations of antagonistic rhizosphere microorganism. Over the past century, evidence has accumulated that such plant-associated microorganisms

M. Kilany (✉)
Biology Department, King Khalid University, Abha, 9004, Saudi Arabia

Microbiology Department, National Organization for Drug Control and Research (NODCAR), Giza, Egypt
e-mail: mona.kilany@yahoo.com

E.H. Ibrahim
Biology Department, King Khalid University, Abha, 9004, Saudi Arabia

Blood Products Quality Control and Research Department, National Organization for Research and Control of Biologicals, Cairo, Egypt
e-mail: essamebrahim@hotmail.com

S. Al Amry • S. Al Roman • S. Siddiqi
Biology Department, King Khalid University, Abha, 9004, Saudi Arabia
e-mail: amri555@yahoo.com; sar200@hotmail.com; kalasaz@yahoo.co.in

M.K. Meghvansi, A. Varma (eds.), *Organic Amendments and Soil Suppressiveness in Plant Disease Management*, Soil Biology 46,
DOI 10.1007/978-3-319-23075-7_9

account for many examples in which susceptible plants remain almost free of infection despite ample exposure to the virulent inoculum of soilborne pathogens. Natural disease-suppressive soils probably are the best examples in which the indigenous microflora effectively protect plants against soilborne pathogens. Suppressive soils initially become apparent because the incidence or severity of disease is lower than expected for the prevailing environment or as compared to that in surrounding soil (Cook and Baker 1983). Suppressive soils have been described for many soilborne pathogens including *Phytophthora cinnamomi*, *Phytophthora infestans*, *Pythium splendens*, *Pythium ultimum*, *Rhizoctonia solani*, and *Ralstonia solanacearum* (Mazzola 2007). Among the fungal diseases, damping-off is a serious disease complex worldwide of a wide range of seedlings in nurseries, glasshouses, gardens, crops, and forests and can damage both germinating seeds and young seedlings. Two types of damping-off diseases were known, preemergence damping-off and postemergence damping-off (Yang 2001). Damping-off is incited by any of a handful of fungal diseases, including several root rots (*Pythium*, *Phytophthora*) and molds (*Sclerotinia* or white mold, *Botrytis* or gray mold) (Agrios 1997). *Pythium* species cause more than 60 % mortality of seedlings both in nursery and in main field (Manoranjitham et al. 2000). Management of *Pythium* damping-off is very difficult due to its wide host range, soilborne nature, and prolonged survival of propagules in the soil. Traditionally, this disease is remediated by the application of synthetic fungicides. But the excessive use of fungicides resulted in the accumulation of residual toxicity and environmental pollution and altered the biological balance in the soil by attacking the beneficial microorganisms besides development of resistance in *Pythium* spp. against fungicides. Therefore, it is necessary to develop an effective, cheap, and environmentally safe nonchemical method for the control of damping-off disease. So, microbial control has been developed successfully as an alternative strategy and became a good promise in the field of microbial control in the past two decades (Muthukumar et al. 2011; Singh and Sachan 2013). Accordingly, biocontrol of *Pythium* damping-off disease with biological control agents (BCAs) including filamentous fungi, bacteria, actinomycetes, and yeasts has been intensively studied involving *Enterobacter cloacae*, *Gliocladium virens*, *Trichoderma harzianum*, *Rhizoctonia* spp., *Pseudomonas* spp., and *Cladorrhinum foecundissimum*, which are considered as ecologically sustainable and safe crop protection solutions (Khare and Upadhyay 2009; Muthukumar et al. 2011). The biological control products are regulated by governmental regulations for registration and use. Suppression of damping-off by biocontrol agents is the consequence of the interactions between soilborne pathogen, plant, and microbial community. The occurrence and development of soilborne diseases depend on several factors affecting either the pathogen or the plant. The complexity of the interactions between a pathogen and its plant host, influenced by biotic and abiotic factors of the environment, makes the control of the diseases often very difficult (Weller et al. 2002). Mycoparasitism, antibiosis by enzymes and secondary metabolites, competition, and induction of plant defense system are typical mechanisms of biocontrol agents (Singh and Sachan 2013). Soil interferes in many ways in the relationships between microorganisms, pathogens, and host plant. It can even modify the interactions among microorganisms themselves. In disease-suppressive soils, disease incidence or severity commonly remains low in spite of the presence of the pathogen, a susceptible host plant, and favorable climatic conditions. Soil suppressiveness to diseases depends on the pathogen itself, its inoculums density and its intrinsic aggressiveness, and also on different soil factors including both biotic and abiotic components. Soil abiotic components such as texture, organic matter content, pH, and temperature and moisture greatly affect the behavior of the pathogens and determine disease incidence or severity. Soil biotic factors that affect on the occurrence and development of soilborne diseases include: autecology of pathogens, interactions between microorganisms and pathogens, and interactions between plants and pathogens (Messiha et al. 2007; Steinberg et al. 2007). Soil physicochemical and biological factors interact to provide rapidly changing ecological niches and microbial components (Cook and Baker 1983). Soil organic matters also have a profound influence on microorganisms in soil, particularly those, including some pathogen, saprophytic and obligate plant parasites. This chapter presents recent advances and findings regarding the role of beneficial microbes in the *Pythium* damping-off disease suppression and the biological aspects highlighting the mechanisms of action of biocontrol process.

9.2 Damping-Off Diseases

Damping-off diseases are worldwide economically significant on numerous agricultural, ornamental, and horticultural crops and can be caused by soilborne plant pathogenic fungi under various environmental conditions (Salman and Abuamsha 2012). The name damping-off usually refers to the disintegration of stem and root tissues at and below the soil line. The plant tissues become water-soaked and mushy, and the seedling wilts and falls over (Fig. 9.1).

Damping-off diseases, however, can have several phases. The fungi that cause these diseases can attack the seed or the seedling below the soil line before it emerges, causing a preemergence damping-off where seeds become soft and mushy, turn dark brown and germinating seedlings shrivel, and may darken. Preemergence damping-off disease is difficult to be diagnosed because the seeds are not visible; consequently, the losses are often attributed to "poor seed" (Baker 1957). If the germinant has not emerged after a considerable period, the seed should be excavated and examined; if the seed contents are decayed, then damping-off fungi may be involved. On the other hand, postemergence damping-off causes death of seedlings after emergence or transplanting at the soil line where stem tissue near the soil line is weakened and decayed, usually causing plants to topple and die.

When only roots are decayed, plants may continue growing but remain stunted, wilt, and eventually die. As seedlings get older, they become more resistant to damping-off pathogens. Most pathogens that cause damping-off diseases are responsible for diseases as the plant grows to maturity. Root rot, crown rot, stem

Fig. 9.1 Damping-off caused by *Pythium* (Courtesy: "Martin Chilvers, Michigan State University." Reproduced with permission)

lesions, basal rot, crater rot, bottom rot, and stem girdling diseases may all be associated with damping-off pathogens attacking mature plants. Generally, damping-off is caused by over 30 species of fungi such as *Pythium*, *Rhizoctonia*, *Fusarium*, *Alternaria*, *Sclerotinia*, *Phytophthora*, *Thielaviopsis*, and *Botrytis* (Flint 1998; Yang 2001). The most common culprits that are associated with damping-off are *Pythium* species (water molds) and *Rhizoctonia solani* (true fungi). *Pythium* is a cosmopolitan and biologically diverse genus. Most species reside in soil inhabitants, although some are aquatic inhabitants. Most *Pythium* spp. are saprobes or facultative or opportunistic plant pathogens causing a wide variety of diseases, including damping-off (Larkin et al. 1995; Sumner et al. 1990). Damping-off diseases caused by *Pythium* species usually begin as root rot. This group of fungi survives as oospores in the soil that germinate to attack root tips and root hairs, causing a progressive deterioration of the root. The seedling may wilt or rot in the ground. *Pythium* species are often responsible for preemergence damping-off (Agrios 1997). The environmental conditions that favor damping-off vary according to the pathogen. *Pythium* spp. tend to be most active during the spring months when soil temperatures are still cool and soil moisture is plentiful (Flint 1998; Yang 2001). Landis et al. (1990) have been reported that although damping-off disease is usually caused by fungi or oomycetes, stresses such as high surface soil temperatures and chemicals can also cause damping-off symptoms.

9.3 Microbial Control of *Pythium* Damping-Off

Strategies to control soilborne diseases are limited because of their extremely broad host range, their ecological behavior, and the high survival rate of resistant forms such as oospores and sclerotia under different environmental conditions, and cultivars with complete resistance are not available (Li et al. 1995). Many pathologists have investigated that biological control agents offer an environmentally friendly alternative to protect plants from soilborne pathogens (Whipps 2001; Weller et al. 2002). Damping-off suppression can operate directly on fungal plant pathogens in the bulk soil, in the rhizosphere, and in some cases in plants. In the bulk soil, antagonistic soil microbes may act directly on resting spores or on active mycelium during a saprotrophic phase of plant pathogens, thus suppressing the plant pathogen directly. This suppression can be either specific or general. Specific disease suppression is caused by one or a few specific microorganisms. General disease suppression is caused by multiple microorganisms, acts against multiple pathogens, and is quickly restored. General disease suppression is directly related to microbial metabolic activity and mediated by availability of nutrients and energy available for growth of the pathogen through the soil. General disease suppression acts mainly in the bulk soil and is therefore largely congruent with pathogen suppression; it is especially effective against pathogens that have a saprotrophic phase. Also, in the rhizosphere, antagonists may suppress pathogens by interfering directly with germination, growth, and infection processes or indirectly through inducing host resistance (Termorshuizen and Jeger 2008). Effective biological control of damping-off requires careful matching of antagonists to pathosystems to achieve any of the three types of biological control: preventative control, eradicative control, or reductive control. Accordingly, biological control agents are more target specific and hence have fewer negative effects on nontarget organisms or even beneficial organisms in the rhizosphere (Cunniffe and Gilligan 2011).

9.3.1 Microbial Diversity and Disease Suppression

BCAs are beneficial organisms acting as naturally occurring enemies against pathogen such as bacteria and fungi. In last three decades, several antagonists were used to provide direct effects on *Pythium* spp., causal agent of damping-off, reducing their growth and preventing establishment in the rhizosphere (Howell 2003; Faltin et al. 2004). However, most of them showed inconsistent in vitro results, and only very few antagonists were analyzed under open field conditions (Grosch et al. 2005). The ability to control disease is more likely related to the production of specific metabolites or other substances than to the ability to produce fungal reproductive propagules (Lewis and Papavizas 1984). A wide range of aerobic microorganisms are involved in this aspect, and their introduction into soil improves fertility and structure and a range of population effects that may lead to suppression of plant pathogens and eventually of disease. However, it is difficult to determine exact suppression mechanisms as compost represents a "microbial community structure rather than a single species" (Boulter et al. 2002).

9.3.1.1 Bacterial Biocontrol Agents

In the past, Broadbent et al. (1971) found difficulty in controlling *Pythium* damping-off. Few actinomycetes were antagonistic to *P. ultimum* than to the other plant pathogenic fungi (Broadbent et al. 1971), as for antagonistic bacteria effectively acted as a biological control agent against *Pythium* damping-off: *Enterobacter cloacae*, *Bacillus* spp., *P. cepacia*, *P. corrugata*, *P. fluorescens*, *P. marginalis*, *P. putida*, *P. syringae*, *P. viridiflava*, and *Erwinia herbicola* (Gravel et al. 2005). *Stenotrophomonas maltophilia* and *Lysobacter enzymogenes* have been exploited to control *P. ultimum* in sugar beet (Palumbo et al. 2005). *Actinoplanes philippinensis* and *Micromonospora chalcea* were also investigated to control damping-off in cucumber (El-Tarabily 2006). Li et al. (2007) concluded that all paenibacilli prevented preemergence damping-off caused by *P. aphanidermatum*. *Serratia entomophila* strain M6 is a suitable candidate for exploitation as biocontrol agent of *P. aphanidermatum* (Chairat and Pasura 2013). Recently, it was observed that *E. faecalis* is a bactericidal agent producing diffusible metabolites which inhibited *P. ultimum* growth in vitro as shown in Fig. 9.2 (Kilany et al. 2015).

Streptomyces rubrolavendulae (Yen) S4 has been described as a biocontrol agent for controlling *Pythium* damping-off disease of the horticultural plant Joseph's coat caused by *P. aphanidermatum* (Loliam et al. 2013).

9.3.1.2 Fungal Biocontrol Agents

Fungi have a broad-spectrum antagonistic activity against *Pythium* damping-off. Biocontrol of preemergence damping-off induced by *Pythium* species is achieved by coating radish and pea seeds with *T. harzianum* or *T. koningii* (El-Katatny et al. 2001). Besides, control of *Pythium* spp. in tobacco, sugar beet, and cauliflower by *T. harzianum* through soil application was recorded (Das et al. 2002). The successful application of *Trichoderma* species for the management of damping-off caused by *Pythium* species in chili and tomato has been reported (Jayaraj et al. 2006; Muthukumar et al. 2011). Two biological control agents, *Pythium nunn* and *T. harzianum* isolate T-95, were combined to reduce *Pythium* damping-off of cucumber in greenhouse (Paulitz et al. 1990). *Gliocladium virens* most consistently and effectively controlled damping-off of zinnia, cotton, and cabbage seedlings caused by *P. ultimum* (Lumsden and Locke 1989). Pre- and postemergence damping-off of wheat caused by *P. diclinum* was successfully controlled by *Gliocladium roseum* or *T. harzianum* (Abdelzaher 2004). Eight isolates of binucleate *Rhizoctonia* spp. from South Australian plant nurseries and potting mix

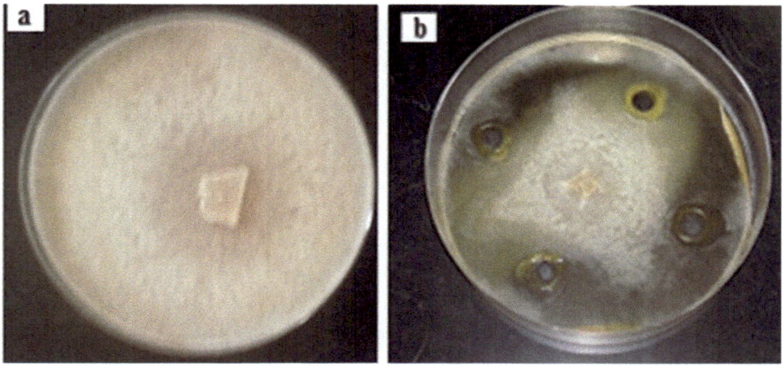

Fig. 9.2 Antifungal activity of *E. faecalis* against *P. ultimum*, where (**a**) refers to the control and (**b**) refers to the sample (Kilany et al. 2015)

suppliers were screened for ability to control damping-off disease caused by *P. ultimum* var. *sporangiiferum* (Harris et al. 1993). *C. foecundissimum* has a considerable potential as a biocontrol agent for damping-off of eggplant and pepper caused by *P. ultimum* (PuZ3). Antagonistic activities of *Aspergillus* species, *Penicillium* species, and *Trichoderma* species against *P. debaryanum* were studied by in vitro dual culture experiment (Hasan et al. 2013).

9.4 Mechanism of Microbial Control of *Pythium* Damping-Off

The antagonists encounter the pathogen either by direct antagonism (physical contact and/or a high degree of selectivity for the pathogen by the mechanism (s) expressed by the BCA(s)) or indirect antagonism (activities that involve stimulating of plant host defense (Pal and Gardener 2006)). *Pythium* damping-off suppression is the consequence of the interactions between the plant, pathogens, and BCAs (parasitism, predation, mutualism, protocooperation, commensalisms, neutralism, and competition), depending on the environmental conditions (Chisholm et al. 2006). Several strategies have been used to study the complex tripartite interaction in order to improve advantageous interactions, enhance the practical application of these beneficial microorganisms, and unravel the mechanisms of biological control (Vinalea et al. 2008; Rey and Schornack 2013). Most described mechanisms of pathogen suppression include the modulation by relative occurrence of other organisms in addition to the pathogen as shown in Fig. 9.3. The most effective BCAs studied appear to antagonize pathogens using multiple mechanisms (Iavicoli et al. 2003).

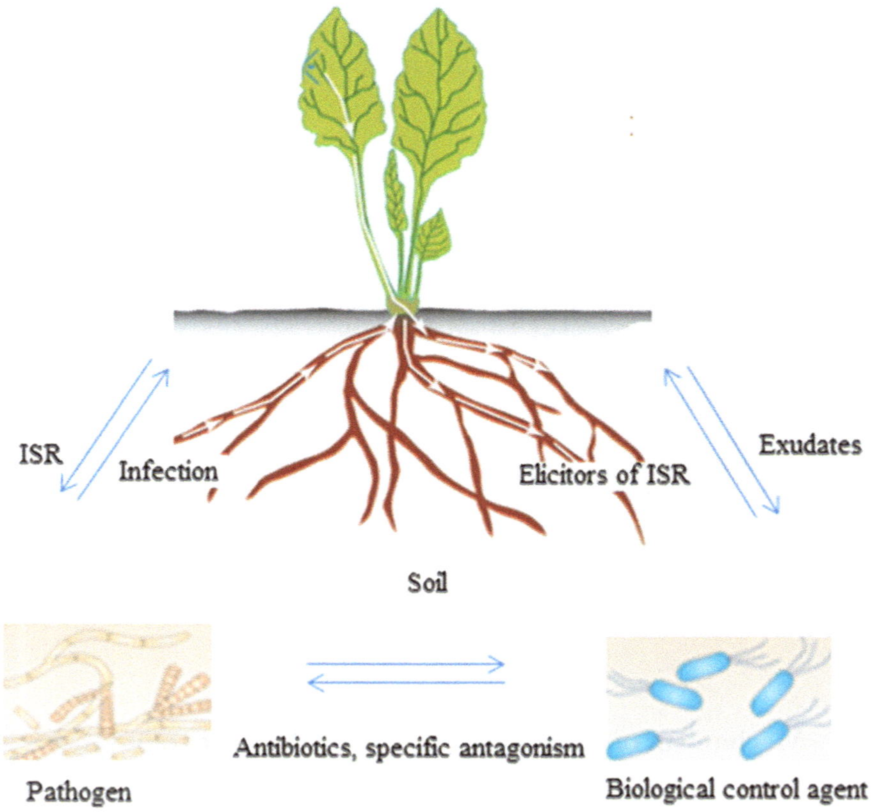

Fig. 9.3 Mechanisms of specific biocontrol agents for controlling plant pathogens

9.4.1 Mycoparasitism

Various microorganisms were recorded as parasites to soilborne pathogenic fungi in many systems (Elad 1995), depending on the production of antibiotics and fungal cell wall-degrading enzymes. It has been reported that antibiotics and hydrolytic enzymes are not only produced together but act synergistically in mycoparasitic antagonism (Schirmböck et al. 1994).

9.4.1.1 Lytic Enzymes

Harman et al. (1980) had suggested that mycoparasitism was the principle mechanism of *Pythium* damping-off when seeds were coated with *Trichoderma hamatum*. Their suggestion was based on evidence that in the presence of *Pythium* spp., *T. hamatum* becomes able to produce hydrolytic enzymes β-1,3-glucanase and cellulase, on observation of hyphal parasitism in vitro on *Pythium* spp. (Elad et al. 1982). *Stenotrophomonas maltophilia* and *Lysobacter enzymogenes* have the potential to antagonize *P. ultimum* infecting sugar beet by production of proteases and glucanases, respectively (Palumbo et al. 2005). The mycoparasitism of *S. rubrolavendulae* S4 against *P. aphanidermatum* was indicated by the degradation of *P. aphanidermatum* mycelium by means of cellulase production which was demonstrated by electron micrographs (Loliam et al. 2013). Moreover, *P. putida* strain N1R provides biocontrol of *P. ultimum* by enzymatic degradation of volatile seed exudates, which would otherwise stimulate the pathogen to cause preemergence damping-off (Paulitz 1991).

9.4.1.2 Antibiotics

Many microbes produce and secrete one or more compounds with antibiotic activity (Shahraki et al. 2009). It has been shown that some antibiotics produced by microorganisms are particularly effective against plant pathogens and the diseases they cause (Islam et al. 2005). Concomitantly, pyoluteorin, a new antibiotic, gliovirin, and gliotoxin had been isolated from *P. fluorescens* Pf-5, *T. virens* (GV-P), and *T. virens* (GL-21), respectively, that are effective antibiotics against damping-off incited by *P. ultimum* (Wilhite et al. 1994). Mutants of *T. harzianum* with altered antibiotic production were found inhibitory to *P. ultimum* (Graeme-Cook and Faull 1991). Further, the mechanism adopted to interpret *Pythium* damping-off biocontrol by *P. fluorescens* was attributed to the production of both antibiotics 2,4-diacetylphloroglucinol and viscosinamide (Thrane et al. 2000). Recently, *T. viride* was found highly inhibitory to *P. indicum* and *P. aphanidermatum*, the causal organisms of damping-off of tomato, by the effect of volatile and nonvolatile metabolites that inhibit the mycelial growth of *P. aphanidermatum* as well as increased the plant growth (Neelamegam 2004; Khare et al. 2010; Muthukumar et al. 2011). Moreover, Leclere et al. (2005) found that *B. subtilis* BBG100 exert profound effect to control damping-off caused by *P. aphanidermatum* by mycosubtilin. The efficacy of *Calothrix elenkenii* against damping-off disease, caused by *P. aphanidermatum* in three vegetable crops, tomato, chili, and brinjal, is due to antifungal compound production (Manjunath et al. 2010). *Chaetomium globosum* control the damping-off in sugar beet caused by *P. ultimum* by production of cheatomin (Lo 1998).

9.4.2 Suppression by Other By-Products

The suppression of *Pythium* damping-off involves production of microbial metabolites such as ethanol, ammonia, siderophore, etc. Toxic metabolites produced by *Trichoderma* spp. on seed coats are the principal mechanism of biological control of *Pythium* damping-off. *E. cloacae* is a potential antagonist against *Pythium* spp.

owing to production of ethanol which is an effective stimulant of sporangium germination, and reductions in ethanol production may reduce or delay sporangium germination of *Pythium* spp. thereby delaying seed colonization (Nelson 1987). Besides, Howell et al. (1988) reported that ammonia produced by *E. cloacae* was involved in the suppression of *P. ultimum*-induced damping-off of cotton. Similarly, *P. aeruginosa* and *P. fluorescens* produced siderophores to control *P. ultimum* damping-off in tomato and potato (Goud and Muralikrishnan 2009). Moreover, glycolipids that are produced by *Pseudomonas* spp. can damage the zoospores that are released from the sporangia of *Pythium* spp. (Stanghellini and Miller 1997). *P. putida* produced volatile metabolites to control *P. ultimum* damping-off in pea and soybean (Lo 1998).

9.4.3 Attachment to Pathogen Surfaces

Biocontrol activity may be requiring the attachment of BCAs to the surface of the host cells. Attachment mechanisms play a vital role in cell-cell interactions between fungi and other microorganisms (Douglas 1987).Cook and Long (1995) successfully used the attachment phenomenon to select potential BCAs among phyllosphere bacteria and yeasts. A common observed feature of the *E. cloacae*-*Pythium* system in vitro was the ability of *E. cloacae* to attach to the hyphae, agglutinate cell wall fragments of *P. ultimum* inhibiting mycelial growth (Nelson et al. 1986). It was suggested that the agglutination of cell wall fragments of *P. ultimum* occurred in the absence of some sugars or in the presence of others. In the absence of sugars, cells of *E. cloacae* attached to the hyphae. This is consistent with studies of phytoplanktonic bacteria where carbon starvation apparently promotes the attachment of bacteria to surfaces (Marshall 1980). On the other hand, in the presence of glucose or sucrose (sugars that block the agglutination of cell wall fragments by blocking available receptor sites), bacteria did not attach to the hyphae (Nelson et al. 1986). Furthermore, *P. fluorescens* provided superior seed protection from *Pythium* damping-off in naturally infested soils by adhering to hyphae of *P. ultimum* leading to fungal growth inhibition (Callan et al. 1990).

9.4.4 Competition

Generally, nutrient and space competition have been believed to play an important role in disease suppression. Biocontrol by nutrient competition can occur when the biocontrol agent decreases the availability of a particular substance, thereby limiting the growth of the pathogen. Soilborne pathogens, such as species of *Pythium*, infecting through mycelial contact, are more susceptible to competition by other soil- and plant-associated microbes than by those germinating directly on plant surfaces which they invade through appressoria and infection pegs. Rhizosphere or phyllosphere BCAs are generally protecting the plant by rapid colonization, thus consuming completely the limited available substrates so that none is left for pathogens to grow. Apparently, it was suggested that competition for nutrients between germinating oospores of *P. aphanidermatum* and bacteria significantly correlated with suppression of damping-off in the greenhouse (Fldd and Chet 1987). It appears more likely that competition is the primary mechanism by which *P. oligandrum* protects seed from infection by *P. ultimum* resulting in protection of sugar beet seeds from damping-off (Martin and Hancock 1987). Furthermore, Green et al. (2001) explained the biological control using *T. harzianum* by competition with *P. ultimum* for substrates from the seed coat and wounded or infected root tissue. Moreover, effective catabolism of nutrients in the spermosphere has been identified as a mechanism contributing to the suppression of *P. ultimum* by *E. cloacae* (van Dijk and Nelson 2000; Kageyama and Nelson 2003).

9.4.5 Role of Host and Disease Suppression

Apparently, host plants possess a little predictive value for the disease that is actually developing. This can be due to host-induced factors, such as induced systemic resistance (ISR), systemically acquired resistance (SAR), and specific disease suppression. The importance of host species in substrate-induced disease suppression has rarely been investigated. Van Rijin (2007) studied the effect of compost on disease suppression of the same isolate of *P. ultimum* using five different host seedlings (pea, cucumber, tomato, carrot, and sugar beet) and six composts mixed with peat. There was a significant interaction between pathosystem and compost type. Since in this experiment the host was the sole source of variation, host-mediated effects must explain this interaction. The genetic and functional diversity of the rhizosphere community is a key factor of specific disease suppression (Weller et al. 2002). This diversity varies according to plant species through the quantity and quality of root exudation and rhizodeposition (Bergsma-Vlami et al. 2005). Therefore, plants evolve strong defense mechanisms to effectively ward off pathogens while supporting development toward useful interactions (Jones and Dangl 2006; Bonneau et al. 2013). Microbe-associated chemical stimuli can induce plant host defenses through biochemical changes that enhance resistance against subsequent infection by a variety of pathogens. Induction of host defenses can be local and/or systemic, depending on the type, source, and amount of stimuli. Induced systemic resistance (ISR) is mediated by jasmonic acid (JA) and/or ethylene, which are attributed to a variety of microorganisms and can result in control of multiple pathogens (Paulitz and Matta 1999). One of the most important biological agents is *S. plymuthica*, currently used in greenhouses which may provide economical prolonged protection against damping-off by sensitizing susceptible cucumber plants to elaborate a wide range of defense mechanisms (Benhamou et al. 2000). Ramamoorthy et al. (2002) recorded that in addition to direct antagonism of

P. fluorescens and plant growth promotion, induction of defense-related enzymes involved in the phenylpropanoid pathway collectively contributed to enhance resistance against invasion of *P. aphanidermatum* in tomato and hot pepper. Moreover, Howell et al. (2000) and Howell (2003) demonstrated that application of *T. virens* to cotton seedling induced the resistance in the host plant by synthesis of much higher concentrations of the terpenoids desoxyhemigossypol (dHG), hemigossypol (HG), and gossypol (G) in developing roots than those found in untreated controls. Some biocontrol strains of *Pseudomonas* sp. and *Trichoderma* sp. are known to strongly induce plant host defenses against *Pythium* damping-off (Harman et al. 2004; Haas and Défago 2005). The mechanism of *T. harzianum* Rifai for controlling maize seedling disease caused by *P. ultimum* Trow was investigated by proteome technique, and the result suggested that *T. harzianum* strain T22 was not only able to promote seedling growth but also induce the plant resistance by protein accumulation (Chen et al. 2005). *B. subtilis* strain BSCBE4 and *P. chlororaphis* strain PA23 obviously reduced the incidence of damping-off of hot pepper incited by triggering the plant-mediated defense mechanism in response to infection by *P. aphanidermatum* (Nakkeeran et al. 2006).

9.4.6 Metabolism of Germination Stimulants

Preemergence damping-off incited by *P. ultimum* in cotton was controlled using *Trichoderma virens*; this was attributable to metabolism of pathogen germination stimulants by the biocontrol agent released by the seed (Chen et al. 1988). In addition, the mechanisms involved in the biocontrol of preemergence damping-off of cotton seedlings incited by *P. ultimum* were studied by Howell (2002; 2003) who found that control by *T. virens* (G6, G6-5) or protoplast fusants of *T. virens* and *T. longibrachiatum* (Tvl-30, Tvl-35) was due to metabolism of germination stimulants released by the cotton seed. These compounds normally induced pathogen propagules to germinate. It is apparent that *T. virens* completely inhibited mycelial growth and sporangium production of *P. aphanidermatum*, the causal agent of Chinese-kale damping-off (Intana and Chamswarng 2007). It is apparent that *P. fluorescens* had the capacity to inhibit the germination of *Pythium* oospores, its growth, and the infection process (Cook and Long 1995; Ellis et al. 1999). One of the more effective bacterial species studied for its *Pythium* suppressiveness is *E. cloacae*. Molecular evidence showed that strain E6 of *E. cloacae* has the potential to inactivate the fatty acid that stimulates *Pythium* sp. germination, consequently protecting seeds from damping-off disease (van Dijk and Nelson 1997). Researchers provided strong evidence to support a mechanism for the suppression of *Pythium* damping-off by *E. cloacae* through which *E. cloacae* metabolize seed exudate fatty acid stimulants of *P. ultimum* sporangium germination resulting in reduction in sporangium germination and subsequent seed infection (van Dijk and Nelson 2000). *E. cloacae* protect the corn and cucumber seeds from *P. ultimum* infections by reducing sporangial activation and germination

(Windstam and Nelson 2008). *E. cloacae* are also effective in inactivation of the stimulatory activity of the seed exudates, thereby reducing *P. ultimum* sporangium germination on carrot, cotton, cucumber, lettuce, radish, tomato, and wheat (Kageyama and Nelson 2003). Suppressive efficiency of bacterial consortia to *P. ultimum* damping-off was attributed to degradation of seed exudate linolenic acid that stimulates the germination of *P. ultimum* sporangia (McKellar and Nelson 2003).

9.4.7 *Soil Dynamics*

Soil physicochemical and biological factors interact to provide hastily changing ecological niches and microbial components. Biological control of soilborne pathogens could be possible through manipulation of soil condition (Cook and Baker 1983). Soil organic substances support the largest numbers and types of microorganisms interacting with each other leading to modification or alteration in soil conditions that greatly influence the microbial community and their activity in soil ecosystem (Boulter et al. 2002). The extent of soilborne pathogen suppression will vary substantially depending on the quantity and quality of organic matter present in soil (Hoitink and Boehm 1999). It appears more likely that the primary mechanism by which *P. oligandrum* protects sugar beet seed from *P. ultimum* damping-off infection is alteration of the quality and quantity of sugar beet seed exudates in the spermosphere (Martin and Hancock 1987). Another aspect of the microbial populations studied was their composition and diversity in relation to disease suppression. Broad-spectrum biological control of diseases caused by *Pythium* requires the supplementation of organic nutrients in soil for survival of biocontrol agents where the decomposition level of organic matter significantly affects the composition of bacterial taxa as well as the populations and activities of biocontrol agents (Hoitink and Boehm 1999). Concomitantly, soil microbial community and carbon and nitrogen availability could be exploited as predictors to the relative growth of *P. ultimum* and *P. aphanidermatum* and the incidence of cotton seedling damping-off (Kowalchuk et al. 2003). The influence of microbial community structures in the different rock wool treatments toward *Pythium* disease suppression was investigated (Postma et al. 2005). Furthermore, the findings obtained by Manici et al. (2004) indicate that the green manures suppress *Pythium* sp. and also induced an increase in total soil microbial activity.

9.5 Commercially Available Biocontrol Agents

Currently, biocontrol of damping-off with bacterial and fungal antagonists is being investigated very intensively. The problems associated with the commercial acceptance of biological control agents of *Pythium* damping-off are discussed, and several methods of improving selection, activity, and use are described. Commercially available biocontrol rhizobacteria include *B. subtilis* strains GB03 (Kodiak; Gustafson), MBI 600 (Subtilex; Becker Underwood), and QST 713 (Serenade; AgraQuest), *B. pumilus* strain GB34 (Yield Shield; Gustafson), *B. licheniformis* strain SB3086 (EcoGuard; Novozymes), a mixture of *B. subtilis* strain GB122 and *B. amyloliquefaciens* strain GB99 (BioYield; Gustafson), several *Bacillus* spp. (yield-increasing bacteria in China), *S. griseoviridis* K61 (Mycostop; AgBio Development), and a few strains of *P. fluorescens*, *P. putida*, and *P. chlororaphis* (Cedomon; BioAgri). These biocontrol bacteria can be applied as dry products (granules or powders), cell suspensions (with or without microencapsulation), or seed coatings (Schisler et al. 2004). Several commercial products of *Trichoderma* like Biocure, Antagon, Bioderma, Trichofit, Dermapack, and Trichosan in India and Binab-7, Azadderma, F-Stop, Trichodex, and Trichodermin abroad have appeared in the market which indicate that bioagents are becoming popular (Kanjanamaneesathian et al. 2003; Khare and Upadhyay 2009). In the field, reproducible cost-effective biological control is rare. Nevertheless, *G. virens*, *P. oligandrum*, *T. harzianum*, and *C. minitans* have been exploited commercially for the control of damping-off disease incited by *Pythium*. Fungal antagonists have been introduced into soil or applied to seeds, and biocontrol of damping-off is sometimes equivalent to standard fungicide applications (Whipps 1997; Fravel et al. 1998). Although the number of biocontrol products is increasing tremendously, these products still represent a low proportion of fungicides: a total share of 3.5 % of the total crop protection markets (Fravel 2005).

9.6 Methods of Application of BCAs

Pythium spp. are effectively controlled by seed treatment because the fungus is active early in the season during seed germination (Heydari and Misaghi 2003). Eventually, application of biological control strategies requires more knowledge-intensive management to be effective. So, there are several methods of application of antagonisms: (1) overall application, (2) application to the infection site, (3) one place application, and (4) occasional application (Heydari et al. 2004).

9.7 Conclusion

Generally, damping-off disease is caused by different species of *Pythium* and it represents a major economic problem. Traditionally, chemical pesticides have been used to control most soilborne fungal diseases, but they are restricted by many hazards they cause. An alternative strategy for damping-off disease management was established by a tremendous number of biocontrol agents including bacteria, actinomycetes, and fungi. Such BCAs became successfully popular for control of

Pythium damping-off diseases and are considered as an important economic tool for protecting the crops. BCAs have different suppressive potentials on *Pythium* damping-off diseases in the same particular ecological niche. The study of the population dynamics and tripartite interaction between the plant, pathogen, and antagonist is crucial to understand the mechanistic pathway of biocontrol agents and, consequently, to develop an appropriate biocontrol strategy. Predominantly, the different mechanisms of antagonism occur across a spectrum of directionality related to the amount of interspecies contact and specificity of the interactions. The most effective type of antagonism is direct antagonism resulting from physical contact and/or a high degree of selectivity for the pathogen by the mechanism expressed by the BCA (e.g., hyperparasitism). Conversely, indirect antagonisms result from activities that do not involve sensing or targeting a pathogen by the BCA through two mechanisms, competition and stimulation of plant host defense. The latter mechanism is more prevalent within the indirect antagonism. Mixed-path antagonism has been observed though some mechanisms involved the production of antibiotics, lytic enzymes, and other by-products as well as suppression of germination. Additionally, some microorganisms exhibited one mechanism, while others may work through several mechanisms. The latter microorganisms are likely to be more robust under extreme conditions. In spite of a plethora of examples in the literature of microbes with biocontrol activity against *Pythium* damping-off diseases, very few have given considerable levels of reproducible control across a number of seasons and sites. Two microbial groups, *Pseudomonas* spp. and *Trichoderma* spp., have given the greatest success. Therefore, BCA applications have been used successfully in combination with each other's. In the future, it is expected that better-performing BCAs will be developed. However, there is still great potential for the discovery of microbes with increased biocontrol abilities and to produce novel bioactive products.

References

Abdelzaher HMA (2004) Occurrence of damping-off of wheat caused by *Pythium diclinum* tokunaga in El-minia, Egypt and its possible control by *Gliocladium roseum* and *Trichoderma harzianum*. Arch Phytopathol Plant Prot 37:147–159

Agrios GN (1997) Induced structural and biochemical defenses, 4th edn, Plant pathology. Academic, London, pp 93–114

Baker KF (1957) Damping-off and related diseases . The UC system for producing healthy container-grown plants. California agriculture experimental station extension service manual 23. Australian Nurserymen's Association, Parramatta, Australia, pp 34–51

Benhamou N, Gagné S, Le Quéré D, Dehb L (2000) Bacterial-mediated induced resistance in cucumber: beneficial effect of the endophytic bacterium *Serratia plymuthica* on the protection against infection by *Pythium ultimum*. Phytopathology 90(1):45–56

Bergsma-Vlami M, Prins ME, Raaijmakers JM (2005) Influence of plant species on population dynamics, genotypic diversity and antibiotic production in the rhizosphere by indigenous *Pseudomonas spp.*. FEMS Microbiol Ecol 52:59–69

Bonneau L, Huguet S, Wipf D, Pauly N, Truong HN (2013) Combined phosphate and nitrogen limitation generates a nutrient stress transcriptome favorable for arbuscular mycorrhizal symbiosis in Medicago truncatula. New Phytol 199:188–202

Boulter JI, Trevors JT, Boland GJ (2002) Microbial studies of compost: bacterial identification, and their potential for turfgrass pathogen suppression. World J Microbiol Biotechnol 18 (7):661–671

Broadbent P, Baker K, Waterworth Y (1971) Bacteria and actinomycetes antagonistic to fungal root pathogens in Australia. Aust J Biol Sci 24:925–944

Callan NW, Mathre DE, Miller JB (1990) Bio-priming seed treatment for biological control of *Pythium ultimum* pre-emergence damping-off in sh-2 sweet corn. Plant Dis 74:368–372

Chairat Y, Pasura A (2013) Isolation and identification of rhizobacteria having inhibitory capability on pathogenic fungi, *Pythium sp.*. J Sci Technol Human 11(2):117–127

Chen W, Hoitink HAJ, Schmitthenner AF, Tuovinen OH (1988) The role of microbial activity in suppression of damping-off caused by *Pythium ultimum*. Phytopathology 78:314–322

Chen R, Harman GE, Afio COMI, Cheng SG (2005) Proteins related to the biocontrol of *Pythium* damping-off in maize with *Trichoderma harzianum*. J Integr Plant Biol 47(8):988–997

Chisholm ST, Coaker G, Day B, Staskawicz BJ (2006) Host-microbe interactions: shaping the evolution of the plant immune response. Cell 124:803–814

Cook R, Baker KF (1983) The nature and practice of biological control of plant pathogens. American Phytopathological Society, St Paul, MN, p 539

Cook DWM, Long PG (1995) Screening of microbes that attach to *Botrytis cinerea* hyphae before tests of biocontrol activity. In: XIII international plant protection congress, The Hague (Eur J Plant Pathol Abstract no 503)

Cunniffe N, Gilligan C (2011) A theoretical framework for biological control of soil-borne plant pathogens: identifying effective strategies. J Theor Biol 278:32–43

Das S, Biswapati M, Maity D, Raj SK (2002) Different techniques of seed treatment in the management of seedling disease of sugar beet. J Mycopathol Res 40(2):175–178

Douglas W (1987) Adhesion to surfaces. In: Rose AH, Harrison JS (eds) The yeast, vol II. Academic, New York

Elad Y (1995) Mycoparasitism. In: Kohmoto K, Singh US, Singh RP (eds) Pathogenesis and host specificity in plant diseases: histopathological, biochemical, genetic and molecular basis, vol 2, Eukaryotes. Elsevier Science, Oxford, pp 289–307

Elad Y, Chet I, Henis Y (1982) Degradation of plant pathogenic fungi by *Trichoderma harzianum*. Can J Microbiol 28:719–725

El-Katatny MH, Gudelj M, Robra KH, Elnaghy MA, Gubitz GM (2001) Characterization of a chitinase and an endo-b-1,3- glucanase from *Trichoderma harzianum* Rifai T24 involved in control of the phytopathogen *Sclerotium rolfsii*. Appl Microbiol Biotechnol 56:137–143

Ellis RJ, Timms-Wilson TM, Beringer JE, Rhodes D, Renwick A, Stevenson L, Bailey MJ (1999) Ecological basis for biocontrol of damping-off disease by *Pseudomonas fluorescens* 54/96. J Appl Microbiol 87(3):454–463

El-Tarabily KA (2006) Rhizosphere-competent isolates of streptomycete and non-streptomycete actinomycetes capable of producing cell-wall degrading enzymes to control *Pythium aphanidermatum* damping-off disease of cucumber. Can J Bot 84:211–222

Faltin F, Lottmann J, Grosch R, Berg G (2004) Strategy to select and assess antagonistic bacteria for biological control of Rhizoctonia solani Kuhn. Can J Microbiol 50:811–820

Fldd Y, Chet L (1987) Possible role of competition for nutrients in biocontrol of *Pythium* damping-off by bacteria. Phytopathology 77:190–195

Flint ML (1998) Pests of the garden and small farm: a grower's guide to using less pesticide. University of California Agriculture and Natural Resources Publication, Oakland, p 3332

Fravel DR (2005) Commercialization and implementation of biocontrol. Annu Rev Phytopathol 43:337–359

Fravel DR, Connick WJJ, Lewis JA (1998) Formulation of microorganisms to control plant diseases. In: Burges HD (ed) Formulation of microbial biopesticides: beneficial microorganisms, nematodes and seed treatments. Kluwer Academic, Dordrecht, pp 187–202

Goud MJP, Muralikrishnan V (2009) Biological control of three phytopathogenic fungi by *Pseudomonas fluorescens* isolated from rhizosphere. Internet J Microbiol 7(2)

Graeme-Cook KA, Faull JL (1991) Effect of ultraviolet-induced mutants of *Trichoderma harzianum* with altered production on selected pathogens *in vivo*. Can J Microbiol 37:659–664

Gravel V, Martinzi C, Antoun H (2005) Antagonist microorganisms with the ability to control Pythium damping-off of tomato seeds in rockwool. BioControl 50:771–786

Green H, Heiberg N, Lejbolle K, Jensen DF (2001) The use of a GUS transformant of *Trichoderma harzianum*, strain T3a, to study metabolic activity in the spermosphere and rhizosphere related to biocontrol of Pythium damping-off and root rot. Eur J Plant Pathol 107:349–359

Grosch R, Faltin F, Lottmann J, Kofoet A, Berg G (2005) Effectiveness of three antagonistic bacterial isolates to suppress *Rhizoctonia solani* Kühn on lettuce and potato. Can J Microbiol 51:345–353

Haas D, Défago G (2005) Biological control of soil-borne pathogens by fluorescent pseudomonads. Nat Rev Microbiol 3:307–319

Harman GE, Chet I, Baker R (1980) *Trichoderma hamatum* on seedling disease induced in radish and pea by *Pythium spp.* or *Rhizoctonia solani*. Phytopathology 70:1107–1172

Harman GE, Howell CR, Vitarbo A, Chet I, Lorito M (2004) *Trichoderma* species - opportunistic, avirulent plant symbionts. Nat Rev Microbiol 2:43–56

Harris AR, Schisler DA, Ryder MH (1993) Binucleate *Rhizoctonia* isolates control damping-off caused by *Pythium ultimum* var. *Sporangiiferum*, and promote growth, in *Capsicum* and *Celosia* seedlings in pasteurized potting medium. Soil Biol Biochem 25(7):909–914

Hasan S, Gupta G, Anand S, Chaturvedi A, Kaur H (2013) Biopotential of microbial antagonists against soilborne fungal plant pathogens. Int J Agric Food Sci Technol 4(2):37–39

Heydari A, Misaghi IJ (2003) The role of rhizosphere bacteria in herbicide-mediated increase in *Rhizoctonia solani*-induced cotton seedling damping-off. Plant Soil 257:391–396

Heydari A, Fattahi H, Zamanizadeh HR, Zadeh NH, Naraghi L (2004) Investigation on the possibility of using bacterial antagonists for biological control of cotton seedling damping-off in greenhouse. Appl Entomol Phytopathol 72:51–68

Hoitink HAJ, Boehm MJ (1999) Biocontrol within the context of soil microbial communities: a substrate-dependent phenomenon. Annu Rev Phytopathol 37:427–446

Howell CR (2002) Cotton seedling pre-emergence damping-off incited by *Rhizopus oryzae* and *Pythium spp.* and its biological control with *Trichoderma spp.*. Biol Control 92(2):177–180

Howell CR (2003) Mechanisms employed by *Trichoderma* species in the biological control of plant diseases: the history and evolution of current concepts. Plant Dis 87(1):4–10

Howell CR, Beier RC, Stipanovic RD (1988) Production of ammonia by *Enterobacter cloacae* and its possible role in the biological control of *Pythium* pre-emergence damping-off by the bacterium. Phytopathology 78:1075–1078

Howell CR, Hanson LE, Stipanovic RD, Puckhaber LS (2000) Induction of terpenoid synthesis in cotton roots and control of *Rhizoctonia solani* by seed treatment with *Trichoderma virens*. Phytopathology 90:248–252

Iavicoli A, Boutet E, Buchala A, Métraux JP (2003) Induced systemic resistance in *Arabidopsis thaliana* in response to root inoculation with *Pseudomonas fluorescens* CHA0. Mol Plant Microbe Interact 16:851–858

Intana W, Chamswarng C (2007) Control of Chinese-kale damping-off caused by *P. aphanidermatum* by antifungal metabolites of *Trichoderma virens*. Songklanakarin J Sci Technol 29(4):919–927

Islam MT, Hashidoko Y, Deora A, Ito T, Tahara S (2005) Suppression of damping-off-disease in host plants by the rhizoplane bacterium *Lysobacter sp.* strain SB-K88 is linked to plant colonization and antibiosis against soil-borne peronosporomycetes. Appl Environ Microbiol 71:3786–3796

Jayaraj J, Radhakrishnan NV, Velazhahan R (2006) Development of formulations of *Trichoderma harzianum* strain M1 for control of damping-off of tomato caused by *Pythium aphanidermatum*. Arch Phytopathol Plant Prot 39:1–8

Jones JD, Dangl JL (2006) The plant immune system. Nature 14:323–329

Kageyama K, Nelson EB (2003) Differential inactivation of seed exudate stimulation of *Pythium ultimum* sporangium germination by *Enterobacter cloacae* influences biological control efficacy on different plant species. Appl Environ Microbiol 69:1114–1120

Kanjanamaneesathian M, Phetcharat V, Pengnoo A, Upawan S (2003) Use of *Trichoderma harzianum* cultured on ground mesocarp fibre of oil-palm as seed treatment to control *Pythium aphanidermatum*, a causal agent of damping-off of Chinese kale seedling. World J Microbiol Biotechnol 19:825–829

Khare A, Upadhyay RS (2009) Induction of mutant strains of *Trichoderma viride* 1433 for biocontrol of *Pythium aphanidermatum*. Environ Biol Conserv 14:21–27

Khare A, Singh BK, Upadhyay RS (2010) Biological control of *Pythium aphanidermatum* causing damping-off of mustard by mutants of *Trichoderma viride1433*. J Agric Technol 6(2):231–243

Kilany M, Amry SA, Hashem M (2015) Evaluation of the role of soil-isolated bacteria in the production of antifungal metabolites against *Arabian journal of science and engineering*. Microbiol (in press)

Kowalchuk GA, van Os GJ, Aartrijk J, Veen JA (2003) Microbial community responses to disease management soil treatments used in flower bulb cultivation. Biol Fertil Soils 37(1):55–63

Landis TD, Tinus RW, McDonald SE, Barnett JP (1990) The container tree nursery manual. The biological component: nursery pests and mycorrhizae, vol 5, Agriculture handbook 674. USDA Forest Service, Washington, DC, p 171

Larkin RP, English JT, Mihail JD (1995) Effects of infection by *Pythium spp* on the root system morphology of alfalfa seedlings. Phytopathology 85:430–435

Leclere V, Beche M, Adam A, Guez JS, Wathelet B, Ongena M, Thonart P, Gancel F, Chollet-Imbert M, Jacques P (2005) Mycosubtilin overproduction by *Bacillus subtilis* BBG100 enhances the organism's antagonistic and biocontrol activities. Appl Environ Microbiol 71:4577–4584

Lewis A, Papavizas GC (1984) A new approach to stimulate population proliferation of *Trichoderma* species and other potential biocontrol fungi introduced into natural soils. Phytopathology 74:1240–1244

Li Z, Pinson SRM, Marchetti MA, Stansel JW, Park WD (1995) Characterization of quantitative trait loci (QTLs) in cultivated rice contributing to field resistance to sheath blight (*Rhizoctonia solani*). Theor Appl Genet 91:382–388

Li B, Ravnskov S, Xie G, Larsen J (2007) Biocontrol of *Pythium* damping-off in cucumber by arbuscular mycorrhiza-associated bacteria from the genus *Paenibacillus*. Biocontrol 52:863–875

Lo CT (1998) General mechanisms of action of microbial biocontrol agents. Plant Pathol Bull 7:155–166

Loliam B, Morinaga T, Chaiyanan S (2013) Biocontrol of *Pythium aphanidermatum* by the cellulolytic actinomycetes *Streptomyces rubrolavendulae* S4. Sci Asia 39:584–590

Lumsden RD, Locke JC (1989) Biological control of damping-off caused by *Pythium ultimum* and *Rhizoctonia solani* with *Gliocladium virens* in soilless mix. Am Phytopathol Soc 79 (3):361–366

Manici LM, Caputo F, Bambini V (2004) Effect of green manure on *Pythium* spp. population and microbial communities in intensive cropping systems. Plant Soil 263:133–142

Manjunath M, Prasanna R, Nain L, Dureja P, Singh R, Kumar A et al (2010) Biocontrol potential of cyanobacterial metabolites against damping-off disease caused by *Pythium aphanidermatum* in solanaceous vegetables. Arch Phytopathol Plant Prot 43:666–677

Manoranjitham SK, Prakasam V, Rajappan K, Amutha G (2000) Control of chilli damping-off using bioagents. J Mycol Plant Pathol 30:225–228

Marshall KC (1980) Bacterial adhesion in natural environments. In: Berckey RCW (ed) Microbial adhesion to surface. Ellisttorwood, Chichester, Great Britain, pp 187–196
Martin FN, Hancock JG (1987) The use of *Pythium oligandrum* for biological control of pre-emergence damping-off caused by *P. ultimum*. Phytopathology 77:1013–1020
Mazzola M (2007) Manipulation of Rhizosphere bacterial communities to induce suppressive soils. J Nematol 39(3):213–220
McKellar ME, Nelson EB (2003) Compost-induced suppression of *Pythium* damping-off is mediated by fatty acid-metabolizing seed-colonizing microbial communities. Appl Environ Microbiol 69:452–460
Messiha NAS, Van Diepeningen AD, Farag NS, Abdallah SA, Janse JD, Van Bruggen AHC (2007) *Stenotrophomonas maltophilia*: a new potential biocontrol agent of *Ralstonia solanacearum*, causal agent of potato brown rot. Eur J Plant Pathol 118:211–225
Muthukumar A, Eswaran A, Sanjeevkumas K (2011) Exploitation of Trichoderma species on the growth of *Pythium aphanidermatum* in chilli. Braz J Microbiol 42:1598–1607
Nakkeeran S, Kavitha K, Chandrasekar G, Renukadevi P, Fernando WGD (2006) Induction of plant defence compounds by *Pseudomonas chlororaphis* PA23 and *Bacillus subtilis* BSCBE4 in controlling damping-off of hot pepper caused by *Pythium aphanidermatum*. Biocontrol Sci Technol 16(4):403–416
Neelamegam R (2004) Evaluation of fungal antagonists to control damping-off of tomato (Lycopersicon esculentum Mill.) caused by *Pythium ultimum*. J Biol Control 18:97–102
Nelson EB (1987) Rapid germination of sporangia of *Pythium* species in response to volatiles from germination seeds. Phytopathology 77:1108–1112
Nelson EB, Chao WL, Norton GT, Harman GE (1986) Attachment of *Enterobacter cloacae* to hyphae of *Pythium ultimum*: possible role in the biological control of *Pythium* pre-emergence damping-off. Phytopathology 76:327–335
Pal KK, Gardener BM (2006) Biological control of plant pathogens. Plant Health Instructor. doi:10.1094/PHI-A-2006-1117-02
Palumbo JD, Yuen GY, Jochum CC, Tatum K, Kobayashi DY (2005) Mutagenesis of beta-1, 3-glucanase genes in *Lysobacter enzymogenes* strain C3 results in reduced biological control activity toward Bipolaris leaf spot of tall fescue and *Pythium damping*-off of sugar beet. Phytopathology 95:701–707
Paulitz TC (1991) Effect of *Pseudomonas putida* on the stimulation of *Pythium ultimum* by seed volatiles of pea and soybean. Phytopathology 81:1282–1287
Paulitz TC, Matta A (1999) The role of the host in biological control of diseases. In: Albajes R, Gullino ML, van Lenteren JC, Elad Y (eds) Integrated pest and disease management in greenhouse crops. Kluwer, Wageningen, pp 394–410
Paulitz TC, Ahmad JS, Baker R (1990) Integration of *Pythium nunn and Trichoderma harzianum* isolate T-95 for the biological control of Pythium damping-off of cucumber. Plant Soil 121:243–250
Postma J, Geraats BPJ, Pastoor R, van Elsas JD (2005) Characterization of the microbial community involved in the suppression of *Pythium aphanidermatum* in cucumber grown on rockwool. Phytopathology 95:808–818
Ramamoorthy V, Raguchander T, Samiyappan R (2002) Enhancing resistance of tomato and hot pepper to *Pythium* diseases by seed treatment with fluorescent pseudomonads. Eur J Plant Pathol 108:429–441
Rey T, Schornack S (2013) Interactions of beneficial and detrimental root-colonizing filamentous microbes with plant hosts. Genome Biol 14(6):121
Salman M, Abuamsha R (2012) Potential for integrated biological and chemical control of damping-off disease caused by *Pythium ultimum* in tomato. BioControl 57:711–718
Schirmböck M, Lorito M, Wang Y-L et al (1994) Parallel formation and synergism of hydrolytic enzymes and peptaibol antibiotics, molecular mechanisms involved in the antagonistic action of *Trichoderma harzianum* against phytopathogenic fungi. Appl Environ Microbiol 60:4364–4370

Schisler DA, Slininger PJ, Behle RW, Jackson MA (2004) Formulation of *Bacillus* spp. for biological control of plant diseases. Phytopathology 94:1267–1271
Shahraki M, Heydari A, Hassanzadeh N (2009) Investigation of antibiotic, siderophore and volatile metabolites production by *Bacillus* and *Pseudomonas* bacteria. Iran J Biol 22:71–85
Singh R, Sachan NS (2013) Review on biological control of soil borne fungi in vegetable crops. Hort Flora Res Spectr 2(1):72–76
Stanghellini ME, Miller RM (1997) Biosurfactants. Their identity and potential efficacy in the biological control of zoosporic plant pathogens. Plant Dis 81:4–12
Steinberg C, Edel-Hermann V, Alabouvette C, Lemanceau P (2007) Soil suppressiveness to plant diseases. In: van Elsas JD, Jansson J, Trevors JT (eds) Modern soil microbiology. CRC, New York, pp 455–478
Sumner DR, Gascho GJ, Johnson AW, Hook JE, Threadgill ED (1990) Root diseases, populations of soil fungi, and yield decline in continuous double-crop corn. Plant Dis 74:704–710
Termorshuizen AJ, Jeger MJ (2008) Strategies of soilborne plant pathogenic fungi in relation to disease suppression. Fungal Ecol 1:108–114
Thrane C, Nielsen TH, Nielsen MN, Sørensen J, Olsson S (2000) Viscosinamide-producing *Pseudomonas fluorescens* DR54 exerts a biocontrol effect on *Pythium ultimum* in sugar beet rhizosphere. Microbiol Ecol 33(2):139–146
van Dijk K, Nelson EB (1997) Inactivation of seed general mechanisms of action of microbial bio-control agents 1 6 5 exudate stimulants of *Pythium ultimum* sporangium germination by strains of *Enterobacter cloacae* and other seed-associated bacteria. Appl Environ Microbiol 63:331–335
van Dijk K, Nelson EB (2000) Fatty acid competition as a mechanism by which *Enterobacter cloacae* suppresses *Pythium ultimum* sporangium germination and damping-off. Appl Environ Microbiol 66:5340–5347
van Rijin E (2007) Disease suppression and phytosanitary aspects of compost. PhD thesis. Wageningen University, Wageningen, The Netherlands
Vinalea F, Sivasithamparamb K, Ghisalbertic EL, Marraa R, Wooa DL, Loritoa M (2008) Trichoderma–plant–pathogen interactions. Soil Biol Biochem 40:1–10
Weller DM, Raaijmakers JM, Gardener BB, Thomashow LS (2002) Microbial populations responsible for specific soil suppressiveness to plant pathogens. Annu Rev Phytopathol 40:309–348
Whipps JM (1997) Developments in the biological control of soil-borne plant pathogens. Adv Botan Res 26:1–134
Whipps JM (2001) Microbial interactions and biocontrol in the rhizosphere. J Exp Bot 52:487–511
Wilhite SE, Lumsden RD, Straney DC (1994) Mutational analysis of gliotoxin production by the biocontrol fungus *Gliocladium virens* in relation to suppression of pythium damping-off. Phytopathology 84:816–821
Windstam S, Nelson EB (2008) Differential interference with *Pythium ultimum* sporangial activation and germination by *Enterobacter cloacae* in the corn and cucumber spermospheres. Appl Environ Microbiol 74(14):4285–4291
Yang XB (2001) Identification of soybean seedling diseases. Integr Crop Manag 486(10):79–80

Chapter 10
Interaction of Rhizobia with Soil Suppressiveness Factors

Kim Reilly

10.1 Introduction

Soil health can be broadly defined as 'the competence with which soil functional processes (e.g. nutrient cycling, energy flow) are able to support viable, self-sustaining (micro) faunal and (micro) floral ecosystems' (Sturz and Christie 2003). In contrast soil quality is defined by its 'suitability for a specific use'. This definition encompasses biological, physical and chemical attributes and is dependent on the soil type and land use context (Griffiths et al. 2010). The balance between chemical, physical and biological components contributes to maintaining soil health and quality (Nautiyal et al. 2010). There has been an increasing awareness of the importance of soil quality and soil health in sustainable agricultural production. The role of rhizobia in N fixation has long been recognised; however, more recently, the role of rhizobia in soil suppressiveness has been explored and is reviewed here. For the purposes of this chapter, soil suppressiveness is defined as 'the phenomenon whereby incidence of crop disease remains low even though a virulent pathogen and a susceptible host are present'. As discussed elsewhere in this volume, both biotic and abiotic factors can contribute to suppressiveness. From a biological perspective, suppressiveness may be general (i.e. nontransferable and attributed to the activity of soil total microbial biomass) or specific (i.e. due to the effect of individual or specific groups of microorganisms and is transferable) (Weller et al. 2002).

The rhizobia are defined as symbiotic nitrogen-fixing soil bacteria which form nodules on the roots of leguminous plants. They are generally aerobic, motile, non-sporulating rods and are chemoheterotrophic (i.e. require preformed organic compounds as a source of carbon and energy) and diazotrophic (i.e. can fix atmospheric nitrogen). N fixation occurs only after forming nodules on the roots

K. Reilly (✉)
Teagasc, Ashtown Food Research Centre, Dublin 15, Ireland
e-mail: Kim.Reilly@teagasc.ie

M.K. Meghvansi, A. Varma (eds.), *Organic Amendments and Soil Suppressiveness in Plant Disease Management*, Soil Biology 46,
DOI 10.1007/978-3-319-23075-7_10

of leguminous plants and not in free-living states. The first rhizobia species *Rhizobium leguminosarum* was described in 1889 (Frank 1889), and subsequently described rhizobia were initially classified within the genus *Rhizobium* (Young et al. 2001). Initially those bacteria capable of forming symbiotic N-fixing nodules were placed in the genus *Rhizobium* and considered as rhizobia, and those which formed tumours or hairy roots in the genus *Agrobacterium*. However, with the advent of molecular methods, the classification of rhizobia has recently undergone extensive and significant change (Weir 2012). Current taxonomy is based on 16S rRNA sequence data and has led to substantial reclassification and renaming. Well-known examples include *Agrobacterium tumefaciens* (now *Rhizobium radiobacter*) and *Sinorhizobium meliloti* (now *Ensifer meliloti*).

To date over 98 bacterial species belonging to 14 genera have been identified as rhizobia (Fikri-Benbrahim and Berrada 2014; Shiraishi et al. 2010; Weir 2012). Most rhizobia belong to the order Rhizobiales within the α-proteobacteria and are predominantly found in the genera *Rhizobium*, *Ensifer* (previously *Sinorhizobium*), *Bradyrhizobium* and *Mesorhizobium*. Not all species within each genus form N-fixing symbiotic root nodules. For example, species such as *Rhizobium radiobacter* (previously *Agrobacterium tumefaciens*), *Ensifer adhaerens* and *Bradyrhizobium betae* are not considered rhizobia although they may cause galls or tumours as described above. Some rhizobial species occur in widely different genera in the β-proteobacteria and γ-proteobacteria and have arisen due to horizontal gene transfer of plasmids carrying symbiotic genes.

The ability to form symbiotic nodules is usually mediated by a plasmid pSym which carries *nod* and *nif* genes responsible for nodule formation and nitrogen fixation, respectively. In contrast pathogenic types which cause tumours or hairy roots harbour pTi or pRi plasmids, which carry *vir* genes responsible for virulence. In some instances some bacterial strains may carry both types of plasmid (Fig. 10.1). The first reported strains of this type were reported by Velazquez et al. (2005) who described *Rhizobium rhizogenes* strains which induced N-fixing symbiotic nodules in *Phaseolus vulgaris* and also caused hairy root or tumour formation in non legume plants (Velazquez et al. 2005). Rhizobial plasmids are commonly large (>500 kb) and may carry essential genes; thus, some authors consider that the rhizobial genome may more properly be considered a multipartite genome with two components—a core genome and an accessory genome. In some instances, for example, within the genus *Bradyrhizobium*, the symbiosis genes are integrated by lateral gene transfer into the main chromosome forming a 'symbiosis island' (SI) (Laranjo et al. 2014) (see Fig. 10.1).

Under the previous classification scheme, rhizobia were classified based on phenotypic criteria and were grouped depending on host specificity (cross inoculation between rhizobia and their host plants) and on growth in culture media. Species were classified as 'fast growing' and 'slow growing', and such descriptors are still used in the literature. Rhizobia such as *Bradyrhizobium japonicum* are categorised as 'slow growing', whilst *Ensifer xingianense* (*Sinorhizobium fredii*) is described as 'fast growing' (Chen et al. 1988; Fikri-Benbrahim and Berrada 2014). Certain rhizobial species or subspecies (i.e. biovars) only nodulate specific legume species,

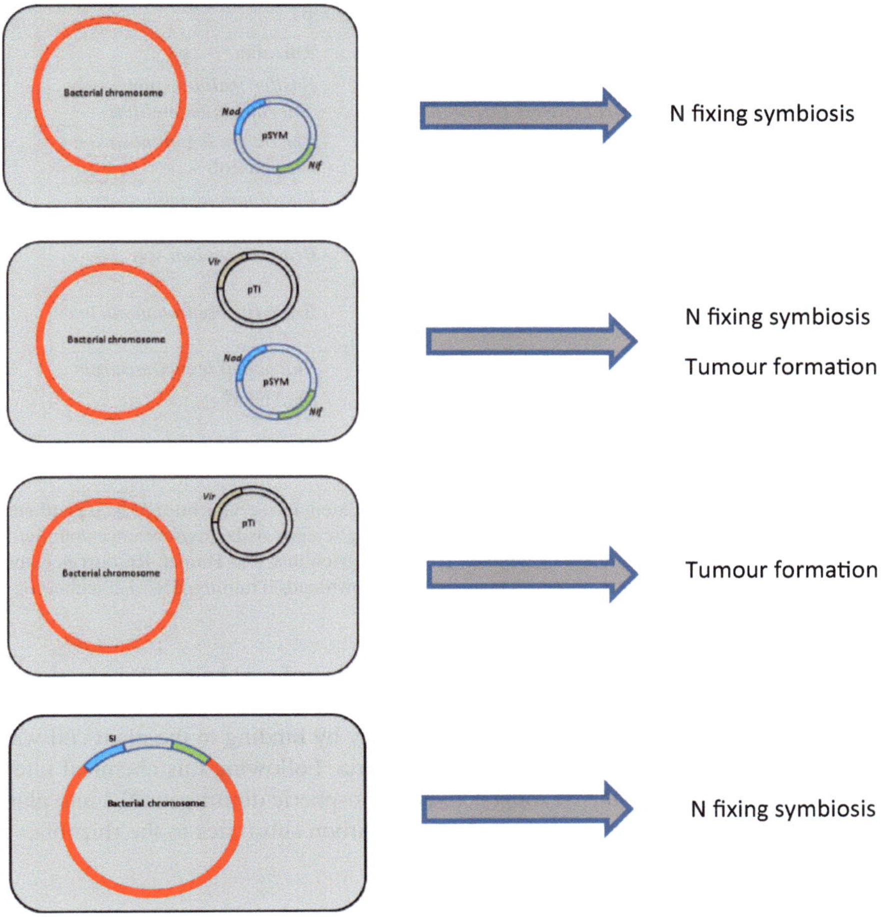

Fig. 10.1 Variability in plasmid composition and functional outcome in symbiotic and pathogenic bacteria

whilst others are promiscuous and can nodulate a range of hosts. Rhizobia and their legume hosts have been categorised into 'cross-inoculation groups'. Each group comprises all of the legume species that will develop nodules if inoculated with rhizobia isolated from any other member of the sample group (Table 10.1). This concept is widely used by famers and extension services to guide formulation of soil rhizobial inoculants.

Formation of N-fixing symbiotic nodules involves a complex interplay of responses between the plant root and the rhizobia bacterium and involves the coordinated expression of both plant and bacterial genes. During symbiotic nodule formation, the legume plant produces flavonoid compounds in the root exudate that are recognised by the bacterium and induce expression of nodulation genes (*nod* genes) in the symbiotic rhizobia bacterium. Nod factors induce responses in the

Table 10.1 Common legume/rhizobia cross-inoculation groups

Cross-inoculation group	Rhizobia
Alfalfa group: Alfalfa (*Medicago* spp.); sweet clovers (*Melilotus* spp.)	*Ensifer meliloti* (previously *Sinorhizobium meliloti*)
Bean group: Beans (*Phaseolus* spp.)	*Rhizobium leguminosarum* bv. phaseoli
Clover group: Clovers (*Trifolium* spp.)	*Rhizobium leguminosarum* bv. trifolii
Cowpea group: Peanut (*Arachis hypogaea*); cowpea (*Vigna unguiculata*)	*Bradyrhizobium* spp.
Soybean group: Soybeans (*Glycine max*)	*Bradyrhizobium japonicum*
Pea and vetch group: Peas (*Pisum* spp.); vetches (*Vicia* spp.); lentils (*Lens culinaris*); faba bean *Vicia faba*)	*Rhizobium leguminosarum* bv. viceae
Chickpea group: Chickpea (*Cicer arietinum*)	*Mesorhizobium ciceri*

Source: Modified and adapted from University of Florida Extension service http://edis.ifas.ufl.edu/aa126, USDA extension services http://efotg.sc.egov.usda.gov/references/public/ia/agronomytechnote11attach.pdfand College of Tropical Agriculture and Human Resources extension service http://www.ctahr.hawaii.edu/bnf/Downloads/Training/BNF%20technology/Rhizobia.PDF

legume host including (1) deformation and branching of root hairs, (2) induction of gene expression and (3) induction of cell division (Atlas and Bartha 1986) (Gage 2004). Host plant lectins play a role in specificity by binding to the plant cell wall and to saccharide moieties on compatible bacteria. Following this chemical interplay within the nodules, the rhizobia convert atmospheric dinitrogen (N_2) into plant available forms and the plant in turn provides carbon substrates to the rhizobia.

10.2 Suppressive Effects of Rhizobia

Rhizobia may occur not only in symbiotic nodules on legume roots but also survive as free-living bacteria prior to nodulation. Some strains are able to colonise and survive in the rhizosphere of non legume crops or in other soil microenvironments, and both free-living and symbiotic forms can have a range of direct and indirect effects which may contribute to soil suppressiveness and prevention of crop disease (Antoun et al. 1998; Avis et al. 2008; Dakora 2003). Suppressive effects of rhizobia have predominantly been reported against fungal and nematode plant pathogens, although other effects are also reported (Avis et al. 2008).

10.2.1 Suppressive Effects Against Fungal Pathogens

Suppressive effects of rhizobia against fungal pathogens have been described since the 1970s. A strain of *B. japonicum* could inhibit sporulation in a range of fungal plant pathogens including *Phytophthora megasperma*, *Pythium ultimum*, *Fusarium oxysporum* and *Ascochyta imperfecta*, whilst 49 different strains of *Sinorhizobium meliloti* were reported to reduce growth of *F. oxysporum* (Antoun et al. 1978; Tu 1979). A number of studies indicated that inoculation of plants such as soybean, common bean, mung bean, sunflower and okra with rhizobia could reduce incidence of root pathogens including *Phytophthora*, *Fusarium*, *Macrophomina* and *Rhizoctonia* (Buonassisi et al. 1986; Ehteshamulhaque and Ghaffar 1993; Tu 1979).

In a recent study, soil amendment with crop debris of wild rocket (*Diplotaxis tenuifolia*) led to suppression of Fusarium wilt in cucumber seedlings and was accompanied by a change in the rhizosphere microbial population. *Rhizobium* spp. were amongst a group of bacterial species whose populations showed a significant increase in the suppressive soil (Klein et al. 2013).

10.2.2 Suppressive Effects Against Nematodes

Nematode suppressive soils have been described in a number of studies and *Rhizobium* species have been consistently implicated in this type of suppression. *Rhizobium* species have been identified as responsible for soil suppressiveness against the plant pathogenic nematode *Heterodera schachtii* (Yin et al. 2003). *H. schachtii* cysts isolated from a suppressive soil could transfer this property to non-suppressive soils, and when rDNA from the cyst-associated bacteria was sequenced, only rDNA sequences from the *Rhizobium* spp. group were consistently associated with high levels of suppressiveness. *Rhizobium* spp. have been shown to suppress juveniles of the soybean cyst nematode *H. glycines*, and in a recent study *Rhizobium* spp. and *Streptomyces* spp. sequences were the dominant bacterial DGGE bands detected in the bacterial communities associated with nematode cysts in a 2-year study on a long-term monoculture nematode suppressive soil (Zhu et al. 2013).

10.2.3 Other Suppressive Effects

Although less widely reported, suppressive effects on other crop pathogens and pests have been noted. For example, during legume/cereal rotations in Africa, populations of the parasitic weed *Striga* (witchweed) that can cause heavy losses in cereal crops have shown significant decreases when legume soybean, groundnut or cowpea was used as the preceding crop (Carsky et al. 2000). This is attributed both to symbiotic N supply and to antagonistic compounds produced by legume rhizobia. Similarly, broomrapes (*Orobanche* spp.) are parasitic weeds which cause severe losses in vegetable, legume and sunflower crops. In a recent study (Bouraoui et al. 2012), germination of *O. foetida* in a co-culture with its plant host faba bean (*Vicia faba*) was reduced by up to 75 % following inoculation with selected *Rhizobium* strains. Suppressive effects of rhizobia against bacterial and viral disease in bean have also been reported (Elbadry et al. 2006; Huang et al. 2007).

10.2.4 Mechanism of Suppressive Effects

Proposed mechanisms include direct effects such as competition with pathogens for nutrients and for preferred colonisation sites on the root; production of antimicrobial, antibacterial or germination inhibition compounds by the rhizobia themselves; as well as indirect mechanisms such as improved plant nutrition and plant growth and elicitation of plant defence responses (induced resistance). Where plant or rhizobial produced compounds are present in the soil as root exudates and/or from crop residue decomposition, they can inhibit soilborne pathogens in the legume crop and also for crops rotated with the legume crop.

10.3 Substances Produced by Soil Rhizobia

10.3.1 Nod Factors

Nod factors are lipooligosaccharide molecules produced by rhizobial bacteria that are involved in symbiotic nodule formation. The structure of the oligosaccharide backbone determines the host specificity and the biological activity of the bacterium (Savoure et al. 1994). Several studies indicate that Nod factors can elicit the production of plant phytoalexin defence compounds, thereby protecting the plant from disease (Dakora 2003) (see Sect. 10.4.1). Nod factors can be perceived by non legume crops (Khaosaad et al. 2010), and this provides a direct mechanism to explain disease suppression in legume and succeeding crops.

Nod factors of some *Rhizobium* spp. have also been demonstrated to increase colonisation of roots by mycorrhizal fungi such as *Glomus*. Application of Nod factors at concentrations of 10^{-9} M could promote the colonisation of legume and non legume plants by AM fungi (Dakora 2003). Application of low concentrations of Nod factors (10^{-7}–10^{-9} M) to soybean roots has been shown to increase root biomass, and foliar application of Nod factors also increased grain yield and photosynthate production in non legumes including rice, bean, canola, apple and grape (Matiru and Dakora 2005; Souleimanov et al. 2002). Recent studies in barley and alfalfa indicate that where rhizobia or Nod factors are applied prior to the AM fungus (rather than co-inoculated), colonisation by the second symbiont was inhibited (Catford et al. 2006; Khaosaad et al. 2010).

10.3.2 Phytohormones and Growth-Promoting Compounds

Rhizobia have long been known to produce plant growth-stimulating hormones such as gibberellins, cytokinins and indole-3-acetic acid (IAA). Direct growth-promoting effects of IAA have been observed in legumes and also in non legume plants such as lettuce (*Lactuca sativa*) and canola (*Brassica campestris*) in response to IAA produced by *R. leguminosarum* (Antoun et al. 1998). Some studies have shown substantial root hair proliferation in rice and other cereals in response to inoculation with rhizobia, the proliferation being attributed to hormones including IAA and gibberellins produced by the rhizobia and resulting in enhanced nutrient uptake capacity by the plant (Yanni et al. 2001).A number of other compounds produced by rhizobia can stimulate plant growth. The compound lumichrome was originally identified from culture filtrates of *E. meliloti*. At low concentration (5 nM), lumichrome increased growth in a range of legume and non legume crops including soybean, maize, millet, sorghum and cowpea. At high concentration, however, growth was inhibited in soybean, millet and cowpea (Matiru and Dakora 2005). Lumichrome may readily be produced in the rhizosphere by breakdown of riboflavin, which may account for its effect on a range of plant types. Rhizobial strains which produce large amounts of lumichrome could thus be of interest as a general crop growth promoter (Matiru and Dakora 2005).

10.3.3 Siderophores and Organic Acids

It has been demonstrated that rhizobia grown in culture secrete siderophores and organic acids that would allow them to obtain nutrients from the soil in free living stages. Organic acids serve to solubilise phosphorus (P) and manganese (Mn). Siderophores are small, high-affinity iron-chelating compounds secreted by micro-organisms such as bacteria and fungi. It is proposed that plants benefit directly from the pool of bacterially solubilised nutrients in the rhizosphere (Dakora 2003). Rhizobia produce a range of siderophores which can modulate plant iron nutrition and in addition play a role in control of soilborne pathogens by sequestering iron from pathogens, thereby suppressing their growth and proliferation (Hamdan et al. 1991). Out of 196 *Rhizobium* species examined, 181 produced siderophores; subsequent studies showed that only siderophore-producing strains of *Sinorhizobium meliloti* could inhibit growth of the pathogen *Macrophomina phaseolina* (Arora et al. 2001).

10.3.4 Hydrogen Gas

Hydrogen (H_2) gas is a by-product of nitrogenase activity during nodule nitrogen fixation. In HUP^+ rhizobia, a hydrogenase uptake system oxidises the H_2 produced; however, in HUP^- strains, the H_2 is released into the soil. The released hydrogen gas has been shown to stimulate growth in a range of cereals, and this growth was accompanied by an increase in soil bacteria able to oxidise nodule produced H_2. It is proposed that the secreted H_2 stimulates the proliferation of plant growth-promoting bacteria resulting in increased plant growth (Dakora 2003).

10.4 Substances Produced by Legumes in Response to Rhizobia

10.4.1 Phytoalexins and Phytoanticipins

Many plants produce phytoalexin defence compounds in response to both biotic (e.g. pathogens) and abiotic (e.g. heavy metals, UV radiation) stresses (Dixon and Lamb 1990). Phytoalexins are defined as low-molecular-weight compounds produced in response to pathogens which can slow or prevent the growth of plant pathogens. The term phytoanticipin is used to describe preformed antimicrobial compounds already present in the plant. Legume phytoalexins are predominantly isoflavonoid phenolic compounds, including pisatin (*Pisum* sp.), glyceollin (*Glycine* sp.), medicarpin (*Medicago* sp.) and coumestrol (*Phaseolus vulgaris*). Non-isoflavonoid compounds such as stilbenes, benzofurans and furanoacetylenes also occur as legume phytoalexins (Dakora and Phillips 1996). There is considerable evidence that inoculating legume plants with rhizobia or treating with Nod factors induces production of phytoalexins that protect the plant against pathogens (Avis et al. 2008; Dakora 2003; Savoure et al. 1994). Legume host plant secretion of phytoalexins into the rhizosphere can occur at significant levels and is a direct mechanism by which rhizobia can contribute to soil disease suppressiveness (Dakora et al. 1993; Parniske et al. 1991).

Some isoflavonoid compounds such as coumestrol act both as nod gene inducers and as phytoalexins, although the majority of nod gene inducers are not phytoalexins (Dakora et al. 1993).

10.4.2 Other Phenolic Compounds

During symbiotic nodule formation, the legume plant produces phenolic compounds (commonly isoflavonoids) in the root exudate which can induce expression of *nod* genes in the symbiotic rhizobia bacterium. For example, in soybean (*Glycine max*) compounds such as the flavonoids coumestrol, daidzein, genistein and isoliquiritigenin are strong inducers of *nod* genes in the soybean symbiotic rhizobia *B. japonicum* (Kape et al. 1992). In addition to their role in *nod* gene induction, such compounds also play other roles. For example, flavonoid compounds can exhibit chemotaxis, attracting the symbiont towards the root. This has been described for fast-growing rhizobia species such as *E. meliloti* previously (*R. meliloti*), a symbiont of alfalfa. However, chemotaxis towards pathogens also occurs, for example, the isoflavonoids genistein and daidzein in soybean root exudate attract zoospores of the soybean pathogen *Phytophthora sojae* (Dakora and Phillips 1996).

Isoflavonoids also affect the growth of both fungi and bacteria although the interaction is complex. Daidzein can stimulate spore germination of the arbuscular mycorrhizal fungus *Glomus*. Formononetin and biochanin A showed inhibitory effects on growth of *Glomus* spp. at low concentration but stimulated growth at high concentration (Dakora and Phillips 1996). In continuous soybean monoculture systems, increases in total soil biomass were correlated with levels of daidzein and genistein in the rhizosphere. Genistein significantly increased arbuscular mycorrhizal (AM) hyphal length and spore density; however, soilborne pathogens were also stimulated by soybean root flavonoids, which inhibited formation of symbiosis (Weller et al. 2002; Wang et al. 2012).

Interestingly some flavonoids can also induce resistance within the symbiotic rhizobia to plant root defence compounds such as phytoalexins. Soybean isoflavonoids such as genistein, daidzein and isoliquiritigenin can induce resistance to bactericidal concentrations of glyceollin (a soybean phytoalexin) (Kape et al. 1992). Early experiments showed that a broad range of soybean rhizobia, although initially susceptible to the soybean phytoalexin glyceollin, could adapt to the presence of the phytoalexin and were able to tolerate previously bactericidal concentrations following exposure to low concentrations of daidzein and genistein. Such induced resistance allows the rhizobia to survive in the root rhizosphere, despite accumulation of significant amounts of bactericidal phytoalexins (Parniske et al. 1991).

10.5 Conclusions

Whilst the benefits of rhizobia for crop growth have long been appreciated, it is clear that the benefits of rhizobia go beyond their role as nitrogen-fixing symbionts in legume crops. There is convincing evidence that rhizobia not only enhance plant growth but also play a role in reducing plant disease, in legume and other crops. The mechanisms underlying these suppressive effects however are complex and interrelated and remain to be fully explored. Unfortunately despite an initial flurry of research in the 1990s, little further exploration has been carried out. Much of this lack of study may be due to (a) the significant reclassification of the rhizobia resulting in lack of clarity in terms, classification and nomenclature of rhizobial strains and (b) the complexity of the interactions between rhizobia, other soil microbes and host and non-host plants. However, given the potential for identification and/or development of rhizobial strains that could be used to boost crop growth as N-fixing symbionts and also as disease suppressants, a renewed focus would be welcome. Questions of interest include: (1) Are there differences in compounds produced and suppressive effects between different cross-inoculation groups and within different strains in the same cross-inoculation group? (2) What is the role of pSYM plasmids in suppressiveness? For example, is plasmid copy number important? How much variation is seen in pSYM sequences, and are some variants more effective at inducing suppressiveness than others? Are some variants more effective at establishing N fixation? (3) Is it possible to engineer optimised inoculants by transforming effective N-fixing strains with additional pSYM plasmids selected for disease suppressiveness? A further exploration of these and other topics would be of future benefit.

References

Antoun H, Bordeleau LM, Gagnon C (1978) Antagonism between *Rhizobium meliloti* and *Fusarium oxysporum* with respect to symbiotic effectiveness. Can J Plant Sci 58:75–78

Antoun H, Beauchamp CJ, Goussard N, Chabot R, Lalande R (1998) Potential of *Rhizobium* and *Bradyrhizobium* species as plant growth promoting rhizobacteria on non-legumes: effect on radishes *(Raphanus sativus* L.). Plant Soil 204:57–67

Arora NK, Kang SC, Maheshwari DK (2001) Isolation of siderophore-producing strains of *Rhizobium meliloti* and their biocontrol potential against *Macrophomina phaseolina* that causes charcoal rot of groundnut. Curr Sci 81:673–677

Atlas R, Bartha R (1986) Microbial ecology. Benjamin/Cummings Publishing, California

Avis TJ, Gravel V, Antoun H, Tweddell RJ (2008) Multifaceted beneficial effects of rhizosphere microorganisms on plant health and productivity. Soil Biol Biochem 40:1733–1740

Bouraoui M, Abbes Z, Abdi N, Hmissi I, Sifi B (2012) Evaluation of efficient *Rhizobium* isolates as biological control agents of *Orobanche foetida* Poir. parasitizing *Vicia faba* L. *minor* in Tunisia. Bulgarian J Agric Sci 18:557–564

Buonassisi AJ, Copeman RJ, Pepin HS, Eaton GW (1986) Effect of *Rhizobium* spp on *Fusarium solani* f. sp.*phaseoli*. Can J Plant Pathol 8:140–146

Carsky RJ, Berner DK, Oyewole BD, Dashiell K, Schulz S (2000) Reduction of *Striga hermonthica* parasitism on maize using soybean rotation. Int J Pest Manage 46:115–120

Catford JG, Staehelin C, Larose G, Piche Y, Vierheilig H (2006) Systemically suppressed isoflavonoids and their stimulating effects on nodulation and mycorrhization in alfalfa split-root systems. Plant Soil 285:257–266

Chen WX, Yan GH, Li JL (1988) Numerical taxonomic study of fast-growing soybean Rhizobia and a proposal that *Rhizobium fredii* be assigned to *Sinorhizobium* gen. nov. Int J Syst Bacteriol 38:392–397

Dakora FD (2003) Defining new roles for plant and rhizobial molecules in sole and mixed plant cultures involving symbiotic legumes. New Phytol 158:39–49

Dakora FD, Phillips DA (1996) Diverse functions of isoflavonoids in legumes transcend anti-microbial definitions of phytoalexins. Physiol Mol Plant Pathol 49:1–20

Dakora FD, Joseph CM, Phillips DA (1993) Common bean root exudates contain elevated levels of daidzein and coumestrol in response to *Rhizobium* inoculation. Mol Plant-Microbe Interact 6:665–668

Dixon RA, Lamb CJ (1990) Molecular communication in interactions between plants and microbial pathogens. Annu Rev Plant Physiol Plant Mol Biol 41:339–367

Ehteshamulhaque S, Ghaffar A (1993) Use of rhizobia in the control of root-rot diseases of sunflower, okra, soybean and mungbean. J Phytopathol 138:157–163

Elbadry M, Taha RM, Eldougdoug KA, Gamal-Eldin H (2006) Induction of systemic resistance in faba bean (*Vicia faba* L.) to bean yellow mosaic potyvirus (BYMV) via seed bacterization with plant growth promoting rhizobacteria. J Plant Dis Protect 113:247–251

Fikri-Benbrahim H, Berrada K (2014) Taxonomy of the Rhizobia: current perspectives. Br Microbiol Res J 4:616–639

Frank B (1889) Uber die Pilzsymbiose der Leguminosen. Ber Deut Bot Ges 7:332–346

Gage DJ (2004) Infection and invasion of roots by symbiotic, nitrogen-fixing rhizobia during nodulation of temperate legumes. Microbiol Mol Biol Rev 68:280–286

Griffiths BS, Ball BC, Daniell TJ, Hallett PD, Neilson R, Wheatley RE, Osler G, Bohanec M (2010) Integrating soil quality changes to arable agricultural systems following organic matter addition, or adoption of a ley-arable rotation. Appl Soil Ecol 46:43–53

Hamdan H, Weller DM, Thomashow LS (1991) Relative importance of fluorescent siderophores and other factors in biological-control of *Gaeumannomyces graminis* var *Tritici* by *Pseudomonas fluorescens* 2–79 and m4-80r. Appl Environ Microbiol 57:3270–3277

Huang HC, Erickson RS, Hsieh TF (2007) Control of bacterial wilt of bean (*Curtobacterium flaccumfaciens* pv. *flaccumfaciens*) by seed treatment with *Rhizobium leguminosarum*. Crop Prot 26:1055–1061

Kape R, Parniske M, Brandt S, Werner D (1992) Isoliquiritigenin, a strong nod gene-inducing and glyceollin resistance-inducing flavonoid from soybean root exudate. Appl Environ Microbiol 58:1705–1710

Khaosaad T, Staehelin C, Steinkellner S, Hage-Ahmed K, Antonio Ocampo J, Manuel Garcia-Garrido J, Vierheilig H (2010) The *Rhizobium* sp. strain NGR234 systemically suppresses arbuscular mycorrhizal root colonization in a split-root system of barley (*Hordeum vulgare*). Physiol Plant 140:238–245

Klein E, Ofek M, Katan J, Minz D, Gamliel A (2013) Soil suppressiveness to fusarium disease: shifts in root microbiome associated with reduction of pathogen root colonization. Phytopathology 103:23–33

Laranjo M, Alexandre A, Oliveira S (2014) Legume growth-promoting rhizobia: an overview on the *Mesorhizobium* genus. Microbiol Res 169:2–17

Matiru VN, Dakora FD (2005) The rhizosphere signal molecule lumichrome alters seedling development in both legumes and cereals. New Phytol 166:439–444

Nautiyal CS, Chauhan PS, Bhatia C (2010) Changes in soil physicochemical properties and microbial functional diversity due to 14 years of conversion of grassland to organic agriculture in semi-arid agroecosystem. Soil Tillage Res 109:55–60

Parniske M, Ahlborn B, Werner D (1991) Isoflavonoid-inducible resistance to the phytoalexin glyceollin in soybean rhizobia. J Bacteriol 173:3432–3439

Savoure A, Magyar Z, Pierre M, Brown S, Schultze M, Dudits D, Kondorosi A, Kondorosi E (1994) Activation of the cell-cycle machinery and the isoflavonoid biosynthesis pathway by active *Rhizobium meliloti* nod signal molecules in *Medicago* microcallus suspensions. EMBO J 13:1093–1102

Shiraishi A, Matsushita N, Hougetsu T (2010) Nodulation in black locust by the Gammaproteobacteria *Pseudomonas* sp and the Betaproteobacteria *Burkholderia* sp. Syst Appl Microbiol 33:269–274

Souleimanov A, Prithiviraj B, Smith DL (2002) The major Nod factor of *Bradyrhizobium japonicum* promotes early growth of soybean and corn. J Exp Bot 53:1929–1934

Sturz AV, Christie BR (2003) Beneficial microbial allelopathies in the root zone: the management of soil quality and plant disease with rhizobacteria. Soil Till Res 72:107–123

Tu JC (1979) Evidence of differential tolerance among some root-rot fungi to rhizobial parasitism *in vitro*. Physiol Plant Pathol 14:171–8

Velazquez E, Peix A et al (2005) The coexistence of symbiosis and pathogenicity-determining genes in *Rhizobium rhizogenes* strains enables them to induce nodules and tumors or hairy roots in plants. Mol Plant-Microbe Interact 18:1325–1332

Wang J, Li X, Zhang J, Yao T, Wei D, Wang Y, Wang J (2012) Effect of root exudates on beneficial microorganisms-evidence from a continuous soybean monoculture. Plant Ecol 213:1883–1892

Weir BS (2012) The current taxonomy of rhizobia. NZ Rhizobia website. http://www.rhizobia.co.nz/taxonomy/rhizobia. Last updated 10 Apr 2012

Weller DM, Raaijmakers JM, Gardener BBM, Thomashow LS (2002) Microbial populations responsible for specific soil suppressiveness to plant pathogens. Annu Rev Phytopathol 40:309–18

Yanni YG, Rizk RY et al (2001) The beneficial plant growth-promoting association of *Rhizobium leguminosarum* bv. *Trifolii* with rice roots. Aust J Plant Physiol 28:845–870

Yin B, Valinsky L, Gao XB, Becker JO, Borneman J (2003) Bacterial rRNA genes associated with soil suppressiveness against the plant-parasitic nematode *Heterodera schachtii*. Appl Environ Microbiol 69:1573–1580

Young JM, Kuykendall LD, Martinez-Romero E, Kerr A, Sawada H (2001) A revision of *Rhizobium* Frank 1889, with an emended description of the genus, and the inclusion of all species of *Agrobacterium* Conn 1942 and *Allorhizobium undicola* de Lajudie et al. 1998 as new combinations: *Rhizobium radiobacte*r, *R. rhizogenes, R. rubi, R. undicola and R. vitis.* Int J Syst Evol Microbiol 51:89–103

Zhu Y, Shi F, Tian J, Liu J, Chen S, Xiang M, Liu X (2013) Effect of soybean monoculture on the bacterial communities associated with cysts of *Heterodera glycines*. J Nematol 45:228–235

Chapter 11
Biocontrol of Plant Parasitic Nematodes by Fungi: Efficacy and Control Strategies

Mohd. Sayeed Akhtar, Jitendra Panwar, Siti Nor Akmar Abdullah, Yasmeen Siddiqui, Mallappa Kumara Swamy, and Sadegh Ashkani

11.1 Introduction

Nematodes are filiform roundworms belonging to phylum Nematoda commonly found in plants, animals, and soil. They have the ability to utilize the various organic sources for the production of energy (Akhtar and Panwar 2011). Some plant parasitic nematodes usually feed on plant cells by choosing and establishing a single feeding site known as sedentary feeders, while others are migratory feeders which means they move from site to site on the root and rarely feed on plant single cell. In general, the plant parasitic nematodes are documented as the utmost vicious

M.S. Akhtar (✉) • S.N.A. Abdullah
Laboratory of Plantation Crops, Institute of Tropical Agriculture, Universiti Putra Malaysia (UPM), 43400 Serdang, Selangor, Malaysia
e-mail: sayeedbot@yahoo.co.in; sitiakmar@gmail.com

J. Panwar
Department of Biological Sciences, Birla Institute of Technology and Science, Centre for Biotechnology, Pilani 333031, India
e-mail: drjitendrapanwar@yahoo.co.in

Y. Siddiqui
Laboratory of Food Crops, Institute of Tropical Agriculture, Universiti Putra Malaysia (UPM), 43400 Serdang, Selangor, Malaysia
e-mail: yasmeen@upm.edu.my

M.K. Swamy
Faculty of Agriculture, Department of Crop Science, Universiti Putra Malaysia (UPM), 43400 Serdang, Selangor, Malaysia
e-mail: swamy.bio@gmail.com

S. Ashkani
Department of Agronomy and Plant Breeding, Yadegar-e-Imam Khomeini RAH Shahre-Rey Branch, Islamic Azad University, Tehran, Iran
e-mail: ashkani.sadegh@gmail.com

M.K. Meghvansi, A. Varma (eds.), *Organic Amendments and Soil Suppressiveness in Plant Disease Management*, Soil Biology 46,
DOI 10.1007/978-3-319-23075-7_11

pests for several economically important crops worldwide. Bowers et al. (1996) reported that the nematode had the ability to alter the root exudates in qualitative and quantitative fashion, which may influence the activity of beneficial and pathogenic microbes in the rhizosphere. The estimated average annual yield loss of various crops by plant parasitic nematodes is about 12.3 % (Sasser and Freckman 1987), but it varies from 8.8 to 14.6 % from developed to developing countries (Nicol et al. 2011; Palomares-Rius and Kikuchi 2013). Among the sedentary feeders, *Meloidogyne* species are predominant and are considered as the most damaging genera throughout the world. About 95 % of the total nematode populations are represented only by four major species such as *M. incognita*, *M. javanica*, *M. arenaria*, and *M. hapla*.

Suppression of plant diseases in the presence of a pathogen, suitable host plant, and favorable climatic conditions is known as soil suppressiveness (Mazzola et al. 2004; Weller et al. 2007). It is directly associated with the nature and fertility level of the soil and the types of soil microorganisms. However, the level of disease suppressiveness is directly proportional to the level of soil microbial activity, meaning the larger the active microbial biomass, the greater the soil capacity to use carbon, nutrients, and energy, thus lowering their availability to pathogens (Kumar et al. 2012). Any treatment to increase the microbial activity in the soil enhanced the suppression of pathogens by increasing competition for nutrients, but overall it is a very tough task to control all types of soilborne pathogens by suppressive soils. To control the diseases caused by plant parasitic nematodes, frequent use of chemical nematicides has been increased in the past few decades globally (Gupta and Dikshit 2010; Leng et al. 2011). But these chemical nematicides possess several toxic effects on the human health, soil microbiota, and environment. Thus, several cultural practices have been adopted for the management of nematodes, but gradually the annual losses observed in the quality and quantity of crop yields revealed that there is a decisive need to develop a new eco-friendly way to control the plant parasitic nematodes. In this regard, biological control strategies provide an alternative tool for management of plant parasitic nematodes over the conventional chemical control strategies (Mazzola 2007). The biological control of nematodes could be achieved either by managing the natural habitats to marmalade by increasing the activity of native fungi or by introducing new beneficial rhizospheric fungi or by the combination of both (Timper 2011). Nevertheless, the augmentation of the beneficial microorganisms in the agricultural fields and their potential benefits on the various crops is feasible through the adoption of various management practices such as reduced tillage, crop rotation, and lowering the micronutrient uses.

The rhizosphere is the immediate microenvironment surrounding the plant roots which provides novel environments for microbes due to change in increased levels of nutrients and intense microbial population (Giri et al. 2005; Gupta et al. 2012; Yadav et al. 2015). The rhizoplane and the surrounding rhizosphere soil are colonized and occupied by a wide range of microorganisms. Of the various microorganisms present, opportunistic fungi and arbuscular mycorrhizal (AM) fungi play a key role in the biocontrol of diseases caused by plant parasitic nematodes. Consequently, the plant parasitic nematode and beneficial rhizospheric fungi share common ecological niche and also influenced the plant growth and yield attributes in various means (Akhtar and Siddiqui 2008; Akhtar and Panwar 2011). Because of multifaceted nature, it is very hard to generalize the overall underground interaction processes taking place between the plant parasitic nematodes, opportunistic fungi, and AM fungi. The aim of this chapter is to provide an overview of the biocontrol potential of opportunistic as well as AM fungi on the growth and improvement of various crop plants and population of plant parasitic nematodes in different pathosystems. The chapter also focuses on the cost-effective technologies used for the mass propagation of opportunistic fungi and AM fungi and their ample application in the expansion of practical control system desired for the sustainable agricultural practices.

11.2 Opportunistic Fungi

Fungi have the immense miscellany in their metabolic pathways and offer numerous important classes of commercial compounds having nematicidal activity (Li et al. 2007; Anke 2010) and limit the nematode densities by the production of nematotoxic compounds due to their parasites and antagonistic or predatory actions between fungi and plant parasitic nematodes (Lopez-Llorca and Jansson 2007; Akhtar and Panwar 2011). Lopez-Llorca and Jansson (2007) found that the opportunistic fungi either directly parasitize the nematodes or secrete some nematicidal metabolites which may affect the viability of one or more stages of the nematode life cycle or having deleterious effects on reproductive structures of a nematode. The secondary reproductive stage of the nematode is highly susceptible against the opportunistic fungi. The obese females are highly prone to fungal attack similarly like the parasitism of egg masses. The opportunistic fungi when come in contact with nematode eggs grow more rapidly and parasitize the eggs during initial embryonic developmental stages. This may reduce the parasitic actions of nematode juveniles. Among the various known opportunistic fungi, *P. lilacinus* and *P. chlamydosporia* have been extensively studied by several previous researchers for their nematophagous knack and biocontrol potentiality (Khan et al. 2004; Kiewnick and Sikora 2006; Siddiqui and Akhtar 2009a, b; Akhtar and Panwar 2011; Azam et al. 2013).

11.2.1 Paecilomyces lilacinus

Paecilomyces lilacinus (Thom) Samson is a mutual Hyphomycetes and is ubiquitously distributed especially in warmer climates (Samson 1974). It is encompassed in the group of frequently tested biocontrol agents against the plant parasitic nematodes (Brand et al. 2010; Pau et al. 2012; Azam et al. 2013). It is basically a saprophyte but could also compete for extensive range of substrates (Holland et al. 2003; Pau et al. 2012).

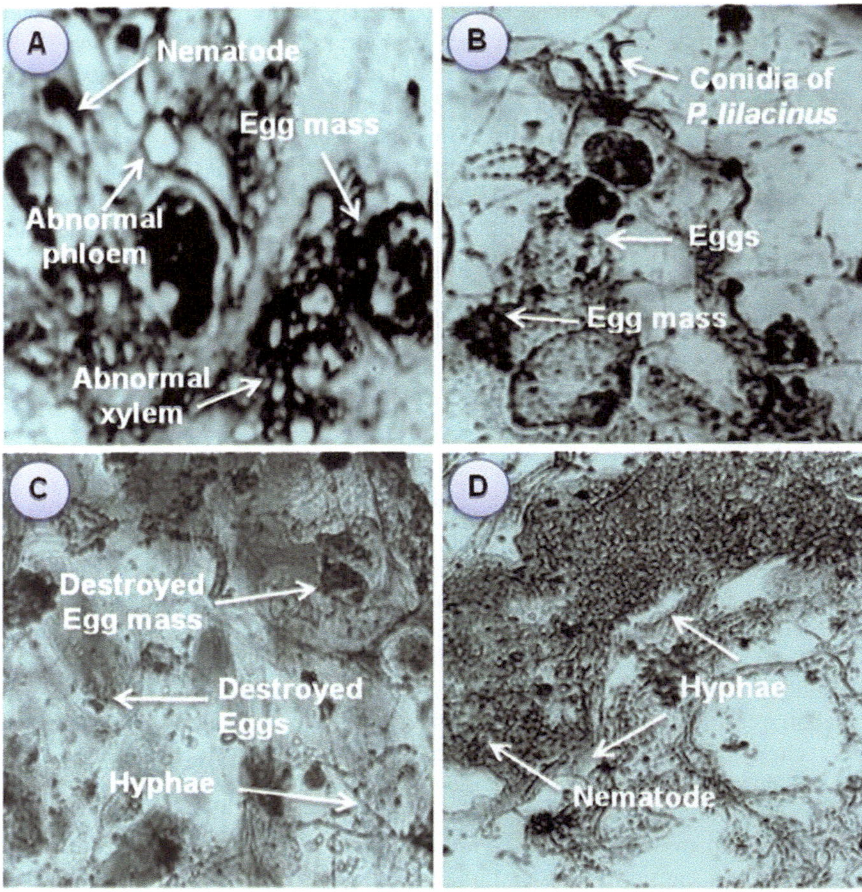

Fig. 11.1 Cross section of tomato root infected with root-knot nematode; (**a**) showing presence of nematode, egg masses, abnormal phloem, and abnormal xylem in the cortical region; (**b**) showing conidia of *P. lilacinus* surrounding the nematode eggs and egg masses; (**c**) disruption of eggs and egg masses by *P. lilacinus* hyphae; (**d**) complete disintegration of nematode eggs by *P. lilacinus* hyphae

Jatala (1986) reported that *P. lilacinus* infects eggs and females of plant parasitic nematodes and destroyed the embryo within 5 days under laboratory conditions. He found that the infection of nematode eggs starts in a gelatinous matrix with the development of fungal hyphae which latter surrounds the entire nematode eggs. The colonization of nematode eggs occurred through the diffusion of egg cuticle by the fungal hyphal network by enzymatic or mechanical actions. His experiments clearly indicated that *P. lilacinus* grow well between 15 and 30 °C. It also had the adaptability to grow in a wide range of soil pH which made it a pretty modest organism in most of the cultivated fields. The suppression of plant parasitic nematode by *P. lilacinus* is ascribed by disintegration of the embryo, inhibition of hatching, and parasitism of adult females (Fig. 11.1). However, after establishment of *P. lilacinus* in soil, it grows faster and spread rapidly within a short span in the introduced area as dominant species. Moreover, the production of secondary metabolites such as chitinases, leucinotoxins, and proteases has also been associated with *P. lilacinus* infection (Park et al. 2004).

11.2.2 Pochonia chlamydosporia

Pochonia chlamydosporia (Goddard) Zare and Gams is a well-known nematophagous fungus and ubiquitously distributed in all parts of the world. It is naturally occurring as a facultative parasite of females, eggs, cyst, and plant parasitic nematodes (Lopez-Llorca et al. 2008; Manzanilla-Lopez et al. 2013). In the rhizosphere, this fungus could settle the host root as endophytes preferably with the plants belonging to families Gramineae and Solanaceae and provide numerous benefits to host plant defense against the soilborne pathogens (Macia-Vicente et al. 2009a, b). *P. chlamydosporia* have been extensively studied for its biocontrol potential against plant parasitic nematodes (Kerry and Hirsch 2011; Manzanilla-Lopez et al. 2013). The efficacy of this potential biological fungus against the plant parasitic nematode is affected by three major factors: (1) the amount of fungus in the rhizosphere, (2) the rate of development of eggs in the egg masses, and (3) the size of galls in which female nematode develops.

The population of *P. chlamydosporia* could be identified on the basis of position and shape of conidia, the plethora of dictyo-chlamydospores, and the development of conidia either in heads or chains (Zare and Gams 2004). *P. chlamydosporia* infects the nematode eggs through the expansion of aspersoria at the tip or lateral position of hyphae, which encompasses tightly to the surface of eggshells (Fig. 11.2), and finally penetrated into eggshells by the formation of an infection peg (Holland et al. 1999). A postinfection bulb leads to the expansion of mycelia within the eggs that caused almost the complete devastation of their contents (Tikhonov et al. 2002; Esteves et al. 2009a). Khan et al. (2004) reported that the eggshells and juvenile cuticles both have been physically disrupted, and the fungal hyphae willingly multiplied inside the eggs and juveniles due to enzymatic activity and biosynthesis of diffusible toxic metabolites. *P. chlamydosporia* are reported to secrete serine, protease, and chitinase responsible for the major structural changes inside the nematode eggs which may result in the disintegration of lipid and vitelline layers. Application of *P. chlamydosporia* as soil inoculants could reduce the natural nematode population up to 90 % under field condition (Bordallo et al. 2002), but the fungus differs in virulence toward nematode competence to colonize the root and production of chlamydospore (Bordallo et al. 2002; Yang et al. 2007; Macia-Vicente et al. 2009a, b). All these specific features make *P. chlamydosporia* a successful biocontrol agent under different pathosystems (van Damme et al. 2005; Rumbos et al. 2006; Esteves et al. 2009b).

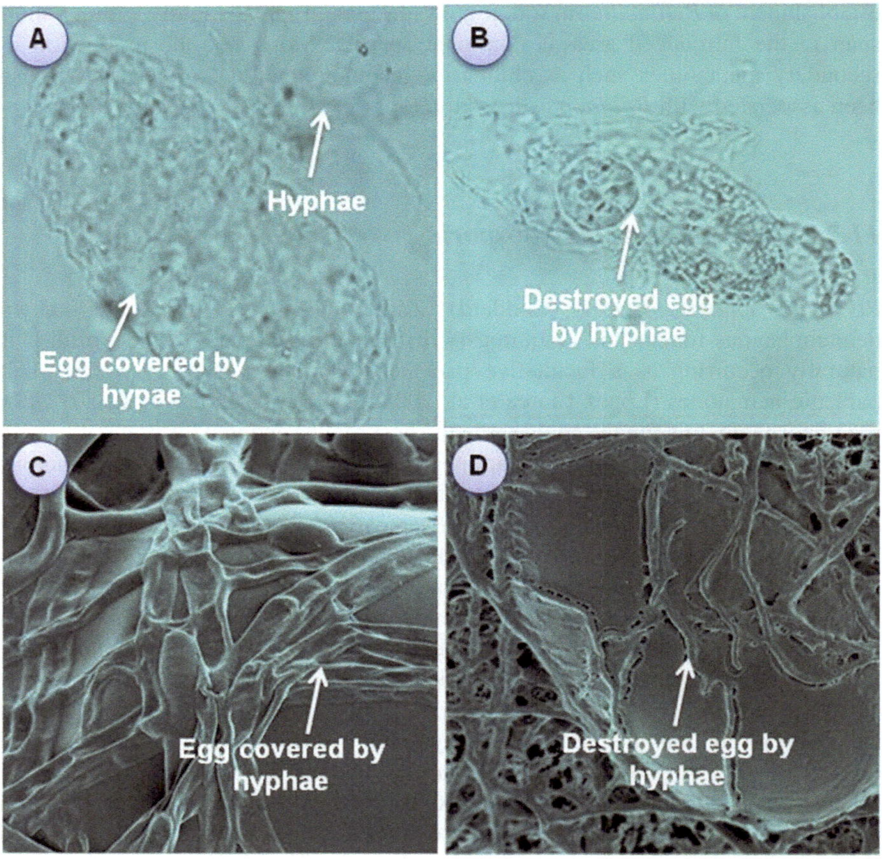

Fig. 11.2 Classical and electron microscopic images of root-knot nematode infected by *P. chlamydosporia*; (**a**) egg of a nematode infected by *P. chlamydosporia* hyphae; (**b**) complete disintegration of nematode egg by *P. chlamydosporia* hyphae; (**c**) electron microscopic view of *P. lilacinus* hyphae covering the nematode egg; (**d**) disruption of nematode egg by *P. chlamydosporia* hyphae

11.3 Arbuscular Mycorrhizal Fungi

Arbuscular mycorrhizal (AM) fungi are the key components of soil microbial populations with ubiquitous distribution in almost all the agroclimatic conditions of the world and form symbiosis with most of the land plants, in any kind of terrestrial ecosystem (Akhtar and Siddiqui 2008). Currently, AM fungi have been cited in the phylum Glomeromycota (Redecker and Raab 2006), and over 200 morphospecies of Glomeromycota have been described (Schüßler 2008). AM fungi have been categorized on the basis of extra-radical mycelium and branched haustoria-like structure within the cortical cells, termed as arbuscules. These

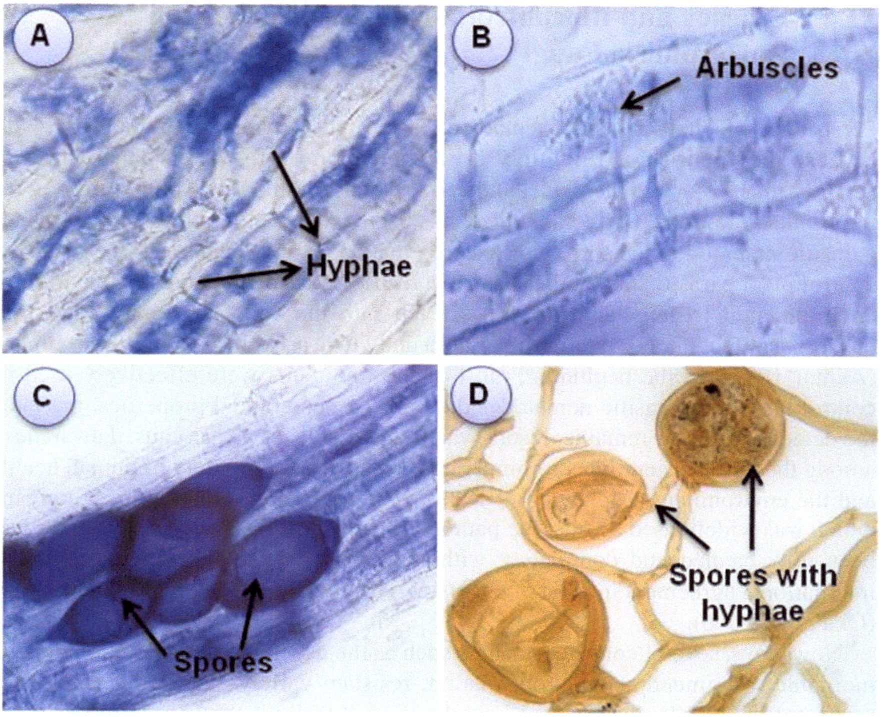

Fig. 11.3 Microscopic view of colonization pattern of AM fungi inside the tomato root; (**a**) showing hyphae of AM fungi; (**b**) showing formation of arbuscles; (**c**) visualization of AM spores inside the cortical tissue; (**d**) AM spores with hyphae stained with Melzer's reagents

arbuscules are the core sites for the nutrient exchange (Fig. 11.3), where the fungi supply water and nutrients like N and P to plants and in turn receive carbon from plants (Bonfante and Genre 2010).

Due to their unique ability and adaptability in different agroclimatic conditions, the AM fungi improved plant health by the acquisition of essential mineral nutrient and water from soil and enhanced production of growth regulations, tolerance toward various abiotic conditions, and mutualistic relationship with additional rhizospheric microorganisms existing in the same ecological niche (Akhtar and Siddiqui 2008; Akhtar 2011).

11.4 Efficacy and Biocontrol Strategies of Beneficial Rhizospheric Fungi

Persistence of plant parasitic nematodes is the most serious problem worldwide because they nourish and multiply their population on live host plants and also actively migrate inside the plants and aerial parts or in the rhizosphere. Among all the available options, chemical control has been extensively used against the plant parasitic nematode, due to its nonselective nature. However, use of chemicals to control plant parasitic nematodes has been restricted in many countries due to their environmental toxicity and ability to leach into the soil. They may cause the hazardous effect on the soil microbial flora and fauna as well as on the environment (Akhtar 1997). In the beginning, most of the fumigants were effectively used to control the plant parasitic nematodes due to their nematicidal properties, but later the detection of their remains in soil, water, and edible crops has caused awareness among the global scientific community concerned about the safety of human health and the environment (Alphey et al. 1988). Methyl bromide was the first fumigant which was widely used against the pathogens causing soilborne diseases, but it has been now banned and completely withdrawn from the market by imposing an international agreement in most of countries worrying about the environment safety (Oka et al. 2000).

Nowadays, several control measures such as the use of green manure, organic or inorganic soil amendments, crop rotation, resistant variety cultivation, unplanted treatment, and biological control have been used to limit the population of plant parasitic nematodes in the soil. But, unfortunately, all these control methods have led to limited success (Barker and Koenning 1998). Integrated pest management provides a working methodology for pest management in sustainable agricultural systems. With the increasing cost of inorganic fertilizers and the environmental and human health hazards associated with the use of pesticides, opportunistic and AM fungi may provide a more suitable and environmentally acceptable alternative for sustainable agriculture. Several comprehensive reviews have been published time to time exploring the possibilities of using AM fungi (Barea et al. 2005; Akhtar and Siddiqui 2008; Smith and Read 2008; Akhtar and Panwar 2011) and opportunistic fungi in the biocontrol of plant diseases (Atkins et al. 2005; Hildalgo-Diaz and Kerry 2008). We have summarized some recently published results of interaction studies between opportunistic fungi, AM fungi, and plant parasitic nematodes in tabular forms (Tables 11.1, 11.2, and 11.3).

Table 11.1 Effect of *Paecilomyces lilacinus* on the plant growth and reproduction of plant parasitic nematodes

Plant	Nematode	Effect on plant growth	Effect on nematode population	References
Faba bean	*M. incognita*	Pre- and posttreatment of plants with *P. lilacinus* increased the shoot dry weight from 11.1 to 13.3 % and 9.1–12.1 %, respectively	Pre- and postinoculation of plant with *P. lilacinus* reduced the number of juveniles from 95.4 to 97.4 % and 91.1–98.9 %, respectively, compared to control	El-Shanshoury et al. (2005)
Tomato	*M. incognita*	Pre- and postinoculation of nematode to *P. lilacinus* significantly reduced the dry weight of plant by 26.15–56.92 %	Pre- and postinoculation of fungus parasitized the nematode eggs by 72.0 % and 68.0 %, respectively	Esfahani and Pour (2006)
Tomato	*M. incognita*	Use of *P. lilacinus* increased the root and shoot weight of plants up to 27.83 % and 46.8 %, respectively	Inoculation of fungus reduced the number of galls per plant, egg masses per root system, and eggs per egg mass up to 44.74 %, 34.23 %, and 16.90%, respectively	Goswami et al. (2006)
Tomato	*M. incognita*	Use of various glucose formulations of *P. lilacinus* increased the shoot weight by 1.83–9.89 % and root weight by 5.0–14.2 % compared to control	Soil treated with fungus reduced root galling, number of egg masses, and final nematode population in the roots by 66 %, 74 %, and 71 %%, respectively, compared to control	Kiewnick and Sikora (2006)
Tomato	*M. incognita*	Use of single or combined application of *P. lilacinus* with bacterial inoculants increased the plant height up to 4.3 %	Treatment with *P. lilacinus* reduced the number of eggs per egg mass up to 18 % compared to untreated control	Anastasiadis et al. (2008)
Tomato	*M. incognita*	Inoculation of *P. lilacinus* increased plant	Use of fungus caused the 44.0 % and 76.0 %	Siddiqui and Akhtar (2008a)

(continued)

Table 11.1 (continued)

Plant	Nematode	Effect on plant growth	Effect on nematode population	References
		length and shoot dry of plants by 42.82 % and 42.25 %, respectively, over nematode-infected plants	parasitism on females and eggs of nematode	
Tomato	*M. incognita*	Enhancement in shoot length (72.66 cm), shoot weight (42.66 g), and root length (36.66 cm) was recorded when *P. lilacinus* was applied in dose 10 g /kg soil compared to control treatment	Inoculation of *P. lilacinus* caused the highest reduction in nematode population, galling, and egg mass per gram root on nematode-infested plants	Kannan and Veeravel (2008)
Tomato	*M. incognita*	Treatment with different spore inoculums of *P. lilacinus* increased the root weight from 37.94 to 65.58 %	The galling is reduced from 89.89 to 97.31 % by the application of different loads of spore inoculum of *P. lilacinus*	Oclarit and Cumagun (2009)
Lettuce	*Meloidogyne* spp.	Application of *P. lilacinus* increased the yield of lettuce by 59.33 % in nematode-infested soil under field conditions	The reduction in galling and nematode population was achieved by 34.89 % and 61.76 % with the application of *P. lilacinus*	Prakob et al. (2009)
Banana	*M. incognita*	Use of *P. lilacinus* significantly increased the plant length (23.09 %) and pseudo stem girth (39.61%) compared to nematode-infected plants	Treatment with *P. lilacinus* reduced the nematode population in soil root by 91.18 % and 81.82 %	Sundararaju and Kiruthika (2009)
Ashwagandha	*M. incognita*	Treatment with *P. lilacinus* increased the shoot dry weight by 84.23% over nematode plants	Use of *P. lilacinus* reduced the root-knot indices approximately up to 50.0 % compared to	Sharma and Pandey (2009)

(continued)

Table 11.1 (continued)

Plant	Nematode	Effect on plant growth	Effect on nematode population	References
			nematode-inoculated plants	
Chickpea	*M. incognita*	Application of *P. lilacinus* caused 26.83 % increase in shoot dry weight of plants as compared to nematode-infested control treatments	Inoculation of *P. lilacinus* caused 42 % and 70 % of re-isolation of females and eggs from a nematode-infested plants	Siddiqui and Akhtar (2009a)
Tomato	*M. javanica*	Simultaneous inoculation of *P. lilacinus* was found better in terms of plant growth than sequential inoculation and causes 41.26 % increase in shoot dry weight of plant compared to control treatments	Concurrent use of *P. lilacinus* reduced the galling, egg masses, egg per egg mass, and final nematode population by 31.44, 33.39, 46.40, and 47.13 %, respectively, compared to control treatments	Ganaie and Khan (2010)
Tomato	*M. incognita*	Results showed that there is no significance difference between the treatments observed in terms of plant growth compared to control under growth chamber experiment	Preplanting soil treatment with *P. lilacinus* reduced the galling, egg masses per root system, and final nematode population by 66 %, 74 %, and 71 %, respectively, compared to the inoculated control under growth chamber experiment	Kiewnick et al. (2011)
Guava	*M. enterolobii*	ND	Application of *P. lilacinus* reduced the egg and egg masses up to 40 % over control treatments	Carneiro et al. (2011)
Okra	*M. incognita*	Application of *P. lilacinus* as soil inoculants with neem cake increased shoot weight up to	Use of various combinations of *P. lilacinus* as seedling treatment and soil inoculants reduced the galling	Kannan and Veeravel (2012)

(continued)

Table 11.1 (continued)

Plant	Nematode	Effect on plant growth	Effect on nematode population	References
		4.17 %, but the results were more pronounced (28.33%) when this combination was applied with seedling dip treatments of *P. lilacinus*	from 26.50 to 64.96 % and juvenile population from 9.96 to 28.18 % over control	
Tomato	*M. incognita*	Application of *P. lilacinus* increased the shoot dry of plants up to 73.19 % over untreated control	Treatment with *P. lilacinus* reduced the galls and egg masses up to 88.23 % and 76.94 %, respectively	Khalil et al. (2012)
Pepper	*M. incognita*	Use of *P. lilacinus* as seed and substrate treatment increased the seedling length from 4.68 to 7.03 %	Seed and substrate treatment with *P.* significantly lowered the root-knot indices from 6.3 to 5.8 %	Rao et al. (2012)
Brinjal	*M. incognita*	Inoculation of *P. lilacinus* reduced the shoot dry weight of nematode-infested plants from 33.70 to 37.76 %	Use of *P. lilacinus* lowered the root-rot indices from 1.20 to 1.28	Usman and Siddiqui (2012)
Tomato	*M. incognita*	ND	Alginate-formulated *P. lilacinus* pellets at 1.6 % (w/w) with soil mixture reduced the root galling by 66.7 %	Aminuzzaman et al. (2013)
Tomato	*M. incognita*	One-week prior inoculation of *P. lilacinus*, nematode increased the shoot dry weight by 57.0 %	One-week prior inoculation of *P. lilacinus*, nematode reduced the root-knot indices and egg-mass indices from 11 to 30 %	Azam et al. (2013)
Tomato	*M. incognita*	Inoculation of *P. lilacinus* increased the root length by 59.49 %	Treatment with *P. lilacinus* reduced the galling up to 58.58 % and egg masses by 65.18 %	Khalil (2013)
Okra	*M. incognita*	Treatment with various concentrations	Application of various concentrations	Mukhtar et al. (2013)

(continued)

Table 11.1 (continued)

Plant	Nematode	Effect on plant growth	Effect on nematode population	References
		of *P. lilacinus* propagules increased the shoot dry weight by 4 to 8 %	of *P. lilacinus* reduced the number of galls from 14 to 37 %, egg masses from 15 to 37 %, nematode reproduction factor from 20 to 52 %	
Tomato	*M. incognita*	Single and twice application of *P. lilacinus* increased the shoot dry weight of plants by 31.40–37.00 %, respectively, over control	Single or twice treatment with *P. lilacinus* reduced the number of galls and egg mass per root system by 52.86–67.71 % and 75.86–87.58 %, respectively	Udo et al. (2013)
Potato	*M. arenaria*	Use of Bio-Nematon (*P. lilacinus* at 10^8 unit/cm^3) increased the plant height, number of leaves, and number of branches by 68.2 %, 106.9 %, and 137.0 %, respectively, compared to control treatments under field condition	Treatment with Bio-Nematon (*P. lilacinus* at 10^8 unit/cm^3) reduced the number of galls and number of egg masses in root system by 77.4 % and 83.3%, respectively, under field condition	Abd-El-Khair and El-Nagdi, (2014)
Chickpea	*M. incognita*	Use of *P. lilacinus* increased the shoot length of plants by 40.62 % compared to control treatments	Application of *P. lilacinus* reduced the number of juvenile in root and galling by 44.42 % and 65.88 %, respectively	Mishra et al. (2014)
Brinjal	*M. incognita*	Treatment with *P. lilacinus* increased the shoot and root length by 45.62 % and 29.41 %, respectively, compared to control treatments	Inoculation of *P. lilacinus* reduced the root-knot indices up to 63.88 % compared to control treatments	Ravindra et al. (2014)

Table 11.2 Effect of *Pochonia chlamydosporia* on the plant growth and reproduction of plant parasitic nematodes

Plant	Nematode	Effect on plant growth	Effect on nematode population	References
Hollyhock Petunia Poppy	*M. incognita*	Root dip treatment with *P. chlamydosporia* increased the flower production by 7–15 % on various tested ornamental plants under field conditions	The frequency of colonization of eggs, egg masses, and females by *P. chlamydosporia* was recorded as 25–29 %, 47–60 %, and 36–41 %, respectively, under field conditions	Khan et al. (2005a)
Chickpea	*Meloidogyne* spp.	Application of *P. chlamydosporia* increased the plant growth by 28 % and yields by 25 % of nematode-infected chickpea plants	Use of *P. chlamydosporia* reduced the galling by 23 % and egg mass production by 18 %	Khan et al. (2005b)
Faba bean	*M. incognita*	Application of *P. chlamydosporia* reduced the population density of nematodes on faba bean	Application of *P. chlamydosporia* reduced the population density of nematodes on faba bean either with post- or preinfection with the range of 97.1 to 98.9 % compared to control	El-Shanshoury et al. (2005)
Cabbage Tomato	*M. incognita*	ND	Use of *P. chlamydosporia* reduced nematodes population by 51–78 % in the tomato compared to cabbage	Tahseen et al. (2005)
Tomato	*M. incognita*	Treatment with *P. chlamydosporia* increased plant length and shoot dry weight by 36.71 % and 36.63 %, respectively, compared to nematode-infested plants	Use of *P. chlamydosporia* caused the parasitism on females and eggs of nematodes by 30.0 % and 67.0 %, respectively	Siddiqui and Akhtar (2008a)
Okra	*M. incognita*	Combined application of *P. chlamydosporia* with neem cake or carbofuran significantly increased the plant growth and yield by 53 % and 64 %, respectively, over non-inoculated control	Use of *P. chlamydosporia* with neem cake or carbofuran reduced the galling, egg production, and nematode population by 89 %, 90 %, and 81 %, respectively	Dhawan and Singh (2009)

(continued)

Table 11.2 (continued)

Plant	Nematode	Effect on plant growth	Effect on nematode population	References
Chickpea	*M. incognita*	Use of *P. chlamydosporia* caused 22.41 % increase in shoot dry weight of plants as compared to nematode-infested control plants	Inoculation of *P. chlamydosporia* caused 28 % and 66 % of re-isolation of females and eggs from nematode-infested plants	Siddiqui and Akhtar (2009a)
Okra	*M. incognita*	Use of *P. chlamydosporia* increased the shoot length, shoot weight, root length, and root weight of plant by 80.9, 74.1, 73.9, and 80 %, respectively, over control treatment under pot conditions	Treatment with *P. chlamydosporia* reduced galls and egg masses per plant and eggs per egg mass by 54.8, 53.7, and 46.5 %, respectively, under pot condition	Dhawan and Singh (2011)
Guava	*M. enterolobii*	ND	Application of *P. chlamydosporia* reduced the disease severity up to 61.5 % as compared to control under glasshouse conditions	Carneiro et al. (2011)
Tomato	*M. javanica*	ND	Among the various tested isolates of *P. chlamydosporia*, isolates 64 and 10 were most efficient in reducing the number of eggs by 72.0 % and 60.0 %, respectively	Dallemole-Giaretta et al. (2012)
Tomato	*M. javanica*	Inoculation of *P. chlamydosporia* Pc123gfp increased the root and shoot growth of plants 20 days after inoculation compared to nematode-inoculated plants	Treatment with *P. chlamydosporia* Pc123gfp reduced the number of galls and egg masses per root system by 53.6 % and 32 %, respectively, compared to control	Escudero and Lopez-Llorca (2012)
Tomato	*M. incognita*	Use of chlamydospore inoculum of *P. chlamydosporia* (strain 4) increased the shoot dry weight up to 12.14 % compared to non-inoculated control treatment	Use of chlamydospore inoculum of *P. chlamydosporia* (strain 4) reduced the number of egg per root system by almost 50 % compared to non-inoculated control treatment	Yang et al. (2012)

(continued)

Table 11.2 (continued)

Plant	Nematode	Effect on plant growth	Effect on nematode population	References
Tomato	*M. incognita*	ND	Use of alginate-formulated *P. chlamydosporia* pellets at 1.6 % (w/w) with soil mixture reduced the nematode density by 90 % on tomato under green-house conditions	Aminuzzaman et al. (2013)
Okra	*M. incognita*	Use of various concentrations of fungal propagules enhanced the shoot dry weight from 5 to 10 %	Treatment with various concentrations of fungal propagules suppressed the number of galls from 12 to 32 %, egg masses from 11 to 30 %, and reproduction factor from 20 to 43 %	Mukhtar et al. (2013)
Tomato	*M. javanica*	Treatment with *P. chlamydosporia* increased the shoot by 7.38 % and root mass by 4.64 %	Application of *P. chlamydosporia* reduced the number of galls per plant by 12.68 % and number of eggs per plant by 17.39 %	Podestá et al. (2013))
Brinjal	*M. incognita*	Use of *P. chlamydosporia* increased the shoot length and root length by 29.46 % and 33.88 %, respectively, compared to control treatments	Inoculation of *P. chlamydosporia* reduced the root-knot indices by 58.33 % compared to control treatments	Ravindra et al. (2014)
French bean	*M. javanica*	Pre- and postinoculation of fungus to nematode in soil increased the shoot dry weight of plant by 43.39–48.36 % and 13.79–29.24 %, respectively	Pre- and posttreatment of plants with fungus to nematode reduced the number of galls per root system up to 55–62.5 % and 2.5–7.5 %, respectively	Sharf et al. (2014)
Cucumber	*M. javanica*	Application of *P. chlamydosporia* to the soil increased cucumber root mass by 12.03 % compared to control plants	The application of *P. chlamydosporia* reduced the number of galls per gram of roots by 49.44 % and the number of eggs per gram of roots by 40.58 %	Viggiano et al. (2014)

Table 11.3 Effect of AM fungi on the plant growth and reproduction of plant parasitic nematodes

Plant	Nematode	Effect on plant growth	Effect on nematode population	References
Chickpea	*M. incognita*	Inoculation of *G. intraradices* increased the shoot and root dry weight up to 9.68 and 14.75%, respectively	Application of AM fungus reduced the galling up to 28.57% and nematode population up to 27.32%	Akhtar and Siddiqui, (2006)
Mentha	*M. incognita*	Use of *G. aggregatum* increased the herb yield up to 16.61 % and oil yield up to 37.25 % compared to control treatments	Inoculation of *G. aggregatum* reduced root-knot indices up to 27.3 % over control treatments	Pandey (2005)
Tomato	*M. incognita*	Application of both isolates of *G. fasciculatum* increased the shoot weight up to 8.20–10.93 % and yield up to 9.75–10.40 %	Treatment with both isolates of *G. fasciculatum* reduced galling up to 41.3–44.7 % and 60.1–63.1 %, respectively	Kantharaju et al. (2005)
Banana	*M. javanica*	ND	Results showed that AM fungus-inoculated plants had 20 % less galling compared to non-mycorrhizal plants	Rodríguez Romero and Jaizme-Vega (2005)
Papaya	*M. incognita*	Inoculation of *G. mosseae* and *G. manihotis* significantly increased the plant growth, but the increase in plant growth was marginal when each AM fungus was compared individually	Inoculation of *G. mosseae* and *G. manihotis* reduced the galling by 84–44–99.59 % and number of nematodes per root by 83.33–99.54 %	Jaizme-Vega et al. (2006)
Tomato	*M. incognita*	Use of *G. mosseae* and *G. margarita* both increased the shoot dry weight of plant by 35.34 % and 31.74 %, respectively, but the results were more pronounced when the AM fungi were used with tested organic manures	Treatment with *G. mosseae* and *G. margarita* both reduced the galling by 60.22 % and 51.14 %, respectively, and nematode population by 60.27 % and 50.41%, respectively, but the results were more pronounced when the AM fungi were used with tested organic manure	Siddiqui and Akhtar (2007)
Chickpea	*M. incognita*	Inoculation of *G. intraradices* increased the shoot dry	Use of *G. intraradices* reduced the galling by 25.0 % and nematode	Akhtar and Siddiqui (2007)

(continued)

Table 11.3 (continued)

Plant	Nematode	Effect on plant growth	Effect on nematode population	References
		weight by 8.5 % compared to nematode-infested control	population by 25.83 % compared to control	
Tomato	*M. incognita*	Inoculation of AM fungus increased the plant dry weight by 34.80 % and yield by 54.54 % compared to nematode-infested plants	Inoculation of AM fungus reduced the galling by 66.09 %, number of egg masses by 66.47 %, and nematode population by 55.20 %	Shreenivasa et al. (2007)
Sunflower	*M. incognita*	Pre- and posttreatment of AM fungi to nematode increased the plant length by 6.02 % and 2.41 %, respectively, compared to nematode-inoculated control treatment	Pre- and postinoculation of AM fungi reduced the nematode infestation by 83.33 and 33.33 %, respectively, compared to nematode-inoculated control treatment	Jalaluddin et al. (2008)
Tomato	*M. incognita*	ND	Inoculation of *G. intraradices* reduced the galling by 24 %, while the results were more pronounced (60 %) with the combination of *R. etli*	Reimann et al. (2008)
Tomato	*M. incognita*	Treatment with AM fungus increased the shoot dry weight by 30.69 % compared to nematode-infested control plants	Inoculation of AM fungus reduced the galling by 30.30 % and nematode population by 38.44%	Siddiqui and Akhtar (2008b)
Cucumber	*M. incognita*	Inoculation of *G. mosseae* and *G. versiforme* significantly increased the shoot dry of plants by 39.38 % and 50.17 %, while the *G. intraradices* was found least effective in terms of plant growth	All the tested AM fungi reduced the galling index by 3.0, 2.4, and 2.0, respectively. However, inoculation with *G. versiforme* decreased the number of galls per gram root by 45 %, while the other two fungi also showed the similar propensity, but the trend was not significant	Zhang et al. (2008)
Chickpea	*M. incognita*	Use of AM fungus increased the shoot dry weight by 15.11 %, grain weight by 16.23 %, and yield by	Use of AM fungus reduced the galling by 27.27 % under field conditions	Akhtar and Siddiqui (2009)

(continued)

Table 11.3 (continued)

Plant	Nematode	Effect on plant growth	Effect on nematode population	References
		15.13 % under field conditions		
Tomato	*M. incognita*	Treatment of AM fungi increased the shoot dry weight of plants by 29.9–30.9 % compared to untreated control	Use of AM fungi reduced the galling up to 26.08–29.71 % and nematode population up to 24.59–33.33 % compared to untreated control	Siddiqui and Akhtar (2009b)
Cucumber	*M. incognita*	Application of both levels of P with *G. intraradices* increased the shoot dry of plants by 25.0 % and 28.42 %, respectively	Use of both levels of P with AM fungus reduced the galling approximately up to 50–54 %	Zhang et al. (2009)
Sweet passion fruit	*M. incognita*	Inoculation of AM fungus stimulated the root biomass of plants up to 35.71 % and 10.94 % in the non-disinfected and disinfected soil	AM fungus-treated plants showed 72.0 % reduction in the number of galls per gram of roots and 87.7 % in egg masses per gram of roots in disinfested soil, while in noninfected soil the number of eggs and galls per root system were recorded 44.0 and 26.5 %, respectively	Anjos et al. (2010)
Cowpea	*M. incognita*	ND	Inoculation of AM fungus suppressed the root galling and nematode reproduction up to 12.80–72.73 % and 24.24–55.43 % on various tested varieties of cowpea in both pot experiments	Odeyemi et al. (2010)
Acacia farnesiana *Acacia saligna*	*M. incognita*	Treatment of AM fungi together with oxamyl increased the shoot dry weight of both plants by 66.57–72.90 and 61.73–65.18 %, respectively	Application of AM fungi together with oxamyl decreased no. of egg masses, eggs per egg mass, final nematode population, and buildup of nematode approximately by 80.40 %, 47.90 %, 79.70%, and 89.80 %, respectively, in both tested plant species	Soliman et al. (2011)

(continued)

Table 11.3 (continued)

Plant	Nematode	Effect on plant growth	Effect on nematode population	References
Pea	*M. incognita*	Use of AM fungus significantly increased the growth (24.54 %), in the nematode-inoculated plants	Inoculation of AM fungus reduced the number of galls and nematode population up to 30.13 % and 32.23 %, respectively	Akhtar and Panwar (2013)
Tobacco	*M. incognita*	Combined inoculation of *G. aggregatum* with neem cake caused the maximum increase in the shoot dry (48.81 %) of plants over nematode-infested soil	Combined inoculation of *G. aggregatum* with neem cake reduced the nematode population by 60.87 % and nematode reproduction rate by 58.96 %	Serfoji et al. (2013)
Mays	*M. incognita*	Use of AM fungus increased shoot weight (17.58 % and 11.63 %) and the yield (64.92 % and 20.07 %) of plants under pot and field conditions	Treatment with AM fungus reduced the galling (47.56 % and 44.81 %) and nematode population (98.23 % and 80.81 %) under pot and field conditions	Odeyemi et al. (2013)
Tomato	*M. incognita*	Results showed that all the tested AM fungi increased the shoot dry weight of plant compared to control, but *G. deserticola* caused the height increase (40.17 %) in shoot dry weight compared to other tested fungi	Among all the tested AM fungi, *G. deserticola* reduced the number of galls per root system by 44.28 % and number of eggs per root system by 72.42 %	Udo et al. (2013)
Potato	*M. arenaria*	Treatments with Stanes symbion vam (mixture of *G. fasciculatum* and *Gigaspora* sp.) increased the plant height, number of leaves, and number of branches by 64.5 %, 82.2 %, and 113.4 %, respectively, compared to control treatments under field condition	Inoculation of Stanes symbion vam (mixture of *G. fasciculatum* and *Gigaspora* sp.) reduced the number of juveniles in soil, eggs, and egg masses on root system by 86.1 %, 69.8 %, and 71.9 %, respectively, compared to control treatments under field condition	Abd-El-Khair and El-Nagdi, (2014)

11.5 Mass Propagation Strategies of Opportunistic Fungi and AM Fungi

11.5.1 Mass Production of Opportunistic Fungi

Several media have been extensively used for the mass production of opportunistic fungi. For the mass production of *P. lilacinus* potato dextrose broth (Rangaswami 1972), Richard's medium, 10 % molasses (Rangaswami 1972), and semi-selective medium (Mitchell et al. 1987) can be used. The highest mycelium weight and spore production were achieved by using the semi-selective medium followed by 10 % molasses medium (Prabhu et al. 2008). Corn meal agar and potato dextrose agar media have also been used for the mass production of *P. lilacinus* (Robl et al. 2009). Similarly, the mass production of *Pochonia* spp. was achieved by using shrimp agar medium (Moosavi et al. 2010). Besides this wheat, bran and barley grain were also used for the mass production of *Pochonia* spp. (de Leij and Kerry 1991; Crump and Irving 1992). For the large-scale commercial production, liquid fermentation method is generally used because of difficulties to improve spore production on solid medium (Khan and Anwer 2011).

11.5.2 Mass Production of AM Fungi

AM fungi have the unique ability to improve the uptake of water and mineral nutrients from the soil and also to guard the plants against the pathogen attack (Smith and Read 2008). AM fungi also scavenge the available P through their extra-radical hyphae and upsurge the secretion of various amino acids (such as serine and isoleucine) and defense-related proteins (Akhtar and Siddiqui 2008; Akhtar et al. 2011), which augments their importance toward the modern and profitable agronomic practices. Due to their obligate nature, the AM fungi could not be cultured in vitro, which may limit the mass production of AM fungal propagules. In the conventional method of propagation, the AM fungi are propagated through the pot or pan culture usually with single spore culture, swiftly spread on the substrate, and finally colonize the root of host plants (Akhtar and Abdullah 2014). This method is quite useful for the production of clean fungal inoculum with high potentiality in a short span of time. Similarly, aeroponic culture systems allow the production of cleaner spores and enable even nourishment of AM fungi-colonized plants (Jarstfer and Sylvia 1999). Propagation of any AM fungal strains on root-organ culture permitted the propagation of monoxenic strains that could be used either directly as inoculum or as a starter inoculum for the mass production of AM fungi. A very simple and low-cost technique of single spore pot culture has been developed by Panwar et al. (2007). It permits undistributed growth of the mutualistic partners and visualization of germinating AM fungal spores and their mass multiplication. Moreover, the mass production of AM fungal inoculum requires control and optimization of both host growth and fungal development. The microscopic sizes of AM fungi, together with the complex identification processes, also contribute to the drawbacks of inoculum propagation.

Nevertheless in vitro bulk production of AM fungal inoculum is a promising approach, offering clean, viable, contamination-free fungal propagules. The cost of in vitro inoculum may appear expensive compared to the greenhouse-propagated fungal inoculum, but its use as starting inoculums is a warranty of purity (Akhtar and Abdullah 2014). The main purpose of this cultural method is to provide pure, clean, and reliable material as starter inoculum for the fundamental and applied research. There were several reports which indicate that mycorrhizologists were able to produce 25 spores/ml in 4 months' incubation time (Chabot et al. 1992), while the other workers claimed for the production of 3250 spores/ml in 7 months (Douds 2002). Recently another work justifies the production of more than 2400 spore/100 g of soil after 120 days from single spore culture (Panwar et al. 2007).

11.6 Conclusions

The present chapter provides an overview on the interactions between opportunistic fungi, AM fungi, and plant parasitic nematodes. Use of opportunistic and AM fungi will not only reduce the load of nematicides in agricultural practices but also increase the plant vigor through the uptake of essential mineral nutrients and also reduce the nematode buildup in the plant and soil. Moreover, use of these biocontrol agents has an eco-friendly approach toward the environment as well as human health. The protection of nematode diseases by the application of these biocontrol agents is a complex process which may depend upon the molecular interactions between hosts, biocontrol agents, and pathogenic microorganisms. Application of single or mixed inoculum of opportunistic fungi, AM fungi were found to be effective in controlling the nematode diseases under greenhouse, pot, and field conditions in various agroclimatic conditions. An overview of the recent cost-effective technologies used for the mass propagation of these beneficial rhizospheric microorganisms is discussed. The success of mass propagation of indigenous biocontrol agents depends upon its selective nature toward edaphic, environment, and other rhizospheric biota, but it is still a challenge to develop these biocontrol agents in the sustainable agricultural practices to understand real underground mechanisms involved between the host, biocontrol agents, and pathogenic microorganisms.

References

Abd-El-Khair H, El-Nagdi WMA (2014) Field application of biocontrol agents for controlling fungal root rot and root-knot nematode in potato. Arch Phytopathol Plant Protect 47:1218–1230

Akhtar M (1997) Current options in integrated management of plant-parasitic nematodes. Integr Pest Manag Rev 2:187–197

Akhtar MS (2011) Biocontrol of root-rot disease complex of chickpea by arbuscular mycorrhizal fungi and other phosphate solubilizing microorganisms. Lambert Academic, Saarbrücken

Akhtar MS, Abdullah SNA (2014) Mass production techniques of arbuscular mycorrhizal fungi: major advantages and disadvantages: a review. Biosci Biotechnol Res Asia 11:1199–1204

Akhtar MS, Panwar J (2011) Arbuscular mycorrhizal fungi and opportunistic fungi: efficient root symbionts for the management of plant parasitic nematodes. Adv Sci Eng Med 3:165–175

Akhtar MS, Panwar J (2013) Efficacy of root-associated fungi and the growth of *Pisum sativum* (Arkil) and reproduction of root-knot nematode *Meloidogyne incognita*. J Basic Microbiol 53:318–326

Akhtar MS, Siddiqui ZA (2006) Effects of phosphate solubilizing microorganisms on the growth and root-rot disease complex of chickpea. Mycol Phytopathol 40:246–254

Akhtar MS, Siddiqui ZA (2007) Biocontrol of a chickpea root-rot disease complex with *Glomus intraradices*, *Pseudomonas putida* and *Paenibacillus polymyxa*. Aust Plant Pathol 36:175–180

Akhtar MS, Siddiqui ZA (2008) Arbuscular mycorrhizal fungi as potential bioprotectants against plant pathogens. In: Siddiqui ZA, Akhtar MS, Futai K (eds) Mycorrhizae: sustainable agriculture and forestry. Springer, Dordrecht, pp 61–98

Akhtar MS, Siddiqui ZA (2009) Effects of phosphate solubilizing microorganisms and *Rhizobium* sp. on the growth, nodulation, yield and root-rot disease complex of chickpea under field condition. Afr J Biotechnol 8:3479–3488

Akhtar MS, Siddiqui ZA, Wiemken A (2011) Arbuscular mycorrhizal fungi and *Rhizobium* to control plant fungal diseases. In: Lichtfouse E (ed) Alternative farming systems, biotechnology, drought stress and ecological fertilisation, vol 6, Sustainable agriculture reviews. Springer, Dordrecht, pp 263–292

Alphey TJW, Robertson WM, Lyon GD (1988) Rishitin, a natural plant product with nematicidal activity. Rev Nematol 11:399–404

Aminuzzaman FM, Xie HY, Duan WJ, Sun BD, Liu XZ (2013) Isolation of nematophagous fungi from eggs and females of *Meloidogyne* spp. and evaluation of their biological control potential. Biocontrol Sci Technol 23:170–182

Anastasiadis IA, Giannakou IO, Prophetou-Athanasiadou DA, Gowen SR (2008) The combined effect of the application of a biocontrol agent *Paecilomyces lilacinus*, with various practices for the control of root-knot. Crop Prot 27:352–361

Anjos ECT, Cavalcante UMT, Gonçalves DMC, Pedrosa EMR, dos Santos VF, Maia LC (2010) Interactions between an arbuscular mycorrhizal fungus (*Scutellospora heterogama*) and the root-knot nematode (*Meloidogyne incognita*) on sweet passion fruit (*Passiflora alata*). Braz Arch Biol Technol 53:801–809

Anke H (2010) Insecticidal and nematicidal metabolites from fungi. In: Hotrichter M (ed) The mycota: industrial applications, 2nd edn. Springer, Berlin

Atkins SD, Clark IM, Pande S, Hirsch PR, Kerry BK (2005) The use of real-time PCR and species-specific primers for the identification and monitoring of *Paecilomyces lilacinus*. FEMS Microbiol Ecol 51:257–264

Azam T, Akhtar MS, Hisamuddin A (2013) Histological interactions of *Paecilomyces lilacinus* with root-knot nematode *Meloidogyne incognita* and their effect on the growth of tomato. Adv Sci Eng 5:335–341

Barea JM, Pozo MJ, Azcon R, Azcon-Aguilar C (2005) Microbial co-operation in the rhizosphere. J Exp Bot 56:1761–1778

Barker KR, Koenning SR (1998) Developing sustainable systems for nematode management. Annu Rev Phytopathol 36:165–205

Bonfante P, Genre A (2010) Mechanisms underlying beneficial plant-fungus interactions in mycorrhizal symbiosis. Nat Commun 1:1–11

Bordallo JJ, Lopez-Llorca LV, Jansson HB, Salinas J, Persmark L, Asensio L (2002) Effects of egg-parasitic and nematode-trapping fungi on plant roots. New Phytol 154:491–499

Bowers JH, Nameth ST, Riedal RM, Rowe RC (1996) Infection sand colonization of potato roots by *Verticillium dahliae* as infected by *Pratylenchus penetrans* and *P. crenatus*. Phytopathology 86:614–621

Brand D, Soccol CR, Sabu A, Roussos S (2010) Production of fungal biological control agents through solid state fermentation: a case study on *Paecilomyces lilacinus* against root-knot nematodes. Micol Apl Int 22:31–48

Carneiro RMDG, Hidalgo-Díaz L, Martins I, de Souza Silva KFA, de Sousa MG, Tigano MS (2011) Effect of nematophagous fungi on reproduction of *Meloidogyne enterolobii* on guava (*Psidium guajava*) plants. Nematology 13:721–728

Chabot S, Becard G, Piche Y (1992) Life cycle of *Glomus intraradices* in root organ culture. Mycologia 84:315–321

Crump DH, Irving F (1992) Selection of isolates and methods of culturing *Verticillium chlamydosporium* and its efficacy as a biocontrol agent of beet and potato cyst nematodes. Nematologica 38:367–374

Dallemole-Giaretta R, Freitas LG, Lopes EA, Pereira OL, Zooca RJF, Ferrazb S (2012) Screening of *Pochonia chlamydosporia* Brazilian isolates as biocontrol agents of *Meloidogyne javanica*. Crop Prot 42:102–107

de Leij FAAM, Kerry BR (1991) The nematophagous fungus *Verticillium chlamydosporium* as a potential biological control agent for *Meloidogyne arenaria*. Rev Nematol 14:157–164

Dhawan SC, Singh S (2009) Compatibility of *Pochonia chlamydosporia* with nematicide and neem cake against root-knot nematode, *Meloidogyne incognita* infesting okra. Indian J Nematol 39:85–89

Dhawan SC, Singh S (2011) Biomanagement of root-knot nematode, *Meloidogyne incognita* by egg parasitic fungus, (*Pochonia chlamydosporia*) on okra. Veg Sci 38:128–134

Douds DDJ (2002) Increased spore production by *Glomus intraradices* in the split-plate monoxenic culture system by repeated harvest, gel replacement, and resupply of glucose to the mycorrhiza. Mycorrhiza 12:163–167

El-Shanshoury AR, El-Sayed SA, Mahmoud YAG, Khalefa DM (2005) Evaluation of *Pochonia chlamydosporia*, *Paecilomyces lilacinus* and *Arthrobotrys dactyloide* as biocontrol agents for *Meloidogyne incognita* under greenhouse condition. Pak J Biol Sci 8:1511–1516

Escudero N, Lopez-Llorca LV (2012) Effects on plant growth and root-knot nematode infection of an endophytic GFP transformant of the nematophagous fungus *Pochonia chlamydosporia*. Symbiosis 57:33–42

Esfahani MN, Pour BA (2006) The effect of *Paecilomyces lilacinus* on the pathogenesis of *Meloidogyne javanica* and tomato plant growth parameters. Iran Agric Res 24:67–76

Esteves I, Peteira B, Atkins SD, Magan N, Kerry BR (2009a) Production of extracellular enzymes by different isolates of *Pochonia chlamydosporia*. Mycol Res 113:867–876

Esteves I, Peteira B, Powers S, Magan N, Kerry B (2009b) Effects of osmotic and matric potential on growth and accumulation of endogenous reserves in three isolates of *Pochonia chlamydosporia*. Biocontrol Sci Technol 19:185–199

Ganaie MA, Khan T (2010) Biocontrol potential of *Paecilomyces lilacinus* on pathogenesis of *Meloidogyne javanica* infecting tomato plant. Eur J Appl Sci 2:80–84

Giri B, Giang PH, Kumari RAP, Oelmuller R, Varma A (2005) Mycorrhizosphere: strategies and function. In: Buscot F, Varma A (eds) Microorganisms in soil: roles in genesis and function, vol 3, Soil biology. Springer, Berlin

Goswami BK, Pandey RK, Rathour KS, Bhattacharya C, Singh L (2006) Integrated application of some compatible biocontrol agents along with mustard oil seed cake and furadan on *Meloidogyne incognita* infecting tomato plants. J Zhej Uni Sci 7:873–875

Gupta S, Dikshit AK (2010) Biopesticides: an ecofriendly approach for pest control. J Biopest 3:186–188

Gupta G, Panwar J, Akhtar MS, Jha PN (2012) Endophytic nitrogen-fixing bacteria as biofertilizer. In: Lichtfouse E (ed) Sustainable agriculture reviews, vol 11. Springer, Dordrecht, pp 183–221

Hildalgo-Diaz L, Kerry BR (2008) Integrated management and biocontrol of vegetable and grain crops nematodes. In: Ciancio A, Mukerji KG (eds) Integrated management and biocontrol of vegetable and grain crops nematodes. Springer, Dordrecht, pp 29–49

Holland RJ, Williams KL, Khan A (1999) Infection of *Meloidogyne javanica* by *Paecilomyces lilacinus*. Nematology 1:131–139

Holland RJ, Williams KL, Nevalainen KMH (2003) *Paecilomyces lilacinus* strain Bioact251 is not a plant endophyte. Aust Plant Pathol 32:473–478

Jaizme-Vega MC, Rodríguez-Romero AS, Barroso Nunez LA (2006) Effect of the combined inoculation of arbuscular mycorrhizal fungi and plant-growth promoting rhizobacteria on papaya (*Carica papaya* L.) infected with the root-knot nematode *Meloidogyne incognita*. Fruits 61:1–7

Jalaluddin M, Hajra NB, Firoza K, Shahina F (2008) Effect of *Glomus callosum*, *Meloidogyne incognita* and soil moisture on growth and yield of sunflower. Pak J Bot 40:391–396

Jarstfer AG, Sylvia DM (1999) Aeroponic culture of VAM fungi. In: Varma A, Hock B (eds) Mycorrhiza: structure, function, molecular biology and biotechnology, 2nd edn. Springer, Berlin, pp 427–441

Jatala P (1986) Biological control of plant parasitic nematodes. Annu Rev Phytopathol 24:453–489

Kannan R, Veeravel R (2008) Effect of an ovi-parasitic fungus *Paecilomyces lilacinus* Samson on *Meloidogyne incognita* in tomato. Ann Plant Prot Sci 16:466–470

Kannan R, Veeravel R (2012) Effect of different dose and application methods of *Paecilomyces lilacinus* (Thom.) Samson against root-knot nematode, *Meloidogyne incognita* (Kofoid and White) Chitwood in okra. J Agric Sci 4:119–127

Kantharaju V, Krishnappa K, Ravichardra NG, Karuna K (2005) Management of root-knot nematode, *Meloidogyne incognita* on tomato by using indigenous isolates of AM fungus, *Glomus fasciculatum*. Indian J Nematol 35:32–36

Kerry BR, Hirsch PR (2011) Ecology of *Pochonia chlamydosporia* in the rhizosphere at the population, whole organisms and molecular scales. In: Davies K, Spiegel Y (eds) Biological control of plant parasitic nematodes building coherence between microbial ecology and molecular mechanisms. Progress in biological control, vol 11. Springer, Dordrecht, pp 71–182

Khalil MS (2013) The potential of five eco-biorational products on the reproduction of root-knot nematode and plant growth. ESci J Plant Pathol 2:4–91

Khalil MSEH, Allam AFG, Barakat AST (2012) Nematicidal activity of some biopesticide agents and microorganisms against root-knot nematode on tomato plants under greenhouse conditions. J Plant Protect Res 50:47–52

Khan MR, Anwer MA (2011) Fungal bioinoculants for plant disease management. In: Ahmad I, Ahmad F, Pichtel J (eds) Microbes and microbial technology: agriculture and environmental applications. Springer, Dordrecht, pp 447–488

Khan A, Williams KL, Nevalainen HKM (2004) Effects of *Paecilomyces lilacinus* protease and chitinase on the eggshell structures and hatching of *Meloidogyne javanica* juveniles. Biol Contr 31:346–352

Khan MR, Khan SM, Mohide F (2005a) Root-knot nematode problem of some winter ornamental plants and its biomanagement. J Nematol 37:198–206

Khan MR, Mohiddin FA, Khan SM, Khan B (2005b) Effect of seed treatment with certain biopesticides on root-knot of chickpea. Nematol Medit 33:107–112

Kiewnick S, Sikora RA (2006) Biological control of root-knot nematode *Meloidogyne incognita* by *Paecilomyces lilacinus* strain 251. Biol Contr 38:179–187

Kiewnick S, Neumann S, Sikora RA, Frey JE (2011) Effect of *Meloidogyne incognita* inoculum density and application rate of *Paecilomyces lilacinus* strain 251 on biocontrol efficacy and colonization of egg masses analyzed by real-time quantitative PCR. Phytopathology 101:105–112

Kumar P, Satyajeet Khare S, Dubey RC (2012) Diversity of *Bacilli* from disease suppressive soil and their role in plant growth promotion and yield enhancement. New York Sci J 5:90–111

Leng P, Zhang Z, Guangtang P, Zhao M (2011) Applications and development trends in biopesticides. Afr J Biotechnol 10:19864–19873

Li GK, Zhang J, Xu JD, Liu Y (2007) Nematicidal substances from fungi. Recent Pat Biotechnol 1:1–22

Lopez-Llorca LV, Jansson HB (2007) Fungal parasites of invertebrates: multimodal biocontrol agents. In: Robson GD, van West P, Gadd GM (eds) Exploitation of fungi. Cambridge University Press, Cambridge, pp 310–335

Lopez-Llorca LV, Macia-Vicente JG, Jansson HB (2008) Mode of action and interactions of nematophagous fungi. In: Ciancio A, Mukerji KG (eds) Integrated management and biocontrol of vegetable and grains crops nematodes. Springer, Dordrecht, pp 51–76

Macia-Vicente JG, Jansson HB, Talbot NJ, Lopez-Llorca LV (2009a) Real-time PCR quantification and live-cell imaging of endophytic colonization of barley (*Hordeum vulgare*) roots by *Fusarium equiseti* and *Pochonia chlamydosporia*. New Phytol 182:213–228

Macia-Vicente JG, Rosso LC, Ciancio A, Jansson HB, Lopez-Llorca LV (2009b) Colonisation of barley roots by endophytic *Fusarium equiseti* and *Pochonia chlamydosporia*: effects on plant growth and disease. Ann Appl Biol 155:391–401

Manzanilla-Lopez RH, Esteves I, Finetti-Sialer MM, Hirsch PR, Ward E, Devonshire J, Hidalgo-Diaz L (2013) *Pochonia chlamydosporia*: advances and challenges to improve its performance as a biological control agent of sedentary endo-parasitic nematodes. J Nematol 45:1–7

Mazzola M (2007) Manipulation of rhizosphere bacterial communities to induce suppressive soils. J Nematol 39:213–220

Mazzola M, Funnell DL, Raaijmakers JM (2004) Wheat cultivar-specific selection of 2,4-diacetylphloroglucinol-producing fluorescent Pseudomonas species from resident soil populations. Microbial Ecol 48:338–348

Mishra KK, Dwivedi S, Pandre PK (2014) Evaluation of fungal bio-agents on plant growth and *M. incognita* infestation on chickpea. Chem Mater Res 6:135–139

Mitchell DJ, Kannwischer-Mitchell ME, Dickson DW (1987) A semi selective medium for the isolation of *Paecilomyces lilacinus* from soil. J Nematol 19:255–256

Moosavi MR, Zare R, Zamanizadeh HR, Fatemy S (2010) Pathogenicity of *Pochonia* species on eggs of *Meloidogyne javanica*. J Invertebr Pathol 104:125–133

Mukhtar T, Hussain MA, Kayani MZ (2013) Biocontrol potential of *Pasteuria penetrans*, *Pochonia chlamydosporia*, *Paecilomyces lilacinus* and *Trichoderma harzianum* against *Meloidogyne incognita* in okra. Phytopathol Medit 52:66–76

Nicol JM, Turner SJ, Coyne DL, den Nijs L, Hockland S, Maafi ZT (2011) Current nematodes threats to world agriculture. In: Jones J, Gheysen G, Fenoll C (eds) Genomics and molecular genetics of plant-nematode interactions. Springer, Dordrecht, pp 21–43

Oclarit EL, Cumagun CJR (2009) Evaluation of efficacy of *Paecilomyces lilacinus* as biological control agent of *Meloidogyne incognita* attacking tomato. J Plant Protect Res 49:337–340

Odeyemi IS, Afolami SO, Sosanya OS (2010) Effect of *Glomus mosseae* (arbuscular mycorrhizal fungus) on host-parasite relationship of *Meloidogyne incognita* (Southern root-knot nematode) on four improved cowpea varieties. J Plant Protect Res 50:320–325

Odeyemi IS, Afolamia SO, Adekoyejoa AB (2013) Integration of *Glomus mosseae* with *Chromolaena odorata* powder for suppression of *Meloidogyne incognita* on maize (*Zea mays* L.). Arch Phytopathol Plant Protect 46:1589–1597

Oka Y, Koltai H, Bar-Eyal M, Mor M, Sharon E, Chet I, Spiegel Y (2000) New strategies for the control of plant-parasitic nematodes. Pest Manag Sci 56:983–988

Palomares-Rius JE, Kikuchi T (2013) Omics fields of study related to plant-parasitic nematodes. J Integr Omics 3:1–10

Pandey R (2005) Field application of bio-organics in the management of *Meloidogyne incognita* in *Mentha arvensis*. Nematol Medit 33:51–54

Panwar J, Tarafdar JC, Yadav RS, Saini VK, Aseri GK, Vyas A (2007) Technique for visual demonstration of germinating AM spore and their multiplication in pots. J Plant Nutri Soil Sci 170:659–663

Park JO, Hargreaves JR, McConville EJ, Stirling GR, Ghisalberti EL, Sivasithamparam K (2004) Production of leucinostatins and nematicidal activity of Australian isolates of *Paecilomyces lilacinus* (Thom) Samson. Lett Appl Microbiol 38:271–276

Pau CG, Leong CTS, Wong SK, Eng L, Jiwan M, Kundat FR, Aziz ZFBA, Ahmed OH, Majid NM (2012) Isolation of indigenous strains of *Paecilomyces lilacinus* with antagonistic activity against *Meloidogyne incognita*. Int J Agric Biol 14:197–203

Podestá GS, Freitas LG, Dallemole-Giaretta R, Zooca RJF, Caixeta LB, Ferraz S (2013) *Meloidogyne javanica* control by *Pochonia chlamydosporia*, *Gracilibacillus dipsosauri* and soil conditioner in tomato. Summa Phytopathol 39:122–125

Prabhu S, Kumar S, Subramanian S (2008) Mass production and commercial formulation of *Paecilomyces lilacinus*. Madra Agric J 95:415–417

Prakob W, Nguen-Hom J, Jaimasit P, Silapapongpri S, Thanunchai J, Chaisuk P (2009) Biological control of lettuce root-knot disease by the used of *Pseudomonas aeruginosa*, *Bacillus subtilis* and *Paecilomyces lilacinus*. J Agric Technol 5:179–191

Rangaswami G (1972) Disease of crop plants in India. Prentice Hall, New Delhi, p 504

Rao MS, Sowmya DS, Chaya MK, Kumar RM, Rathnamma K, Gavaskar J, Priti K, Ramachandran N (2012) Management of nematode induced wilt disease complex in capsicum using *Pseudomonas fluorescens* and *Paecilomyces lilacinus*. Nematol Medit 40:101–105

Ravindra H, Sehgal M, Pawan AS, Archana BS, Shruti SA, Narasimhamurty HB (2014) Eco-friendly management of root-knot nematodes using acacia compost and bioagents in brinjal. Pak J Nematol 32:33–38

Redecker D, Raab P (2006) Phylogeny of the Glomeromycota (arbuscular mycorrhizal fungi): recent developments and new gene markers. Mycologia 98:885–895

Reimann S, Hauschild R, Hildebrandt U, Sikora RA (2008) Interrelationships between *Rhizobium etli* G12 and *Glomus intraradices* and multitrophic effects in the biological control of the root-knot nematode *Meloidogyne incognita* on tomato. J Plant Dis Protect 115:108–113

Robl D, Sung LB, Novakovich JH, Marangoni PRD, Zawadneak MAC, Dalzoto PR, Gabardo J, Pimentel IC (2009) Spore production in *Paecilomyces lilacinus* (Thom.) Samson strains on agro-industrial residues. Braz J Microbiol 40:296–300

Rodríguez Romero AS, Jaizme-Vega MC (2005) Effect of the arbuscular mycorrhizal fungus *Glomus manihotis* on the root-knot nematode, *Meloidogyne javanica* in banana. Nematol Medit 33:217–221

Rumbos C, Reimann S, Kiewnick S, Sikora RA (2006) Interactions of *Paecilomyces lilacinus* strain 251 with the mycorrhizal fungus *Glomus intraradices*: implications for *Meloidogyne incognita* control on tomato. Biocontrol Sci Technol 16:981–986

Samson RA (1974) *Paecilomyces* and some allied Hyphomycetes. Study Mycol 6:1–119

Sasser JN, Freckman DW (1987) A world perspective on nematology: the role of the society. In: Veech JA, Dickson DW (eds) Vistas on nematology. Society of Nematologists, Hyattsville, pp 7–14

Schüßler A (2008) Glomeromycota species list. http://www.lrz-muenchen.de/~schuessler/amphylo/. Assessed on June 2014

Serfoji P, Smithra P, Saravanan K, Durai Raj K (2013) Management of *Meloidogyne incognita* on tobacco through *Glomus aggregatum* and oil cakes. Int J Pharma Biol Arch 4:451–454

Sharf R, Shiekh H, Abbasi S, Akhtar A, Robab MI (2014) Interaction between *Meloidogyne incognita* and *Pochonia chlamydosporia* and their effects on the growth of *Phaseolus vulgaris*. Arch Phytopathol Protect 47:622–630

Sharma P, Pandey R (2009) Biological control of root-knot nematode; *Meloidogyne incognita* in the medicinal plant; *Withania somnifera* and the effect of biocontrol agents on plant growth. Afr J Agric Res 4:564–567

Shreenivasa KR, Krishnappa K, Ravichandra NG (2007) Interaction effects of arbuscular mycorrhizal fungus *Glomus fasciculatum* and root-knot nematode, *Meloidogyne incognita* on growth and phosphorous uptake of tomato. Karn J Agric Sci 20:57–61

Siddiqui ZA, Akhtar MS (2007) Effects of AM fungi and organic fertilizers on the reproduction of the nematode *Meloidogyne incognita* and on the growth and water loss of tomato. Biol Fertil Soils 43:603–609

Siddiqui ZA, Akhtar MS (2008a) Synergistic effects of antagonistic fungi and a plant growth promoting rhizobacterium, an arbuscular mycorrhizal fungus, or composted cow manure on populations of *Meloidogyne incognita* and growth of tomato. Biocontrol Sci Technol 18:279–290

Siddiqui ZA, Akhtar MS (2008b) Effects of fertilizers, AM fungus and plant growth promoting rhizobacterium on the growth of tomato and on the reproduction of root-knot nematode *Meloidogyne incognita*. J Plant Inter 3:263–271

Siddiqui ZA, Akhtar MS (2009a) Effect of plant growth promoting rhizobacteria, nematode parasitic fungi and root-nodule bacterium on root-knot nematodes *Meloidogyne javanica* and growth of chickpea. Biocontrol Sci Technol 19:511–521

Siddiqui ZA, Akhtar MS (2009b) Effects of antagonistic fungi and plant growth-promoting rhizobacteria on growth of tomato and reproduction of the root-knot nematode, *Meloidogyne incognita*. Aust Plant Pathol 38:22–28

Smith SE, Read DJ (2008) Mycorrhizal symbiosis, 3rd edn. Academic, London

Soliman AS, Shawky SM, Omar MNA (2011) Efficiency of bioagents in controlling root-knot nematode on Acacia plants in Egypt. American-Eurasian J Agric Environ Sci 10:223–229

Sundararaju P, Kiruthika P (2009) Effect of biocontrol agent, *Paecilomyces lilacinus* along with neemcake and botanicals for the management of *Meloidogyne incognita* on banana. Indian J Nematol 39:201–206

Tahseen Q, Clark IM, Atkins SD, Hirsch PR, Kerry BR (2005) Impact of the nematophagous fungus *Pochonia chlamydosporia* on nematode and microbial populations. Commun Agric Appl Biol Sci 70:81–86

Tikhonov VE, Lopez-Llorca LV, Salinas J, Jansson HB (2002) Purification and characterization of chitinases from the nematophagous fungi *Verticillium chlamydosporium* and *V. suchlasporium*. Fun Gen Biol 35:67–78

Timper P (2011) Utilization of biological control for managing plant-parasitic nematodes. In: Davies K, Spiegel Y (eds) Biological control of plant-parasitic nematodes: building coherence between microbial ecology and molecular mechanisms. Springer, New York, pp 259–289

Udo IA, Uguru MI, Ogbuji RO (2013) Pathogenicity of *Meloidogyne incognita* race 1 on tomato as influenced by different arbuscular mycorrhizal fungi and bioformulated *Paecilomyces lilacinus* in a dysteric cambisol soil. J Plant Protect Res 53:71–78

Usman A, Siddiqui MA (2012) Effect of some fungal strains for the management of root-knot nematode (*Meloidogyne incognita*) on eggplant (*Solanum melongena*). J Agric Technol 8:213–218

van Damme V, Hoedekie A, Viaene N (2005) Long-term efficacy of *Pochonia chlamydosporia* for the management of *Meloidogyne javanica* for glasshouse crops. Nematology 7:727–736

Viggiano JR, de Freitas LG, Lopes EA (2014) Use of *Pochonia chlamydosporia* to control *Meloidogyne javanica* in cucumber. Biol Control 69:72–77

Weller DM, Landa BB, Mavrodi OV, Schroeder KL, De la Fuente L, Bankhead SB, Molar RA, Bonsall RF, Mavrodi DV, Thomashow LS (2007) Role of 2,4-diacetylphloroglucinol-producing fluorescent Pseudomonas spp. in the defense of plant roots. Plant Biol 9:4–20

Yadav K, Akhtar MS, Panwar J (2015) Rhizospheric plant microbe interactions: key factor to soil fertility and plant nutrition. In: Arora NK (ed) Plant microbe symbiosis: applied facets. Springer, India, pp 127–145

Yang J, Tian B, Liang L, Zhang K (2007) Extracellular enzymes and the pathogenesis of nematophagous fungi. Appl Microbiol Biotechnol 75:21–31

Yang J, Loffredo A, Borneman J, Becker JO (2012) Biocontrol efficacy among strains of *Pochonia chlamydosporia* obtained from a root-knot nematode suppressive soil. J Nematol 44:67–71

Zare R, Gams W (2004) A monograph of *Verticillium* section Prostrata. Rostaniha 3:1–188

Zhang L, Zhang J, Christie P, Li X (2008) Pre-inoculation with arbuscular mycorrhizal fungi suppresses root knot nematode (*Meloidogyne incognita*) on cucumber (*Cucumis sativus*). Biol Fertil Soils 45:205–211

Zhang L, Zhang J, Christie P, Li X (2009) Effect of inoculation with the arbuscular mycorrhizal fungus *Glomus intraradices* on the root-knot nematode *Meloidogyne incognita* in cucumber. J Plant Nutr 32:967–979

Chapter 12
Soil Suppressive Microorganisms and Their Impact on Fungal Wilt Pathogens

M.K. Mahatma and L. Mahatma

12.1 Introduction

Saprophytic microorganisms are indispensible members of food chain and play vital role in recycling of carbon and nutrients by decomposition in the ecosystem. Availability of the nutrients creates favorable conditions for the growth of plants. Starvation of saprophytes poses natural selection pressure and adaption to it leads to the evolution of parasitism. Some of these microorganisms evolved complex mechanisms in response to the host defense and adapted to utilize nutrients from the living organisms, and gradually facultative parasite, facultative saprophyte, obligate parasite, and hyperparasitism developed making the soil ecosystem highly complex. In soil, many microorganisms occur in close proximity, and they interact in a unique way. The sum total of all of the individual interactions establishes the equilibrium population. Odum (1959) proposed seven relationships between the different living organisms in the equilibrium as follows: (a) *neutralism*, in which two organisms behave entirely independently; (b) *symbiosis*, the two symbionts relying upon one another and both benefiting by the relationship; (c) *protocooperation*, an association of mutual benefit to the two species but without the cooperation being obligatory for the existence or for their performance of some reaction; (d) *commensalism*, in which only one species derives benefit while the other is unaffected; (e) *competition*, a condition in which there is a

M.K. Mahatma (✉)
ICAR-Directorate of Groundnut Research, Ivnagar Road, P. Box # 5, Junagadh 362001, GJ, India
e-mail: maheshmahatma@gmail.com

L. Mahatma
Department of Plant pathology, N.M. College of Agriculture, Navsari Agricultural University, Navsari 396 450, GJ, India
e-mail: mahatmalalit@yahoo.co.in

M.K. Meghvansi, A. Varma (eds.), *Organic Amendments and Soil Suppressiveness in Plant Disease Management*, Soil Biology 46,
DOI 10.1007/978-3-319-23075-7_12

suppression of one organism as the two species struggle for the limiting quantities of nutrients, O_2, space, or other common requirements; (f) *amensalism*, in which one species is suppressed while the second is not affected, often the result of toxic production; and (g) *parasitism* and *predation*, the direct attack of one organism upon another. Existence of these relationships and their predominance characterize the soil. Conveniently, the soil has been classified in to two different categories, viz., conducive and suppressive soil. If in the soil, plant pathogenic microorganisms develop well and provide congenial conditions for the severe diseases, it is known as conducive soil. To be conducive, there should be the appropriate population density of the particular pathogen in the soil. Whereas, soils in which the pathogen does not establish, or establishes but causes little or no damage, or establishes and causes disease for a while but thereafter the disease is less important, although the pathogen may persist in the soil is known as suppressive soil (Baker and Cook 1974). Suppressive soil provides hostile environmental conditions for the pathogen to build up inoculum potential and penetration. Numerous biotic and/or abiotic factors cumulatively make the soil suppressive. Many antagonistic, pathogenic, as well as unapparent microorganisms remain in equilibrium proportion in the soil which predominately determines its characteristics. As long as the equilibrium remains ideal or shifted towards the antagonistic microorganisms by selectively favoring its activities, the soil suppresses the disease and support good crop. However, if the equilibrium shifts towards the pathogenic microorganisms and increases its potentiality, it becomes conducive soil. Range of the suppressiveness has been observed, and there may be intermediate or ideal suppressive soil. Suppressive soils have been described for many soil-borne pathogens, viz., *Gaeumannomyces graminis* var. *tritici* (take-all of wheat, which causes blackening of the plant base, stunting, and, in severe cases, white inflorescence with shrivelled grains and no yield); *Fusarium oxysporum* (wilt diseases of tomato, radish, banana, and others); *Phytophthora cinnamon* (root rot of eucalyptus); *Pythium* spp. and *Rhizoctonia solani* (damping-off of seedlings of several crops, including sugar beet and radish); *Thielaviopsis basicola* (black root rot of tobacco, bean, cherry trees, and others); *Streptomyces scabies* (bacterial potato scab; i.e., lesions on potato tubers); *Ralstonia solanacearum* (bacterial wilt of tomato, tobacco, and others); *Meloidogyne incognita* (root swelling and root-knot galls on several crops, mostly in tropical and subtropical countries). In the present chapter, different aspects of microbial soil suppressiveness and their impact on wilt disease have been discussed in detail.

12.2 Historical Landmark

Soil suppressiveness and microorganisms in the suppression of disease were first time realized by Sanford (1926) while working on potato scab disease caused by *S. scabies*. He observed that the incidence of potato scab caused by *S. scabies* is reducing in the green manuring crop in Canada. Attention of Millard in England was drawn to this observation and soon reported that the reduced disease incidence was due to the presence of inhibitory effect of nonpathogenic bacteria (Millard and Taylor 1927). Henery (1932) from the University of Alberta, Canada, reported that with the increasing temperature, the infection curve for *G.graminis* on wheat seedling was downwards in unsterilized but upward in sterilized soil. Decrease in the infection with the raise in the temperature in the unsterilized soil was also due to enhanced effect of soil mycroflora. The result was confirmed by Garrett (1934) of Waite Institute, Adelaide. This has guided to focus on different soil inhabiting microorganisms for the management of devastating diseases. Weindling (1932) showed that *Trichoderma* sp., a common saprophytic fungus, was able to parasitize the mycelia of other fungi. The first report of fusarium wilt suppressive soil was made by Stover (1962). Suppression of fusarium wilt of radish by growth-promoting effects of fluorescent pseudomonads was first published by Kloepper and Schroth (1978) and later by Geels et al. (1985). Substantial work on the biological control and suppressive soil has been done by the different scientists throughout the word. Weller et al. (2002) thoroughly reviewed the microbial populations responsible for specific soil suppressiveness to plant pathogens.

12.3 Fungal Wilt Disease

Vascular wilt characterized by the presence of pathogen in the vessels of angiosperm is one of the most destructive diseases. Four genera of fungi, viz., *Fusarium*, *Verticillium*, *Ceratocystis*, and *Ophiostoma* cause the vascular wilt. Among the different genera, *Fusarium* and *Verticillium* are most important and cause disease to the wide range of plants. *Verticillium* sp., a cold loving fungus that thrives best in heavy soils, does not require injury for infection whereas *Fusarium* sp., found in the tropical and subtropical region, grows best in sandy soil and causes more damage when root-knot, reniform, or sting nematodes injure the roots. *Fusarium* sp. prefers acidic condition and can be transmitted internally in seed, while *Verticillium* prefers alkaline conditions and is not transmitted internally in the seeds. High nitrogen fertilizer, excessive soil moisture, thin stands, and deep cultivation during the growing season favor wilt disease. Both fungi survive long periods in soil in the absence of a cultivated host. Fusarium wilt was first recognized in the nineteenth century by Atkinson (1892) and was later described for other soils around the globe. *Verticillium* (*Verticillium albo-atrum* and *V. dahliae*) causes vascular wilts of vegetables, flowers, field crops, perennial ornamentals, and fruit and forest trees in the temperate region. All vascular wilts have certain disease symptoms in common and are almost similar to the physiological drought, however, are irreversible. In cross sections of infected stems and twigs, discolored brown areas appear as a complete or interrupted ring consisting of discolored vascular tissues. Vessels may be clogged with mycelium, spores, or polysaccharides produced by the fungus. Clogging is increased further by gels and gums formed by the breakdown products of plant cells by the enzymatic action of the fungi. In some hosts, balloon-like tyloses are produced by parenchyma cells adjoining some xylem vessels (Agrios 2005).

12.4 Classification of Suppressive Soils

Two different categories, viz., general and specific suppressiveness, are most commonly used by many scientists to classify soil suppressiveness. The widespread but limited ability of soils to suppress the growth or activity of soil-borne pathogens has been referred to as "general suppression." Nonspecific antagonism or biological buffering terminologies (Weller et al. 2002) have also been used to simplify the nomenclature; however, they are less accepted. Accordingly, specific suppressiveness is due to antagonistic effect of individual or selected groups of microorganisms during some stage in the life cycle of a pathogen. Though the general and specific suppressiveness is most widely used terminology, it seems ambiguous and represents the notation of general suppressiveness encompassing wide range of pathogens. Similarly, specific suppressiveness gives the notation of specific suppression of the disease. However, general suppressiveness is also effective against the given class of pathogen only. This is clearly illustrated for the soil suppressive to the fusarium wilt which is not even suppressive to the disease caused by *F. solani, F. roseum*, and other soil-borne pathogens (Alabouvette 1986; Deacon and Berry 1993; Steinberg et al. 2007). Suppressiveness is so specific and sometimes is cultivar specific. Hopkins et al. (1987) from Florida in a long-term monoculture of watermelon cultivar observed that most of the cultivars wilted severely after 4–5 years regardless of previously described levels of resistance to *Fusarium oxysporium* f. sp. *niveum*. Only the resistance in Smokylee and Crimson Sweet was stable in the monoculture, and only Crimson Sweet continued to have acceptable level of yields throughout the monoculture. Crimson Sweet only moderately resistant to fusarium wilt in greenhouse tests had a unique resistance that was effective throughout the 7 years monoculture. Instead, if the horizontal suppressiveness and vertical suppressiveness are used to notify the general and specific suppressiveness, respectively, it would be more comprehensive. The classification can be metaphorically comprehended as horizontal resistant (horizontal suppressive) and vertical resistant (vertical suppressive). However, in the present chapter old terminologies, viz., general and specific suppressiveness, are used. Various characteristics of suppressive soils are given in Table 12.1 and described suitably in the different subheading in the chapter.

Table 12.1 Characteristics of nonspecific and specific suppressive soils

Sr. No.	Characteristics	General suppressive	Specific suppressive	References
1	Synonym	Nonspecific suppressive soil Nonspecific antagonist or biological buffering, horizontal suppressive soil	Specific antagonist or biological buffering, vertical suppressive soil	Baker and Cook (1974), Alabouvette (1986), Weller et al. (2002)
2	Number of microorganisms associated	Many	One or few	Alabouvette (1986), Weller et al. (2002)
3	Effect of soil organic matter	Enhanced on addition	Not affected much	Hopkins et al. (1987), Weller et al. (2002)
4	Transferability	Less	More	Menzies (1959), Cook and Rovira (1976), Weller et al. (2002)
5	Inducibility	No	Yes	Cook and Rovira (1976), Alabouvette (1986), Weller et al. (2002)
6	Effect of edaphic, climatic, and agronomic conditions	More	Less	Cook and Rovira (1976)
7	Duration	Retain from the longer period	Retain from the shorter period	Hopkins et al. (1987)
8	Effectiveness in the absence of plants	Not affected	Affected	Hopkins et al. (1987), Weller et al. (2002)
9	Reversibility	Difficult to convert in conducive soil	Easy to convert in conducive soil	Cook and Rovira (1976), Larkin et al. (1993)

12.5 Wilt Suppressive Soils

Among the different wilts, fusarium wilt suppressive soil has only been observed and studied extensively. Suppressive soil has been reported from the four places, viz., in the Salinas Valley, California, United States; the Chateaurenard region, near Cavaillon, France; the Canary Islands and the Broye Valley, Switzerland. Among these, the Chateaurenard soils in France and the Salinas Valley soil in California are known for their natural suppressiveness to fusarium wilt diseases (Louvet et al. 1976; Kloepper et al. 1980; Scher and Baker 1980). Monoculture-induced suppressiveness to fusarium wilt of watermelon was studied at Central Florida Research and Education Centre, Leesburg, Florida (Larkin et al. 1993). Alabouvette

(1986) extensively worked on the fusarium wilt suppressive soils from the Châteaurenard region and reviewed the results. They coined the concept of soil receptivity to soil-borne pathogens while working on fusarium wilt in melon which reflects the capacity of a soil to allow a pathogen to establish, develop, persist, and express its pathogenicity on host plants (Alabouvette et al. 1982). Study reveled that the absence of disease could not always be accounted for the absence of the pathogen (*F. oxysporum* f. sp. *melonis*). This was demonstrated by introducing into various soils increasing amounts of a given pathogen; at similar inoculum densities, severity of disease on a population of susceptible host plants varies significantly according to soils indicating the various degrees of soil receptivity to Fusarium wilt (Alabouvette et al. 1982). It is thus possible to identify disease suppressiveness.

12.6 Microorganisms in Soil Suppressiveness and Its Mechanisms

Suppressiveness of soil is mainly related to its biological properties; however, physical, chemical, and meteorological factors affect the biological factors and thereby indirectly affect the suppressiveness of the soil. Both general and specific suppression are eliminated by autoclaving and gamma radiation which support the biological basis of disease suppression. General suppression is reduced but not eliminated by soil fumigation, and 70 ± C moist heat (Cook and Rovira 1976). The specific suppressiveness was eliminated by pasteurization (Shipton et al. 1973; Scher and Baker 1982; Alabouvette 1986; Raaijmakers and Weller 1998; Westphal and Becker 2000). Numerous kinds of antagonistic microorganisms have been found to increase in suppressive soils; most commonly, however, pathogen and disease suppression has been shown to be caused by fungi, such as *Trichoderma* sp., *Penicillium* sp., and *Sporidesmium* sp., or by bacteria of the genera *Pseudomonas* sp., *Bacillus* sp., and *Streptomyces* sp. However, populations of nonpathogenic *F. oxysporum* and fluorescent *Pseudomonas* spp. have been repeatedly shown to be involved in suppression of fusarium wilts in naturally occurring disease suppressive soils. Other antagonistic microorganisms have been proposed having lesser roles in the suppression of fusarium wilts (Alabouvette 1990; Larkin et al. 1996). Suppressiveness to *F. oxysporum* f. sp. *melonis* (Scher and Baker 1980) and *F. oxysporum* f. sp. *niveum* (Hopkins et al. 1987; Larkin et al. 1993) was induced following continuous cropping of melon and watermelon, respectively. Interestingly, the induction of suppressiveness in these cases was associated with continuous cropping of partially resistant cultivars, whereas induction of suppressiveness against other soil-borne pathogens normally involves monoculture of susceptible cultivars (Whipps 1997). Evidence of a similar induction of suppression in the early 1900s was reviewed by Kommedahl et al. (1970), in which long-term monoculture of cultivar resistant to flax wilt resulted in a marked decline in disease following several years of increases at Ventura Count, California, USA. Whereas, cropping to susceptible cultivars resulted in complete wilt (100 %) every year. Schneider (1982) also observed islands of healthy celery plants in fields uniformly devastated by wilt. In both of these cases, the organisms responsible for suppressiveness were non-pathogenic *F. oxysporum*. Transfer of suppressiveness to a raw conducive, fumigated, or sterilized soil by addition of 0.1–10 % or less (w/w) of the suppressive soil further consolidated the role of microorganisms in suppressiveness. Mechanisms in suppression of fusarium wilt by microorganisms may involve competition for substrate and root surface, antagonism, PGPR activities, and cytological modification of host plant holistically.

12.6.1 Competition for Nutrients and Root Surface

After the germination of the pathogen, it has to travel to some distance before it comes in the contact of the host surface, and host parasite relationship is established. Till the distance is travelled, the pathogen need to remain dependent on some other source of nutrients. Presence of other microorganisms may exert competition for the nutrients and site of infection which is a general phenomenon regulating the population dynamics of microorganisms sharing the same ecological niche and having the same physiological requirements (Alabouvette et al. 2009). Carbon, nitrogen, and phosphorous are the important nutrients required for the growth of fungi. Among the different nutrients, competition for the carbon is most significant and is responsible for the inhibition of germination and subsequent growth of the fusarium wilt pathogen in Châteaurenard region (Bouches-du-Rhône, France) in melon and cotton field (Alabouvette et al. 1977; Sivan and Chet 1989; De Boer et al. 2003). Larkin et al. (1996) isolated 400 different microorganisms including actinomycetes, bacteria, and fungi from watermelon root growing in the suppressive and non-suppressive soil to fusarium wilt of watermelon and concluded that nonpathogenic *F. oxysporium* was the primary antagonist responsible for the disease suppressiveness. Other than *F. oxysporum, Trichoderma* spp., Arbuscular mycorrhizal fungi (AMF), fluorescent *Pseudomonas spp., Bacillus* spp., *Alcaligenes* sp., etc. have been reported to control *fusarium* diseases in different crops (Park et al. 1988; Duijff et al. 1991; Lemanceau and Alabouvette 1991; Chen et al. 1995; Tanwar et al. 2013). *Trichoderma* spp. and *Pseudomonas spp.* colonize near to the root surface and exert multiple effects on the pathogen by competing for the nutrient & infection site and parasitizing the pathogen either directly or indirectly by secreting many growth limiting metabolites (Perelló et al. 2003). In artificially developed suppressive soil, application of a combination of biocontrol agents is likely to more closely mimic the natural situation and may, therefore, represent a more viable control strategy of wilt diseases in many crops. Lemanceau et al. (1992, 1993) described increased suppression of fusarium wilt of carnation by combining *P. putida* WCS358 with nonpathogenic *F. oxysporum* Fo47. The enhanced disease suppression by this combination is due to siderophore-mediated competition for iron by *P. putida* WCS358, which makes the pathogenic *F. oxysporum* strain more sensitive to competition for glucose by the nonpathogenic strain *F. oxysporum* Fo47. Furthermore, Leeman et al. (1996) showed that combining strains of nonpathogenic *Verticillium lecanii*, *Acremonium rutilum*, or *F. oxysporum* with the fluorescent *Pseudomonas* spp. strains WCS358, WCS374, or WCS417 resulted in significantly better suppression of fusarium wilt of radish compared to the single organisms. This mechanism was proved using a GUS-marked strain of pathogenic *F. oxysporum* f. sp. *lini* and a *pvd-in a Z*-marked derivative *of P. putida* WCS358. The study confirmed that suppression by the nonpathogenic *Fusarium* sp., strain is related to reductions in both population density and metabolic activity of the pathogen on the root surface, and that competition for iron contributes to the suppression by *Pseudomonas* sp., and enhances the biological activity of the nonpathogenic *F. oxysporum* strain. The significant role for pyoverdine production by *P. putida* WCS358 in this interaction was ascertained as the siderophore deficient mutant did not enhance disease control achieved by use of the nonpathogenic *F. oxysporum* alone (Duijff et al. 1999).

Competition for the infection court by quantifying root colonization by a nonpathogenic and a pathogenic strain of *F. oxysporum* was observed by Eparvier and Alabouvette (1994). Glucuronidase activity of the GUS-transformed pathogen was reduced in the presence of the protective strain and concluded that these strains were competing for root colonization. It is evident that stable suppressiveness such as suppressiveness of the soil from the Salinas Valley or Chateaurenard is based on the collective effects of several microorganisms and mechanisms (Schippers 1992).

12.6.2 Antagonism

Antagonism which involves the destruction or inhibition of the growth of the pathogen by other microorganisms is a well-known phenomenon in the ecosystem. The parasitic activity of strains of *Trichoderma spp*. towards various pathogens has been studied and reviewed thoroughly (Harman et al. 2004; Motlagh and Samimi 2013; El-Rahman and Mohamed 2014; Lelavthi et al. 2014). Chitin and β-1,3-glucan are the main structural components of fungal cells walls, except those from members of the class oomycetes, which contain β-1,3-glucan and cellulose. Antagonism by the *Trichoderma spp.* involves specific recognition between the antagonist and its target pathogen and triggers cell wall-degrading enzymes, viz., β-(1,3)- glucanases, chitinases, lipases, and proteases. These enzymes penetrate the hyphae of the pathogen resulting into death of the target organism (De la Cruz et al. 1992; Sivan and Chet 1989). In addition, they produce some lytic enzymes during the parasitic interaction between *Trichoderma* spp. and some pathogenic fungi (Haran et al. 1996). Other cell wall-degrading enzymes, including hydrolyzing minor polymers (proteins, β-1,6-glucans, α-1,3-glucans, etc.), may be involved in the effective and complete degradation of mycelial or conidial walls of phytopathogenic fungi by *Trichoderma* spp. Mycoparasitism describes the type of biotrophic interactions in which organisms benefit at the expense of the fungi (Druzhinina et al. 2011). Partial degradation of the host cell wall is normally observed in later stages of the parasitic process. Initially, the mycoparasite grows directly towards its host and often coils around it or attaches to it by forming hook-like structures and apressoria. Following these interactions, *Trichoderma* spp. sometimes penetrate the host mycelium, apparently by partially degrading its cell walls (Elad et al. 1984). Heterotrimeric G-proteins and mitogen-activated protein (MAP) kinases affected biocontrol-relevant processes such as the production of hydrolytic enzymes and antifungal metabolites and the formation of infection structures. MAPK signaling was also found to be involved in induction of plant systemic resistance in *T. virens* and in the hyperosmotic stress response in *T. harzianum*. *Trichoderma* mycoparasitism combines processes such as nutrient competition (Chet 1987), the secretion of antifungal metabolites (Lorito et al. 1996), and formation of morphological changes such as coiling around the host and development of appressorium-like structures (Lu et al. 2004).

Antibiosis is also a very common phenomenon of antagonism of many biocontrol agents (BCAs) such as fluorescent *Pseudomonas* spp., *Bacillus* spp., *Streptomyces* spp., and *Trichoderma* spp. Various secondary metabolites have been reported from these microorganism with their role in the suppression of several plant pathogens (Weller and Thomashow 1993; Alabouvette et al. 2009). Production of antibiotics including phenazine-1-carboxylic acid, 2,4 diacetylphloroglucinol (2,4- DAPG), pyoluteorin, and pyrrolnitrin play an important role in the biological control of soil-borne pathogens by certain strains of fluorescent *Pseudomonas* spp. that produce these antibiotics (Keel et al. 1992; Kraus and Loper 1995). There are some evidences of the activity of phenazines and anthranilate in the antagonism of *Pseudomonas aeruginosa* toward *F. oxysporum* (Anjaiah et al. 1998).

12.6.3 PGPR Activities

Microbial activities are 10–1000 times higher in the vicinity of plant roots than in unplanted soil (Lugtenberg and Bloemberg 2004). Plant Growth Promoting Rhizobacteria (PGPR), viz., *Azotobacter* spp., *Azospirillum* spp., *Acetobacter* spp., *Rhizobium* spp., *Bacillus* spp., AMF, *Trichoderma* spp., *Pseudomonas* spp., etc., competitively colonize near plant roots and stimulate plant growth and/or inhibit the pathogenic activities. Signal molecules secreted by the root surface of the susceptible plant activate the germination of prologues pathogenic fungi; however, presence of PGPR and its colonization prevents subsequent growth of pathogenic microorganisms near to the root surface. Supply of nutrients either by fixing or solubilizing, production of phytohormones (such as auxin and cytokinin), and volatile growth stimulants such as ethylene and 2,3-butanediol help plants in growing better and controlling diseases (Haas and Défago 2005; Ayed et al. 2006;

Daami-Remadi et al. 2006; Chowdappa et al. 2013). Efficiency and level of disease suppression depend upon efficacy, population dynamics, and location of these microorganisms. Biodegradation activities of PGPR, through the action of ACC deaminase activity that hydrolyzes ACC into ammonia and α-ketobutyrate, prevent the synthesis of plant growth inhibiting levels of ethylene in the roots (Viterbo et al. 2010). ACC deaminase has previously been reported for *Pseudomonas spp.*, and its activity has been associated with an increase in root elongation due to the reduced inhibition caused by ethylene (Avis et al. 2008). *Trichoderma* strains colonize the plant roots and influencing the synthesis of chloroplast enzymes that increases rate of photosynthesis (Abo-Ghalia and El-Khallal 2005) or establishing chemical communication and systemically altering the expression of numerous plant genes that alter plant physiology and photosynthetic efficiency (Harman et al. 2004; Hermosa et al. 2012). Further, *Trichoderma* sp. has the ability to increase the solubility of nutrients with low solubility like phosphates and other micronutrients like zinc, copper, iron, and manganese (Altomare et al. 1999), and the soluble form of phosphorus was easily absorbed by the extensive plant roots. Thus, through an increased nutrient uptake, bioagents compensate for the losses caused by pathogen attack. Biocontrol potential of AMF could be explained in terms of its ability to change root architecture, improved nutrient uptake, competition with the pathogen for infection site, activation of plant defense enzymes (chitinase, chitosanase, β-1,3-glucanase, and superoxide dismutase), phenolic and phytoalexin production (Avis et al. 2008).

12.6.4 Induced Systemic Resistance

Induced systemic resistance (ISR) is the process whereby the detrimental effect of a pathogen on plant is induced by prior treatment with an elicitor, either an organism or chemical. It has been proposed that, in suppressive soils, plant roots are associated with microbial communities that have an overall beneficial effect on plant health. ISR allows plants to withstand pathogen attack to the leaves or roots, without offering total protection (Harman et al. 2004). Many effective PGPMs elicit ISR, irrespective of antibiotic production (Zehnder et al. 2001; Ongena et al. 2004). Systemic induced resistance (SAR) by *P. fluorescens* WCS417r was established in carnation, radish, Arabidopsis tomato (Van Peer et al. 1991; Leeman et al. 1995; Pieterse et al. 1996; Duijff et al. 1998). Indeed, a mutant of this bacterial strain reduced endophytic root colonization and a lower ability to induce systemic resistance (Duijff et al. 1997). The effects of three different strains of *Pseudomonas* spp. mediating ISR in *Arabidopsis thaliana* have been investigated through transcriptome analysis of plants with roots that were colonized by one of these strains (*P. fluorescens*WCS417r, *P. thivervalensis*, or *P. fluorescens* CHA0). Studies with *A. thaliana* mutants indicate that the jasmonate/ethylene-inducible defense pathway is important for ISR, whereas the salicylate-inducible pathway mediating SAR seems to be less important. Total six classes of antibiotic compounds, viz., phenazines, phloroglucinols, pyoluteorin, pyrrolnitrin, cyclic lipopeptides (all of which are diffusible), and hydrogen cyanide (HCN; which is volatile) were reported from *P. fluorescens* which suppress root disease. The modes of action of these secondary metabolites are partly understood. These antibiotics exert inhibition of electron transport chain and fungal respiratory chains and cause membrane damage (Reviewed by Haas and Défago 2005). In bean, ISR elicited by a *P. putida* strain was associated with elevated levels of hexenal, which is a volatile antifungal compound, and with enhanced expression of enzymes that are involved in hexenal synthesis (Ongena et al. 2004). The ability of nonpathogenic *F. oxysporum* to induce resistance has been shown in carnation, cucumber, chickpea, and tomato (Kroon et al. 1991; Mandeel and Baker 1991; Hervás et al. 1995; Fuchs et al. 1997). However, the efficacy of the induced resistance varies according to the fungal biocontrol strain (Olivain et al. 1995). The spatial separation between the biocontrol strains used to induce resistance and the challenging pathogen in the split root system led to the conclusion that the reduction of the disease incidence by the inducing microorganisms was plant mediated (Hoffland et al. 1996). Further, inoculation with nonpathogenic *F. oxysporum strain* Fo47 increased chitinase, β-1,3-glucanase, and β-1,4-glucosidase activity in plants, confirming the ability of Fo47 to induce resistance in tomato Fuchs et al. (1997). This study suggests that Fo47 may act as an inducer of resistance through a classic SAR-like mechanism and induces PR proteins. *T. harzianum* strain T-39 also found to induce resistance and made leaves of bean plants resistant to diseases that are caused by the fungal pathogens *Botrytis cinerea* and *Colletotrichum lindemuthianum*, even though T-39 was present only on the roots and not on the foliage Bigirimana et al. (1997).

12.6.5 Cytological Modification

Induction of cytological modification in response to the presence and activities of nonpathogenic, antagonistic, plant growth promoting microorganisms tends to make the root surface incompatible for the penetration and subsequent establishment. Treating tomato plants with *Trichoderma* species has resulted in the formation of hemispherical cell wall appositions and the occlusion of some intercellular spaces by an amorphous material (Hibar 2007). Similarly, Benhamou and Thériault (1992) showed that treating tomato plants with *Pythium oligandrum* before inoculation with *F. oxysporum f.* sp. *radicis-lycopersici* has entailed cytological changes, mainly characterized by the elaboration of structural barriers, cell wall thickenings, and plugging of most intercellular spaces. Bao and Lazarovits (2001) observed reduced wilt disease incidence due to cell wall thickening in tomato plants after treatment of nonpathogenic strain of *F. oxysporum* (70T01).

12.7 Techniques to Study the Soil Suppressiveness

Haas and Défago (2005) discussed that the complexity of the disease suppression and observed four phenomena: First, certain suppressive soils when pasteurized (e.g., by wet heat at 70 °C for 30 min) lose their suppressiveness, and other harsher antimicrobial treatments (e.g., gamma radiation or autoclaving) have the same effect (Shipton et al. 1973; Scher and Baker 1980). Second, suppressiveness can be transferable: an inoculum of 0.1–10 % of a specific suppressive soil introduced into a conducive soil can establish disease suppression (Menzies 1959; Cook and Rovira 1976; Weller et al. 2002). Third, when the pH of a fusarium wilt suppressive soil is lowered from 8 to 6 by the addition of H_2SO_4, the soil looses suppressiveness (carnation to the wilt disease) because of the change in the soil environment. Fourth, several years of monoculture can induce disease suppression in some soils. The best-studied example is suppressiveness to *F. oxysporum* f. sp. *melonis* (Scher and Baker 1980) and *F. oxysporum* f. sp. *niveum* (Hopkins et al. 1987; Larkin et al. 1993) which was induced following continuous cropping of melon and watermelon, respectively. All these decisively establish that microorganisms are invariably associated with the soil suppressiveness; however, the soil environmental conditions also play role in making the soil suppressive either directly or indirectly by making the environment conducive for the antagonistic microorganism. To study the soil suppressiveness, these four phenomena should be systematically studied by using various techniques. Even with the advent of the advanced soil monitoring techniques, the nature of the soil microbiota, its dynamics, activities, and interactions are still largely enigmatic. One or few microorganisms may primarily be responsible for the suppressiveness, but interactions with other members of the rhizosphere community can significantly modulate its degree. Moreover, the phenomenon of disease suppression might be related to specific functions or activities of soil microorganisms rather than the simple presence or abundance of particular populations in the soil. Traditional approaches to study microbial communities in soils were based on culture-dependent techniques. These approaches were useful for isolation purposes, but were very limited in their scope to understand microbial communities and diversity. Recent developments in new types of media and methods have led to considerable advances in this composition and diversity of soil microbial communities; however, still less than 1 % of the microorganisms present in soil may be readily isolatable whereas remaining 99 % microorganisms viable but nonculturable (VBNC) stage (Torsvik et al. 1996; Kuske et al. 1997; Oliver 2005). It is generally admitted that disease suppressiveness is related to a global increase in soil microbial biomass. A large biomass would create a competitive environment deleterious for the pathogens (Janvier et al. 2007). To overcome the dependence on the culture dependence techniques and expand our understanding, culture-independent techniques to "first identify and then recover" important antagonists are extensively useful. These are holistic, high throughput, accurate, and comprehensive techniques; however, they have not been used to study the soil suppressiveness. For better understanding, it is recommended to use combination of culture-dependent and culture-independent techniques (Liesack et al. 1997).

12.7.1 Culture-Dependent Techniques

Culture-dependent techniques involves soil sampling, isolation of bacteria, and determination of colony forming units; screening of isolates for in vitro antagonistic activity towards pathogen; screening of antagonists for production of siderophores and cell-wall degrading enzymes; and identification of the isolated microorganism based on various biochemical and molecular techniques. The identification techniques include biochemical as well as nucleic acid based identifications which are quite accurate and reproducible. A thorough understanding of the mechanisms of action is needed to maximize consistency. In the *F. oxysporum* wilt suppressive soil, many studies dealing with nonpathogenic *F. oxysporum* have proven that not all the nonpathogenic strains are effective in controlling wilts. Since there is currently no known genetic marker to identify these strains, the only available and reliable method to screen for efficient strains is a bioassay in which the potential biocontrol agents are confronted with the pathogen in the presence of the host plant and disease incidence or severity is monitored. In general, the closer the screening method is to the production system, the greater the chances are for success.

12.7.2 Culture-Independent Techniques

Culture-independent techniques allow the study of a much greater part of the soil microflora. These techniques may be biochemical or molecular depending upon the test performed (Table 12.2). Biochemical techniques involve different assays, viz., Ability of microbial communities to degrade different carbon substrates (BIOLOG); Phospholipid fatty acid (PLFA); Fatty acid methyl ester (FAME); Enzyme activities and Metabolites (volatile and nonvolatile) profiling. Molecular techniques involves ITS/IGS or NTS sequencing ITS/IGS sequencing; Terminal-restriction fragment length polymorphism (T-RFLP); Denaturing gradient gel electrophoresis (DGGE and PCR-DGGE); RAPD and Gene-specific primers; 16S rRNA microarray probes; etc.

12.7.2.1 Biochemical Techniques

The community level physiological patterns established using the BIOLOG systems have been used to detect differences in the ability of microbial communities to degrade different carbon substrates (Garland and Mills 1991). Pérez-Piqueres

Table 12.2 Various techniques to identify microorganism in suppressive soil

Sr. No.	Techniques	Instruments	Suppressive soil to pathogen/disease	Microorganism Identified/specific method	References
1.0	Biochemical				
1.1	Carbon utilization	Biolog	South-East Asian isolates Take-all disease and *Rhizoctonia solani*	*Trichoderma* spp. *Pseudomonas corrugata*	Kubicek et al. (2003) Barnett et al. (1999)
1.2	FAME and PLFA)	GC	Take-all disease, *Rhizoctonia solani* *Verticillium dahliae* Kleb *Fusarium* crown and root rot of asparagus	*Pseudomonas corrugata* *Pseudomonas putida* Bacteria, Fungi, and Mycorrhiza	Barnett et al. (1999) Berg et al. (2002) Hamel et al. (2005)
1.3	Enzyme activity	Spectrophotometer/ substrate degradation in medium	Soil suppressiveness to seedling blight of barley (*Fusarium culmorum*) Suppressive soil to *F. oxysporum* on melon plants	Higher activities of β-glucosidase and cellobiohydrolase Higher phosphatase and β-glucosidase	Rasmussen et al. (2002) Ros et al. (2005)
1.4	Metabolites DAPG and other	HPLC	*Fusarium oxysporum* f. sp. *Lycopersici* and *F. oxysporum* f. sp. *albedinis*	*Pseudomonas chlororaphis* (phenazine carboxylic acid, 2-hydroxy phenazine carboxylic acid, and 2-hydroxy phenazine)	Mezaache-Aichour et al. (2012)
1.5	Volatile organic antifungal molecules	GC-MS	*Sclerotinia sclerotiorum*	*Pseudomonas* spp.	Fernando et al. (2005)
2.0	Molecular				
2.1	ITS/IGS or NTS sequencing	PCR	*Fusarium oxysporum* f. sp. *lycopersici* and *F. oxysporum* f. sp. *albedinis* Biocontrol isolates from different countries Fusarium Wilt of cucumber and other crops	*Pseudomonas chlororaphis* (16S rDNA and sequencing) *Trichoderma* spp. (ITS1 region of rDNA) Nonpathogenic *Fusarium oxysporum* (FIGS11/FIGS12 primers of IGS region	Mezaache-Aichour et al. (2012) Hermosa et al. (2004)

				of rDNA produced 500-bp DNA fragment)	Wang et al. (2013)
2.2	T-RFLP DGGE	PCR	Effect of Biocontrol Agent *Pseudomonas fluorescens* 2P24	Soil Fungal Community in Cucumber Rhizosphere using T-RFLP and DGGE	Gao et al. (2012)
2.3	3.RAPD M13 and D7 primers	PCR	Take-all disease suppressive soil *Fusarium oxysporum* f. sp. *radicis-lycopersici*	DAPG-producing *Pseudomonas spp.*	Raaijmakers and Weller (2001)
2.4	Gene-specific primers	PCR	Bio control isolates *Verticillium dahliae* Kleb (Isolated from potato, oilseed rape, strawberry, and from bulk soil)	*Trichoderma* spp. (tef1 gene) *P. putida* (phlD and chiA genes)	Hermosa et al. (2004) Berg et al. (2002)
2.5	5.16S rRNA probes	Microarray	(*Tobacco basicola*) tobacco black root rot suppressive soil	Fluorescent Pseudomonas (with Pseudomonas probe:Pseu1, PseuD, PseubC2BC3-2, and PseubC2-10)	Kyselkova et al. (2009)

et al. (2006) compared the BIOLOG profiles of different soil mixes suppressive to *R. solani* from the non-amended highly conducive control soil. Similarly, Benizri et al. (2005) compared the BIOLOG profiles of the bacteria inhabiting two healthy and one sick soil, mimicking peach tree replant disease. Analysis separated the soil bacteria isolated from healthy soils from those isolated from sick soils. Kubicek et al. (2003) identified *Trichoderma* spp. and *Pseudomonas corrugita* from the suppressive soil to Take-all disease from South East Asian Isolates. Barnett et al. (1999) characterized a collection of 14 spontaneous phenotype variants, derived from in vitro and in vivo cultures (wheat roots) of *P. corrugata* 2140, using fatty acid methyl ester profiles (GC-FAME), carbon substrate utilization (BIOLOG), and in vitro inhibition against seven soil microorganisms. All three phenotype profiles indicated marked differences between some variants and the parent isolate. Some variant types were classified taxonomically by GC-FAME as different species to their wild-type parent, and up to a Euclidian distance of 11 from their parent. Taxonomic identification by the BIOLOG assay was more consistent than others. Phospholipid-derived fatty acids (PLFA) are chemotaxonomic markers of bacteria and other organisms. Phospholipids are primary lipids found in cell membranes that are saponified, releasing fatty acids contained in their diglyceride tail. Phospholipids are extracted from the whole soil and analyzed. Once the phospholipids of an unknown sample are saponified, the composition of the resulting PLFA can be compared to the PLFA)of known organisms to determine the identity of the sample organism. Many fatty acids have been isolated and are representative of specific microbial groups, making PLFA analysis a useful tool to describe microbial diversity and structure (Bossio et al. 1998; Ibekwe and Kennedy 1998). Various fatty acid biomarkers have been reported for microorganism identification, viz., PLFA C18:2ω6 was taken as indicator of fungal biomass (Frostegard and Baath 1996); C16:1ω5, as indicator of extra radical mycorrhizal hyphae and spores (Olsson 1999); while 16:0 and 16:1 (equivalent proportions) along with 18:1ω7c/ω9t/ω12t fatty acids as biomarkers for *Pseudomonas* spp. (Piotrowska-Seget and Mrozik 2003).

The types and proportions of fatty acids present in cytoplasm membrane and outer membrane (gram negative microorganisms) lipids of cells are major phenotypic traits. FAME is a type of fatty acid ester that is derived by transesterification of fats with methanol. Since every microorganism has its specific FAME fingerprint, it can be used as a tool for microbial source tracking (MST). FAME microbial markers would be a useful indicator of soil health and that the soil odd number fatty acid proportion changed due to organic amendment, which also reduced the disease incidence (Cai et al. 2003). From Fusarium wilt suppressive soil of Chateaurenard, France, total 37 species of bacteria with 71 antagonists were identified using FAME and/or 16S rRNA gene sequencing. A high proportion of the antagonists isolated from this soil produced siderophores (94 % of 71) and chitinase activity (46 %). Interestingly, suppressive soil of Chateaurenard, France, displayed higher diversity of antagonistic bacteria (Adesina et al. 2007). Soil enzymes and metabolites play vital roles for the maintenance of soil ecology and soil health. Enzymatic activities in the soil are mainly of microbial origin; therefore, microorganisms are acting as the indicators of soil health and can be used as measures of microbial activity and characteristics of the soil. The potential enzymes playing major roles in maintaining soil health are—amylase, arylsulphatase, β-glucosidase, cellulase, chitinase, dehydrogenase, phosphatase, protease, and urease. These enzymes and other metabolites can be studied by the spectrophotometric techniques. Higher activities of β-glucosidase, cellobiohydrolase, phosphatase, and β-glucosidase was observed by in the soil suppressive to seedling blight of barley *(F. culmorum)* and to *F. oxysporum* on melon plants (Rasmussen et al. 2002; Ros et al. 2005). Phenazine carboxylic acid, 2-hydroxy phenazine carboxylic acid, and 2-hydroxy phenazine have been observed by HPLC in the *Fusarium oxysporum* f. sp. *lycopersici* and *F. oxysporum* f. sp. *albedinis* suppressiveness (Mezaache-Aichour et al. 2012).

12.7.2.2 Molecular Techniques

All the molecular techniques are based on the nucleic acid of the microbial communities which involves amplification of the DNA and sometimes its sequencing to validate the result with higher precision. Depending upon the specificity of the DNA fragment and primers used for the amplification of the DNA, various techniques have been named. Random Amplified Polymorphic DNA (RAPD) employ short primers (8–12 nucleotides) to amplify large template of genomic DNA without its prior knowledge, expecting that fragments will amplify. This makes the method popular for comparing the DNA of biological systems that have not been resolved. Other PCR uses gene-specific primers sets from the different part of the DNA. Gene-specific primers (phlD and phz) for the biosynthesis genes 2,4-diacetylphloroglucinol (2,4- DAPG) and phenazine-1-carboxylic acid (PCA) in pseudomonads in soils have been used to characterize wilt suppressive soil (Raaijmakers et al. 1997). Internal transcribed spacer (ITS) is a piece of nonfunctional RNA situated between 5′ external transcribed sequence (5′ ETS), 18S rRNA, ITS-1, 5.8S rRNA, ITS-2, 28S rRNA, and finally the 3′ ETS. During rRNA maturation, ETS and ITS pieces are spliced. Genes encoding ribosomal RNA and spacers occur in tandem repeats that are thousands of copies long, each separated by regions of non-transcribed DNA termed intergenic spacer (IGS) or non-transcribed spacer (NTS). Sequence of the ITS region is highly conserved because of low evolutionary pressure and widely used in taxonomy. Isolation of these from the soil samples is easy as they are in high copy number. Several taxon-specific primers have been described that allow selective amplification of fungal sequences. By using oligonucleotide primers targeted to conserved regions in the 16S and 23S genes, RISA (Ribosomal intergenic spacer analysis) fragments can be generated from most of the dominant bacteria in the soil sample. Amplification results in complex banding pattern that provides a community-specific profile where each DNA band corresponds to a bacterial population on the original assemblage. Majority of the rRNA operon serves a structural function; portions of the 16S-23S intergenic region can encode tRNAs depending on the bacterial species. *P. chlororaphis, Trichoderma* spp., nonpathogenic *F.oxysporum*, and many more biocontrol agents have been identified by RISA (Mezaache-Aichour et al. 2012; Hermosa et al. 2004; Wang et al. 2013). Terminal Restriction Fragment Length Polymorphism (T-RFLP) is a molecular tool for the profiling of microbial communities based on the position of a restriction site closest to a labeled end of an amplified gene. The method is based on digesting a mixture of PCR-amplified variants of a single gene using one or more restriction enzymes and detecting the size of each of the individual resulting in terminal fragments using a DNA sequence. Muyzer et al. (1993) described a technique based on the separation of all the same length PCR-amplified fragments coding for 16S rRNA, by denaturing gradient gel electrophoresis (DGGE). DGGE analysis of different microbial communities demonstrated the presence of up to 10 distinguishable bands in the separation pattern, which were most likely derived from as many different species constituting these populations, and thereby generated a DGGE profile of the populations. These techniques allow the analysis of both culturable and nonculturable microorganisms and provide a rapid method for observing changes in community structure in response to different environmental factors. Besides total bacterial and fungal communities, the structure of specific subgroups can also be assessed (Garbeva et al. 2006). In a soil having received pig slurry or compost and showing an increased suppressiveness to *R. solanacearum* biovar 2 on potato, PCR-DGGE revealed differences in the bacterial community structure (Schonfeld et al. 2003; Gorissen et al. 2004). These amendments resulted in the appearance of several novel bands and different relative intensities of bands common to the treated and non-treated soils. In the case of compost amendment, several discriminate DGGE bands and PCR products were cloned and/or sequenced in order to identify the corresponding microorganisms; but their involvement in disease suppressiveness remains to be tested. Nevertheless, even if the microorganisms are not directly responsible, these DNA markers might serve as indicators of these treatments and thus as indicator of the *R. solanacearum*-suppressive status of soil. Comparing bacterial DGGE patterns of soils receiving different treatments, Kowalchuk et al. (2003) found that except for a sterilized and then amended soil, all DGGE patterns from the treated and control soils were highly similar. The same samples were also examined by fungal PCR-DGGE. The profiles obtained were much simpler than those obtained for bacteria. Once again the sterilized and amended soil was very different from the others. Yang et al. (2001) compared DGGE fingerprinting of rhizospheric bacterial communities associated with healthy or *Phytophthora cinnamomi* infected avocado roots. An assay clearly revealed that bacterial communities from healthy roots, both of control trees or trees treated with biocontrol bacteria, were highly similar, but different from the communities on infected roots. Gao et al. (2012) studied soil fungal community in cucumber rhizosphere using T-RFLP and)DGGE and observed *Pseudomonas fluorescens* 2P24 as biocontrol agent. Pérez-Piqueres et al. (2006) used the T-RFLP method to characterize microbial communities. Correspondence analyses clearly separated both fungal and bacterial community structures of the most suppressive amended soil from the other treatments. All these results demonstrate that the microbial community structure and diversity are often sensitive to the phytopathological status of soils, but until now, no microbial component was identified as potential indicator of disease suppression from such studies. Indeed, after the whole community fingerprinting, it is necessary to select the discriminating markers and to identify the microorganisms "hidden" behind.

DNA microarray technique is accurate and helps in handling large number of samples. Kyselkova et al. (2009) assessed 64 16S rRNA microarray probes whose signals correlated with tobacco black-root-rot (*Tobacco basicola*) suppressiveness in greenhouse analyzed to discriminate suppressive from conducive soils under field conditions. Rhizobacterial communities of tobacco and wheat sampled in 2 years from four farmers' fields of contrasted suppressiveness status were compared. The 64 previously identified indicator probes correctly classified 72 % of 29 field samples, with 9 probes for *Azospirillum*, *Gluconacetobacter*, *Sphingomonadaceae*, *Planctomycetes*, Mycoplasma, *Lactobacillus crispatus*, and *Thermodes ulforhabdus* providing the best prediction. The whole probe set (1033 probes) revealed strong effects of plant, field location and year on rhizobacterial community composition, and a smaller (7 % variance) but significant effect of soil suppressiveness status. Study signifies the use of subset of 16S rRNA probes targeting diverse rhizobacteria as indicator of suppressiveness under field conditions.

12.8 Conclusion

In soil, many microorganisms occur in close proximity and interact in a unique way. Soils in which the pathogen does not establish, or establishes but causes little or no damage, or establishes and causes disease for a while but thereafter the disease is less important, although the pathogen may persist in the soil, are known as suppressive soils. Two different categories, viz., general or horizontal (widespread but limited ability of soils to suppress the growth or activity of soil-borne pathogens) and specific or vertical (due to antagonistic effect of individual or selected groups of microorganisms during some stage in the life cycle of a pathogen) suppressiveness is most commonly observed. Wilt suppressive soils have been reported from the four places, viz., in the Salinas Valley, California, United States; the Chateaurenard region, near Cavaillon, France; the Canary Islands and the Broye Valley, Switzerland. Among these, the Chateaurenard soil in France and the Salinas Valley soil in California are known for their natural suppressiveness to Fusarium wilt diseases. Numerous kinds of antagonistic microorganisms have been found to increase in suppressive soils; most commonly, however, pathogen and disease suppression has been shown to be caused by fungi, such as *Trichoderma* sp., *Penicillium* sp., and *Sporidesmium* sp., or by bacteria of the genera *Pseudomonas* sp., *Bacillus* sp., and *Streptomyces* sp. Populations of nonpathogenic *F. oxysporum* and fluorescent *Pseudomonas* spp. have been repeatedly shown to be involved in suppression of fusarium wilts in naturally occurring disease suppressive soils. Mechanisms in suppression of fusarium wilt by microorganisms are may be involving; competition for substrate and root surface; antagonism; PGPR activities; and cytological modification of host plant holistically. Less than 1 % of the microorganisms present in soil may be readily isolatable whereas remaining 99 % microorganism viable but nonculturable (VBNC) stage. To overcome the dependence on the culture dependence techniques and expand our understanding, culture-independent techniques to "first identify and then recover" important antagonists are extensively useful. For better understanding, it is recommended to use combination of culture-dependent and culture-independent techniques. Culture-independent techniques allow the study of a much greater part of the soil microflora. These techniques may be biochemical or molecular depending upon the test performed.

References

Abo-Ghalia H, El-Khallal SM (2005) Alleviation of heavy metal stress by arbuscular mycorrhizal fungi and jasmonic acid in maize plants. Egypt J Bot 45:55–77

Adesina MF, Lembke A, Costa R, Speksnijder A, Smalla K (2007) Screening of bacterial isolates from various European soils for *in vitro* antagonistic activity towards *Rhizoctonia solani* and *Fusarium oxysporum*: site dependent composition and diversity revealed. Soil Biol Biochem 39:2818–2828

Agrios GN (2005) Plant pathology, 5th edn. Elsevier, New York

Alabouvette C (1986) Fusarium wilt suppressive soils from the Chateaurenard region: review of a 10-year study. Agronomie 6:273–284

Alabouvette C (1990) Biological control of Fusarium wilt pathogens in suppressive soils. In: Hornby D (ed) Biological control of soilborne plant pathogens. CAB International, Wallingford, pp 27–43

Alabouvette C, Rouxel F, Louvet J (1977) Recherches sur la résistance des sols aux maladies. III. Effets du rayonnement a sur lamicroflore d'un sol et sa resistance a la fusariose vasculaire du melon. Ann Phytopathol 9:467–471

Alabouvette C, Couteaudier Y, Louvet J (1982) Comparaison dela réceptivité de différents sols et substrats de culture aux fusarioses vasculaires. Agronomie 2:1–6

Alabouvette C, Olivain C, Migheli Q, Steinberg C (2009) Microbiological control of soil-borne phytopathogenic fungi with special emphasis on wilt-inducing *Fusarium oxysporum*. New Phytol 184:529–544

Altomare C, Norvll WA, Bjrkman T, Harman GE (1999) Solubilization of phosphates and micronutrients by plant growth promoting and biocontrol *Trichoderma harzianum* strain 1295–22. Appl Environ Microbiol 65:2926–2933

Anjaiah V, Koedam N, Nowak-Thompson B, Loper E, Hofte M, Tambong LT, Comelis P (1998) Involvement of phenazines and anthranilate in the antagonism of *Pseudomonas aeruginosa* PNAI and Tn5 derivatives toward *Fusarium* spp. and *Pythium* spp. Mol Plant Microbe Interact 9:847–854

Atkinson GF (1892) Some diseases of cotton. Al Agric Exp Stn Bull 41:65

Avis TR, Gravel V, Antoun H, Tweddell RJ (2008) Multifaceted beneficial effects of rhizospheric microorganisms on plant health and productivity. Soil Biol Biochem 40:1733–1740

Ayed F, Daami-Remadi M, Jabnoun-Khiareddine H, El-Mahjoub M (2006) Potato vascular Fusarium wilt in Tunisia: incidence and biocontrol by *Trichoderma* spp. Plant Pathol J 5:92–98

Baker KF, Cook RJ (1974) Biological control of plant pathogens. Freeman, San Francisco, p 433

Bao JR, Lazarovits G (2001) Differential colonization of tomato roots by non-pathogenic and pathogenic *Fusarium oxysporum* may influence Fusarium wilt control. Phytopathology 91:449–456

Barnett SJ, Alami Y, Singleton I, Ryder MH (1999) Diversification of *Pseudomonas corrugata* 2140 produces new phenotypes altered in GC-FAME, BIOLOG, and *in vitro* inhibition profiles and taxonomic identification. Can J Micobiol 45:287–298

Benhamou N, Thériault G (1992) Treatment with chitosan enhances resistance of tomato plants to the crown and root rot pathogen *Fusarium oxysporum* f. sp. *radicis-lycopercisi*. Physiol Mol Plant Pathol 41:33–52

Benizri E, Piutti S, Verger S, Pages L, Vercambre G, Poessel JL, Michelot P (2005) Replant diseases: bacterial community structure and diversity in peach rhizosphere as determined by metabolic and genetic fingerprinting. Soil Biol Biochem 37:1738–1746

Berg G, Roskot N, Steidle A, Eber L, Zock A, Smalla K (2002) Plant-dependent genotypic and phenotypic diversity of antagonistic rhizobacteria isolated from different Verticillium host plants. Appl Environ Microbiol 68:3328–3338

Bigirimana J, De Meyer G, Poppe J, Elad Y, Höfte M (1997) Induction of systemic resistance on bean (*Phaseolus vulgaris*) by *Trichoderma harzianum*. Med Fac Landbouww Univ Gent 62:1001–1007

Bossio DA, Scow KM, Gunapala N, Graham KJ (1998) Determinants of soil microbial communities: effects of agricultural management, season and soil type on phospholipid fatty acid profiles. Microb Ecol 36:1–12

Cai Y, Zhao S, LiaoZ He C, Zhuang X (2003) Effect of fertilization on tomato bacterial wilt biocontrolling and soil health restoration using FAME analysis. Agric Sci China 2:779–785

Chen C, Bauske EM, Musson G, Rodríguez-Kábana R, Kloepper JW (1995) Biological control of Fusarium wilt on cotton by use of endophytic bacteria. Biol Control 5:83–91

Chet I (1987) Innovative approaches to plant disease control, Trichoderma - application, mode of action, and potential as a biocontrol agent of soilborne pathogenic fungi. Wiley, New York, pp 137–160

Chowdappa P, Kumar SPM, Lakshmi MJ, Upreti KK (2013) Growth stimulation and induction of systemic resistance in tomato against early and late blight by *Bacillus subtilis* OTPB1 or *Trichoderma harzianum* OTPB3. Biol Control 65:109–117

Cook RJ, Rovira AD (1976) The role of bacteria in the biological control of *Gaeumannomyces graminis* by suppressive soils. Soil Biol Biochem 8:269–273

Daami-Remadi M, Hibar K, Jabnoun- Khiareddine H, Ayed F, El Mahjoub M (2006) Effect of two *Trichoderma* species on severity of potato tuber dry rot caused by Tunisian *Fusarium* complex. Int J Agric Res 1:432–441

De Boer W, Verheggen P, Klein Gunnewiek PJA, Kowalchuk GA, Van Veen JA (2003) Microbial community composition affects soil fungistasis. Appl Environ Microbiol 69:835–844

De la Cruz J, Hidalgo-Gallego A, Lora JM, Benítez T, Pintor-Toro JA, Llobell A (1992) Isolation and characterization of three chitinases from *Trichoderma harzianum*. Eur J Biochem 206:856–867

Deacon JW, Berry LA (1993) Biocontrol of soil-borne plant-pathogens–concepts and their application. Pestic Sci 37:417–426

Druzhinina IS, Seidl-Seiboth V, Herrera-Estrella A, Horwitz BA, Kenerley CM, Monte E, Mukherjee PK, Zeilinger S, Grigoriev IV, Kubicek CP (2011) *Trichoderma*: the genomics of opportunistic success. Nat Rev Microbiol 9:749–759

Duijff BJ, Bakker PAHM, Schippers B (1991) Suppression of fusarium wilt of carnation by Pseudomonas in soil: mode of action. In: Keel C, Knoller B, Defago G (eds) Plant growth-promoting rhizobacteria, progress and prospect, vol 14. International Organization for Biological and Integrated Control of Noxious Animals and Plants/West Palaearctic Regional Section, Interlaken, pp 152–157

Duijff BJ, Gianinazzi-Pearson V, Lemanceau P (1997) Involvement of the outermembrane lipopolysaccharides in the endophytic root colonization of tomato roots by biocontrol *Pseudomonas fluorescens* strain WCS417r. New Phytol 135:325–334

Duijff BJ, Pouhair D, Olivain C, Alabouvette C, Lemanceau P (1998) Implication of systemic induced resistance in the suppression of fusarium wilt of tomato by *Pseudomonas fluorescens* WCS417r and by nonpathogenic *Fusarium oxysporum* Fo47. Eur J Plant Pathol 104:903–910

Duijff BJ, Recorbet G, Bakker PAHM, Loper JE, Lemanceau P (1999) Microbial antagonism at the root level is involve in the suppression of fusarium wilt by the combination of nonpathogenic *Fusarium oxysporum* Fo47 and *Pseudomonas putida* WCS358. Phytopathology 89:1073–1979

Elad Y, Barak R, Chet I (1984) Parasitism of sclerotia of *Sclerotium rolfsii* by *Trichoderma harzianum*. Soil Biol Biochem 16:381–386

El-Rahman SSA, Mohamed HI (2014) Application of benzothiadiazole and Trichoderma harzianum to control faba bean chocolate spot disease and their effect on some physiological and biochemical traits. Acta Physiol Plant 36:343–354

Eparvier A, Alabouvette C (1994) Use of ELISA and GUS-transformed strains to study competition between pathogenic and non-pathogenic *Fusarium oxysporum* for root colonization. Biocontrol Sci Technol 4:35–47

Fernando WGD, Ramarathnam R, Krishnamoorthy AS, Savchuk SC (2005) Identification and use of potential bacterial organic antifungal volatiles in biocontrol. Soil Biol Biochem 37:955–964

Frostegard A, Baath E (1996) The use of phospholipid fatty acid analysis to estimate bacterial and fungal biomass in soil. Biol Fertil Soils 22:59–65

Fuchs JG, Moenne-Loccoz Y, Défago G (1997) Non-pathogenic *Fusarium oxysporum* strain Fo47 induces resistance to Fusarium wilt in tomato. Plant Dis 81:492–496

Gao G, Yin D, Chen S, Xia F, Yang J, Qing L, Wang W (2012) Effect of biocontrol agent *Pseudomonas fluorescens* 2p24 on soil fungal community in cucumber rhizosphere using T-RFLP and DGGE. PLoS ONE 7(2), e31806

Garbeva P, Postma J, Van Veen JA, Van Elsas JD (2006) Effect of above-ground plant species on soil microbial community structure and its impact on suppressionn of *Rhizoctonia solani* AG3. Environ Microbiol 8:233–246

Garland JL, Mills AL (1991) Classification and characterization of heterotrophic microbial communities on the basis of patterns of community-level sole-carbon-source utilization. Appl Environ Microbiol 57:2351–2359

Garrett SD (1934) Factors affecting the pathogenicity of cereal root-rot fungi. Biol Rev 9:351–361

Geels FP, Schmidt EDL, Schippers B (1985) The use of 8- hydroxyquinoline for the isolation and prequalification of plant growth-stimulating rhizosphere pseudomonads. Biol Fertil Soils 1:167–173

Gorissen A, Van Overbeek LS, Van Elsas JD (2004) Pig slurry reduces the survival of Ralstonia solanacearum biovar 2 in soil. Can J Microbiol 50:587–593

Haas D, Défago G (2005) Biological control of soil-borne pathogens by fluorescent pseudomonads. Nat Rev Microbiol 3:307–319

Hamel C, Vujanovic V, Jeannotte R, Nakano-Hylander A, St- Arnaud M (2005) Negative feedback on a perennial crop: Fusarium crown and root rot of asparagus is related to changes in soil microbial community structure. Plant Soil 268:75–87

Haran S, Schickler H, Chet I (1996) Molecular mechanisms of lytic enzymes involved in the biocontrol activity of *Trichoderma harzianum*. Microbiology 142:2321–2331

Harman GE, Howell CR, Viterbo A, Chet I, Lorito M (2004) *Trichoderma* spicies.-Opportunistic avirulent plant symbionts. Nat Rev Microbiol 2:43–56

Henery MD (1932) A study of scab resistance in the potato. J Agric Res 54:305–317

Hermosa MR, Keck E, Chamorro I, Rubio B, Sanz L, Vizaino JA, Grondona I, Monte E (2004) Genetic diversity shown in *Trichoderma* biocontrol isolates. Mycol Res 108:879–906

Hermosa R, Viterbo A, Chet I, Monte E (2012) Plant-beneficial effects of *Trichoderma* and of its genes. Microbiology 158:17–25

Hervás A, Trapero-Casas JL, Jiménez-Díaz RM (1995) Induced resistance against Fusarium wilt of chickpea by nonpathogenic races of *Fusarium oxysporum* f. sp. *ciceris* and nonpathogenic isolates of *F. oxysporum*. Plant Dis 79:1110–1116

Hibar K (2007) Induction of resistance in tomato plants against *Fusarium oxysporum* f. sp. *radicis-lycopersici* by *Trichoderma* spp. Tunis J Plant Prot 2:47–58

Hoffland E, Hakulinen J, Van Pelt JA (1996) Comparison of systemic resistance induced by avirulent and non-pathogenic *Pseudomonas* species. Phytopathology 86:757–762

Hopkins DL, Larkin RP, Elmstrom GW (1987) Cultivar-specific induction of soil suppressiveness to fusarium wilt of water melon. Phytopathology 77:607–611

Ibekwe AM, Kennedy AC (1998) Phospholipid fatty acid profiles and carbon utilization patterns for analysis of microbial community structure under field and greenhouse conditions. FEMS Microbiol Ecol 26:51–163

Janvier C, Villeneuve F, Alabouvette C, Edel-Hermann V, Mateille T, Steinberg C (2007) Soil health through soil disease suppression: which strategy from descriptors to indicators. Soil Biol Biochem 39:1–23

Keel C, Schnider U, Maurhofer M, Voisard C, Laville J, Burger P, Wirthner P, Haas D, Défago G (1992) Suppression of root diseases by *Pseudomonas fluorescens* CHAO: importance of the secondary metabolite 2,4-diacetylphloroglucinol. Mol Plant Microbe Interact 5:4–13

Kloepper JW, Schroth MN (1978) Plant growth-promoting rhizobacteria on radishes. In: Proceedings of the 4th international conference on plant pathogenic bacteria, vol 2. Station de Pathologie Végétale et de Phytobactériologie, INRA, Angers, France, pp 879–882

Kloepper JW, Leong J, Teintze M, Schroth M (1980) *Pseudomonas* siderophores: a mechanism explaining diseasesuppressive soils. Curr Microbiol 4:317–320

Kommedahl T, Christensen JJ, Frederickson RA (1970) A half century of research in Minnesota on flax wilt caused by *Fusarium oxysporum*. Minn Agric Exp Stn Tech Bull 273:35

Kowalchuk GA, Van Os GJ, Aartrijk J, Veen JA (2003) Microbial community responses to disease management soil treatments used in flower bulb cultivation. Biol Fertil Soils 37:55–63

Kraus J, Loper JE (1995) Characterization of a genomic region required for production of the antibiotic pyoluteorin by the biological control agent *Pseudomonas fluorescens* Pf-5. Appl Environ Microbiol 61:849–854

Kroon BAM, Scheffer RJ, Elgersma DM (1991) Induced resistance in tomato plants against Fusarium wilt invoked by *Fusarium oxysporum* f. sp. *dianthi*. Neth J Plant Pathol 97:401–408

Kubicek CP, Bissett J, Druzhinina I, Kullnig-Gradinger C, Szakacs G (2003) Genetic and metabolic diversity of *Trichoderma*: a case study on South-East Asian isolates. Fungal Gen Biol 38:310–319

Kuske CR, Barns SM, Busch JD (1997) Diverse uncultivated bacterial groups from soils of the arid southwestern United States that are present in many geographic regions. Appl Environ Microbiol 63:3614–3621

Kyselkova M, Kopecky J, Frapolli M, Defago G, Sagova-Mareckova M, Grundmann GL, Moenne-Loccoz Y (2009) Comparison of rhizobacterial community composition in soil suppressive or conducive to tobacco black root rot disease. ISME J 3:1127–1138

Larkin RP, Hopkins DL, Martin FN (1993) Effect of successive watermelon plantings on *Fusarium oxysporum* and other microorganisms in soils suppressive and conducive to fusarium-wilt of watermelon. Phytopathology 83:1097–1105

Larkin RP, Hopkins DL, Martin FN (1996) Suppression of Fusarium wilt of watermelon by nonpathogenic *Fusarium oxysporum* and other microorganisms recovered from a disease-suppressive soil. Phytopathology 86:812–819

Leeman M, Van Pelt JA, Den Ouden FM, Heinsbroek M, Bakker PAHM, Schippers B (1995) Induction of systemic resistance against fusarium wilt of radish by lipopolysaccharides of *Pseudomonas fluorescens*. Phytopathology 85:1021–1027

Leeman M, den Ouden FM, van Pelt JA, Cornelissen C, Matamala-Garros A, Bakker PAHM, Schippers B (1996) Suppression of fusarium wilt of radish by co-inoculation of fluorescent *Pseudomonas* spp. and root-colonizing fungi. Eur J Plant Pathol 102:21–31

Lelavthi MS, Vani L, Pscal R (2014) Antimcrobial activity of *Trichodema harzinum* against bacteria and fungi. Int J Curr Micobiol Appl Sci 3:96–103

Lemanceau P, Alabouvette C (1991) Biological control of fusarium diseases by fluorescent Pseudomonas and nonpathogenic *Fusarium*. Crop Prot 10:279–286

Lemanceau P, Bakker PAHM, Dekogel WJ, Alabouvette C, Schippers B (1992) Effect of pseudobactin 358 production by *Pseudomonas putida* WCS358 on suppression of fusarium wilt of carnations by nonpathogenic *Fusarium oxysporum* Fo47. Appl Environ Microbiol 58:2978–2982

Lemanceau P, Bakker PAHM, Dekogel WJ, Alabouvette C, Schippers B (1993) Antagonistic effect of nonpathogenic *Fusarium oxysporum* Fo47 and pseudobactin 358 upon pathogenic *Fusarium oxysporum* f. sp. Diant. Appl Environ Microbiol 59:74–82

Liesack W, Janssen PH, Rainey FA, Ward-Rainey NL, Stackenbrandt E (1997) Microbial diversity in soil: the need for a combined approach using molecular and cultivation techniques. In: van Elsas JD, Trevors JT, Wellington EMH (eds) Modern soil microbiology. Marcel Dekker, New York, pp 375–439

Lorito M, Woo SL, D'Ambrosio M (1996) Synergistic interaction between cell wall degrading enzymes and membrane affecting compounds. Mol Plant Microbe Interact 9:206–213

Louvet J, Rouxel F, Alabouvette C (1976) Recherches sur la resistance des sols aux maladies. I. Mise en evidence de la nature microbiologique de la resistance d'un sol au développement de la fusariose vasculaire du melon. Ann Phytopathol 8:425–436

Lu Z, Tombolini R, Woo S et al (2004) *In vivo* study of *Trichoderma*-pathogen-plant interactions, using constitutive and inducible green fluorescent protein reporter systems. Appl Environ Microbiol 70:3073–3081

Lugtenberg BJJ, Bloemberg GV (2004) Life in the rhizosphere. In: Ramos JL (ed) Pseudomonas, vol 1. Kluwer Academic/Plenum, New York, pp 403–430

Mandeel Q, Baker R (1991) Mechanism involved in biological control of fusarium wilt of cucumber with strain of non-pathogenic *Fusarium oxysporum*. Phytopathology 81:462–469

Menzies JD (1959) Occurrence and transfer of a biological factor in soil that suppresses potato scab. Phytopathology 49:648–652

Mezaache-Aichour S, Guechi A, Nicklin J, Drider D, Prevost HRN, Strange RN (2012) Isolation, identification and antimicrobial activity of pseudomonads isolated from the rhizosphere of potatoes growing in Algeria. J Plant Pathol 94:89–98

Millard WA, Taylor CB (1927) Antagonism of microorganisms as the controlling factor in the inhibition of scab by green manuring. Ann Appl Biol 14:202–215

Motlagh MRS, Samimi Z (2013) Evaluation of Trichoderma spp., as biological agents in some of plant pathogens. Ann Biol Res 4:173–179

Muyzer G, De Waal EC, Uitterlinden AC (1993) Profiling of complex microbial populations by denaturing gradient gel electrophoresis analysis of polymerase chain reaction-amplified genes for 16S rRNA. Appl Environ Microbiol 55:695–700

Odum EP (1959) Fundamentals of ecology. W.B. Saunders Co, Philadelphia

Olivain C, Steinberg C, Alabouvette C (1995) Evidence of induced resistance in tomato inoculated by nonpathogenic strains of *Fusarium oxysporum*. In: Manta M (ed) Environmental biotic factors in integrated plant disease control 3rd EFPP conference, Poznan, pp 427–430

Oliver JD (2005) The viable but nonculturable state in bacteria. J Microbiol 43:93–100

Olsson PA (1999) Signature fatty acids provide tools for determination of the distribution and interactions of mycorrhizal fungi in soil. FEMS Microbiol Ecol 29:303–310

Ongena M, Duby F, Rossignol F, Fauconnier ML, Dommes J, Thonart P (2004) Stimulation of the lipoxygenase pathway is associated with systemic resistance induce in bean by a nonpathogenic *Pseudomonas* strain. Mol Plant Microbe Interact 17:1009–1018

Park CS, Paulitz TC, Baker R (1988) Biocontrol of fusarium wilt of cucumber resulting from interactions between *Pseudomonas putida* and non-pathogenic isolates of *Fusarium oxysporum*. Phytopathology 78:190–194

Perelló A, Mónaco C, Sisterna M, Dal Bello G (2003) Biocontrol efficacy of Trichoderma isolates for tan spot of wheat in Argentina. Crop Prot 22:1099–1106

Pérez-Piqueres A, Edel-Hermann V, Alabouvette C, Steinberg C (2006) Response of soil microbial communities to compost amendments. Soil Biol Biochem 38:460–470
Pieterse CMJ, Van Wees SCM, Hoffland E, Van Pelt JA, Van Loon LC (1996) Systemic resistance in *Arabidopsis* induced by biocontrol bacteria is independent of salicylic acid accumulation and pathogenesis-related gene expression. Plant Cell 8:1225–1237
Piotrowska-Seget Z, Mrozik A (2003) Signature lipid biomarker (SLB) analysis in determining changes in community structure of soil microorganisms. Pol J Environ Stud 12:669–675
Raaijmakers JM, Weller DM (1998) Natural plant protection by 2,4-diacetylphloroglucinol producing *Pseudomonas*spp. in take-all decline soils. Mol Plant Microbe Interact 11:144–152
Raaijmakers JM, Weller DM (2001) Exploiting genotypic diversity of 2,4-Diacetylphloroglucinol-producing Pseudomonas spp.: characterization of superior root-colonizing P. fluorescens strain Q8r1-96. Appl Environ Microbiol 67:2545–2554
Raaijmakers JM, Weller DM, Thomashow LS (1997) Frequency of antibiotic producing *Pseudomonas* spp. in natural environments. Appl Environ Microbiol 63:881–887
Rasmussen PH, Knudsen IMB, Elmholt S, Jensen DF (2002) Relationship between soil cellulolytic activity and suppression of seedling blight of barley in arable soils. Appl Soil Ecol 19:91–96
Ros M, Hernandez MT, Garcia C, Bernal A, Pascual JA (2005) Biopesticide effect of green compost against Fusarium wilt on melon plants. J Appl Microbiol 98:845–854
Sanford GB (1926) Some factors affecting the pathogenicity of *Actinomyces scabies*. Phytopathology 16:525–547
Scher FM, Baker R (1980) Mechanism of biological control in a Fusarium- suppressive soil. Phytopathology 70:412–417
Scher FM, Baker R (1982) Effect of *Pseudomonas putida* and a synthetic iron chelator on induction of soil suppressiveness to fusarium wilt pathogens. Phytopathology 72:1567–1573
Schippers B (1992) Prospects for management ofnatural suppressiveness to control soilborne pathogens. In: Tjarnos EC, Papavizas GC, Cook RJ (eds) Biological control of plant diseases. Plenum, New York, pp 21–34
Schneider RW (1982) Suppressive soils and plant disease. American Phytopathology Society, St. Paul, MN, p 88
Schonfeld J, Gelsomino A, Overbeek LSV, Gorissen A, Smalla K, Elsas JDV (2003) Effects of compost addition and simulated solarisation on the fate of *Ralstonia solanacearum* biovar 2 and indigenous bacteria in soil. FEMS Microbiol Ecol 43:63–74
Shipton PJ, Cook RJ, Sitton JW (1973) Occurrence and transfer of a biological factor in soil that suppresses takeall of wheat in eastern Washington. Phytopathology 63:511–517
Sivan A, Chet I (1989) The possible role of competition between *Trichoderma harzianum* and *Fusarium oxysporum* on rhizosphere colonization. Phytopathology 79:198–203
Steinberg C, Edel-Hermann V, Alabouvette C, Lemanceau P (2007) Soil suppressiveness to plant diseases. In: van Elsas JD, Jansson J, Trevors JT (eds) Modern soil microbiology, 2nd edn. CRC, Boca Raton, pp 455–501
Stover RH (1962) Fusarial wilt (Panama disease) of bananas and other Musa species. Commonwealth Mycological Institute, p 117
Tanwar A, Aggarwal A, Panwar V (2013) Arbuscular mycorrhizal fungi and *Trichoderma viride* mediated *Fusarium* wilt control in tomato. Biocontrol Sci Technol 23:485–498
Torsvik V, Sorheim R, Goksoyr J (1996) Total bacterial diversity in soil and sediment communities – a review. J Ind Microbiol 17:170–178
Van Peer R, Niemann GJ, Schippers B (1991) Induced resistance and phytoalexin accumulation in biological control of Fusarium wilt of carnation by *Pseudomonas* sp. strain WCS417r. Phytopathology 81:728–734
Viterbo A, Landau U, Kim S, Chernin L, Chet I (2010) Characterization of AAC deaminase from the biocontrol and plant growth-promoting agent Trichoderma asperellum T203. FEMS Microbiol Lett 305:42–48

Wang C, Lin Y, Lin Y, Chung W (2013) Modified primers for the identification of nonpathogenic *Fusarium oxysporum* isolates that have biological control potential against fusarium wilt of cucumber in Taiwan. PLoS ONE 8, e65093
Weindling R (1932) *Trichoderma lignorum* as a parasite of other soil fungi. Phytopathology 22:837–845
Weller DM, Thomashow LS (1993) Microbial metabolites with biological activity. In: Lumsden RD, Vaughn JL (eds) Pest management: biologically based technologies. American Chemical Society, Washington, DC, pp 173–180
Weller DM, Raaijmakers JM, Gardener BBM, Thomashow LS (2002) Microbial populations responsible for specific soil suppressiveness to plant pathogens. Annu Rev Phytopathol 40:309–348
Westphal A, Becker JO (2000) Transfer of biological soil suppressiveness against *Heterodera schachtii*. Phytopathology 90:401–406
Whipps JM (1997) Developments in the biological control of soil-borne plant pathogens. Adv Bot Res 26:1–134
Yang C, Crowley DE, Menge JA (2001) 16S rDNA fingerprinting of rhizosphere bacterial communities associated with healthy and *Phytophthora* infected avocado roots. FEMS Microbiol Ecol 35:129–136
Zehnder GW, Murphy JF, Sikora EJ, Kloepper JW (2001) Application of rhizobacteria for induced resistance. Eur J Plant Pathol 107:39–50

Part III
Concepts in Plant Disease Management Involving Organic Amendments

Chapter 13
Anaerobic Soil Disinfestation and Soilborne Pest Management

Erin N. Rosskopf, Paula Serrano-Pérez, Jason Hong, Utsala Shrestha, María del Carmen Rodríguez-Molina, Kendall Martin, Nancy Kokalis-Burelle, Carol Shennan, Joji Muramoto, and David Butler

13.1 Introduction

Anaerobic soil disinfestation (ASD; also referred to as biological soil disinfestation (BSD)) is a preplant soil treatment method developed to control plant disease and manage yield decline in many crop production systems (Blok et al. 2000; Shinmura 2000). The practice involves induction of anaerobic soil conditions by increasing microbial respiration through incorporation of easily decomposable, carbon-rich organic amendments into moist soil and by preventing the resupply of oxygen

E.N. Rosskopf (✉) • J. Hong • N. Kokalis-Burelle
USDA, ARS, United States Horticultural Research Laboratory, 2001 S. Rock Rd., Fort Pierce, FL 34945, USA
e-mail: erin.rosskopf@ars.usda.gov; Jason.hong@ars.usda.gov; nancy.burelle@ars.usda.gov

P. Serrano-Pérez • M.d.C. Rodríguez-Molina
Centro de InvestigacionesCientíficas y Tecnológicas de Extremadura, Instituto de InvestigacionesAgrariasFinca La Orden-Valdesequera, Autovía A-5, km 372-06187, Guadajira, Badajoz, Spain
e-mail: paulaserrano@outlook.com; carmen.rodriguez@gobex.es

U. Shrestha • D. Butler
Department of Plant Sciences, Organic, Sustainable & Alternative Crop Production, University of Tennessee, 2431 Joe Johnson Dr., Knoxville, TN 37996, USA
e-mail: ushrest1@vols.utk.edu; dbutler@utk.edu

K. Martin
Department of Biology, William Paterson University, Wayne, NJ 07470, USA
e-mail: martin31@wpunj.edu

C. Shennan • J. Muramoto
Environmental Studies Department, Center for Agroecology and Sustainable Food Systems, University of California Santa Cruz, 1156 High St, Santa Cruz, CA 95064, USA
e-mail: Cshennan@ucsc.edu; joji@ucsc.edu

M.K. Meghvansi, A. Varma (eds.), *Organic Amendments and Soil Suppressiveness in Plant Disease Management*, Soil Biology 46,
DOI 10.1007/978-3-319-23075-7_13

through the soil surface by coverage with plastic film for a period of time, as short as 2 weeks or as long as 15 weeks.

ASD research is increasing in the USA (Rosskopf et al. 2010; Butler et al. 2012b, c; McCarty et al. 2014; Shennan et al. 2014), the Netherlands (Blok et al. 2000; Messiha et al. 2007; Korthals et al. 2014), and Japan (Momma et al. 2013; Mowlick et al. 2014). Several different approaches and inputs have been tested with variable levels of pest control associated with different amendments and application techniques (Table 13.1).

The current practice in Florida, for example, utilizes two easily obtained agricultural waste products, composted broiler litter, and feed-grade blackstrap molasses obtained from the sugar processing industry. These inputs are incorporated into prepared planting areas, either in a broadcast application, typical of cut-flower production (Rosskopf et al. 2009), or in a preformed raised bed that is characteristic of vegetable systems in the southeast (Lamont 1996). After incorporation, the bed or flat ground is covered using either clear, UV-stabilized solarization film that is later replaced or with totally impermeable polyethylene film (TIF) which can remain in the field during crop production. Research in Florida has established that for fall production on sandy soils, 5 cm of water applied via a double drip tape under the polyethylene mulch is adequate for the development of anaerobic conditions (Butler et al. 2012b). Similar approaches were pioneered in California, principally without the addition of composted animal waste as a nitrogen source, using locally available agricultural waste, such as rice bran (Muramoto et al. 2014).

Although ASD does not necessarily require either high temperature (Ludeking et al. 2010; Runia et al. 2012; McCarty et al. 2014) or long-term incubation (Momma et al. 2010; Butler et al. 2012b, c), combining ASD with soil solarization can improve the efficacy of each separate component and overcome the limitations of each treatment when applied alone (Butler et al. 2012b). Reduction in disease incidence resulting from the application of organic amendments with solarization has been described in numerous cases (Paulitz and Bélanger 2001; Bailey and Lazarovits 2003), and some authors refer to this combination of techniques as "biosolarization or biodisinfection" (Bello et al. 2008; García Ruíz et al. 2009; Martínez et al. 2011; Núñez-Zofío et al. 2011; Domínguez et al. 2014). Although redox potential was not specifically monitored in these studies, anaerobic conditions achieved in soil may be implicated in the disease control observed. When these organic amendments are *Brassicaceae* species cover crops or seed meal, most authors use the term "biofumigation." With this technique, developed by Kirkegaard et al. (1993), the amount of irrigation water is not likely to be enough to induce anaerobic conditions but is enough to ensure optimal soil moisture for glucosinolate (GSL) hydrolysis. In this case, the main mechanism of control is the accumulation of toxic compounds in the soil atmosphere; the *Brassicaceae* species used are high-GSL-content varieties, and toxicity of the resulting isothiocyanates is critical for the success of the technique (Matthiessen and Kirkegaard 2006). While ASD could be performed using *Brassicaceae* species, this is not the basis of the technique. The objective of this contribution is to summarize the research that has been conducted on ASD around the world and to suggest research areas that are of

Table 13.1 Plant pathogen and nematode control research utilizing anaerobic soil disinfestation (ASD) and biological soil disinfestation (BSD)

Region	Technique name	Conditions	Organic input(s)	Rate	Plastic	Tarp period (weeks)	Soil temp (°C)	Crop	*Pathogen*	Infestation	Author and year
Argentina	Soil reductive sterilization (SRS)	Greenhouse	Wheat bran	10 t/ha	Transparent polyethylene	3	~30 °C	Carnation	*Fusarium oxysporum (Fox)*	Natural	Yossen et al. (2008)
Japan	Biological soil disinfestation (BSD)	Greenhouse	Wheat bran	0.008–0.016 g/g soil	Plastic film	9 days–2 weeks	28 °C	–	*Fox* f. sp. *lycopersici*	Inoculum	Momma et al. (2005)
		Glass bottle	Wheat bran	0.01 g/g soil	Parafilm	2	28 °C	–	*Ralstonia solanacearum*	Inoculum	Momma et al. (2006)
		Chamber/field	Ethanol/wheat bran	0.5–2 % (v/v) 1 kg/m^2	Hermetic pots Plastic film	2–15 days	30–33 °C	Tomato	*Fox* f. sp. *lycopersici*	Inoculum	Momma et al. (2010)
		Field	Ethanol	1 % (v/v) diluted with water	Plastic film	3	–	Tomato	*Fox* f. sp. *lycopersici*	Re-infested after treatment	Momma et al. (2011)
		Chamber	*Brassica juncea* wheat bran	6 kg/m^2 1 kg/m^2	Hermetic pots	18 days	30–31 °C	–	*Fox*	Inoculum	Mowlick et al. (2012)
		Greenhouse	Wheat bran	1 kg/m^2	Double plastic film	3	24–34 °C	Spinach	*Fox* f. sp. *spinaciae*	Natural	Mowlick et al. (2013a, b, c))
		Greenhouse	*Brassica juncea radish roots* wheat bran	2–10 kg/m^2	Double plastic film	3	19.7–39.4 °C	Spinach	*F. ox* f. sp. *spinaciae*	Inoculum	Mowlick et al. (2014)
		Field	*Brassica juncea Avena sativa*	3.29–10.4 kg/m^2	Transparent PE	3	18.1–35.5 °C	Spinach	*F. oxy* f. sp. *spinaciae*	Natural	Mowlick et al. (2013b)
Netherlands	–	Field	Broccoli	3.8–4 kg fresh/m^2	3 layers airtight plastic or uncovered	15	–	–	*F. oxysporum* f. sp. *asparagi* *Rhizoctonia solani* *Verticillium dahliae*	Inoculum in nylon bags	Blok et al. (2000)

(continued)

Table 13.1 (continued)

Region	Technique name	Conditions	Organic input(s)	Rate	Plastic	Tarp period (weeks)	Soil temp (°C)	Crop	*Pathogen*	Infestation	Author and year
	BSD	Field	Grass	42, 62, or 102 t of grass/ha	Airtight plastic (used for ensiling)	–	–	Asparagus	*F. redolens* *R. tuliparum* *V. dahliae*	Inoculum packets	Blok et al. (2005)
	BSD	Field	Italian ryegrass (*Lolium multiflorum*)	40 or 54 t/ha (two locations)	3 layers airtight plastic (used for ensiling)	13	–	Norway maple Southern catalpa	*V. dahliae* *Pratylenchus fallax* *Trichodorus*	Potato stems covered with microsclerotia	Goud et al. (2004)
	ASD	Field	Mixture of ryegrass (fresh organic matter)	50 t/ha	VIF	12	–	Potato, lily, and carrots	*V. dahliae* *P. penetrans*	Natural	Korthals et al. (2014)
	ASD	Field	Green manure crop	40 t/ha	VIF	6	Summer	Several	*Meloidogyne* *Globodera pallida*	–	Lamers et al. (2010)
	BSD	Chamber	Organic by-products	2, 4, and 6 g raw protein/l of soil	Hermetic pots	2, 4, or 8 weeks	16 °C	–	*V. dahliae*	Inoculum	Ludeking et al. (2010)
	BSD	Beaker/field	Plant debris	Various	3 layers airtight plastic (used for ensiling)	6 and 31 days	28 °C	Potato	*R. solanacearum*	Inoculum	Messiha et al. (2007)
Spain	Biosolarization	Field	Biofence/olive pomace/manure/sugar beet vinasse/manure + *Trichoderma*	Various	0.05 mm transparent PE	4	20.2–32.8 °C	Strawberry	*Macrophomina* *Pythium* *Rhizoctonia* *Cylindrocarpon* *Fusarium* *Colletotrichum acutatum* *Phytophthora cactorum.*	Natural	Domínguez et al. (2014)
	Greenhouse	Manure/sugar beet vinasse	0.75–1.5 l/m^2	Transparent PE	–	–	Pepper	*P. capsici*	–		Lacasa et al. (2010)
Spain	Biodisinfection	Greenhouse	(Biofence)/*Sinapis alba* Manure/CPL	Variable	0.05 mm transparent PE	6	-	Pepper	*P. capsici*	Pepper root balls	Nuñez-Zofio et al. (2012)

USA	ASD	Pots in greenhouse	*Cover crops/molasses*	3.1–5.8 t/ha 14.6 t/ha	Transparent PE	3	20–30 °C (25.1 °C)	–	*Fox* f. sp. *lycopersici* *Sclerotium rolfsii*	Inoculum packets	Butler et al. (2012a, 2012b))
		Pots in growth chamber	*Cover crops*	<1 mg C/g soil	(0.025 mm) Black PE	3	15–24 °C	Common bean	*Sclerotinia sclerotiorum*	Inoculum packets	Butler et al. (2014a, b))
		Field	Molasses/CPL	8.2–26 t dry matter/ha	Transparent PE	3	30–40 °C	Pepper and eggplant	*P. capsici* *Fox* f. sp. *lycopersici* *Belonolaimus*	Inoculum packets	Butler et al. (2012b)
		Growth chamber	Composted manure/ plant debris/mustard/ rice bran/ethanol	4.9–40 t/ha Ethanol 10 %	Double layer of two gas-impermeable transparent bags	2	18–24 °C	Apple	*R. solani* *P. ultimum* *F. oxysporum* *P. penetrans*	Petri plate (volatiles)	Hewavitharana et al. (2014)
		Field	Mustard/cover crops/ dried molasses	0.86–1.99 mg C/g soil	0.032 mm black PE	3	~21 °C	Tomato and pepper	*R. solani*	Natural	McCarty et al. (2014)
		Field	Molasses Rice bran	20 t/ha	Plastic film	3	<20–23 °C	Strawberry	*V. dahliae*	Natural	Muramoto et al. (2014)
		Field	CPL/molasses	26–16 t dry matter/ha + 8.2/ha	Transparent PE	3	30–40 °C	Pepper and eggplant	*P. capsici*	Inoculum packets	Rosskopf et al. (2010)
		Field	CPL/molasses	224 kg N/ha + 8.2 t/ha	Transparent PE	3	–	Strawberry	*F. oxysporum* *M. phaseolina*	Inoculum packets	Rosskopf et al. (2014)
		Field	Rice bran	20 t/ha	Plastic film	3	–	Strawberry	*V. dahliae*	Natural	Shennan et al. (2013)

interest and importance for the future. Topics of discussion include the impact that amendment choice and temperature have on generating anaerobic conditions, how the process of ASD changes soil chemistry, changes in the microbial community as a result of ASD and the role microbes play in anaerobicity, and what is currently known about creating a disease-suppressive soil using this method.

13.2 Organic Amendments for Anaerobic Soil Disinfestation

Many different amendments have been tested as inputs for ASD, specifically for the management of fungal plant pathogens (*Fusarium* spp., *Rhizoctonia* spp., *Sclerotinia sclerotiorum*, *Verticillium dahliae*, *Macrophomina phaseolina*, see Table 13.1) (Blok et al. 2000, 2005; Butler et al. 2012a, b, c; Momma et al. 2013; McCarty et al. 2014; Rosskopf et al. 2014), oomycetes (*Phytophthora* spp. and *Pythium*) (Rosskopf et al. 2010), and plant pathogenic bacteria (*Ralstonia solanacearum*) (Messiha et al. 2007).

Cereal bran (wheat *(Triticum aestivum)* or rice *(Oryza sativa)*) is one of the popular plant materials incorporated in soil for treatment using ASD (Momma et al. 2006). Wheat bran has been tested as a carbon source to control spinach wilt caused by *Fusarium oxysporum* f. sp. *spinaciae*, reducing the disease incidence to 21.1 % when tested under greenhouse conditions (Mowlick et al. 2013c). Momma et al. (2005) used wheat bran for reducing viability of *F. oxysporum* f. sp. *lycopersici* chlamydospores under controlled conditions. In addition, the severity of bacterial wilt of tomato caused by *R. solanacearum* was also decreased using wheat bran (1 % w/w) in ASD treatment under laboratory conditions (Momma et al. 2006). However, Momma et al. (2010) did not achieve similar results with *F. oxysporum* f. sp. *lycopersici* inoculum when ASD was applied under field conditions over a 15-day period using 1 kg wheat bran per m^2 of soil. This was attributed to lack of uniformity in the distribution of the amendment and the use of additional material would likely overcome this limitation (Momma et al. 2010). In contrast, in Argentina, Yossen et al. (2008) reported effective control with the same dose of wheat bran mentioned above, with similar soil temperatures (over 30 °C), in a carnation greenhouse naturally infested with *F. oxysporum*. ASD using 2 kg of rice bran per m^2 as a carbon source was evaluated in California strawberry fields and reduced the number of *V. dahliae* microsclerotia in naturally infested soil by 85–100 % providing soil temperatures were above 17 °C (Shennan et al. 2013, 2014). Moreover, strawberry yields using this treatment were comparable to soil fumigation using 1,3-dichloropropene plus chloropicrin (Shennan et al. 2013, 2014). However, Muramoto et al. (2014) suggested reducing the amount of rice bran and mixing it with molasses (1 kg $m^{-1}$of each material) in order to prevent high nitrogen addition to the soil (~400 kg total $N \cdot ha^{-1}$) associated with the use of a high rate of rice bran. In a previous study, Daugovish et al. (2011) reported up to

94 % reduction in *V. dahliae* inoculum density in soils heavily infested with the pathogen (20–30 microsclerotiag^{-1} soil) after ASD with rice bran under clear mulch.

Solid materials, such as cereal bran, are easily incorporated, but it is often difficult to achieve sufficient disinfestation deep in the soil profile because their effect is limited to the depth of carbon source incorporation, approximately 20–30 cm depending upon the method of incorporation and bed formation (Shennan et al. 2014). This is not the case when using liquid amendments such as molasses or ethanol that are applied with irrigation water and penetrate deeper into the soil (Momma 2008). Furthermore, these materials have advantages over cereal bran in their low N content (0.5 %) (Muramoto et al. 2014). Molasses has been used successfully as a carbon source for ASD in Japan (Shinmura 2004) and Florida (Rosskopf et al. 2010, 2014; Butler et al. 2012b, 2014a). In Florida, ASD using composted broiler litter and heavy blackstrap molasses provided excellent control of *Phytophthora capsici*, *F. oxysporum* f. sp. *lycopersici,* and *M. phaseolina* (Rosskopf et al. 2010) in raised-bed vegetable crop production including pepper and eggplant (Butler et al. 2012b, 2014a), tomato, cucumbers, and strawberries and for the production of cut flowers on flat ground (Rosskopf et al. 2012; Shennan et al. 2014). In a 2-year field study Butler et al. (2012b) reported, after ASD treatment with diluted blackstrap molasses (20 $Mg \cdot ha^{-1}$) combined with solarization during summer in Florida, mortalities of *F. oxysporum* and *P. capsici* inoculum buried in soil were equivalent to methyl bromide soil fumigation. The effectiveness in *P. capsici* control was associated with the solarization effect, although the addition of molasses increased the temperature achieved in solarized plots (Butler et al. 2012b). Similarly, it was noted that in field experiments of biosolarization with sugar beet vinasse and *Brassica juncea* pellets applied as organic amendments, with soil irrigation to field capacity and covered with plastic, significant reductions in strawberry plant mortality and incidence of charcoal rot caused by *M. phaseolina* were observed (Domínguez et al. 2014). Similar results were found in the Florida strawberry system where the combination of composted broiler litter and molasses resulted in a significant reduction in charcoal rot and near-complete mortality of introduced inoculum of *M. phaseolina* (Rosskopf et al. 2014). However, in California use of molasses alone as a carbon source for ASD was not as effective as rice bran ASD in strawberry production and did not induce shifts in soil microbial communities, whereas distinctly different microbial communities were observed after ASD with rice bran (Zavatta et al. 2014).

Kobara et al. (2007) and Uematsu et al. (2007) proposed the use of ethanol as a carbon source for ASD because the redox potential dropped significantly when 1 % ethanol dilution was used in saturated soil (Kobara et al. 2007). Uematsu et al. (2007) reported that inoculum viability of *F. oxysporum* f. sp. *cucumerinum* and *R. solanacearum* was significantly decreased in soil saturated with an ethanol (0.5–1.0 %, v/v) treatment (Momma et al. 2010, 2013). Under controlled laboratory conditions, ASD with wheat bran required a minimum of 9 days for mortality of chlamydospores of *F. oxysporum* f. sp. *lycopersici* (Momma et al. 2005), while with ethanol, only 3 days for 2 % (v/v), 6 days for 1 %, and 9 days for 0.5 % were needed to achieve undetectable levels of the pathogen (Momma et al. 2010). Ethanol, besides the advantage of its easy application, is a pure substance, so its composition is uniform and stable for long periods of time (Momma et al. 2013). It is widely used in the irrigation systems in commercial fields and plastic houses in Japan (Shennan et al. 2014). In field trials, added inoculum with chlamydospores and bud cells of *F. oxysporum* f. sp. *lycopersici* and natural infestations of *F. oxysporum* were both significantly decreased in soil saturated with 1 % ethanol solution at a rate of 200 $l \cdot m^{-1}$, while wheat bran (1 $kg \cdot m^{-1}$) treatment was not as effective (Momma et al. 2010). In the laboratory, fewer viable propagules of this pathogen survived at 30 °C than at 20 °C, regardless of the ethanol concentration.

Cover crops, crop residues, and *Brassicaceae* seed meal have also been evaluated as C sources in ASD. The efficacy of grass residues *(Dactylis glomerata)* and *Brassica juncea* seed meal as C sources for control of apple seedling root infection by *Pythium* spp. was assessed by Hewavitharana and Mazzola (2013) in growth chamber experiments. Both amendments significantly reduced *Pythium* root infection. Butler et al. (2012c) investigated the effectiveness of warm-season cover crops cowpea (*Vigna unguiculata*); sunn hemp (*Crotalaria juncea*); pearl millet (*Pennisetum glaucum*), and sorghum-Sudan grass (*Sorghum bicolor* x *S. bicolor* var. *sudanense*), in monoculture, and cowpea mixed with pearl millet or sorghum-Sudan grass compared to molasses as C sources for ASD treatment to control *F. oxysporum* f. sp. *lycopersici* and *S. rolfsii* inoculum introduced to greenhouse pots. Mortality of *F. oxysporum* was similar between the molasses control and cover crop C sources (except pearl millet), although the effect varied depending on the trial, with reductions by more than 97 % in all treatments with an added C source (cover crop or molasses) in one trial. In control of *S. rolfsii*, the effects of the C source treatments on survival of inoculum were inconsistent. Although warm-season cover crops have potential to serve as C source, more research is needed to improve their efficacy and consistency to control diseases using ASD (Butler et al. 2012c).

Utilizing the technique in areas where warm-season vegetable production eliminates the potential for combining ASD with soil solarization requires significantly more work to define appropriate inputs. In field experiments to control natural populations of *Rhizoctonia solani*, McCarty et al. (2014) evaluated dried molasses and several cool-season cover crops, including Indian mustard (*Brassica juncea*) and white mustard (*Sinapis alba*) seeded with arugula (*Eruca sativa*) and cereal rye (*Secale cereale*), applied alone or in combination with a low rate of molasses. These inputs were compared to biofumigation with mustard seed meal and an untreated control. Accumulated anaerobic conditions in ASD treatments were greater than in the untreated and biofumigated control. Populations of *R. solani* were lowest for ASD treatments with cereal rye and mustard with arugula treatments and were equivalent to the biofumigant control. Butler et al. (2014b) also evaluated the effect of several cool-season cover crops (crimson clover (*Trifolium incarnatum*), hairy vetch (*Vicia villosa*), cereal rye (*Secale cereale*), wheat (*Triticum aestivum*), mustard (*B. juncea* and *S. alba*), and arugula (*Eruca sativa*)) as C source for ASD on the viability of *S. sclerotiorum* and on the incidence of *Fusarium* root rot of common bean in a growth chamber study using soil temperatures typical of spring in Tennessee. ASD treatments, where C source rates were low (less than 1 mgg^{-1}), did not consistently decrease viability of *S. sclerotiorum* or incidence of *Fusarium* root rot.

Broccoli (*B. oleracea* L. convar. *botrytis*) and perennial ryegrass (*Lolium perenne*), with and without plastic cover, were assayed as organic amendments in field conditions in the Netherlands to control introduced inoculum of *F. oxysporum* f. sp. *asparagi*, *R. solani*, and *V. dahliae* (Blok et al. 2000). In amended treatments with plastic cover, anaerobic conditions developed and continued during the treatment that lasted 15 weeks (from midsummer to early fall), resulting in the inactivation of the inoculum of the three pathogens. Also in the Netherlands, Goud et al. (2004) confirmed the effectiveness of ASD for the management of *V. dahliae,* in *Acer platanoides* and *Catalpa bignonioides* trees, using 40–54 t of fresh residues/ha of Italian ryegrass (*Lolium multiflorum*) as the carbon source during 10–13-week treatment. Soil inoculum levels of *V. dahliae* were reduced by 85 %, after ASD, compared with the non-treated control. The *V. dahliae* population did not increase for the next 4 years. The results showed that ASD could be an effective, economically profitable, and environmentally safe control method for tree nurseries (Goud et al. 2004).

A study examining ASD methods for control of potato brown rot, a disease caused by the quarantine bacterium *R. solanacearum* race 3 biovar 2, was conducted by Messiha et al. (2007) in the Netherlands. The effect of ASD on the survival of inoculum was determined at three different scales: glass mesocosm, microplots, and in a naturally infested commercial field. The population of the pathogen was reduced by greater than 92 % compared to control treatments in a microplot trial using grass leaves. In the field experiments, anaerobic conditions could not be maintained for long periods, due to damage to the plastic tarp. Nevertheless, pathogen populations were still significantly reduced.

Management of some plant parasitic nematodes using ASD has also been shown. Goud et al. (2004) in their work on *V. dahliae* in tree crops reported a significant reduction in native soil populations of *Pratylenchus fallax* resulting from BSD) treatment ($p < 0.01$) as well as by plastic application ($p = 0.05$). This reduction in soil populations translated into a significant reduction in root infection. *Trichodorus* spp. were also reduced and these reductions persisted over multiple years. Similar results have been reported from the Netherlands for other species of *Pratylenchus* (Lamers et al. 2010).

Significant reductions in root-knot nematode (*Meloidogyne* spp.) and potato cyst nematode (*Globodera pallida*) have also been reported from the Netherlands (Lamers et al. 2010; Thoden et al. 2011). Recent work in that country, referred to as "advanced ASD," has been conducted using a commercial ASD input (Herbie®, Thatchtec, Wageningen, NL) which has a protein content of approximately 30 % and is composed of plant material. Similar results were found using traditional ASD and advanced ASD, both of which resulted in high levels of control of tested phytopathogens, including plant parasitic nematodes (van Overbeek et al. 2014).

Research in Florida on nematode control with ASD has focused on root-knot and sting nematode (*Belonolaimus* spp.) as well as the survival of nonpathogenic free-living nematodes. Factorial field trials were conducted to determine the most critical inputs for control of root-knot nematode using ASD. Application of molasses or molasses with composted broiler litter combined with irrigation resulting in the application of 5 and 10 cm of water caused significant reductions in naturally occurring populations of root-knot nematodes in soil as well as eggplant roots in the second cropping season when compared to soil solarization without amendments or irrigation (Butler et al. 2012b). Root galling was most significantly reduced in treatments containing molasses when compared to solarization alone. In the same study, numbers of free-living nematodes in the soil were increased with the application of composted broiler litter. A similar increase in nonpathogenic nematodes was seen with ASD application using both molasses and CPL with 5 cm of water in a strawberry production system, although counts of these organisms were highly variable over time (Rosskopf et al. 2014).

In the Florida cut-flower system, the success of ASD for root-knot nematode management was highly dependent upon the host susceptibility to the nematode. Tested in three different cut-flower crops using molasses, CPL, and 5 cm of water covered with clear polyethylene, ASD treatment resulted in commercial yield equivalent to methyl bromide application but did not provide season-long root-knot nematode control, demonstrated by heavily galled roots of the highly nematode-susceptible crop, snapdragon (Rosskopf et al. 2012).

13.3 Temperature and Anaerobiosis

Shrestha et al. (2014) proposed that identifying a suitable C/N ratio could be most critical for effective ASD. Soil temperature is, in addition to carbon source, a factor affecting the efficacy of ASD treatment in disease control (Shennan et al. 2013; Stapleton et al. 2010). Butler et al. (2014b) suggested that C source rate higher than $4\ mg \cdot g^{-1}$ is needed when soil temperatures during ASD treatment are low (15–25 °C), and Shennan et al. (2013) indicated that soil temperature needs to be above 17 °C for at least a week to control *V. dahliae* in strawberry fields. The relationship between the C/N ratio and soil temperature requires additional clarification in order to provide suitable recommendations for quantities of material necessary for effective anaerobic conditions in different regions.

In ASD, the application of soil amendments leads to reductions in redox potential (Eh) implying that oxygen is being consumed (Momma 2008). As an indicator of anaerobic intensity, several authors use "cumulative anaerobicity," referring to the hourly accumulation of anaerobic soil conditions calculated from average redox potentials below a critical redox potential indicative of anaerobic conditions (Butler et al. 2012b). Determining the level of anaerobiosis needed for pathogen control at different temperature regimes is recommended. Temperature plays a significant role in the survival of pathogens under anaerobic conditions.

Ebihara and Uematsu (2014) showed that the survival period of a pathogen under anaerobic conditions became shorter as the incubation temperature became higher, although the sensitivity to anaerobic conditions apparently differed among species. Generally, *V. dahliae* was eradicated quickly, but *F. oxysporum* f. sp. *fragariae* survived longer, and *Phytophthora cactorum* was intermediate between the two. The ability of *F. oxysporum* to grow under anoxic conditions by performing ammonia fermentation (Zhou et al. 2002) has likely contributed to the long survival of *F. oxysporum* f. sp. *fragariae* (Ebihara and Uematsu 2014). However, in the case of *S. sclerotiorum*, no significant relationship between accumulated anaerobic conditions and total germination of sclerotia or disease was observed in two ASD trials with cover crops (Butler et al. 2014b). In a 2-year field study in Florida, Butler et al. (2012b) reported accumulated anaerobic values ($mV \cdot h^{-1}$) for a molasses ASD treatment below 5000 $mV \cdot h^{-1}$ the first year and near 30,000 $mV \cdot h^{-1}$ the second year. A similar pattern of increase in the accumulation of anaerobic conditions was observed by McCarty et al. (2014). Adaptation of the soil microbial community to large inputs of labile carbon may have an influence on the increase of anaerobic activity in the second year (McCarty et al. 2014). In California, Shennan et al. (2011) showed that a threshold of 50,000 $mV \cdot h^{-1}$ at 25 °C soil temperature is necessary for control of *V. dahliae*.

13.4 Anaerobic Soil Disinfestation and Soil Chemistry

The precise mechanisms through which ASD works have not yet been well defined (Momma 2008). ASD effectiveness for soilborne disease management is likely due to a variety of factors: creation of anaerobic conditions (Messiha et al. 2007; Ebihara and Uematsu 2014), accumulation of crop-specific toxic compounds liberated during the breakdown of plant material and crop-nonspecific fermentation products (Blok et al. 2000; Momma et al. 2013; Hewavitharana et al. 2014), higher soil temperatures for tarped amended soil (Butler et al. 2012c), and an increase in disease control resulting from microbial changes in the soil environment (Momma et al. 2011; Mazzola 2010, 2011; Mazzola and Manici 2012; Mazzola et al. 2012a; Mowlick et al. 2012, 2013a, b, c). Importantly, ASD does not result in a biological vacuum, and instead of soil becoming disease conducive, it can become suppressive to various plant pathogens (Goud et al. 2004; Mazzola et al. 2012b; Muramoto et al. 2014).

Relatively high rates of organic amendments followed by irrigation and plastic mulching (transparent or opaque) induce anaerobic (reduced) conditions in ASD-treated soils that not only affect soilborne pests but also potentially impact a range of soil chemical, physical, and biological properties (Inglett et al. 2005; Butler et al. 2014a). Soil chemical changes induced by ASD treatment are especially important due to potential impacts on soil fertility and crop nutrition, as well as potential environmental impacts through nutrient export and greenhouse gas emissions.

13.4.1 Soil pH

Soil pH in anaerobic soils is affected by the chemical nature of reduction processes. In saturated acidic soils without large C inputs, soil pH potentially increases as various reduction processes consume protons (Inglett et al. 2005), most notably iron oxyhydroxides (Faulkner and Richardson 1989) which are abundant in many soil types. Other common reactions in anaerobic soils that are proton consuming include reduction of NO_3^-, MnO_2, and SO_4^{2-} (Kögel-Knabner et al. 2010). In contrast, soil pH potentially decreases in alkaline soils as organic acids are generated and dissolved carbon dioxide accumulates (Inglett et al. 2005). Increases in soil pH in anaerobic acidic soils are likely less typical in ASD treatment, where substantial amounts of labile C can lead to significant production of organic acids during anaerobic decomposition (Momma et al. 2006; Runia et al. 2014). Momma et al. (2006) demonstrated that soil pH decreased following ASD treatment for at least 2 weeks due to production of organic acids. Other studies have shown that soil pH response is highly influenced by the type of amendments, anaerobic conditions, temperature, and soil types. Amendments such as rice bran, mustard seed meal, ethanol, and orchard grass residues (Hewavitharana et al. 2014), various warm-season cover crops (Butler et al. 2012c), molasses (Butler et al. 2012b), and wheat bran (Momma et al. 2006) used during ASD have been reported to decrease soil pH. In a 2-year study by McCarty et al. (2014), soil pH was minimally affected by ASD treatment (molasses or cover crop residues), likely owing to lessened anaerobic conditions compared to some other studies. Decomposition of organic amendments can potentially increase soil pH through the release of basic cations (e.g., Marschner and Noble 2000; Xu et al. 2006), which could be a dominant process if significant organic acids are not created during anaerobic decomposition, indicating the importance of strong and sustained reducing conditions driven by adequate amendment of labile C in order to obtain pH reductions conducive to soilborne pest control. Soil type is also an important determinant of the effect of ASD treatment on soil pH, as the high buffering capacity of soils high in organic matter is likely to limit impact of ASD treatment on soil pH (Bohn et al. 1985), whereas sandy soils with low buffering capacity are likely to see a strong soil pH response to ASD treatment or organic matter additions (e.g., Butler et al. 2012a).

In terms of pesticidal activity, it is also important to consider that even in highly disturbed plasticulture horticultural production systems, soils are relatively heterogeneous, leading to microsites of varying anaerobic activity and soil pH during ASD. It is expected that during ASD treatment, microsites of reducing activity initiated by the presence of labile C are likely to have soil pH values less than the bulk soil pH (Strong et al. 1997) which can still contribute to effective pesticidal activity (Katase et al. 2009). Recovery of soil pH following ASD treatment termination and soil reoxidation is likely to be relatively rapid, as organic acids in soils generally have a relatively short half-life (<12 h) as they are a relatively labile source of C for soil microbial biomass, and many of the organic acids created during ASD (e.g., acetate, lactate, propionate) are weakly sorbed to soil mineral surfaces (Jones et al. 2003).

The production of an unpleasant odor is often associated with an effective ASD treatment due to organic acid production by anaerobic bacteria activated under a reduced soil condition (Momma 2008). During ASD treatment, organic acids or volatile fatty acids (VFAs), primarily acetic and butyric acids, are released into the soil solution as the organic matter decomposes anaerobically (Momma 2008; Huang et al. 2015), followed by small amounts of isovaleric acid and propionic acid (Huang et al. 2014). These short-chain)VFAs and others, such as valeric and caproic acids, are metabolic products of bacterial anaerobic fermentation (Guenzi and Beard 1981), and they have been shown to be toxic to a wide range of soil pathogens (Okazaki and Nose 1986; Tenuta et al. 2002; Conn et al. 2005). Momma et al. (2006) reported conidia of *F. oxysporum* f. sp. *lycopersici* were completely killed in 0.02 % acetic acid or 0.02 % butyric acid solutions, while for chlamydospores, doses as high as 0.14 % acetic acid or 0.2 % butyric acid solutions were necessary. Huang et al. (2015) confirmed that the organic acids generated with a 2 % maize ASD amendment suppressed *F. oxysporum*, *R. solani*, and *R. solanacearum*. Increasing the temperature has the effect of altering the dissociation constant (pKa) of weak acids such as VFAs resulting in greater amounts of non-ionized forms (Schwarzenbach et al. 2002; Conn et al. 2005). Importantly, only the non-ionized forms of VFAs (e.g., acetic acid, not acetate) are toxic to *V. dahliae* microsclerotia (Tenuta et al. 2002) and other microorganism (Wallace et al. 1989; Lazarovits et al. 2003). Besides temperature, the critical factor regulating disease control activity of VFAs is soil pH. For example, at pH 3, acetic acid is in its non-ionized toxic form, whereas at pH 6.0 it is in the non-active acetate form. In this way, small changes in pH can cause very large changes in the quantity of each of these chemical products, thus influencing disease control efficacy (Conn et al. 2005). Besides VFAs, a large diversity and amount of volatile organic compounds such as alcohols, organic sulfides, esters, ketones, hydrocarbons, and isothiocyanates (ITCs) were detected during the anaerobic phase of ASD using ethanol, grass residues, or *Brassica juncea* seed meal as the carbon input (Hewavitharana et al. 2014). Each ASD carbon source produced a unique spectrum of volatile compounds (Hewavitharana et al. 2014).

13.4.2 Soil Organic Matter

Plasticulture systems used for high-value specialty crop production are typified by high levels of disturbance by intensive tillage (typically rotovation) which can lead to increased microbial decomposition of soil organic matter pools. Incorporation of amendments for ASD treatment can potentially mitigate this impact, if amendments contain a portion of more recalcitrant C to build soil organic matter (such as lignin from plant biomass or composts of plant materials or animal manures) in addition to the large pool of labile C used to drive the ASD treatment process. While some labile C is likely to be incorporated in organic matter pools through microbial activity, it may also stimulate decomposition of passive pools of organic matter through priming effects (Kuzyakov and Domanski 2000).

13.4.3 Soil Nitrogen

Treatment of soil with ASD is likely to cause substantial changes in soil nitrogen (N) status. The labile C in organic amendments provides a substrate for rapid growth and respiration of indigenous soil microbes. Largely dependent on the ratio of C to N (C/N ratio) in the organic soil amendment (Whitmore 1996), this microbial activity will lead to mineralization of organic N forms in the amendment to ammonium (NH_4^+; Fig. 13.1) when C/N ratios are low (approximately <20:1), whereas potential immobilization of available soil inorganic N into soil microbial

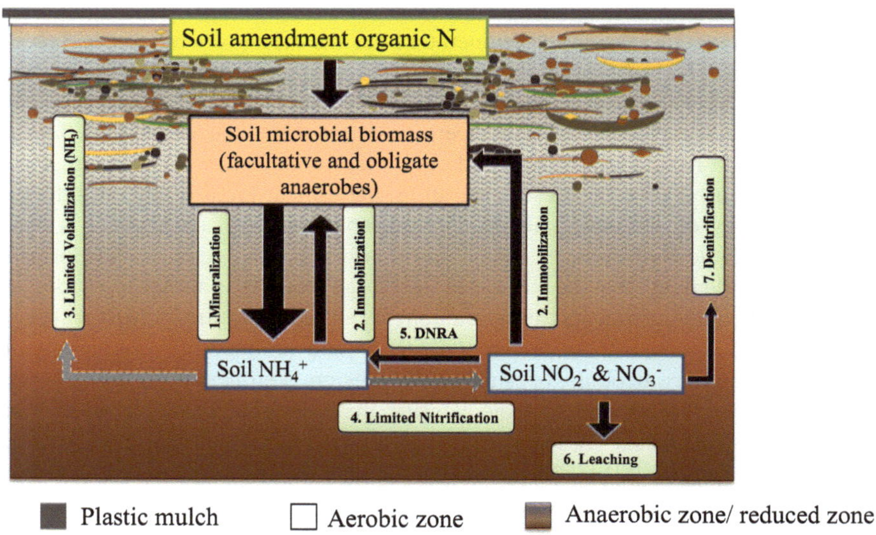

Fig. 13.1 Proposed nitrogen (N) cycle during anaerobic phase of anaerobic soil disinfestation (ASD). Following addition of organic amendment to drive ASD treatment, soil microbes decompose organic compounds leading to (1) mineralization of organic N to NH_4^+ if amendment C/N ratio is moderate to low. Soil ammonium (NH_4^+) and nitrate (NO_3^-) can potentially be immobilized (2) into microbial biomass, especially if the amendment C/N ratio is high. Soil NH_4^+ can be (3) volatilized to ammonia, although this process is likely limited during ASD. Soil NH_4^+ under aerobic conditions would typically be (4) nitrified to NO_2^- and then NO_3^- by nitrifying soil bacteria (aerobes), but this process is limited under anaerobic conditions. The reverse of this process is (5) dissimilatory nitrate reduction to ammonium (DNRA), which is performed by various anaerobic bacteria and may be significant in some ASD systems. Soil NO_3^- at the beginning of ASD treatment is potentially vulnerable to loss via (6) leaching with irrigation applied for treatment or (7) denitrification, the extent of which likely depend on site conditions and management

biomass is possible with higher C/N ratios (approximately >20:1). If the soil is alkaline, soil NH_4^+ can volatize to ammonia (NH_3) and accumulate below the plastic layer and in soil pores; however, these conditions are not typical in ASD-treated soils. Nitrification is likely limited during anaerobic phases of ASD treatment due to the absence of oxygen and limitation in activity of nitrifying soil bacteria (aerobes) but can quickly proceed as treated soils transition to an aerobic state. Nitrifying soil bacteria are also known to be sensitive to other soil disinfestation treatments, such as fumigation by methyl bromide and solarization (Chen et al. 1991), although there has been no research to determine the sensitivity of these bacteria to ASD treatment. Nitrate (NO_3^-) present in aerobic soils prior to treatment is vulnerable to loss by leaching as treatments are initiated if excess irrigation water is applied. During ASD treatment, NO_3^- is also vulnerable to loss via denitrification due to the anaerobic condition present during treatment. The denitrification potential under ASD may be high given the transient nature of the pH changes (i.e., a denitrifier community not heavily influenced by long-term pH selection) and the abundance of carbon in ASD (Šimek and Cooper 2002), although this will also depend upon how much nitrate is present prior to treatment and populations of denitrifying bacteria which can be relatively low in highly disturbed soils (Doran 1980). However, N_2O is also produced during nitrification which may well be elevated as the system returns to aerobic conditions and the accumulated NH_4^+ is nitrified (Khalil et al. 2004) or in microsites of aerobic activity where nitrification proceeds and then diffuses to anaerobic zones during treatment phases dominated by anaerobiosis (e.g., Nielsen et al. 1996). There is data from agricultural soils in California that a significant amount of N_2O is emitted following fertilizer additions via ammonia (NH_3) oxidation pathways (nitrifier nitrification, nitrifier denitrification, and nitrification-coupled denitrification) as opposed to heterotrophic denitrification of existing nitrate (Zhu et al. 2013).

Chemical analyses from various field and laboratory experiments with ASD showed a rapid depletion of nitrate (NO_3^-) with a concurrent increase of NH_4^+ (Butler et al. 2014a). This could be due to a process known as dissimilatory nitrate reduction (Broadbent and Stojanovic 1952). Also, some classes of bacteria known to be important in ASD treatment (e.g., *Clostridia*, *Pseudomonas*) are known to perform DNRA (Burgin and Hamilton 2007). It is possible that the transient decrease in pH shifts the balance toward dissimilatory nitrate reduction during the relevant initial period of anaerobiosis (Stevens et al. 1998). While this process has typically been thought to be relatively minor compared to denitrification in most soils, research is generally lacking (Rütting et al. 2011). However, disappearance of nitrate and accumulation of ammonium could also be due in part to a combination of denitrification and an increase in ammonium from mineralization of the added C source and the priming effect on SOM decomposition. Primary considerations for management of the soil N cycle during ASD treatment include the C/N ratio of the selected organic amendment(s), the rate of N applied in amendments, and initial soil nitrate levels. With large amounts of N applied in relatively low C/N ratio amendments, substantial inorganic N can be present in soils following treatment (e.g., >150 mg N kg^{-1} soil; Butler et al. 2014a), which must be managed effectively to provide adequate plant nutrition and prevent environmental degradation. With amendments having a relatively high C/N ratio (e.g., molasses), immobilization of soil N in microbial biomass can be an issue for crop productivity, as N fertilization must be increased posttreatment to improve crop nutrition (Butler et al. 2014a).

13.4.4 Soil Phosphorus, Calcium, Magnesium, and Sulfur

Although soil phosphorus (P) is relatively insoluble in most soils, reduction processes can increase P solubility in acidic soils. However, Butler et al. (2012b, 2014a) observed no differences in extractable P (Mehlich 1984) at the termination of ASD treatment, likely due to reoxidation of the treated soil by this time and possibly the sandy soils in which the trials were conducted. However, in soils with high clay content, McCarty et al. (2014) also observed minimal impact of ASD treatment on soil P status posttreatment. Changes in soil P status during ASD are unlikely to have a significant impact on plant nutrition or environmental degradation. A more important consideration of ASD impact on soil P status (and other nutrients) is the relative amount of nutrients added in soil amendments used to provide labile C. Many organic soil amendments contain high amounts of P, calcium (Ca), potassium (K), magnesium (Mg), and sulfur (S), which can potentially reduce fertilizer costs but must be effectively managed as a part of a farmer's overall soil fertility program. For example, at an 8 Mg ha^{-1} rate (dry matter) of molasses application in a Florida ASD study, Butler et al. (2014a) reported relatively high increases in exchangeable soil K (from <50 mg K kg^{-1} soil to >500 mg K kg^{-1} soil), soil Ca (from ~200 mg Ca kg^{-1} soil to >300 mg Ca kg^{-1} soil), and soil Mg (from ~20 mg Mg kg^{-1} soil to >mg Mg 50 kg^{-1} soil to), which impacted crop leaf tissue nutrient concentrations. While no published ASD research to date has detailed treatment impacts on soil S or plant S uptake, SO_4^- is used as an electron acceptor in strong anaerobic conditions which leads to the formation of gaseous S forms such as hydrogen sulfide (Runia et al. 2014), which can potentially be removed from the system, as well as contributing pesticidal effects within soil pores under the plastic mulch.

There is an accumulation of Fe^{2+} and Mn^{3+} ions in soil solution in treated soil with fresh plant material (van Bruggen and Blok 2014). Previously, Momma et al. (2011) showed that creation of Fe^{2+} and Mn^{2+} in reduced soils might be one of the mechanisms of ASD. In a recent study, Cao et al. (2014) suggested that the suppression in mycelial growth and zoospore germination of *P. capsici* were caused by the higher concentration of NH_4^+ and humic substances of anaerobically digested pig slurry. In addition, Núñez-Zofío et al. (2011) observed reduction in disease incidence and *P. capsici* oospore survival by application of organic amendments followed by soil plastic mulching. These authors hypothesized that the success was, at least, partially attributed to the production of NH_3 and to the increase in soil microbial activity. Under controlled conditions, Runia et al. (2012) reported the production of CO_2, NH_3, H_2S, CH_4, and N_2O during ASD treatment depending on the type of organic material, characteristics of the soil, temperature, dosage, and exposure time.

Soil treatment by ASD has potential to greatly impact soil fertility status, which is an important consideration for crop production as well as minimization of negative environmental impacts of crop nutrients. Farmers should consider existing soil fertility status, irrigation method, ASD amendment composition, and post-treatment management in order to effectively adjust current practices to a production system utilizing ASD. Farmers and researchers alike should also consider the impacts of ASD treatment on soil properties when comparing the relative merits of soil disinfestation practices. Either positive or negative crop performance following alternative soil disinfestation practices compared to an existing standard may not necessarily be due to treatment impacts on soilborne pests, if treatment impacts on soil chemical, physical, and biological properties are not also considered.

13.5 Microbial Mechanisms of Pathogen Inactivation

Currently, mechanistic studies of ASD are focused on changes in microbial communities, both bacterial and fungal, under ASD treatments (Hong et al. 2014; Mowlick et al. 2012, 2013a, b, 2014; Messiha et al. 2007; Momma 2008; Momma et al. 2010; Rosskopf et al. 2014). Failed applications of ASD have been associated with heavily fumigated soils (Rosskopf, personal observation), where the addition of labile carbon has not resulted in the development of anaerobic conditions. ASD is analogous to fermentation processes that transform raw ingredients such as milk, fruits, or grains into cheese, alcohols, and breads. In part, the success of the food industry in consistently delivering products that meet quality standards is dependent on creating an environment conducive for the microorganisms of choice to grow and produce their products and by-products. In turn, in order to consistently control and suppress pathogens, important microorganisms key to ASD must be identified, which would allow for the environmental factors needed to be defined for optimal pathogen control.

The shift to a microbial community well adapted to persistent anaerobic conditions in the soil would draw on the resident pool of bacteria that can take advantage of this loss of oxygen and the progressive reduction of soil minerals. Anaerobic metabolism can support significant microbial populations in an otherwise aerobic soil. There are microsites in soil aggregates that are effectively anaerobic (Sexstone et al. 1985), and the worm gut drives ingested soil through a period of anaerobiosis before releasing the castings to the ambient aerobic conditions (Horn et al. 2003). Also, many rain events will temporarily shift the soil to a primarily anaerobic condition (Linn and Doran 1984). For bacterial populations that are primarily competitive in aerobic conditions, however, ASD removes all such niches and negates any competitive advantages they may have. Bacteria originating from soil amendments would undergo similar selection. Much of the effect of ASD on microbial community structure may be based on simple shifts in competitiveness from one group of bacteria to another.

Once a soil becomes anaerobic, a series of changes in redox can occur; depletion of oxygen is just the first change. Subsequent reduction of the soil matrix begins with compounds with the highest reduction potential, such as NO_3^- and other nitrogen oxides, MnO_2, Fe_3^+, and organic acids (Kögel-Knabner et al. 2010). This will change the solubility of many minerals and cause a shift in the range of small organic molecules present (Momma et al. 2011). Many bacteria have unique capacities that allow them to thrive in the wide array of different niches available in a soil shifting toward lower Eh (Kögel-Knabner et al. 2010). There is little potential for growth for obligate aerobes, and it is likely that many facultative anaerobes are not sufficiently competitive in an increasingly anaerobic soil.

As stated previously, ASD has been used in the Netherlands, Japan, and various parts of the USA including California, Florida, Washington, Tennessee, North Carolina, and Michigan (Momma et al. 2013; Yoder 2014). Pre- and posttreatment soil microbial communities have been characterized to identify population shifts resulting from ASD application in both field and greenhouse studies. The majority of these findings have focused on changes observed in bacterial populations. Quantification of bacterial communities in post-ASD soil resulted in the identification of an increase in bacterial populations belonging to the Firmicutes phylum, which includes members of the Clostridia and Bacilli classes (Momma et al. 2010; Mowlick et al. 2012, 2013a, b; Stremińska et al. 2014). Fungal community changes after ASD treatment showed increases in some fungal populations as well, including yeasts (Mazzola et al. 2012a, b), total fungi (Stremińska et al. 2014), and an increase of *Trichoderma* spp. colonization of *S. rolfsii* sclerotia (Shrestha et al. 2013).

The types of soil and soil amendments can both have an effect on the microbial population and the efficacy of ASD. A recent study on the use of ASD for control of potato cyst nematode (PCN) compared six soil types, including an artificial soil that did not contain any organic matter (Runia et al. 2014). By day 28, hatching of PCN eggs was reduced in all ASD-treated soils. It was observed that PCN declined more rapidly in three soil types: glacial sand, marine loam, and peat. These three soils had higher total N, total P, and organic matter content prior to the addition of the organic amendment. No differences were detected in O_2 depletion or the accumulation of other gases or organic acids. The authors hypothesize that the control of PCN was biologically based.

Various amendments can be added to enhance the performance of ASD if the soil or environmental conditions are less than ideal. In the previously mentioned paper, Runia et al. (2014), again using the six soils, added organic matter to each soil type and found that the population of Firmicutes, measured by qPCR with Firmicutes-specific primers, was consistently greater in soils that had the carbon amendment compared to those lacking it. The pathogen was significantly reduced 7 days after treatment for the soils that were treated with the amendment and had the increased Firmicutes population. Soils with the amendment had an inactivation of

PCN by >99.5 % by day 28. As discussed previously, it was shown that under moderate temperatures, of 15–20 °C, an increase of carbon, up to four times the standard amount, controlled *S. rolfsii* better than the traditional method (Shrestha et al. 2013). This phenomenon could be attributed to the microbes requiring a more abundant and readily available carbon source in colder temperatures. Shennan et al. (2014) compared the effect of various soil treatments, including non-amended, chloropicrin, methyl bromide/chloropicrin, and ASD with rice bran, molasses, or a combination of the two as carbon source, on fungal communities. Using terminal restriction fragment length polymorphism (T-RFLP), the fungal communities of the chemically treated soil grouped together by multivariate analysis and distinct from the ASD and non-amended samples. The communities of the soil treated with rice bran and the combination of rice bran and molasses were closely related. The communities found in molasses-treated soil grouped together and were most similar to the non-amended samples. Rice bran alone significantly increased total fungi.

In California, soil bacterial populations from posttreatment ASD plots, identified using T-RFLP, were significantly different than non-treated soil (Mazzola et al. 2012a, b). In Japan, an increase in the *Clostridia* and *Bacilli* groups, including an increase of obligate anaerobes (Mowlick et al. 2013b), was detected. In another ASD experiment in Florida, this time on strawberries, soil dilution plating for native soil fungal populations from mid- and late-season soil sampling observed a significant increase of *Trichoderma* species (Rosskopf et al. 2014). Several *Trichoderma* species have been observed to be biocontrol agents against fungal plant pathogens, and plants can benefit from direct interactions with some *Trichoderma* species (Bae et al. 2011). ASD was applied in the same strawberry fields the following year, yet the *Trichoderma* species count was similar or lower than the control, non-treated plots, for both sampling dates. Instead of isolating *Trichoderma* spp. on the semi-selective plates, the plates were dominated by similar bacterial colonies, presumably a *Pseudomonas* species. To note though, anaerobicity and strawberry yield for the ASD-treated field the second year was significantly higher than the control and the first year ASD was applied.

To understand the changes in the microbial community during ASD, a few experiments have focused on sampling the soil throughout the ASD treatment (Momma et al. 2010; Mowlick et al. 2012; Stremińska et al. 2014). These experiments have been performed in greenhouses or growth chambers with the temperatures set at 20–30 °C. Destructive sampling took place at various time periods to sample the soil. Total DNA was extracted from the soil samples and PCR-based detections were used. Momma et al. (2010) performed an experiment to test the effectiveness of ASD in managing *F. oxysporum* f. sp. *lycopersici*. In this study ethanol at different dilutions was used as the carbon source. Autoclaved soil was not effective in reducing the pathogen 14 days post-ASD application. However, the pathogen was not detected in non-autoclaved soil with 2.0 % ethanol ASD treatment, again indicating pathogen control could be biologically based. Plating soil samples taken every 3 days revealed that the anaerobic bacterial population peaked after day 3 and was significantly higher than the bacterial population in the control.

By day 15 the anaerobic population did not differ from the control and the various dilutions of ethanol. Using polymerase chain reaction/denaturing gradient gel electrophoresis (PCR-DGGE) to determine the microbial population dynamics of the soil, the authors compared pretreated soil to soil sampled at 15 days post-treatment. Based on the PCR-DGGE gel, the pretreatment communities were similar. Comparing the posttreatment samples, the two control samples, watered and non-watered soil, were similar, while the ASD-treated samples had unique bands.

In another study, wheat bran and *B. juncea* were used as the carbon source and soil samples were taken every 3 days for 18 days (Mowlick et al. 2013c). In order to understand the microbial community in this study, the soil samples were both plated traditionally and total DNA was extracted from the soil samples. Universal bacterial primers specific to the 16 s region were used for PCR-DGGE and cloning. Unique bands observed from the PCR-DGGE gels of the ASD-treated soil were extracted and sequenced. Based on these sequences, early in ASD treatment, days 3–9, there seemed to be an increase of Firmicutes, specifically members of the Bacilli and Clostridia classes from both carbon sources. While later in the treatment, 15–18 days after treatment, these populations were less abundant. Based on the PCR-DGGE)results, samples from seven soils were selected for creating clone libraries, which included a pretreatment sample and samples from the two carbon sources and the control at days 9 and 18 posttreatment. The control cloned libraries presented highly diversified populations for the three dates, with the most dominant group being members belonging to the phylum Proteobacteria. The bacterial populations in the ASD-treated soil were dominated by the phylum Firmicutes (between 58 and 74 % of total). Within the Firmicutes population, 32–62 % was composed of *Clostridium* spp. at day 9 for both carbon sources. The Proteobacteria population at day 9 for both carbon sources ranged from 10 to 16 %, while for the control soil at day 9, the Proteobacteria was the major population, constituting 36 % of the total population. Posttreatment sampling of the ASD-treated soil revealed a reduction of the *Firmicutes* population, yet it was still the dominant phylum detected.

In another study, as mentioned earlier, six different types of soils were used for ASD, and one of the soils was an artificial soil that lacked organic matter (Stremińska et al. 2014). The carbon source in this study was a commercially available product (Herbie® 7022, Thatchtec BV, Netherlands). Destructive soil sampling took place at 3, 7, 14, and 28 days post-ASD initiation. Non-amended soil was used as a control. Total DNA was extracted from the soil samples, and the abundance of bacteria, fungi, Firmicutes, and sulfate-reducing bacteria (SRB) was characterized using group-specific primers and qPCR. The total abundance of bacteria by day 3 was significantly higher for the soils with the carbon amendment compared to the control. The Firmicutes population was greater for the amended soil throughout the entire study compared to the control. By day 3, the relative abundance of the Firmicutes population was higher than the controls; however, they were not statistically different between artificial and river clay soils. The Firmicutes population accounted for up to 67 % of the total bacterial population for the

ASD-treated soil early in the process. By day 28, the relative abundance of the Firmicutes decreased. The SRB are anaerobic bacteria that use acetic, butyric, and propionic acids as carbon sources, acids produced by some members of Firmicutes. SRB could potentially be useful biomarkers for identifying organic acid production. SRB were detected in four of the six soils 3 days after treatment and in all of the soils by day 14. The fungal population also increased in all of the ASD-treated soils when measured 3 days into treatment. However, by the end of the study, both the control and the ASD-treated soils had a similar abundance of fungi. The authors suggest that the fungal populations they detected were facultative anaerobic yeasts, supporting the observation of Mazzola et al. (2012a, b). However, more recent work by Shennan et al. (2013) shows significant, longer-term changes in fungal communities following ASD with rice bran.

With advances in molecular biology, bioinformatics, and the decreasing cost of sequencing, identification of microbes is quicker and easier than ever before. Previously microbes were identified by first culturing and then describing their phenotypic traits. In fact Bergey's Manual, in 1923, stated that no organism could be classified without first being cultured (Society of American Bacteriologists 1923). A discrepancy was observed between dilution plating and microscopy, in which some plate counts and estimated viable cells could differ by a magnitude of 4–6 (Handelsman 2004). It was estimated that only 0.1–1.0 % of soil bacteria are culturable using common media and standard practices. DNA-DNA hybridization was used to show relatedness among bacteria (Johnson and Ordal 1968), but it was not until Pace and Campbell (1971) and Woese (1987) who showed that 16S rRNA could be used to infer phylogenetic relationships and to identify the unculturable bacteria. This ushered in an era where identification of bacteria was based predominantly on 16SrRNA sequences. However, this approach is not applicable to all scenarios. Type strains *Bacillus globisporus* and *B. psychrophilus* share >99.5 % 16SrRNA sequence similarity, yet comparison of their genomes exhibits only a 23–50 % relatedness in reciprocal DNA-DNA hybridization reactions (Fox et al. 1992). Next-generation sequencing has created a method to obtain many sequences, ~40,000 amplicons, for a fraction of the time and cost of cloning. Currently, many researchers are trying to circumvent the inherent biases of PCR amplification of a single gene by using whole-genome shotgun (WGS) approaches to estimate the composition of a microbial community (Poretsky et al. 2014). WGS consists of extracting DNA and sequencing, assembling the reads into contiguous sequences of DNA (contigs) and annotating the contigs. These predicted genes are then searched against a database of all sequenced bacterial and archaeal genomes. In comparison to WGS, 16SrRNA can determine broad changes in the bacterial community over time yet is limited in resolution and sensitivity (Poretsky et al. 2014).

13.6 Anaerobic Soil Disinfestation and Disease Suppression

This is the golden age of biology. Advances in molecular microbial ecology are being made more quickly than ever before. Combining these new technologies with field applications to increase plant health and yield stands at the intersection of basic and applied science. In order to optimize ASD for different soils, regions, and targeted plant pathogen control (Weller et al. 2002), a thorough understanding of the role that microbes play in the mechanism is critical. At this stage in the research, organisms have been identified that contribute to the development of the anaerobic condition, but their role, if any, in direct disease suppression has not been established. Few studies have determined whether suppressive soils, by definition (Baker and Cook 1974), have been created with the method. Work by Blok et al. (2000) and Goud et al. (2004) established that treatment with BSD) did not result in a disease-conducive soil when pathogens were added to previously treated soil. The goal of their approach was not to establish that a defined specific suppressive nature of soil could be maintained after the treatment but that the approach did not create a biological void in which the introduction of the pathogen would result in an increase in disease compared to an untreated soil. Recent work by Mazzola et al. (2012b) has established that ASD-treated soil did not prevent *Pythium* spp. associated with root infection from colonizing soil, but it did result in disease suppression. Similarly, work in CA resulted in reductions of *V. Dahlia* microsclerotia almost 2 years after ASD treatment using rice bran as carbon source, despite tillage and production of a cover crop followed by a lettuce crop (Shennan and Muramoto, personal observation). Whether disease suppression in these systems is associated with specific organisms and can be transferred to other soils (specific suppression) or is a general suppression that cannot be transferred (Weller et al. 2002) has yet to be established.

One potential approach to understand the mechanism of ASD, rather than elucidating the entire compositional change in the microbial community, is to quantify the presence or increase in the presence of genes associated with acid production (Fujita et al. 2007) or biological control of plant diseases (Joshi and McSpadden Gardener 2006). As previously mentioned, failed trials of ASD have been associated with heavily fumigated soils. Preliminary data by the authors has indicated that by applying amendments rich in Firmicutes, such as composted broiler litter, in heavily fumigated soils creates a more diverse soil bacterial population posttreatment, and these ASD treatments have been successful in managing weeds and phytopathogens. The combination of advanced molecular techniques with traditional approaches will allow for the identification of specific organisms that are responsible for the various phases of ASD, including the development of anaerobicity and resulting disease control. Using these techniques will also better define how specific organisms, such as members of the Bacilli, contribute to disease suppression in this system.

13.7 Conclusion

Although it is a relatively new approach to soil pest management, research on ASD has identified several critical components that are necessary for successful application. While the overall goal is to increase sustainability of the production system by utilizing locally available agricultural waste products as carbon sources, each input can generate different organic compounds as well as having different decomposition rates, subsequently resulting in different changes in soil microbial communities. These changes may be associated with the generation of the anaerobic condition, the production of organic acids, direct or indirect biological control processes, and ultimately disease suppression. It is clear that inputs used in one location may not have the same effects when used in another soil type or under different environmental conditions. Soil temperature plays a significant role in the success of ASD, but exactly how temperature and carbon source interact to impact metabolites produced has not yet been well defined. Many of these interactions among these components will need to be investigated for their impact on the soil chemistry, the microbial community, and how each of these changes influences both the short-term control and long-term suppression of plant disease.

Acknowledgment The authors wish to thank Ariena HC van Bruggen, Yuso Kobara, and Noriaki Momma for the introduction to anaerobic and biological soil disinfestation and to Wesley Schonborn for assistance in manuscript preparation.

Disclaimer Mention of trade names or commercial products in this publication is solely for the purpose of providing specific information and does not imply recommendation or endorsement by the USDA. The USDA is an equal opportunity provider and employer.

References

Bae H, Roberts DP, Lim HS, Strem MD, Park SC, Ryu CM, Melnick RL, Bailey BA (2011) Endophytic *Trichoderma* isolates from tropical environments delay disease onset and induce resistance against *Phytophthora capsici* in hot pepper using multiple mechanisms. Mol Plant Microbe Interact 24:336–351

Bailey KL, Lazarovits G (2003) Suppressing soil-borne diseases with residue management and organic amendments. Soil Till Res 72:169–180

Baker KF, James Cook R (1974) Biological control of plant pathogens. WH Freeman and Company, San Francisco, CA

Bello A, Porter I, Díez-Rojo MA, Rodríguez-Kábana R (2008) Soil biodisinfection for the management of soil-borne pathogens and weeds. In: 3rd international biofumigation symposium, Canberra, Australia, p 35

Blok WJ, Coenen TCM, Termorshuizen AJ, Lamers JG (2005) The potential of biological soil disinfestation to manage Fusarium foot and root rot in asparagus. Acta Hortic 776:135–144

Blok WJ, Lamers JG, Termorshuizen AJ, Bollen GJ (2000) Control of soilborne plant pathogens by incorporating fresh organic amendments followed by tarping. Phytopathology 90:253–259

Bohn HL, McNeal BL, O'Connor GA (1985) Soil chemistry. Wiley, New York
Broadbent FL, Stojanovic BF (1952) The effect of partial pressure of oxygen on some soil nitrogen transformations. Soil Sci Soc Am J 16:359–363
Burgin AJ, Hamilton SK (2007) Have we overemphasized the role of denitrification in aquatic ecosystems? A review of nitrate removal pathways. Front Ecol Environ 5:85–96
Butler DM, Eichler Inwood SE, McCarty DG, Sams CE, Wszelaki AL, Dee ME, Ownley BH (2012a) Optimizing anaerobic soil disinfestation for Tennessee. In: Proceedings of annual international research conference of methyl bromide alternatives and emissions reductions, Orlando, FL, pp 13.11–13.14
Butler DM, Kokalis-Burelle N, Muramoto J, Shennan C, McCollum TG, Rosskopf EN (2012b) Impact of anaerobic soil disinfestation combined with soil solarization on plant–parasitic nematodes and introduced inoculum of soilborne plant pathogens in raised-bed vegetable production. Crop Prot 39:33–40
Butler DM, Rosskopf EN, Kokalis-Burelle N, Albano JP, Muramoto J, Shennan C (2012c) Exploring warm-season cover crops as carbon sources for anaerobic soil disinfestation (ASD). Plant Soil 355:149–165
Butler DM, Kokalis-Burelle N, Albano JP, McCollum TG, Muramoto J, Shennan C, Rosskopf EN (2014a) Anaerobic soil disinfestation (ASD) combined with soil solarization as a methyl bromide alternative: vegetable crop performance and soil nutrient dynamics. Plant Soil 378: 365–381
Butler DM, Shrestha U, Dee ME, EichlerInwood SE, McCarty DG, Ownley BH, Rosskopf EN (2014b) Low carbon amendment rates during anaerobic soil disinfestation (ASD) at moderate soil temperatures do not decrease viability of *Sclerotinia sclerotiorum* sclerotia or fusarium root rot of common bean. In: VIII international symposium on chemical and non-chemical soil and substrate disinfestation, vol 1044, Torino, Italy, pp 203–208
Cao Y, Chang Z, Wang J, Ma Y, Yang H, Fu G (2014) Potential use of anaerobically digested manure slurry to suppress phytophthora root rot of chilli pepper. Sci Hortic 168:124–131
Chen Y, Gamliel A, Stapleton JJ, Aviad T (1991) Chemical, physical, and microbial changes related to plant growth in disinfested soils. In: Katan J, DeVay JE (eds) Soil solarization. CRC, Boca Raton
Conn KL, Topp E, Lazarovits G (2005) Factors influencing the concentration of volatile fatty acids, ammonia, and other nutrients in stored liquid pig manure. J Environ Qual 36:440–447
Daugovish O, Shennan C, Muramoto J, Koike S (2011) Anaerobic soil disinfestation for southern California strawberry. In: Proceedings of annual international research conference on methyl bromide alt emissions reductions, San Diego, CA, pp 2-1/2-4
Domínguez P, Miranda L, Soria C, de los Santos B, Chamorro M, Romero F, Daugovish O, López-Aranda JM, Medina JJ (2014) Soil biosolarization for sustainable strawberry production. Agron Sustain Dev 34:821–829
Doran JW (1980) Soil microbial and biochemical changes associated with reduced tillage. Soil Sci Soc Am J 44:765–771
Ebihara Y, Uematsu S (2014) Survival of strawberry-pathogenic fungi *Fusarium oxysporum* f. sp. *fragariae*, *Phytophthora cactorum* and *Verticillium dahliae* under anaerobic conditions. J Gen Plant Pathol 80:50–58
Faulkner SP, Richardson CJ (1989) Physical and chemical characteristics of freshwater wetland soils. In: Hammer DA (ed) Constructed wetlands for wastewater treatment: municipal, industrial, agricultural. CRC, Boca Raton
Fox GE, Wisotzkey JD, Jurtshuk P Jr (1992) How close is close: 16S rRNA sequence identity may not be sufficient to guarantee species identity. Int J Syst Bacteriol 42:166–170
Fujita Y, Matsuoka H, Hirooka K (2007) Regulation of fatty acid metabolism in bacteria. Mol Microbiol 66:829–839
García Ruíz A, Tello JC, Avilés M, Ordovás J (2009) Fusariosis del clavel (*Fusarium oxysporum f. sp. dianthi*). In: Agrotécnicas SL (ed) Últimos avances en su control. Madrid, p 275

Goud JC, Termorshuizen AJ, Blok WJ, van Bruggen AH (2004) Long-term effect of biological soil disinfestation on verticillium wilt. Plant Dis 88:688–694

Guenzi WD, Beard WE (1981) Volatile fatty acids in a redox-controlled cattle manure slurry. J Environ Qual 10:479–482

Handelsman J (2004) Metagenomics: application of genomics to uncultured microorganisms. Microbiol Mol Biol Rev 68:669–685

Hewavitharana SS, Mazzola M (2013) Carbon source-dependent efficacy of anaerobic soil disinfestation (ASD) in suppression of rhizoctonia root rot of apple. Phytopathology 103:60–60

Hewavitharana SS, Ruddell D, Mazzola M (2014) Carbon source-dependent antifungal and nematicidal volatiles derived during anaerobic soil disinfestation. Eur J Plant Pathol 140:39–52

Hong JC, Martian KJ, Kokalis-Burelle N, Butler D, Rosskopf EN (2014) Comparison of soil bacterial communities from fields treated with Anaerobic Soil Disinfestation located on Florida's east to west coast. Phytopathology 11:S518

Horn MA, Schramm A, Drake HL (2003) The earthworm gut: an ideal habitat for ingested N_2O-producing microorganisms. Appl Environ Microbiol 69:1662–1669

Huang X, Wen T, Zhang J, Meng L, Zhu T, Cai Z (2015) Toxic organic acids produced in biological soil disinfestation mainly caused the suppression of *Fusarium oxysporum* f. sp. *cubense*. BioControl 60:113–124

Inglett PW, Reddy KR, Corstanje R (2005) Anaerobic soils. In: Hillel D (ed) Encyclopedia of soils in the environment. Elsevier, Amsterdam

Johnson JL, Ordal EJ (1968) Deoxyribonucleic acid homology in bacterial taxonomy: effect on incubation temperature on reaction specificity. J Bacteriol 95:893–900

Jones DL, Dennis PG, Owen AG, van Hees PAW (2003) Organic acid behaviour is soils – misconceptions and knowledge gaps. Plant Soil 248:31–41

Joshi R, McSpadden Gardener BB (2006) Identification and characterization of novel genetic markers associated with biological control activities in Bacillus subtilis. Phytopathology 96: 145–154

Katase M, Kubo C, Ushio S, Ootsuka E, Takeuchi T, Mizukubo T (2009) Nematicidal activity of volatile fatty acids generated from wheat bran in reductive soil disinfestation. Nematol Res 39: 53–62

Khalil K, Mary B, Renault P (2004) Nitrous oxide production by nitrification and denitrification in soil aggregates as affected by O_2 concentration. Soil Biol Biochem 36:687–699

Kirkegaard JA, Angus JF Gardner PA, Cresswell HP (1993) Benefits of brassica break crops in the Southeast wheatbelt. In: Proceedings of 7th Australian agronomy conference, Adelaide, pp 19–24

Kobara Y, Uematsu S, Tanaka-Miwa C, Sato R, Sato M (2007) Possibility of the new soil fumigation technique with ethanol solution. In: Proceedings of annual international research conference on methyl bromide alt emissions reductions, San Diego, CA, pp 114-1/114-2

Kögel-Knabner I, Amelung W, Cao Z, Fiedler S, Frenzel P, Jahn R, Kalbitz K, Kölbl A, Schloter M (2010) Biogeochemistry of paddy soils. Geoderma 157(1):1–14

Korthals GW, Thoden TC, van den Berg W, Visser JHM (2014) Long-term effects of eight soil health treatments to control plant-parasitic nematodes and *Verticillium dahliae* in agro-ecosystems. Appl Soil Ecol 76:112–123

Kuzyakov Y, Domanski G (2000) Carbon inputs by plants into the soil. Rev J Plant Nutr Soil Sci 163:421–431

Lacasa CM, Guerrero MM, Ros C, Martínez V, Lacasa A, Fernández P, Núñez-Zofío M, Larregla S, Martínez MA, Díez-Rojo MA, Bello A (2010) Efficacy of biosolarization with sugar beet vinasses for soil disinfestation in pepper greenhouses. Acta Hortic 883:345–352

Lamers JG, Runia WT, Molendijk LPG, Bleeker PO (2010) Perspectives of anaerobic soil disinfestation. Acta Hortic 883:277–283

Lamont WJ Jr (1996) What are the components of a plasticulture vegetable system? Hortic Technol 6:150–154

Lazarovits G, Conn KL, Abbasi PA, Tenuta M (2003) Control of soilborne plant pathogens with organic amendments: challenges and possibilities. In: Proceedings of eighth international congress plant pathology, vol 2, p 273

Linn DM, Doran JW (1984) Effect of water-filled pore space on carbon dioxide and nitrous oxide production in tilled and nontilled soils. Soil Sci Soc Am J 48:1267–1272

Ludeking DJW, Paternotte SJ, Runia WT, Molendijk LPG (2010) Biological soil disinfestation with organic fermentation products. In: I international symposium on organic greenhouse horticulture, vol 915, pp 133–139

Marschner B, Noble AD (2000) Chemical and biological processes leading to the neutralization of acidity in soil incubated with litter materials. Soil Biol Biochem 32:805–813

Martínez MA, Martínez MC, Bielza P, Tello J, Lacasa A (2011) Effect of biofumigation with manure amendments and repeated biosolarization on *Fusarium* densities in pepper crops. J Ind Microbiol Biotechnol 38:3–11

Matthiessen JN, Kirkegaard JA (2006) Biofumigation and enhanced biodegradation: opportunity and challenge in soilborne pest and disease management. Crit Rev Plant Sci 25:235–265

Mazzola M (2010) Management of resident soil microbial community structure and function to suppress soilborne disease development. Clim Change Crop Prod 1:200–218

Mazzola M (2011) Potential of biofumigation of soilborne pest control in strawberry. In: Proceedings of annual international research conference on methyl bromide alt emissions reductions, San Diego, CA, pp 47-1/47-2

Mazzola M, Manici LM (2012) Apple replant disease: role of microbial ecology in cause and control. Annu Rev Phytopathol 50:45–65

Mazzola M, Muramoto J, Shennan C (2012a) Transformation of soil microbial community structure in response to anaerobic soil disinfestation for soilborne disease control in strawberry. Phytopathology 102:S4.77

Mazzola M, Shennan C, Muramoto J (2012b) Application sequence and soil biology influence anaerobic soil disinfestation induced disease suppression. In: Proceedings of annual international research conference on methyl bromide alternatives and emissions reductions, Orlando, pp 57-1/57-3

McCarty DG, Inwood SEE, Ownley BH, Sams CE, Wszelaki AL, Butler DM (2014) Field evaluation of carbon sources for anaerobic soil disinfestation in tomato and bell pepper production in Tennessee. HortSci 49:272–280

Mehlich A (1984) Mehlich 3 soil test extractant: a modification of Mehlich 2 extractant. Commun Soil Sci Plant Anal 15:1409–1416

Messiha NAS, van Diepeningen AD, Wenneker M, van Beuningen AR, Janse JD, Coenen TGC, Termorshuizen AJ, van Bruggen AHC, Blok WJ (2007) Biological soil disinfestation (BSD), a new control method for potato brown rot, caused by *Ralstonia solanacearum* race 3 biovar 2. Eur J Plant Pathol 117:403–415

Momma N (2008) Biological soil disinfestation (BSD) of soilborne pathogens and its possible mechanisms. Japan Agric Res Q 42:7–12

Momma N, Usami T, Amemiya Y, Shishido M (2005) Factors involved in the suppression of *Fusarium oxysporum* f. sp. *lycopersici* by soil reduction. Soil Microorg 59:27–33

Momma N, Yamamoto K, Simandi P, Shishido M (2006) Role of organic acids in the mechanisms of biological soil disinfestation (BSD). J Gen Plant Pathol 72:247–252

Momma N, Momma M, Kobara Y (2010) Biological soil disinfestation using ethanol: effect on *Fusarium oxysporum* f. sp. *lycopersici* and soil microorganisms. J Gen Plant Pathol 76: 336–344

Momma N, Kobara Y, Momma M (2011) Fe2+ and Mn2+, potential agents to induce suppression of *Fusarium oxysporum* for biological soil disinfestation. J Gen Plant Pathol 77:331–335

Momma N, Kobara Y, Uematsu S, Kita N, Shinmura A (2013) Development of biological soil disinfestations in Japan. Appl Microbiol Biotechnol 97:3801–3809

Mowlick S, Hirota K, Takehara T, Kaku N, Ueki K, Ueki A (2012) Development of anaerobic bacterial community consisted of diverse clostridial species during biological soil disinfestation amended with plant biomass. Soil Sci Plant Nutr 58:273–287

Mowlick S, Inoue T, Takehara T, Kaku N, Ueki K, Ueki A (2013a) Changes and recovery of soil bacterial communities influenced by biological soil disinfestation as compared with chloropicrin-treatment. AMB Express 3:46

Mowlick S, Takehara T, Kaku N, Ueki K, Ueki A (2013b) Proliferation of diversified clostridial species during biological soil disinfestation incorporated with plant biomass under various conditions. Appl Microbiol Biotechnol 97:8365–8379

Mowlick S, Yasukawa H, Inoue T, Takehara T, Kaku N, Ueki K, Ueki A (2013c) Suppression of spinach wilt disease by biological soil disinfestation incorporated with *Brassica juncea* plants in association with changes in soil bacterial communities. Crop Prot 54:185–193

Mowlick S, Inoue T, Takehara T, Tonouchi A, Kaku N, Ueki K, Ueki A (2014) Usefulness of Japanese-radish residue in biological soil disinfestation to suppress spinach wilt disease accompanying with proliferation of soil bacteria in the Firmicutes. Crop Prot 61:64–73

Muramoto J, Shennan C, Koike ST, Bolda MP, Daugovish O, Dara SK, Klonsky K, Mazzola M, Zavatta M (2014) Optimizing anaerobic soil disinfestation for California strawberries. In: VIII international symposium chemical and non-chemical soil and substrate disinfestation, vol 1044, Torino, Italy, pp 215–220

Nielsen TH, Nielsen LP, Revsbech NP (1996) Nitrification and coupled nitrification-denitrification associated with a soil-manure interface. Soil Sci Soc Am J 60:1829–1840

Núñez-Zofío M, Larregla S, Garbisu C (2011) Application of organic amendments followed by soil plastic mulching reduces the incidence of *Phytophthora capsici* in pepper crops under temperate climate. Crop Prot 30:1563–1572

Nuñez-Zofío M, del Palacio SL, Garbisu C (2012) Repeated biodisinfection controls the incidence of Phytophthora root and crown rot of pepper while improving soil quality. Span J Agric Res 10:794–805

Okazaki H, Nose K (1986) Acetic acid and n-butyric acid as causal agents of fungicidal activity of glucose-amended flooded soil. Nihon Shokubutsu Byori Gakkaiho Ann Phytopathol Soc Japan

Pace B, Campbell LL (1971) Homology of ribosomal ribonucleic acid of diverse bacterial species with *Escherichia coli* and *Bacillus stearothermophilus*. J Bacteriol 107:543–547

Paulitz TC, Bélanger RR (2001) Biological control in greenhouse systems. Annu Rev Phytopathol 39:103–133

Poretsky R, Rodriguez-R LM, Luo C, Tsementzi D, Konstantinidis KT (2014) Strengths and limitations of 16S rRNA gene amplicon sequencing in revealing temporal microbial community dynamics. PLoS ONE 9, e93827

Rosskopf EN, Burelle NK, Mcsorely R, Skvarch, E (2009) Optimizing alternative fumigant applications for ornamental production in Florida. EDIS. ENY-901 (IN818). Available at http://edis.ifas.ufl.edu/in818

Rosskopf EN, Kokalis-Burelle N, Butler D, Muramoto J, Shennan C (2010) Development of anaerobic soil disinfestation for Florida vegetable and flower production. In: Proceedings of annual international research conference on methyl bromide alternatives and emissions reductions, Orlando, pp 15-1/15-2

Rosskopf EN, Hong J, Kokalis-Burelle N, Butler D (2012) Status of ASD development in Florida. In: Proceedings of annual international research conference on methyl bromide alternatives and emissions reductions, Orlando, pp 84-1/84-2

Rosskopf EN, Burelle N, Hong J, Butler DM, Noling JW, He Z, Booker B, Sances F (2014) Comparison of anaerobic soil disinfestation and drip-applied organic acids for raised-bed specialty crop production in Florida. In: VIII international symposium chemical and non-chemical soil and substrate disinfestation, vol 1044, Torino, Italy, pp 221–228

Runia WT, Molendijk LPG, Ludeking DJW, Schomaker CH (2012) Improvement of anaerobic soil disinfestation. Commun Appl Biol Sci 77:753–762, Ghent University

Runia WT, Thoden TC, Molendijk LPG, van den Berg W, Termorshuizen AJ, Stremińska MA, van der Wurff AWG, Feil H, Meints H (2014) Unravelling the mechanism of pathogen inactivation during anaerobic soil disinfestation. In: VIII international symposium chemical and non-chemical soil and substrate disinfestation, vol 1044, Torino, Italy, pp 177–193

Rütting T, Boeckx P, Müller C, Klemedtsson L (2011) Assessment of the importance of dissimilatory nitrate reduction to ammonium for the terrestrial nitrogen cycle. Biogeosciences 8: 1779–1791

Schwarzenbach R, Gschwend P, Imboden D (2002) Environmental organic chemistry, 2nd edn. Wiley, New York

Sexstone AJ, Revsbech NP, Parkin TB, Tiedje JM (1985) Direct measurement of oxygen profiles and denitrification rates in soil aggregates. Soil Sci Soc Am J 49:645–651

Shennan C, Muramoto J, Koike S, Bolda M, Daugovish O, Mochizuki M, Rosskopf EN, Kokalis-Burelle N, Butler DM (2011) Anaerobic soil disinfestation for suppressing *Verticillium dahliae* in strawberry production in California. HortSci 46:S174–S175

Shennan C, Muramoto J, Fennimore S, Mazzola M, Lazarovits G (2013) Non-fumigant strategies for soilborne disease control in California strawberry production systems. In: Cal strawberry commission annual production research report 2011–2012, pp 145–159

Shennan C, Muramoto J, Lamers J, Rosskopf EN, Kokalis-Burelle N, Mazzola M, Momma N, Butler DM, Kobara Y (2014) Anaerobic soil disinfestation for soil borne disease control in strawberry and vegetable systems: current knowledge and future directions. In: VIII international symposium on chemical and non-chemical soil and substrate disinfestation, vol 1044, pp 165–175

Shinmura A (2000) Causal agent and control of root rot of welsh onion. PSJ Soilborne Dis Workshop Rep 20:133–143

Shinmura A (2004) Principle and effect of soil sterilization method by reducing redox potential of soil. PSJ Soilborne Dis Workshop Rep 22:2–12

Shrestha U, Ownley BH, Rosskopf EN, Dee ME, Butler DM (2013) Optimization of amendment C:N ratio in anaerobic soil disinfestation for control of *Sclerotium rolfsii*. In: Proceedings of annual international research conference on methyl bromide alternatives and emissions reduction, San Diego, CA, pp 14.1–14.3

Shrestha U, Bruce A, Ownley BH, Butler DM (2014) Optimizingamendment C:N ratio for *Fusarium oxysporum* f. sp. *lycopersici* suppression under Anaerobic Soil Disinfestation. In: Proceedings of annual international research conference on methyl bromide alternatives and emissions reductions, Orlando, pp 53-1/53-3

ŠImek M, Cooper JE (2002) The influence of soil pH on denitrification: progress towards the understanding of this interaction over the last 50 years. Eur J Soil Sci 53:345–354

Society of American Bacteriologists (1923) Bergey's manual of determinative bacteriology. Williams and Wilkins, Baltimore

Stapleton JJ, Summers CG, Mitchell JP, Prather TS (2010) Deleterious activity of cultivated grasses (Poaceae) and residues on soilborne fungal, nematode and weed pests. Phytoparasitica 38: 61–69

Stevens RJ, Laughlin RJ, Malone JP (1998) Soil pH affects the processes reducing nitrate to nitrous oxide and di-nitrogen. Soil Biol Biochem 30:1119–1126

Stremińska MA, Runia WT, Termorshuizen AJ, Feil H, van der Wurff AWG (2014) Anaerobic soil disinfestation in microcosms of two sandy soils. Commun Agric Appl Biol Sci 79:15–18

Strong DT, Sale PWG, Helyar KR (1997) Initial soil pH affects the pH at which nitrification ceases due to self-induced acidification of microbial microsites. Soil Res 35:565–570

Tenuta M, Conn KL, Lazarovits G (2002) Volatile fatty acids in liquid swine manure can kill microsclerotia of *Verticillium dahliae*. Phytopathology 92:548–552

Thoden TC, Korthals GW, Termorshuizen AJ (2011) Organic amendments and their influences on plant-parasitic and free-living nematodes: a promising method for nematode management? Nematology 13:133–153

Uematsu S, Tanaka-Miwa C, Sato R, Kobara Y, Sato M (2007) Ethyl alcohol as a promising material of reductive soil disinfestation for controlling root knot nematode and soilborne plant diseases. In: Proceedings of annual international research conference on methyl bromide alternatives and emissions reductions, San Diego, CA, pp 75-1/75-3

van Bruggen A, Blok W (2014) Soil disinfestation and food safety issues. In: VIII international symposium on chemical and non-chemical soil and substrate disinfestation, vol 1044, pp 437–445

van Overbeek L, Runia W, Kastelein P, Molendijk L (2014) Anaerobic disinfestation of tare soils contaminated with Ralstonia solanacearum biovar 2 and Globodera pallida. Eur J Plant Pathol 138:323–330

Wallace RJ, Falconer ML, Bhargava PK (1989) Toxicity of volatile fatty acids at rumen pH prevents enrichment of *Escherichia coli* by sorbitol in rumen contents. Curr Microbiol 19: 277–281

Weller DM, Raaijmakers JM, McSpadden Gardener BB, Thomashow LS (2002) Microbial populations responsible for specific soil suppressiveness to plant pathogens. Annu Rev Phytopathol 40:309–348

Whitmore AP (1996) Modelling the release and loss of nitrogen after vegetable crops. Neth J Agric Sci 44:73–86

Woese CR (1987) Bacterial evolution. Microbiol Rev 51:221–271

Xu JM, Tang C, Chen ZL (2006) The role of plant residues in pH change of acid soils differing in initial pH. Soil Biol Biochem 38:709–719

Yoder AJ (2014) Assessing the impact of biofumigation and anaerobic soil disinfestation on soil biology, nitrogen cycling, crop establishment and yield in vegetable cropping systems. Michigan State University

Yossen V, Zumelzu G, Gasoni L, Kobayashi K (2008) Effect of soil reductive sterilisation on *Fusarium* wilt in greenhouse carnation in Córdoba, Argentina. Austr Plant Pathol 37:520–522

Zavatta M, Shennan C, Muramoto J, Mazzola M (2014) Evaluating C-sources for anaerobic soil disinfestation. In: 2014 proceedings of annual international research conference on methyl bromide alternatives and emissions reductions, Orlando, pp 12-1/12-4

Zhou A, Takaya N, Nakamura A, Yamaguchi M, Takeo K, Shoun H (2002) Ammonia fermentation, a novel anoxic metabolism of nitrate by fungi. J Biol Chem 277:1892–1896

Zhu X, Burger M, Doane TA, Horwath WR (2013) Ammonia oxidation pathways and nitrifier denitrification are significant sources of N2O and NO under low oxygen availability. PNAS 110:6328–6333

Chapter 14
Bio-intensive Management of Fungal Diseases of Fruits and Vegetables Utilizing Compost and Compost Teas

Yasmeen Siddiqui, Yuvarani Naidu, and Asgar Ali

14.1 Introduction

As modern agriculture struggles to support the booming global population, plant diseases contribute to a major setback in quantity and quality of food including vegetable and fruit production worldwide. The losses may be catastrophic or chronic but estimated to be more than 40% of the total production. Crop losses tend to be greatest in tropical countries where environmental conditions are particularly favorable and knowledge and investments in crop health management are minimal.

Diseases specifically caused by fungal pathogens affect plants right from the planting stage to harvesting and storage of produce. Largely, farmers rely heavily on chemical fungicides to minimize the disease pressure. Modern fungicides, however, are organic compounds, with a high degree of specificity toward their target organism. They also generally exhibit low overall toxicity and have little immediate impact on the environment. Despite the positive results of the use of modern fungicides, concern continues to be expressed about the wisdom of using large quantities of chemicals in the environment. Methyl bromide fumigation, for example, not only destroyed beneficial microorganisms, such as mycorrhizae, biocontrol agents, and plant growth-promoting microorganisms but also is a potent contributor in ozone layer depletion. For this reason it and many more are scheduled to be phased out internationally under the Montreal protocol (UNDP 2003). The

Y. Siddiqui (✉) • Y. Naidu
Laboratory of Food Crops, Institute of Tropical Agriculture, University Putra Malaysia, Serdang 43400, Selangor, Malaysia
e-mail: yasmeen@upm.edu.my

A. Ali
Faculty of Science, School of Bioscience, The University of Nottingham Malaysia Campus
Jalan Broga, Semenyih 43500, Selangor Darul Ehsan, Malaysia

M.K. Meghvansi, A. Varma (eds.), *Organic Amendments and Soil Suppressiveness in Plant Disease Management*, Soil Biology 46,
DOI 10.1007/978-3-319-23075-7_14

recent drift to near-zero market tolerance for pesticide residues in fresh leafy vegetables and fruits provides an additional stimulus to search for nonchemical means to control pests and diseases (Reuveni et al. 2002). These issues are motivating increased interest in the disease-suppression benefits of organic products including compost.

Composting has long been recognized as one of the most cost-effective area in agricultural biotechnology, which not only minimizes organic waste production but also is an environmentally sound alternative for recycling of substrates (Siddiqui et al. 2009). From decades, compost is known for its outstanding fertilizer and soil-conditioner characteristics (Hoitink et al. 1993; Lamondia et al. 1998; Siddiqui et al. 2008b). A possible application about which little was known scientifically is the use of water extracts from compost to control plant diseases and as inoculants to restore or enhance soil and leaf microflora.

The biological control of leaf diseases and emphasis on antifungal properties of watery extracts of compost is evident since 1986 (Weltzien and Ketterer 1986). In addition, it has been reported that compost teas improved soil fertility and quality by altering the physical and chemical properties of the soil, such as increasing organic matter content, water-holding capacity, and diversity of microbes and providing available micro- and macronutrients essential for plant growth and ultimately improve the yield (Stoffella et al. 1997; Scheuerell and Mahaffee 2004; Siddiqui et al. 2008a, 2009).

Based on several studies it is well established that the introduction of compost and compost teas can be merged with integrated biocontrol strategies, offering an alternative and attractive approach for disease control to minimize the negative impact of chemicals and maintain a sustainable productivity in intensive vegetable and fruit production systems. This chapter highlights the potentiality of harnessing microbial diversity utilizing compost and compost teas for mitigation of fungal diseases of fruits and vegetables in an eco-friendly manner.

14.2 Compost and Compost Teas for Plant Disease Suppression

14.2.1 *Disease-Suppressive Compost*

The importance of compost in suppression of soilborne diseases in container media was first documented by Hoitink et al. (1977). These initial findings triggered the cascade of studies worldwide in search for the different types of suppressive compost (Hadar and Mandelbaum 1986; Craft and Nelson 1996; Ryckeboer 2001).

Compost prepared from heterogeneous organic wastes (vegetable fruit and garden materials) may have highly suppressive effects against a range of diseases that cause severe losses in many crops and are difficult to control (Postma and Kok 2003). However, not all composts suppress plant diseases with similar efficacies.

For instance, olive and grape marc compost consistently suppressed *Fusarium oxysporum* f. sp. *dianthi* and f. sp. *lycopersici* with high degree, whereas *Rhizoctonia solani* was suppressed moderately. In contrast, cork compost suppressed *R. solani* with high degree, while Fusarium wilt was suppressed with moderate intensity (Borrero et al. 2004, 2009; Trillas et al. 2006). Therefore, compost producers should aim for high-quality "tailored" compost, targeting specific cropping system with high degree of suppression. Vice versa, combinations of diverse types of materials such as manure, lignin-containing materials, and green wastes could be utilized for the development of compost, aiming for broad-spectrum product, in view of growers. This will also minimize the economic pressure of otherwise excessive waste material. Summaries of some compost studied on their effect on pathogens and diseases are listed in Table 14.1.

14.2.2 Compost Teas

An increasing body of experimental evidence indicates that in addition to compost, plant disease suppression could also be achieved by applying a variety of water-based compost preparations (Weltzien 1991).

Compost tea is an aqueous solution that results from the extraction of micro-organisms, fine particulate organic matter, and soluble chemical components of compost (NOSB 2006). Water extracts from compost are recognized by organic growers and researchers through proliferation of preparation methodologies and terminologies (Brinton 1995), though majority referred the end product as compost tea. The first experiment involving the direct application of compost tea on above-ground plant parts was reported by Weltzien and Ketterer (1986). They treated detached grapevine leaves with extracts from horse manure compost. When leaves were later inoculated with suspension of sporangia of downy mildew fungus of grapevines, *Plasmopara viticola*, they showed a highly significant reduction in the diseased area. Subsequently, the potential of compost teas for plant disease suppression and control was attempted more systematically, however, with different response mechanisms (Mcquilken et al. 1994; Yohalem et al. 1996).

It was suggested that watery fermentation extracts of well-composted organic materials reduced disease incidence and severity in various host-pathogen combinations, if applied prophylactically to plant surfaces. When primary leaves of barley were pretreated with the compost extract from horse manure and then inoculated with conidia from powdery mildew (*Erysiphe graminis*), infection levels were reduced by an average of 55% (Weltzien and Ketterer 1986; Budde and Weltzien 1988). Further, detailed study on powdery mildew of sugar beet (*Erysiphe betae*) and of cucumber (*Sphaerotheca fuliginea*) showed that the stages of fungal development were heavily affected by the types of compost extracts. Conidia germination was equal to the control but the formation of secondary hyphae was reduced by more than 50% (Samerski and Weltzien 1988a, b).

Table 14.1 Summary of some compost studied on their effect on plant pathogens and diseases

Plant disease	Pathogen	Crop	Compost type	References
Damping-off	*Rhizoctonia solani*	Radish	Broiler litter and leaf compost; dairy manure with leaf compost; steer/ horse manure compost; Promix	Ringer et al. (1997)
		Cucumber	Vegetable fruit and garden waste	Tuitert et al. (1998)
			Cork, olive marc, grape marc, and spent mushroom compost	Trillas et al. (2006)
		Cabbage	Manure, bark, vermicompost, yard trimmings	Scheuerell et al. (2004)
Damping-off	*Pythium ultimum*	Peas	Garden organics/garden waste and biowastes bark and grape marc	Erhart et al. (1999)
		Cucumber	Peat with different levels of decomposition and bark	Inbar et al. (1991)
			Peat mixtures with different levels of decomposition	Boehm and Hoitink (1992)
			Peat moss amended with composted swine wastes at different weeks of maturity	Diab et al. (2003)
			Manure, bark, vermicompost, yard trimmings	Scheuerell et al. (2004)
	Pythium aphanidermatum	Cucumber	Composted licorice roots	Hadar and Mandelbaum (1986)
	Pythium irregulare		Manure, bark, vermicompost, yard trimmings	Scheuerell et al. (2004)
Phytophthora root rot	*Phytophthora cinnamomi*	Avocado plantation mulch	Organic much (oat straw + mature chicken manure) applied in soil	You and Sivasithamparam (1995)
	Phytophthora nicotianae	Citrus	Composted municipal waste amendment of citrus soils	Widmer et al. (1998)
Phytophthora crown rot and leaf blight	*Phytophthora capsici*	Cucumber	Compost of sawdust and cow manure	Khan et al. (2004)
Phytophthora root and crown rot		Bell pepper	Composted sewage sludge with garden organic; wood chips; commercial humate; crab shell waste; composted MSW; composted paper; composted perennial	Kim et al. (1997)

(continued)

Table 14.1 (continued)

Plant disease	Pathogen	Crop	Compost type	References
			peanuts; composted seed peanuts separately	
Fusarium wilt	*Fusarium oxysporum* f. sp. *conglutinans*	Radish	Hardwood bark	Trillas-Gay et al. (1986)
	Fusarium oxysporum f. sp. *lycopersici*	Tomato	Grape marc and cork compost	Borrero et al. (2004)
			Commercial compost made from mixture of vegetable and animal market wastes and sewage sludge in tunnel system	Cotxarrera et al. (2002)
	Fusarium oxysporum f. sp. *radicis-lycopersici*	Tomato	Pulp and paper mill	Pharand et al. (2002)
	Fusarium oxysporum f. sp. *melonis*	Melon	Compost from tomato plants and cow manure	Saadi et al. (2010)
Fusarium root and stem rot	*Fusarium oxysporum* f. sp. *radicis-cucumerinum*	Cucumber	Dairy solids composted in windrows, dairy solids composted by worms and vegetable refuse composted aerobically	Kannangara et al. (2000)
Verticillium wilt	*Verticillium* spp.	Tomato	Cork compost and light peat	Borrero et al. (2002)
Southern blight	*Sclerotinia rolfsii*	Bean	Mature biosolid compost (sewage sludge and yard waste)	Danon et al. (2007)
		Tomato and soya bean	Powders of kudzu (*Pueraria lobata*), velvet bean (*Mucuna deeringiana*), and pine bark	Blum and Rodríguez-Kábana (2004)
		Beans	Composted grape marc, cattle manure	Gorodecki and Hadar (1990)
Collar spot		Chickpea	Composted grape marc, cattle manure	Gorodecki and Hadar (1990)

In farm trials the effects of compost teas from different sources were tested on a variety of crops. No effect of compost tea application on early blight of tomato was observed, whereas lettuce damping-off incidence was reduced in the summer but not in the spring crop. Postharvest fruit rot of blueberries was significantly reduced. Spinach yield decreased, but broccoli yield increased (Granatstein 1999). It is apparent that impacts on plant health and yield can be crop specific and general

Table 14.2 Summary on the efficacy of compost water extracts or teas in suppressing foliar and soilborne diseases of vegetable and fruit crops

Plant disease	Crop	Pathogen	Type/Source	References
Apple scab	Apple	*Venturia inaequalis*	Spent mushroom	Cronin et al. (1996)
			Spent mushroom and cattle manure	Andrews (1993)
			Spent mushroom	Yohalem et al. (1994, 1996)
Bacterial spot	Tomato	*Xanthomonas vesicatoria*	Cow manure, composted pine bark	Al-Mughrabi et al. (2008)
Wet rot	Okra	*Choanephora cucurbitarum*	Empty fruit bunches of oil palm and rice straw compost	Siddiqui et al. (2008a, 2009)
Late blight	Potato	*Phytophthora infestans*	Thermal compost, static wood chip compost, and vermin castings	Al-Mughrabi (2007)
Damping-off	Cucumber	*Pythium ultimum*	Yard trimmings, vermin compost, and tea compost	Scheuerell and Mahaffee (2004)
			Bovine, sheep, chicken manure, shrimp, and seaweed composts	Dionne et al. (2012)
		Pythium aphanidermatum	Solid olive mill wastes, *Posidonia oceanica*, and chicken manure	Jenana et al. (2009)
Phytophthora blight	Pepper	*Phytophthora capsici*	Pig, cow, and poultry manure, sawdust, livestock waste, dregs of oil and lees	Sang et al. (2010)
Anthracnose	Pepper	*Colletotrichum coccodes*	Pig, cow, and poultry manure, sawdust, livestock waste, dregs of oil and lees	Sang and Kim (2011)
	Cucumber	*Colletotrichum orbiculare*	Pig, cow, and poultry manure, sawdust, livestock waste, dregs of oil and lees	Sang and Kim (2011)
Downy mildew	Grapes	*Plasmopara viticola*	Horse-straw soil	Ketterer (1990)
			Fresh cow dung soil	Achimu and Schlösser (1992)
	Grapes	*Uncinula necator*	Horse manure and cattle manure	Sackenheim (1993)

(continued)

Table 14.2 (continued)

Plant disease	Crop	Pathogen	Type/Source	References
Powdery mildew	Cucumber	*Sphaerotheca fuliginea*	Not stated	Samerski and Weltzien (1988a)
	Sugar beet	*Erysiphe betae*	Not stated	Samerski and Weltzien (1988b)
	Barley	*Erysiphe graminis*	Animal-manure-straw compost	Weltzien (1989)
	Melon	*Erysiphe cichoracearum* DC.	Empty fruit bunches of oil palm	Naidu et al. (2012, 2013)
	Bean	*Erysiphe polygoni*	Not stated	Ketterer and Schwager (1992)
	Tomato		Composted market and garden wastes	Segarra et al. (2009)
	Apple	*Podosphaera leucotricha*	Not stated	Pscheidt and Wittig (1996)
Gray mold	Lettuce	*Botrytis cinerea*	Horse bedding, chicken litter	McQuilken et al. (1994)
	Geranium		Various	Scheuerell and Mahaffee (2006)
	Bean		Cattle, horse manure, horse-straw soil	Urban and Trankner (1993)
			Horse bedding, chicken litter	McQuilken et al. (1994)
	Strawberry		Cattle, chicken manure	Welke (2004)
Gray mold	Grape	*Botrytis cinerea*	Horse-straw soil	Ketterer et al. (1992)
	Grape berries		Horse-straw soil	Ketterer et al. (1992)
	Tomato		Sheep manure	Koné et al. (2010)
	Grape berries		Horse, sheep, cattle manures and plant source (olive)	Hmouni et al. (2006)
	Tomato		Grape marc, cattle manure	Elad and Shtienberg (1994)

inferences about disease suppression or yield cannot be made. A summary of few selected studies was done on the efficacy of compost water extracts or teas in suppressing foliar and soilborne diseases of vegetable and fruit crops as tabulated (Table 14.2).

14.3 Factors Involved in the Suppressive Efficacy of Compost and Compost Teas

Regardless of the various efforts to find elements of disease suppressiveness, the general understanding of what determines the suppressiveness of compost is still in its infancy. Nevertheless, it is expected that disease suppressiveness of compost is most likely due to the interaction of various biotic and abiotic factors. The following sections will discuss few of the characteristics of compost and compost teas which may have a role in disease-suppressive efficacy.

14.3.1 Composting Process and Compost Maturity

Composting can be defined as the biological decomposition and stabilization of organic substrates, under conditions that allow development of thermophilic temperatures ranging from 35 to 75 °C as a result of biologically produced heat (Metacalf and Eddy 1991). The composting process is often divided into three phases signifying the microbial succession. The first phase of rapid composting is characterized by high temperatures usually 40–50 °C, when sugars and easily biodegradable substances are degraded. During the second phase, when high temperature 55–77 °C prevails, less biodegradable substances are destroyed. Thermophilic microorganisms predominate during this phase of the process. The heat generated during this high-temperature phase kills plant pathogens and weed seeds (Bollen 1993; Farrell 1993). This is followed by curing and maturation phase where the temperature gradually drops to environmental temperature and the compost is recolonized with mesophilic bacteria and fungi; decomposition continues but at a very slow rate.

Appropriate curing is essential not only to stabilize the compost and to eliminate or to reduce negative plant responses but also is crucial in determining rate of disease suppression. Compost maturity refers to the phytotoxicity associated with the compost and is defined as the degree of biodegradation at which composts generally release higher levels of soluble mineral nutrients, phytotoxic organic acids, and heavy metals than immature materials (Griffin and Hutchinson 2007). Some of these phytotoxic compounds include salts, ammonia, heavy metals, and organic acids that affect the growth of agricultural crops and predispose them to pest and pathogen attack (Hoitink and Boehm 1999).

It is well established that compost must be of steady quality to be used successfully in biological control of horticultural crops especially in container media (Inbar et al. 1993). Hadar and Mandelbaum (1986) demonstrated that the immature compost was ineffective in suppressing damping-off caused by *Pythium aphanidermatum* in cucumber, whereas mature compost could. The immature compost does not support biocontrol activities (Hoitink et al. 1991), even when inoculated with the best strains. High concentrations of free nutrients (glucose, amino acids, etc.) in fresh crop residues suppress the production of enzymes such as chitinase, cellulose, β-1,3-glucanase, etc. required for parasitism by biocontrol agents such as *Trichoderma* sp. (Hoitink et al. 1993; Chung et al. 1998). Green composted hardwood bark (CHB), which is high in cellulose content, has been shown to be conducive to *Rhizoctonia* damping-off, even though it may be colonized by 10^8 colony-forming units (cfu) g^{-1} of dry weight of antagonistic *T. harzianum* since lytic enzymes responsible for parasitic activity was repressed due to high glucose content and does not exert biological control over *R. solani*, whereas in mature CHB (low in cellulose), the same antagonist renders the medium suppressive. On the contrary, excessively cured composts may lack or have inconsistent disease-suppressive properties. They may also be excessively high in salts and have inferior physical structures, which ultimately will affect the efficacy of the compost (Nelson et al. 1983). Compost used for tea production should be certified free of human pathogens and residual herbicides and is fully mature and cured (Pan et al. 2012). The effectiveness of the compost tea also depends on the raw materials of the compost as well as on the extraction conditions that affect the microbial population density and end product.

14.3.2 Beneficial Microorganisms

Composts are usually pathogen-free due to buildup of high temperatures during thermophilic phase of composting process. Not only pathogens but also beneficial organisms are also either killed or inactivated (Noble and Roberts 2004). Therefore, ability to suppress pathogens and/or diseases is usually induced during curing since most of the biocontrol agents also recolonize compost. This fact has been introduced from the very beginning by Hoitink and colleagues, who observed that suppressive efficacy was reduced or eliminated by heating the compost at 60 °C or by gamma irradiation (Trillas-Gay et al. 1986). However, the suppressive potential could be restored by reintroducing the mixture of microorganisms and a specific organism or amendment of suppressive compost (Trillas et al. 2006; Dukare et al. 2011).

Similarly, microbial composition and the presence of pathogen-suppressive microbial metabolites are the most reported factor influencing the efficacy of compost teas in inhibiting the development of plant pathogens (Koné et al. 2010). Despite their importance, there is very limited understanding of the microbial composition of compost teas and how these organisms can survive on plant surfaces (Scheuerell and Mahaffee 2002). In general, the dominant functional groups isolated from microbial-enriched compost tea were from the genera *Bacillus* sp., *Pseudomonas* sp., *Micrococcus*, *Staphylococcus*, *Burkholderia*, and *Clavibacter*, lactic acid bacteria (*Lactobacillus*), other bacterial species, (Naidu et al. 2010) actinomycetes, yeast, *Trichoderma* sp., *Aspergillus* sp., *Penicillium* sp., and other fungal species (Siddiqui et al. 2009; Naidu et al. 2012).

The study carried out by Siddiqui et al. (2009) demonstrated the role of microbial community in compost tea on suppression of *Choanephora cucurbitarum* causing wet rot of okra. The findings indicated that inhibitory efficacy of compost tea produced from rice straw (RST) and empty fruit bunch (EFB) compost was reduced significantly when the teas were subjected to Millipore membrane filters or heat sterilization. In comparison, the mycelial growth of *C. cucurbitarum* was reduced by 100% in plates amended with both the non-sterilized compost tea. It is hard to determine the involvement of specific microbes in the suppression of phytopathogens by compost teas since a consortium of microbial community is involved rather than a single species (Naidu et al. 2010).

14.3.3 Brewing of Compost Tea

Two principal approaches being endorsed in compost tea production are aerated compost tea (ACT) and non-aerated compost tea (NCT), depending on the degree of aeration given to the system (Scheuerell and Mahaffee 2002).

An array of experimental methods has been utilized, namely, in vitro inhibition, seedling assay, detached leaves, growth chamber, green houses, and field studies to determine their efficacy. For instance, ACTs and NCTs produced from plant residues (rice ash, bean straw, and vegetative fruit waste) and chicken manure significantly reduced in vitro conidial germination and fungal growth of early blight (*Alternaria solani*) in tomato and purple blight (*A. porri*) in onion. Moreover, field evaluations conducted over 2 years resulted in obvious suppression of Alternaria blight infection by NCT treatment compared to ACT treatment. The authors claimed that NCT contained denser biodiversity of microbial biomass than ACT which could be the reason of better performance by NCT in field trials (Haggag and Saber 2007). Correspondingly, non-aerated compost inhibited the in vitro mycelial growth of tomato pathogens, namely, *Alternaria solani*, *B. cinerea*, and *Phytophthora infestans* when compared to the water control (Koné et al. 2010).

In more recent findings, Siddiqui et al. (2008b) observed that disease severity of *Choanephora* wet rot disease on okra was lowest in plants treated with aerated *Trichoderma*-fortified rice straw compost extracts and simultaneously reduced the disease incidence. It was demonstrated that foliar application of ACT eradicated 100% naturally occurring powdery mildew pathogen (*Erysiphe polygoni*) on tomato leaves (Segarra et al. 2009). Similar findings were reported by Naidu et al. (2013), whereby foliar application of microbial-enriched compost tea (ACT) resulted in the reduction of powdery mildew (*Golovinomyces cichoracearum* DC.) severity on melon crops. Conversely, Pscheidt and Wittig (1996) did not observe significant control of powdery mildew of apple or grape, apple scab, pear scab, brown rot of peach, peach leaf curl, and cherry leaf spot when aerated compost tea was applied in the field at regular intervals. Only brown rot blossom blight of sweet cherry caused by *Monilinia laxa* was significantly reduced. The authors concluded that storing the aerated compost tea for 12–15 h might have negatively influenced the observed level of suppression for all host-pathogen combinations.

Investigations on the effectiveness of compost teas showed that the extraction time and the compost-to-water ratio also have a significant effect on its biological activity against plant pathogens. Numerous studies have indicated that suppressive activities of NCTs were increased with fermentation time to a maximum and then decline (Scheuerell and Mahaffee 2000). However, most scientists worked with extraction times between 3 and 10 days. More recently, Hmouni et al. (2006) demonstrated that compost tea significantly reduced the severity of gray mold on tomato as compared to the control with fermentation period of 7 and 15 days. This duration period was in line with Elad and Shtienberg (1994) who stated that the optimal fermentation time was longer than 10 days. In addition, the best effects were noticed with the concentration of 1:2, compost to water (Sackenheim 1993). A relationship of 1:5 was found good for economical and practical purposes in past studies.

14.3.4 Additives

Additives are usually mixtures of different amounts of various microorganisms, mineral nutrients, or readily available forms of carbon, enzymes, and pH-balancing compounds that are meant to enhance microbial activity (Himanen and Hänninen 2009). The primary goal of disease-suppressive compost and compost tea production is to increase the microbial populations. The final balance between bacteria and fungi in compost tea can be achieved by providing additives for the microbes at the beginning/curing of composting or during/after the tea fermentation process (Weltzien 1991; Ingham 2000b). The fermentation nutrients can be classified into two different classes: bacteria additive and fungal additive. Basically, molasses, fruit pulp, juices, proteins, and fish emulsion or fish hydrolysate are commonly termed as bacterial additives, whereas sloughed root cells and dead plant tissue which often supply the more complex carbon substrates that fungi require such as humic acids, seaweed extract (kelp powder), and rock dust are reputed to increase fungal population (Ingham 2000a, b).

The most consistent formulation of ACT)was obtained with kelp, humic acid, and rock dust for the suppression of damping-off caused by *P. ultimum* on cucumber seedlings (Scheuerell and Mahaffee 2004). The authors concluded that the bacterial populations in the ACT were significantly enhanced with addition of nutrient additives. Similarly, compost tea prepared with the addition kelp, humates, rock dusts, grain, and soluble plant sugar sources and liquefied fish prior to brewing process significantly increases the number of stems produced and also significantly inhibits *Helminthosporium solani* and *Rhizoctonia solani*, causal agents of diseases on potato tuber (Al-Mughrabi 2006).

14.4 Mechanisms Involved in Disease Suppression

Compost provides natural biological control for several plant diseases and its water extracts as substitutes for synthetic fungicides (Zhang et al. 1998). Therefore, understanding of mechanism for disease control by compost or its water extracts is crucial to enhance the suppressive effect. In general, biological characteristics of disease suppression can include one or a combination of mechanisms such as competition for nutrients, production of antibiotics or antibiosis, production of lytic and other extracellular enzymes and compounds, predation and direct parasitism, and induced plant resistance (Ketterer 1990; Lorito et al. 1994; Brinton 1995).

Beneficial microorganisms including bacteria (*Bacillus*, *Pseudomonads*), actinomycetes, and fungi (*Trichoderma*, *Gliocladium*) present in compost and its extracts can induce all the four mechanisms associated with disease suppression. For example, fluorescent *Pseudomonads* are the most frequently used plant growth-promoting *Rhizobacteria* that function by suppressing the growth of detrimental rhizosphere microflora present in most soils (Laha et al. 1992). Production of antifungal metabolites, such as antibiotics and siderophore-mediated iron competition, are primary mechanisms by which these bacteria suppress diseases. Siderophores are biosynthetic compounds that are produced under iron-limiting conditions. They serve to chelate the ferric ion (Fe^{3+}) from the environment into microbial cells and reduce the iron availability for the pathogens (Kloepper et al. 1999; Siddiqui et al. 2009). The presence of siderophores was detected in various grape marc aerated compost tea and their suppressive effect on nine selected soilborne pathogens was investigated by Diánez et al. (2006). They concluded that the presence of microorganisms in grape marc compost secreted siderophores into the agar medium that was responsible for inhibiting the growth of the nine tested fungi. Siderophores produced by this microflora play a vital role in nutrient competition among plant pathogens and beneficial microorganisms for the infection site.

Direct inhibition of both conidial germination and mycelium growth of various plant pathogens by beneficial microorganisms that belong to different functional groups, such as *Bacillus*, *Pseudomonas*, lactic acid bacteria, actinomycetes, and fungi (predominantly *Trichoderma* spp. and *Penicillium* spp.) present in the water extract of compost, is well documented by numerous researchers (McQuilken et al. 1994; El-Masry et al. 2002; Siddiqui et al. 2009; Naidu et al. 2010). For instance, Sang and Kim (2011) elucidated that compost water extracts significantly inhibited in vitro conidial germination and appressorium formation of *Colletotrichum coccodes* and *C. orbiculare*, the causal pathogens of anthracnose on pepper and cucumber, respectively. It was suggested that increased populations of beneficial microorganisms could more effectively compete for phylloplane nutrients and niches, leading to a reduction in pathogen infection (Blakeman 1975). These findings were in line with those who elucidated that the competition for nutrients and space by microorganisms in EFB and RST compost teas was likely to be the reason for the greater inhibition of *C. cucurbitarum*, as percentage inhibition of radial growth (% PIRG) was significantly reduced by filter sterilizing the teas. Moreover, this phenomenon has also prevented the formation of a germ tube and led to the lysis of *C. cucurbitarum* conidia (Siddiqui et al. 2009). Similarly, the beneficial microorganisms present in microbial-enriched compost tea contributed to the in vitro conidial germination inhibition of *G. cichoracearum* DC. as reported by Naidu et al. (2012). An in vitro study conducted using various compost extracts in suppressing the radial growth of some phytopathogenic fungi, namely, *Sclerotium bataticola*, *F. oxysporum* f. sp. *lycopersici*, *F. solani*, *F. graminearum*, *Alternaria* sp., *C. coccodes*, *B. cinerea*, *Sclerotinia sclerotiorum*, *A. niger*, *Rhizoctonia solani*, *R. bataticola*, *Pythium* sp., and *Verticillium dahliae*, has been reported (El-Masry et al. 2002; Kerkeni et al. 2007).

In addition, some strains of *Trichoderma* may produce nonvolatile antibiotics that inhibit and, presumably, predispose host hyphae to infection before contact occurs (Merrill and McKeon 2001). As *Trichoderma* recognizes the host, it attaches itself to the host and then either grows along the host hyphae or coils around them and secretes lytic enzymes (chitinase and hydrolases). It has been shown that chitinolytic enzymes isolated from *T. harzianum* inhibit spore germination and hyphal (germ tube) elongation in several plant pathogens (Harman et al. 1993; Claudia et al. 1997).

Several studies have also determined that antibiosis could be the mechanism of suppression based on observations that filter- or heat-sterilized compost teas retain suppressive qualities (Elad and Shtienberg 1994; Yohalem et al. 1994; Cronin et al. 1996). Cronin et al. (1996) elucidated that antibiosis was the mechanism of inhibition for the in vitro conidia germination of *Venturia inaequalis* by spent mushroom compost extracts. When the compost was sterilized and then fermented, no suppressive activity was found. However, fermented non-sterilized compost extracts had equally suppressive activity after 0.1 μm filtration, and most of the suppressive activity was maintained after autoclaving. Using micro-concentrators, the major inhibitory agents were determined to be a low-molecular-weight (<3 kDa), heat-stable, nonprotein metabolite produced by microorganisms during fermentation. On the contrary, Siddiqui et al. (2009) demonstrated the heat sterilization of compost teas (RST and EFB) completely loses suppressive activity against *C. cucurbitarum*. However, micromembrane filtration of the teas maintains the suppressive activity but is of less significant efficacy when compared to non-sterilized compost teas.

Removing the microbial component of compost extracts can have negative impact on the suppressive properties. Filter or heat sterilization results in the loss of disease suppression; hence, it has been concluded that microbial competition for nutrients or space is the mode of action. Plant-pathogen systems demonstrating experimental evidence to support this conclusion include *Uncinula necator* and *Plasmopara viticola* on grapes (Weltzien 1989), *Phytophthora infestans* on tomato and potato (Weltzien and Ketterer 1986), *Botrytis cinerea* on beans (Stindt 1990) and strawberries (Urban and Trankner 1993), and *C. cucurbitarum* on okra (Siddiqui et al. 2009). However, it is not clear whether pathogen inhibition is due to parasitism, competition for nutrients and colonization sites, or production of antibiotics after establishing on the plant surface.

14.4.1 Induction of Plant Disease Resistance

All plants possess resistant mechanism, which can be enhanced as a reaction of plants to various biotic (pathogens, non-pathogens, and beneficial microorganisms) or abiotic (chemicals, metabolites produced by the beneficial organisms, physical stress, etc.) stimuli and is called induced resistance (Sticher et al. 1997). It remains unknown how compost induced resistance in plants. However, plants in compost-amended substrates are colonized by a variety of microorganisms from which strains capable of inducing resistance in plants have been described (Wei and Kloepper 1991; Maurhfiofer et al. 1994; Liu et al. 1995; De Meyer et al. 1998). Such specific strains must be present above a certain threshold population size in the rhizosphere to induce this effect. Once resistance is induced, the population size apparently may decline without affecting resistance (Raaijmakers et al. 1995).

The literature has suggested that compost may induce resistance as an additional biocontrol mechanism against both foliar and root diseases (Zhang et al. 1996, 1998; Kavroulakis et al. 2005; Ntougias et al. 2008). It has also been postulated that the defensive genes encode for several enzymes responsible for induction of resistance in plants including peroxidase (PO) and polyphenol oxidase (PPO) that catalyzes the formation of lignin, while phenylalanine ammonia lyase (PAL) is involved in phytoalexin and phenolic biosynthesis. Goldstein (1998) reported that composts and compost extracts activate disease resistance genes in plants. These genes are usually activated in response to the presence of a pathogen; they mobilize chemical defense against the pathogen invasion. Inducible enzymes such as PO, PPO, and PAL activities were observed to increase in okra plants pretreated with non-sterilized compost tea and challenge inoculated with the pathogen. The induction was correlated with the delay in the development of *Choanephora* wet rot, confirming the possible involvement of induced resistance (Siddiqui et al. 2009). The authors elucidated that PPO induction was a consequence of the biological and chemical activity of the compost teas that might have resulted in the formation of quinones due to oxidation of phenolic compounds, which are more toxic to pathogen than the original phenolics (Kazana et al. 1998).

On the other hand, Zhang et al. (1998) observed the induction of systemic acquired resistance (SAR) in cucumber and *Arabidopsis* by using compost and compost extracts. The authors proposed that β-D-glucuronidase (GUS) activity was induced either with topical sprays of compost water extract or salicylic acid (SA) in non-inoculated cucumber plants with pathogen, namely, *Colletotrichum orbiculare*, suggesting that compost-induced disease suppression more than likely involved in the potentiation of resistance responses, rather than their activation, and also compost-induced SAR differed from SAR induced by pathogens, SA, or compost water extract.

Table 14.3 Summary of compost and compost tea studies on their effect as inducer of plant disease resistance

Compost/compost tea material	Causal pathogen	Disease suppression	Observed effects	References
Spruce and pine bark	*Colletotrichum orbiculare*, *Pythium ultimum*, and *Pythium aphanidermatum*	Foliar anthracnose of cucumber *Pythium* root rot	Reduced root rot severity in split-root plant grown in compost than those produced in peat. Increased Peroxidase activity and enhanced peroxidase isozyme levels in plants	Zhang et al. (1996)
Pine bark fortified with *Trichoderma hamatum* 382 and *Pantoea agglomerans* E278As and their water extract in (1:1 v/v)	*Colletotrichum orbiculare*	Foliar anthracnose of cucumber	Induced SAR. Peroxidase, β-1,3-glucanase and GUS activities were higher after challenge inoculation with pathogen. Induced SAR might be different from salicylic acid-induced resistance	Zhang et al. (1998)
Extracted olive press cakes and grape marc (GM)	*Septoria lycopersici*	Tomato leaf spot	Increased expression of PR genes in the roots of tomato plants, even in the absence of any pathogen. The expression of the PR genes may be triggered by the microflora of the compost or could be associated with abiotic factors of the compost	Kavroulakis et al. (2005)
Rice straw and empty fruit bunch of oil palm water extracts	*Choanephora cucurbitarum*	Wet rot of okra	Induction peroxidase (PO), polyphenol oxidase (PPO), and phenylalanine ammonia lyase (PAL) enzymes in plant. Induction was correlated with delay in disease development	Siddiqui et al. (2009)
Tomato-plant residues mixed with a coarse fraction of separated cattle manure	*Fusarium oxysporum* f. sp. *melonis* *Botrytis cinerea*	Cucumber and melon wilt	Side-grafted split-root system revealed the phenomenon of induced resistance against soilborne pathogens, possibly because of the direct effect of compost microflora	Yogeva et al. (2010)

(continued)

Table 14.3 (continued)

Compost/compost tea material	Causal pathogen	Disease suppression	Observed effects	References
Commercial compost (pig manure, cow manure, poultry manure, sawdust, zeolite) water extract	*Colletotrichum coccodes* *Colletotrichum orbiculare*	Pepper anthracnose Cucumber	Enhanced PR gene expression, defense-related enzyme production, and hydrogen peroxide generation post-pathogen inoculation	Sang and Kim (2011)
Olive marc and olive tree leaves	*Botrytis cinerea*	*Arabidopsis*	178 genes were differently expressed, with a fold change cutoff of 1, of which 155 were upregulated and 23 were downregulated in compost-grown, against perlite-grown, plants. Compost triggered a plant response that shares similarities with both systemic acquired resistance and ABA-dependent/ independent abiotic stress responses	Segarra et al. (2013)
Olive marc	*Botrytis cinerea*	Gray mold of Tomato	The salicylic acid (SA) pathway/abscisic acid (ABA) is involved in compost-induced systemic resistance. Instead, perlite enriched with *Trichoderma asperellum* T34 is not linked to SA pathway/ ABA	Fernández et al. (2014)

Elsewhere, compost water extracts significantly reduced the disease incidence and severity of root and foliar infection on pepper caused by *Phytophthora capsici* compared with the controls. Furthermore, direct inhibition of development and population of *P. capsici* for root infection, as well as indirect inhibition of foliar infection through ISR with broad-spectrum protection, might have contributed to the suppression process (Sang et al. 2010). The authors explained that the enhanced expression of the *pathogenesis-related* (PR) genes, namely, *CABPR1*, *CABGLU*, *CAChi2*, *CaPR-4*, *CAPO1*, or *CaPR-10* as well as β-1,3-glucanase, chitinase, and peroxidase activities, resulted in enhanced plant defense against *P. capsici* in pepper plants. Moreover, compost water extracts also enhanced the chemical and structural defenses of the plants, including H_2O_2 generation in the leaves and lignin accumulation in the stems. A study on various compost and compost teas as an inducer of disease resistance has been summarized in Table 14.3.

14.5 Conclusive Remarks

Though there are existing huge-scale studies on disease suppressiveness of compost and its products, there is still deficient awareness into the general principles of disease suppression by compost in relation to its quality. There are many potential instances where compost and compost teas proved promising; however, their efficacy in field application is still in its infancy. Success in biological control of diseases with compost and compost teas is possible only if all factors involved in the production and utilization of composts are defined and kept consistent. It will be unrealistic to demand a system where compost is able to suppress all pathogens in different situations. "Tailor-made" compost for the suppression of specific pathosystem is perhaps required. Since microbial communities present in the compost are considered to be one of the major driving forces for suppressive efficacy of compost, a better knowledge of the microbial interactions and the enrichment nutrients will enhance the potential suppressiveness of compost and its watery extracts. The use of composts for disease suppression has potential benefits both ecologically and economically. Although the use of composts may not control all diseases to a level that allows the elimination of fungicide use, integration of suitable compost into current disease management practice can reduce fungicide use and associated problems.

References

Achimu P, Schlösser P (1992) Control of *Plasmopara viticola* with compost filtrates. Med Fac Landbouww Rijksuniv Gent 56:171–178

Al-Mughrabi KI (2006) Antibiosis ability of aerobic compost tea against foliar and tuber potato diseases. Biotechnology 5(1):69–74

Al-Mughrabi KI (2007) Suppression of *Phytophthora infestans* in potatoes by foliar application of food nutrients and compost tea. Aust J Basic Appl Sci 1(4):785–792

Al-Mughrabi KI, Berthélémé C, Livingston T, Burgoyne A, Poirier R, Vikram A (2008) Aerobic compost tea, compost and a combination of both reduce the severity of common scab (*Streptomyces scabiei*) on potato tubers. J Plant Sci 3(2):168–175

Andrews JH (1993) Compost extracts and the biological control of foliar plant disease. Grant Report Project # LNC 91-31 Madison, Wisconsin

Blakeman JP (1975) Germination of *Botrytis cineria* conidia *in vitro* in relation to nutrient conditions on leaf surfaces. Trans Br Mycol Soc 65:239–247

Blum LEB, Rodríguez-Kábana R (2004) Effect of soil organic amendments on sclerotial germination, mycelial growth, and *Sclerotium rolfsii*-induced diseases. Fitopatol Bras 29:066–074

Boehm MJ, Hoitink HAJ (1992) Sustenance of microbial activity in potting mixes and its impact on severity of *Pythium* root rot of poinsettia. Phytopathology 82:259–264

Bollen GJ (1993) Factors involved in inactivation of plant pathogens during composting of crop residues. In: Hoitink HAJ, Keener HM (eds) Science and engineering of composting design, environmental microbiological and utilization aspects. Renaissance Publications, Worthington, OH, pp 301–318

Borrero C, Castillo C, Ordovás J, Tello JC, Avilés M (2002) Composted cork as suppressive substrate to *Verticilium* wilt of tomato. In: Michel FC Jr, Rynk RF, Hoitink HAJ (eds) Proceedings of 2002 International Symposium Composting and Compost Utilization, Columbus, OH, 6–8 May 2002, pp 1342–1352

Borrero C, Trillas MI, Ordovás J, Tello JC, Avilés M (2004) Predictive factors for the suppression of fusarium wilt of tomato in plant growth media. Phytopathology 94:1094–1101

Borrero C, Trillas MI, Avilés M (2009) Carnation Fusarium wilt suppression in four composts. Eur J Plant Pathol 123:425–433

Brinton WF (1995) The control of plant pathogenic fungi by use of compost teas. Biodynamics 197:12–15

Budde K, Weltzien HC (1988) PhytosanitareWirkungen von Kompostextrakten und substratenimErreger- System Gerste –EchterMehltau (*Erysiphegraminis* DC f. sp. horder Marchai). Rijksuniv Gent 53:363–371

Chung YR, Hoitink HAJ, Dick WA, Herr LJ (1998) Effects of organic matter decomposition level and cellulose amendment on inoculum potential of *Rhizoctonia solani* in hardwood bark media. Phytopathology 78:836–840

Claudia C, McLean M, Berjak P (1997) *In vitro* studies on the potential for biological control of *Aspergillus flavus* and *Fusarium moniliforme* by *Trichoderma* spp. Mycopathologia 137: 115–124

Craft M, Nelson EB (1996) Microbial properties of composts that suppress damping-off and root rot of creeping Bentgrass caused by *Pythium graminicola*. Appl Environ Microbiol 62(5): 1550–1557

Cronin MJ, Yohalem DS, Harris RF, Andrews JH (1996) Putative mechanism and dynamics of inhibition of the apple scab pathogen (*Venturia inaequalis*) by compost tea. Soil Biol Biochem 28:1241–1249

Cotxarrera L, Trillas-Gay MI, Steinberg C, Alabouvette C (2002) Use of sewage sludge compost and *Trichoderma asperellum* isolates to suppress fusarium wilt of tomato. Soil Biol Biochem 34:467–476

Danon M, Zmora-Nahum S, Chen Y, Hadar Y (2007) Prolonged compost curing reduces suppression of *Sclerotium rolfsii*. Soil Biol Biochem 39:1936–1946

De Meyer G, Bigrimana G, Elad Y, Hofte M (1998) Induced systemic resistance in *T. harzianum* T396 biocontrol of *Botrytis cinerea*. Eur J Plant Pathol 104:279–286

Diab HG, Hu S, Benson DM (2003) Suppression of *Rhizoctonia solani* on Impatiens by enhanced microbial activity in composted swine waste-amended potting mixes. Phytopathology 93(9): 1115–1123

Diánez F, Santos M, Boix A, de Cara M, Trillas I, Avilés M, Tello JC (2006) Grape marc compost tea suppressiveness to plant pathogenic fungi: the role of siderophores. Compost Sci Util 14(1): 48–53

Dionne A, Tweddell RJ, Antoun H, Avis TJ (2012) Effect of non-aerated compost teas on damping-off pathogens of tomato. Can J Plant Pathol 34(1):51–57

Dukare AS, Prasanna R, Dubey SC, Nain L, Chaudhary V, Singh R, Saxena AK (2011) Evaluating novel microbe amended composts as biocontrol agents in tomato. Crop Prot 30:436–442

Elad Y, Shtienberg D (1994) Effect of compost water extracts on grey mould (*Botrytis cinerea*). Crop Prot 2:109–114

El-Masry MH, Khalil AI, Hassouna MS, Ibrahim HAH (2002) *In situ* and *in vitro* suppressive effect of agricultural composts and their water extracts on some phytopathogenic fungi. World J Microbiol Biotechnol 18:551–558

Erhart E, Burian K, Hartl W, Stich K (1999) Suppression of *Pythium ultimum* by biowaste composts in relation to compost microbial biomass, activity and content of phenolic compounds. J Phytopathol 147:299–305

Farrell JB (1993) Fecal pathogen control during composting. In: Hoitink HAJ, Keener HM (eds) Science and engineering of composting: design, environmental, microbiological and utilization aspects. Renaissance Publications, Worthington, OH, pp 282–300

Fernández E, Segarra G, Trillas MI (2014) Physiological effects of the induction of resistance by compost or *Trichoderma asperellum* strain T34 against *Botrytis cinerea* in tomato. Biol Control 78:77–85

Granatstein D (1999) Suppressing plant diseases with compost. The compost connection for western agriculture, No. 8, January 1999

Griffin TS, Hutchinson M (2007) Compost maturity effects on nitrogen and carbon mineralization and plant growth. Compost Sci Util 15:228–236

Goldstein J (1998) Compost suppresses disease in the lab and on the fields. BioCycle 39:62–64

Gorodecki B, Hadar Y (1990) Suppression of *Rhizoctonia solani* and *Sclerotium rolfsii* in container media containing composted separated cattle manure and composted grape marc. Crop Prot 9:271–274

Hadar Y, Mandelbaum R (1986) Suppression of *Pythium aphanidermatum* damping-off in container media containing composted liquorice roots. Crop Prot 5:88–92

Haggag WM, Saber MSM (2007) Suppression of early blight on tomato and purple blight on onion by foliar sprays of aerated and non-aerated compost teas. J Food Agric Environ 5:302–309

Harman GE, Hayes CK, Lorito M, Broadway RM, Di Pietro A, Peterbauer C, Tronsmo A (1993) Chitinolytic enzymes of *Trichoderma harzianum*: purification of chitobiosidase and endochitinase. Phytopathology 83:313–318

Himanen M, Hänninen K (2009) Effect of commercial mineral-based additives on composting and compost quality. Waste Manag 29:2265–2273

Hmouni A, Mouria A, Douira A (2006) Biological control of tomato grey mould with compost water extracts, *Trichoderma* sp. and *Gliocladium* sp. Phytopathol Mediterr 45:110–116

Hoitink HAJ, VanDoren DM, Schmitthenner AF (1977) Suppression of *Phytophthora cinnamomi* in a composted hardwood bark potting medium. Phytopathology 67:561–565

Hoitink HAJ, Inbar Y, Boehm MJ (1991) Status of compost-amended potting mixes naturally suppressive to soilborne diseases of floricultural crops. Plant Dis 75:869–873

Hoitink HAJ, Boehm MJ, Hadar Y (1993) Mechanism of suppression of soil borne plant pathogen in compost-amended substrates. In: Hoitink HAJ, Keener HM (eds) Science and engineering of composting: design, environmental, microbiological and utilization aspects. Renaissance Publication, Worthington, OH, pp 601–621

Hoitink HAJ, Boehm MJ (1999) Biocontrol within the context of soil microbial communities: a substrate-dependent phenomenon. Annu Rev Phytopathol 37:427–446

Inbar Y, Boehm MJ, Hoitink HAJ (1991) Hydrolysis of fluorescein diacetate in sphagnum peat container media for predicting suppressiveness to damping-off caused by *Pythium ultimum*. Soil Biol Biochem 23:479–483

Inbar Y, Chen Y, Hoitink HAJ (1993) Properties for establishing standards for utilization of composts in container media. In: Hoitink HAJ, Keener HM (eds) Science and engineering of composting: design, environmental, microbiological and utilization aspects. Renaissance Publications, Worthington, OH, pp 668–694

Ingham ER (2000a) Brewing compost tea. Kitchen Gard 29:16–19

Ingham ER (2000b) The compost tea brewing manual. Unisun Communications, Corvallis, OR

Jenana RKBE, Haouala R, Ali M, Triki JJG, Hibar K, Ben M (2009) Composts, compost extracts and bacterial suppressive action on *Pythium aphanidermatum* in tomato. Pak J Bot 41:315–327

Kannangara T, Utkhede RS, Paul JW, Punja ZK (2000) Effects of mesophilic and thermophilic composts on suppression of Fusarium root and stem rot of greenhouse cucumber. Can J Microbiol 46:1021–1028

Kazana K, Goulter KC, Way HM, Manners JM (1998) Expression of pathogenesis-related peroxidase of Stylosanthes humilis in transgenic tobacco and canola and its effect on disease development. Plant Sci 136:207–217

Kavroulakis N, Ehaliotis C, Ntougias S, Zervakis GI, Papadopoulou KK (2005) Local and systemic resistance against fungal pathogens of tomato plants elicited by a compost deriving from agricultural residues. Physiol Mol Plant Pathol 66:163–174

Kerkeni A, Daami-Remadi M, Tarchoun N, Khedher MB (2007) *In vitro* assessment of the antifungal activity of several compost extracts obtained from composted animal manure mixtures. Int J Agric Res 2(9):786–794

Ketterer N (1990) Studies on the effect of a compost extraction on the leaf infections of potatoes and tomatoes by *Phytophthora infestans* as well as on the infection of grape vines by *Plasmopara viticola*, *Pseudopeziza tracheiphila* and *Uncinula necator*. PhD Thesis, Bonn

Ketterer N, Schwager L (1992) The effect of compost extracts on disease attack and phyllosphere population in bush bean and tomato leaves. Mededelingen van de Faculteit Landbouwwetenschappen Universiteit Gent 57 (2–3 Part A–B):411–421

Ketterer N, Fisher B, Weltzien H (1992) Biological control of *Botrytis cinerea* on grapevine by compost tea and their microorganisms in pure culture. In: Verhoeff K, Malatharakis N, Williamson B (eds) Recent advances in *Botrytis* research. Proceedings 10th international *Botrytis* symposium, Heraklion, Crete, 5–10 April 1992. Pudoc Scientific Publishers, Wageningen, pp 179–186

Khan J, Ooka JJ, Miller SA, Madden LV, Hoitink HAJ (2004) Systemic resistance induced by *Trichoderma hamatum* 382 in cucumber against Phytophthora crown rot and leaf blight. Plant Dis 88(3):280–286

Kim KD, Nemec S, Musson G (1997) Effects of composts and soil amendments on soil microflora and *Phytophthora* root and crown rot of bell paper. Crop Prot 16(2):165–172

Kloepper JW, Rodriguez-Ubana R, Zehnder GW, Murphy JF, Sikora E, Fernandez C (1999) Plant root-bacterial interactions in biological control of soilborne diseases and potential extension to systemic and foliar diseases. Australas Plant Pathol 28:21–26

Koné SB, Dionne A, Tweddell RJ, Antoun H, Avis TJ (2010) Suppressive effect of non-aerated compost teas on foliar fungal pathogens of tomato. Biol Control 52:167–173

Laha GS, Singh RP, Verma JP (1992) Biological control of *Rhizoctonia solani* in cotton by fluorescent Pseudomonas. Indian Phytopathol 45(4):417–415

Lamondia JA, Gent MPN, Ferrandino FJ, Elmer WH, Stoner KA (1998) The effect of compost amendment or straw mulch on potato early dying disease development and yield. Plant Dis 83(4):661–666

Liu L, Kloepper JW, Tuzun S (1995) Induction of systemic resistance in cucumber against Fusarium wilting plant growth-promoting rhizobacteria. Phytopathology 85:695–8

Lorito M, Peterbauer C, Hayes CK, Harman GE (1994) Synergistic interaction between fungal cell wall degrading enzymes and different antifungal compounds enhances inhibition of spore germination. Microbiology 140:623–629

Maurhfiofer M, Flace C, Meuwly P, Defagu G (1994) Induction of systemic resistance in cucumber by Pseudomonas syringae pv. Syringae. Physiol Mol Plant Pathol 38:223–235

McQuilken MP, Whipps JM, Lynch JM (1994) Effects of water extract of a composted manure-straw mixture on the plant pathogen *Botrytis cinerea*. World J Microbiol Biotechnol 10:20–26

Merrill R, McKeon J (2001) Apparatus design and experimental protocol for organic compost teas. Org Farm Res Found Inf Bull 9:9–15

Metacalf and Eddy (1991) Wastewater engineering treatment, disposal and reuse, 3rd edn. Wiley, New York

Naidu Y, Sariah M, Jugah K, Siddiqui Y (2010) Microbial starter for the enhancement of biological activity of compos tea. Int Agric Biol 12:51–56

Naidu Y, Meon S, Siddiqui Y (2012) *In vitro* and *in vivo* evaluation of microbial-enriched compost tea on the development of powdery mildew on melon. Biol Control 57(6):827–836

Naidu Y, Meon S, Siddiqui Y (2013) Foliar application of microbial-enriched compost tea enhances growth, yield and quality of muskmelon (Cucumis melo L.) cultivated under fertigation system. Sci Hortic 159:33–40

Nelson EB, Kuter GA, Hoitink HAJ (1983) Effects of fungal antagonists and compost age on suppression of rhizoctonia damping-off in container media amended with composted hardwood bark. Phytopathology 73:1457–1462

Noble R, Roberts SJ (2004) Eradication of plant pathogens and nematodes during composting: a review. Plant Pathol 53:548–568

NOSB (2006) Formal recommendation by the NOSB to the National Organic Program. 9 Nov 2006. National Organic Standards Board. Available via http://www.ams.usda.gov/NOSB/FinalRecommendations/Oct06/NOSBFNLCompostGuidanceCombined10192006.pdf

Ntougias S, Papadopoulou KK, Zervakis GI, Kavroulakis N, Ehaliotis C (2008) Suppression of soil-borne pathogens of tomato by composts derived from agro-industrial wastes abundant in Mediterranean regions. Biol Fertil Soils 44:1081–1090

Pan I, Dam B, Sen SK (2012) Composting of common organic wastes using microbial inoculants. Indian J Biotechnol 2:127–134

Pharand B, Carisse O, Benhamou N (2002) Cytological aspects of compost-mediated induced resistance against *Fusarium* crown and root rot in tomato. Phytopathology 92:424–438

Postma J, Kok CJ (2003) Effect of compost application on the soil microflora. In: Proceedings, applying compost – benefits and needs, Brussel, pp 157–161

Pscheidt J, Wittig H (1996) Fruit and ornamental disease management testing program. Extension plant pathology. O.S.U., Corvallis, OR. Available via http://www.science.oregonstate.edu/bpp/Plant_Clinic/Fungicidebooklet/personnel.htm

Raaijmakers JM, Leeman M, Van Oorschot MMP, Van der Sluis I, Schippers B, Bakker PAHM (1995) Dose-response relationships in biological control of Fusarium wilt of radish by *Pseudomonas* spp. Phytopathology 85:1075–1081

Ringer CE, Millner PD, Teerlinck LM, Lyman BW (1997) Suppression of seedling damping-off disease in potting mix containing animal manure composts. Compost Sci Util 5(2):6–14

Reuveni R, Raviv M, Krasnovsky A, Freiman L, Medina S, Bar A, Orion D (2002) Compost induces protection against *Fusarium oxysporum* in sweet basil. Crop Prot 21:583–587

Ryckeboer J (2001) Biowaste and yard waste composts: microbiological and hygienic aspects-suppressiveness to plant diseases. PhD Thesis, Katholieke Universiteit Leuven, Belgium, pp 1–245

Saadi I, Laor Y, Medina S, Krassnovsky A, Raviv M (2010) Compost suppressiveness against *Fusarium oxysporum* was not reduced alter one-year storage under various moisture and temperature conditions. Soil Biol Biochem 42:626–634

Sackenheim R (1993) Untersuchungenuber Wirkungen von wassrigenmikrobilogischaktiven-Extraktenauskompostiertensubstraten auf den Befall der weinrebe (*Vitis vinifera*) mit *Plasmopara viticola*, *Uncinula necator*, *Botrytis cinerea* and *Pseudopeziza tracheiphila*. Dissertation, Inivesiy of Bonn

Samerski C, Weltzien HC (1988a) Studies on the activity and action mechanisms of compost extracts in the pathosystem cucumber powdery mildew *Sphaerotheca fuliginea*. Mededelingen van de Faculteit Landbouwwetenschappen Universiteit Gent 53 (2 Part A):373–378

Samerski C, Weltzien HC (1988b) Investigations on the mode of action of compost extracts in the host-parasite system sugar beet-powdery mildew. J Plant Dis Prot 95:176–181

Sang MK, Kim JG, Kim KD (2010) Biocontrol activity and induction of systemic resistance in pepper by compost water extracts against *Phytophthora capsici*. Phytopathology 100:774–783

Sang MK, Kim KD (2011) Biocontrol activity and primed systemic resistance by compost water extracts against anthracnoses of pepper and cucumber. Phytopathology 101:732–740

Scheuerell SJ, Mahaffee WF (2000) Assessing aerated and non-aerated watery fermented compost and *Trichoderma harzianum* T-22 for control of powdery mildew of rose in the Willamette Valley, Oregon (Abstract). Phytopathology 90:69

Scheuerell SJ, Mahaffee WF (2002) Compost tea: principles and prospects for plant disease control. Compost Sci Util 10:313–318

Scheuerell SJ, Mahaffee WF (2004) Compost tea as a container medium drench for suppressing seedling damping-off caused by *Pythium ultimum*. Phytopathology 94:1156–1163

Scheuerell SJ, Sullivan DM, Mahaffee WF (2004) Suppression of seedling damping-off caused by *Pythium ultimum*, *P. irregulare*, and *Rhizoctonia solani* in container media amended with a diverse range of Pacific Northwest compost sources. Phytopathology 95:306–315

Scheuerell SJ, Mahaffee WF (2006) Variability associated with suppression of gray mold (*Botrytis cinerea*) on geranium by foliar application of non-aerated and aerated compost teas. Plant Dis 90(9):1201–1208

Segarra G, Reis M, Casanova E, Trillas I (2009) Control of powdery mildew (*Erysiphe polygoni*) in tomato by foliar applications of compost tea. J Plant Pathol 91:683–689

Segarra G, Elena G, Trillas I (2013) Systemic resistance against *Botrytis cinerea* in *Arabidopsis* triggered by an olive marc compost substrate requires functional SA signalling. Physiol Mol Plant Pathol 82:46–50

Siddiqui Y, Sariah M, Razi I (2008a) *Trichoderma*-fortified compost extracts for the control of *Choanephora* wet rot in okra production. Crop Prot 27(3–5):385–390

Siddiqui Y, Sariah M, Razi I, Mawardi R, Asgar A (2008b) Bio-efficiency of compost extracts on the wet rot incidence, morphological and physiological growth of okra (*Abelmoschus esculentus* [(L.) Moench]). Sci Hortic 117(1):9–14

Siddiqui Y, Sariah M, Razi I, Mawardi R (2009) Bio-potential of compost tea from agro-waste to suppress *Choanephora cucurbitarum* L. the causal pathogen of wet rot of okra. Biol Control 49(1):38–44

Sticher L, Mauch-Mani B, Metraux JP (1997) Systemic acquired resistance. Phytopathology 35: 235–270

Stindt A (1990) Untersuchungen zur wirkung und zu den wirkungmechanism von kompostextraten auf Botrytis cineria Pers. exNocca and Balb an Erdbeeren kopfsalat und boschbohnen. Dissertation, University of Bonn

Stoffella PJ, Li YC, Roe NE, Ozores-Hampton M, Graetz DA (1997) Utilization of organic waste composts in vegetable crop production systems. In: Morris RA (ed) Managing soil fertility for intensive vegetable production systems in Asia. Asian Vegetable Research and Development Center, Shanhua

Trillas-Gay MI, Hoitink HAJ, Madden LV (1986) Nature of suppression of Fusarium wilt of radish in a container medium amended with composted hardwood bark. Plant Dis 70:1023–1027

Trillas MI, Casanova E, Cotxarrera L, Ordovás J, Borrero C, Avilqs M (2006) Composts from agricultural waste and the *Trichoderma asperellum* strain T-34 suppress *Rhizoctonia solani* in cucumber seedlings. Biol Control 39:32–38

Tuitert G, Szczech M, Bollen GJ (1998) Suppression of *Rhizoctonia solani* in potting mixtures amended with compost made from organic household waste. Phytopathology 88:764–773

UNDP (2003) United Nations Development Programme bureau for development policy MONTREAL PROTOCOL UNIT

Urban J, Trankner A (1993) Control of grey mould (*Botrytis cinerea*) with fermented compost/water extracts. WPRS Bull 16:8–11

Wei G, Kloepper J, Tuzun S (1991) Induction of systemic resistance of cucumber to *Colletotrichum orbiculare* by select strains of plant growth promoting rhizobacteria. Phytopathology 81:1508–1512

Welke SE (2004) The effect of compost extract on the yield of strawberries and the severity of *Botrytis cinerea*. J Sustain Agric 25(1):57–68

Weltzien HC, Ketterer N (1986) Control of downy mildew, *Plasmopara viticola* (de Bary) Berlese et de Toni, on grapevine leaves through water extracts form composted organic wastes. J Phytopathol 116:186–188

Weltzien HC (1989) Some effects of composted organic materials on plant health. Agric Ecosyst Environ 27:439–446

Weltzien HC (1991) Biocontrol of foliar fungal diseases with compost extracts. In: Andrews JH, Hirano SS (eds) Microbial ecology of leaves. Springer, New York, pp 430–450

Widmer TL, Graham JH, Mitchell DJ (1998) Composted municipal waste reduces infection of citrus seedlings by *Phytophthora nicotianae*. Plant Dis 82:683–688

Yohalem DS, Harris RF, Andrews JH (1994) Aqueous extracts of spent mushroom substrate for foliar disease control. Compost Sci Util 2:67–74

Yohalem DS, Nordheim EV, Andrews JH (1996) The effect of water extracts of spent mushroom compost on apple scab in the field. Phytopathology 86:914–922

Yogeva A, Raviv M, Hadar Y, Cohen R, Wolf S, Gil L, Katan J (2010) Induced resistance as a putative component of compost suppressiveness. Biol Control 54:46–51

You MP, Sivasithamparam K (1995) Changes in microbial populations of an avocado plantation mulch suppressive to *Phytophthora cinnamomi*. Appl Soil Ecol 2:33–43

Zhang W, Han DY, Dick WA, Davis KR, Hoitink HAJ (1998) Compost and compost water extract-induced systemic acquired resistance in cucumber and *Arabidopsis*. Phytopathology 88:450–455

Zhang W, Dick WA, Hoitink HAJ (1996) Compost-induced systemic acquired resistance in cumber to Pythium root rot and anthracnose. Phytopathology 86:1066–1070

Chapter 15
Suggested Mechanisms Involved in Suppression of *Fusarium* by Vermicompost Products

Yurdagul Simsek-Ersahin

15.1 Introduction

Fungal plant pathogens, the second major pest group of all plant pests, are responsible for 13 % of preharvest loses due to all plant pests in agricultural crops (Pimentel 1997). More than 10,000 species of fungi can cause disease in plants, some attacking more than one plant species (Agrios 2005). Plant pathogenic fungi could be classified within five major groups as Deuteromycetes (fungi imperfecti), Phycomycetes, Zygomycetes, Ascomycetes, and Basidiomycetes. The imperfect fungal genus *Fusarium*, with perfect states in *Calonectria*, *Gibberella*, *Microneciriella*, and *Nectria*, forms robust survival structures, chlamydospores. This enables *Fusarium* as one of the most ubiquitous and prevalent plant pathogenic fungi group with over a thousand species (Agrios 2005). The genus *Fusarium* contains a number of economically important plant pathogenic species, some causing considerable loses. Both its persistence in soil and high pathogenicity level give *Fusarium* a great deal of importance by both plant pathologists and the plant breeders (Booth 1971; Nelson et al. 1983).

The genus *Fusarium* collectively represents one of the most important groups of fungal plant pathogens, causing various diseases on nearly every economically important plant species. In addition, a plethora of *Fusarium* mycotoxins, fumonisins and trichothecenes, pose health hazards to humans and livestock. Of the equal phytopathogenic and toxigenic importance, species of *Fusarium* also serve as key model organisms for biological and evolutionary research. In 2002, the *F. graminearum* sequencing project was funded by the National Research Initiative within the US Department of Agriculture's National Institute of Food and

Y. Simsek-Ersahin (✉)
Faculty of Science, Department of Biology, Cankiri Karatekin University, Cankiri 18100, Turkey
e-mail: yurdagulersahin@gmail.com

M.K. Meghvansi, A. Varma (eds.), *Organic Amendments and Soil Suppressiveness in Plant Disease Management*, Soil Biology 46,
DOI 10.1007/978-3-319-23075-7_15

Agriculture (Anonymous 2014a). The *F. graminearum* sequencing project represents a partnership between the Broad and the International *Gibberella zeae* Genomics Consortium (IGGR). Meanwhile a *Fusarium* comparative project was founded through the same agency to improve gene annotation and to identify functional noncoding elements of *F. graminearum*. Later, two additional *Fusarium* species, *F. oxysporum* and *F. verticillioides*, have been included in IGGR to assess evolutionary biology among these closely related but biologically distinct *Fusarium* species. Results indicated that genomes of *F. oxysporum* and *F. verticillioides* are diverged from *F. graminearum*. Recently *F. oxysporum* has also emerged as a model for soilborne fungal diseases with *Arabidopsis* and tomato as hosts. *F. oxysporum* has a broad host range and has caused some of the world's most dramatic and economically devastating plant disease epidemics. *F. verticillioides* is a cosmopolitan pathogen of maize and sorghum and produces carcinogenic mycotoxins, the fumonisins. A three-way genome comparison of *F. oxysporum*, *F. verticillioides*, and *F. graminearum* offers powerful synergy in pathogenicity studies and virulence factors and their evolution within this genus (Anonymous 2014a).

Members of the *F. oxysporum* species complex exhibit extraordinary genetic flexibility and cause some of the most destructive diseases across a diverse spectrum of hosts, including many economically important crops, such as banana, cotton, canola, melons, and tomato. The primary solution to control such diseases is through the development of disease-resistant plant cultivars. However, due to its great genetic flexibility and persistence in the soil, it is just a matter of time before the pathogen can adapt and overcome the newly deployed resistance (Kistler and Rep 2010). Consequently, *Fusarium* has been chosen as a model for biological and evolutionary research, as well as research on soilborne fungal diseases. In the *Fusarium* comparative genomics project, the genomes of three phenotypically diverse species, *F. graminearum*, *F. verticillioides*, and *F. oxysporum* f. sp. *lycopersici*, were compared (Anonymous 2014a). This study evaluated the genetic composition and evolutionary origin of lineage-specific chromosomes among a set of carefully selected strains, representing those three species that capture the pathogenic and phenotypic diversity. The *Fusarium* genomics project revealed the existence of lineage-specific chromosomes between otherwise genetically isolated strains, explaining the polyphyletic origin of host specificity and occurrence of new pathogenic lineages in *F. oxysporum*. These results put the evolution of fungal pathogenicity into a new perspective (Ma et al. 2010).

After the Second World War, intensive conventional farming practices add great emphasis on the use of agrochemicals particularly pesticides. However, application of excessive and repeated doses of pesticides has posed many impacts, such as increased pathogen/pest resistance and decreased beneficial organisms in all ecosystems that created the vicious cycle of increased amounts of pesticide use in each coming year to obtain the last year's yield. Most fungicides can cause acute, even chronic toxicity in, primarily, beneficial organisms (Goldman 2008). The International Labour Organization (ILO) reported that up to 14 % of all occupational injuries resulted from exposure to pesticides and other agrochemicals (ILO 1996).

Beginning with "Silent Spring" in 1960s, the modern society has become familiar with detrimental, sometimes irreversible, effects of excessive application of pesticides toward human and environmental health issues (Gold 2007). Increasing social awareness has encouraged scientists for development of environmentally friendly and sustainable farming systems. Sustainable farming systems/good agricultural practices promote the use of organic amendments, e.g., composts (thermophilic)/vermicompost (mesophillic) products, for plant fertilization and protection. The use of both, primarily thermophilic, compost products has increasingly been practiced during the last three decades (Hoitink et al. 1997, 1998; Edwards 1995, 1998; Edwards and Burrows 1988; Edwards and Arancon 2004; Edwards et al. 2006; Simsek-Ersahin 2011). Early studies on implementation of, solid or aqueous, vermicompost products for plant protection targeted soilborne disease control, because maintaining disease suppression effectiveness against soilborne pathogens in soil environment is much difficult than foliar plant pathogens (Szczech 1999; Szczech et al. 1993; Szczech and Smolinska 2001; Nakasone et al. 1999; Orlikowski 1999; Rodriguez et al. 2000; Zaller 2006; Reddy et al. 2012).

A number of studies on sustainable management of economically important soilborne diseases caused by the genus *Pythium*, *Phytophthora*, *Rhizoctonia*, and *Fusarium* by vermicompost products have been carried out increasingly during the last two decades. However, complexity of microbial interactions between soil microflora and those indigenous to vermicastings, various soil properties, and particularly high genetic plasticity of the genus *Fusarium* complicates obtaining accurate understanding of suppression mechanisms of soilborne diseases by vermicompost products. Therefore, more comprehensive and detailed studies are needed to reveal the key factors in mechanisms of vermicompost-mediated suppression toward phytopathogens. This chapter presents an overview of the current understanding of the influence of vermicompost products, solid or liquefied, on *Fusarium* diseases.

15.2 Suppression of Plant Diseases by Vermicompost Products

The early pioneering research on solid vermicompost products for plant protection was aimed to determine their suppression efficacy mostly as pot substrates, few as soil amendments, on soilborne plant pathogens such as *Pythium*, *Phytophthora*, *Rhizoctonia*, *Fusarium*, and *Verticillium* (Edwards and Arancon 2004; Edwards et al. 2006; Szczech et al. 1993; Szczech 1999; Szczech and Smolinska 2001; Rodriguez et al. 2000; Chaoui et al. 2002). The solid vermicomposts are derived from various waste streams, i.e., animal manures (Szczech and Smolinska 2001), separated dairy solids (Kanangara et al. 2000), cattle manure (Szczech 1999), municipal sewage sludge (Szczech and Smolinska 2001), and a mixture of vegetable wastes, bark (*Salix* sp.), and cattle manure (Simsek-Ersahin et al. 2009). The first vermicompost studies evaluated suppression potential of the vermicomposts on the fungal phytopathogens such as *Plasmodiophora brassicae* (clubroot on brassica) (Szczech et al. 1993; Nakamura 1996), *Phytophthora nicotianae* var. *nicotianae* (root rot on tomato) (Szczech et al. 1993; Szczech and Smolinska 2001), *F. oxysporum* f. sp. *lycopersici* (tomato Fusarium wilt) (Szczech 1999), *Pythium* (damping-off), *Rhizoctonia* (root rot), *Verticillium* (wilts) (Chaoui et al. 2002), *Sclerotium* (white rot) (Pereira et al. 1996), and the sugar beet cyst nematode (*Heterodera schachtii*) (Szczech et al. 1993). Rodriguez et al. (2000) also reported significant reduction in disease incidence of fungal pathogens such as *Rhizoctonia solani*, *Phytophthora drechsleri*, and *F. oxysporum* on gerbera via use of solid vermicomposts as growth media (pot substrates).

Given unlabored and versatile application options, producers use liquid compost/vermicompost products for plant protection and fertilizer management particularly in certified organic production systems. Extracts from thermophilic compost have long been proved to be effective against various fungal diseases of leaves and fruits especially when applied prophylactically (Weltzien 1989; Scheuerell and Mahaffee 2002). Studies experimenting the efficacy of aqueous vermicompost-mediated suppressivity on phytopathogens have been started long after those of thermophilic compost-mediated suppressivity. Early studies on implementation of aqueous vermicompost products for plant protection targeted extensively to control soilborne diseases as the case with solid vermicomposts (Szczech et al. 1993; Nakasone et al. 1999; Orlikowski 1999; Rodriguez et al. 2000; Zaller 2006; Reddy et al. 2012). Pioneering studies on suppression efficacy of liquid vermicompost (extracts) against a plant pathogen (*Phytophthora cryptogea*) were reported by Orlikowski (1999) and on a plant parasitic nematode by Arancon et al. (2002). Nakasone et al. (1999) also reported that aqueous vermicomposts inhibited the mycelial growth of *Botrytis cinerea*, *Sclerotinia sclerotiorum*, *Corticium rolfsii*, *R. solani*, and *F. oxysporum*. The researchers in Ohio State University made significant contributions to determination of disease suppression efficacy of vermicompost products in plant disease and pest control (Edwards 1995; Edwards and Arancon 2004; Edwards et al. 2006; Chaoui et al. 2002). Research team in Ohio State University explored disease suppressivity of small applications of some commercially produced vermicomposts on diseases caused by fungus *Pythium* in cucumbers, *Rhizoctonia* in radishes in the greenhouse, *Verticillium* in strawberries, and *Phomopsis* and *Sphaerotheca fulginae* in grapes infield (Edwards and Arancon 2004; Edwards et al. 2006). Studies of aqueous vermicompost products aimed to determine aqueous vermicompost-mediated suppression efficiency toward diseases of aerial plant parts, such as powdery mildews (Singh et al. 2003). Fokkema (1993) also showed that vermicompost products have ability to induce systemic resistance. Researchers in Ohio State University exhibited that aerated vermicompost "teas" suppressed the plant diseases caused by *Fusarium*, *Verticillium*, *Plectosporium*, and *Rhizoctonia* to the same extent as the solid vermicomposts (Edwards et al. 2006).

Liquid compost/vermicompost products offer a multifaceted, yet easy way for implementation of organic products that effectively inoculate to both below- and aboveground plant surfaces with potentially beneficial microbes. However, there is scarce amount of solid and unequivocal scientific study on disease suppression effect of liquid vermicompost products unlike composts (Edwards et al. 2006). Although liquid vermicompost, such as liquid composts, are proven to be effective in certain cases in greenhouse and field studies, respected results exhibit a significant variation. The variation in the results was likely derived from feedstocks used for vermicomposting, composting method, the earthworm species involved, the amendment rate applied, and pathogen and the respected host crop (Szczech and Smolinska 2001; Fernández-Gómez et al. 2012; Jack et al. 2011). As of present, there are two important tasks needed to be fulfilled by researchers working on vermicompost-mediated disease suppression. The first is that reliable, fast, and cost-effective quality disease control parameters are required for manifesting whether a vermicompost product has any disease suppression potential. The second is that vermicompost-mediated suppression mechanisms for each pathogen need to be revealed.

15.3 Suppression of *Fusarium* Diseases by Solid Vermicompost Products

The need for alternative and effective products certified for plant disease and pest management has greatly increased for the last two decades, especially in organic production systems. Therefore, either solid or liquid compost/vermicompost products have received a great deal of attention by growers as well as researchers. By the early 1990s, various researchers have challenged the efficacy of solid vermicomposts derived from different material types for suppression of primarily soilborne plant diseases (Edwards 1995; Szczech 1999; Szczech and Smolinska 2001; Chaoui et al. 2002; Arancon et al. 2002). Several studies demonstrated sufficient levels of solid vermicompost-mediated suppression against diseases caused by *F. oxysporum*, *F. proliferatum* (Moody et al. 1996; Nakamura 1996; Szczech 1999), and *F. oxysporum* f. sp. *cucumerinum* (Simsek-Ersahin 2007). Disease suppression effect of the vermicomposts mostly increased in proportion to rate of vermicompost application, specifically those applied against the pathogens such as *Rhizoctonia* with high parasiticity and endurance in the soil. Studies on disease-suppressive vermicompost products predominantly reported the loss of protective effect after sterilization or heating, implying the significant role of resident microbial population in suppression (Szczech 1999; Simsek-Ersahin et al. 2009).

In a recent study in India, Kumar et al. (2013) evaluated efficacy of biological agents (*Trichoderma harzianum+Pseudomonas fluorescens*) and organic amendments (cow dung manure, spent compost, and vermicompost) against wilt of lentil diseased by *F. oxysporum* f. sp. *lentis*, causing great loses in yield. Vermicompost application provided better disease incidence than cow dung manure and spent compost, as well as the higher grain yield (Kumar et al. 2013). A similar study by Malathi and Mohan (2013) examined efficacy of the joint use of biocontrol agents

Pseudomonas fluorescens and *T. harzianum*, inoculated to organic amendments such as vermicompost, against basal rot incidence caused by *F. oxysporum* f. sp. *cepae* in onion. Application of consortial formulations of organic amendments inoculated with biocontrol agents induced production of defense enzymes in onion and reduced the disease incidence in plants.

15.3.1 In Vitro Antimicrobial Activity of Liquid Vermicompost Products Toward Fusarium

The research on the role of earthworms on microbial community of vermicompost indicated that coelomic fluid released in the decaying biomass by earthworms may have antibacterial properties that kill pathogens, such as *Salmonella*, *Serratia marcescens*, and *Escherichia coli* (Prabha 2009). Earthworms also promote microbial activity and diversity in organic wastes to the levels even greater than those of thermophilic composts (Hoitink et al. 1997, 1998; Hoitink and Boehm 1999) that stimulates increased potential for variability of microbial populations antagonistic against plant pathogens (Doube et al. 1994, 1995). There are researches reporting the hypothesis in that solid or liquid vermicompost products provided plant disease control in vivo or inhibited fungal phytopathogens in vitro (Szczech et al. 1993; Rodriguez et al. 2000; Szczech and Smolinska 2001; Edwards and Arancon 2004; Edwards et al. 2006; Zaller 2006; Simsek-Ersahin et al. 2009; Jack et al. 2011).

There are few studies performed in vitro that examined the antifungal characteristics of vermicompost products against phytopathogens. Singh et al. (2003) reported that the aqueous vermicompost had potential to inhibit spore germination of several fungi from *Alternaria*, *Curvularia*, and *Helminthosporium* genera. Gopalakrishnan et al. (2010) examined the suppression efficacy of vermiwash, extracted from vermicompost made from three different herbs, against three pathogens of chickpea and sorghum, including *F. oxysporum* f. sp. *ciceri* (FOC). They suggested that inhibition of fungal colony growth by biowash toward the phytopathogens, including FOC, has resulted from the presence of secondary metabolites. Manandhar and Yami (2008) have tested the suppression efficacy of compost and vermicompost teas as aerated vermicompost tea (ACTV), non-aerated vermicompost tea (NCTV), aerated compost tea (ACTC), and non-aerated compost tea (NCTV) against foot rot disease of rice caused by *Fusarium moniliforme* Sheldon–*Gibberella fujikuroi* in a field experiment. Among all treatments, ACTV provided maximum control effect in field plants and seeds infected by *F. moniliforme*.

A recent study by Grantina-Ievina et al. (2014) evaluated antifungal activity of 12 vermicompost samples produced from cow manure, sewage sludge, and starchless potato pulp together with composted grass on 27 genera representing plant pathogenic fungi and nonpathogenic fungi associated with seeds or plant growth-promoting feature (i.e., *Trichoderma*). The antifungal activity was observed against fungi *Pseudeurotium*, *Beauveria*, *Nectria*, and *Fusarium* in decreasing order. Researchers suggested that width of the inhibition zones was positively correlated with the pH of the vermicompost extracts and negatively with the ratio E2/E3. High E2/E3 ratios have been related to low aromaticity percentages of natural humic matter as well as high content of low molecular weight substances of fulvic acid solutions (Carvalho et al. 2008).

15.4 Suggested Mechanisms for Vermicompost-Mediated Suppression of Plant Diseases Caused by *Fusarium* sp.

Possible mechanisms of plant disease suppression by organic materials have mainly been defined within two groups as microbially mediated suppression and induced systemic resistance. In general, soil microbes, sometimes abiotic factors, may induce plant resistance called induced, acquired, or induced systemic resistance which is effective against a broad spectrum of phytopathogens and reported to be effective under field conditions as well (Singh et al. 2003; Stone et al. 2004). Microbially mediated suppression is facilitated by specific or general suppression, resulting from microbial competition for nutrients in the rhizosphere, antibiosis, and hyperparasitism (Cook and Baker 1983; Hoitink and Grebus 1997). Specific suppression is derived from activity of narrow range of microorganisms and general suppression from interactions of a wide range of microorganisms. General suppression is a more common context in implementation of vermicomposts for control of soilborne pathogens, due to the rich variety of indigenous microbial populations likely to have antagonistic traits to various pathogens (Arancon et al. 2006). Suppression of phytopathogens such as *Pythium* and *Phytophthora* with high saprophytic capability is usually explained by general suppression mechanism (Cook and Baker 1983; Chen et al. 1988), whereas suppression of *Rhizoctonia* with high parasitic capacity is elucidated by "specific suppression" (Hoitink et al. 1997). Nonpathogenic *Fusarium* sp. and fluorescent *Pseudomonas* sp. were known to have a significant role in soils that are naturally suppressive to Fusarium wilts (Stone et al. 2004).

The phytopathogen genus *Fusarium* sp. has good saprophytic abilities and *Fusarium* populations increase after addition of organic amendment. Even though *F. oxysporum* termed as a soil inhabitant for its ability to persist in soil, tolerance to antagonism, and potential for colonization in organic substrates, many *Fusarium* sp. are poor competitors and cannot colonize organic substrates, previously colonized by other organisms (Park 1958). Pre-colonization of soils/organic matter with two nonpathogenic *F. oxysporum* isolates reduced *F. solani* f. sp. *pisi* growth and infection of pea (Oyarzun et al. 1994). Other fungal genera and bacterial species mildly inhibited *Fusarium* colonization. Expectedly, *Burkholderia cepacia*, an antibiotic-producing bacterial species, strongly inhibited *Fusarium* colonization (Toyota et al. 1996).

Despite a vast amount of studies that examined the efficiency of vermicompost-mediated suppression on plant diseases (Szczech 1999; Edwards and Arancon 2004; Sahni et al. 2008; Simsek-Ersahin 2011), limited number of studies were conducted on the mechanisms of vermicompost-mediated suppression or their impact on plant-associated microbial communities specifically in rhizosphere zone (Jack et al. 2011). One of the widely accepted scientific explanations behind the vermicompost-mediated disease suppression is the "soil-foodweb" concept, defined by Dr. Elaine Ingham of Corvallis, Oregon, USA (Anonymous 2014b). The "soil-foodweb" concept highlights the presence of ergonomically beneficial and indigenous microbial populations in vermicompost that protects plants by outcompeting plant pathogens for available food resources, i.e., by starving them and also by blocking their access to plant roots by occupying all the available sites. An early pioneering study by Szczech (1999) supported this concept. Szczech (1999) indicated that suppressiveness of vermicompost on tomato Fusarium wilt was purely microbially mediated, since suppressiveness was lost when the vermicompost was autoclaved. Edwards and Arancon (2004) also reported that the ability of pathogen suppression disappeared when the vermicompost was sterilized, convincingly indicating that the biological mechanism of disease suppression involved was "microbial antagonism."

In the literature of both suppressive soils (Fravel et al. 2003) and compost-amended container mixes (Hoitink and Boehm 1999), suppression of diseases caused by *F. oxysporum* has been generally considered to be due to suppression generated through activities of one (specific suppression) or several antagonistic microbial populations (general suppression) (Cook and Baker 1983). Therefore, it is not surprising that organic amendments and plant residues, known to enhance both diversity and biomass of microbial community, could suppress diseases caused by *F. oxysporum* in soilless container mixes (Chen et al. 1988), field soils incubated in containers (Serra-Wittling et al. 1996), and unincubated field soils (Lodha 1995). General suppression (Serra-Wittling et al. 1996), specific antagonists (Trillas-Gay et al. 1987), propagule lyses (Oritsejafor and Adeniji 1990), induced resistance (Pharand et al. 2002), and nonbiotic factors (Kai et al. 1990) have been implicated in organic matter-mediated suppressiveness of Fusarium wilts. Other suppression mechanisms suggested for Fusarium wilt suppressive soils are competitive colonization of substrate and roots within the context of OM-mediated disease suppression. However, little is known about the relationships between organic matter quality and suppression of diseases caused by *F. oxysporum* in container systems and field soils. It is worth to note that OM-mediated suppression, e.g., vermicomposts, is widely considered as facilitated by multiple mechanisms at one case (Stone et al. 2004).

There are specific mechanisms that are regarded in biological and OM-mediated suppression of plant diseases, such as microbiostasis, microbial colonization or destruction of pathogen propagules, antibiosis, competition for substrate colonization, and competition for root infection sites (Stone et al. 2004). Microbiostasis or fungistasis, repression of fungal spore germination, is considered for *Fusarium* sp. Small fungal propagules, e.g., conidia and the chlamydospores of *Fusarium*, require an external source of energy for germination that appears to be restricted by limited energy in soil environment (Lockwood 1990). In another assay, forest floor litter is reported to be responsible for induced germination and subsequent lysis of chlamydospores and macroconidia of *F. oxysporum* (Stone et al. 2004). Another example for suppression mechanism facilitated by destruction of pathogen propagules is prevention and decrease of chlamydospore germination of *F. oxysporum* f. sp. *raphani* by *Pseudomonas stutzeri* and *Pimelobacter* sp., isolated from a manure-amended field soil (Toyota and Kimura 1993).

A comprehensive recent study by Yogev et al. (2006) explored the compost-mediated suppression toward four formae speciales of *F. oxysporum*: *melonis*, *bacillici*, *radicis-lycopersici*, and *radicis-cucumerinum*. They evaluated suppression effectiveness of three different composts, produced from mainly plant residues, on four formae speciales of *Fusarium* in melon, tomato, and cucumber. They concluded that different mechanisms may be involved in compost-mediated suppression of four formae speciales of *Fusarium*, revealed by different degree of pathogen population decline in each suppressive compost. The decline in pathogen population was manifested by decrease in inoculum density that may be resulted from a direct effect on the pathogen, for example, by lysis or predation; however, proposed suppression mechanisms do not specially require the elimination of the pathogen in compost-amended container substrate unless necessary inquiry is performed. Yogev et al. (2006) also explored induced systemic resistance mechanism toward formae speciales of *Fusarium* in respected hosts. The study demonstrated that compost-mediated suppressiveness toward a wide range of pathogens is rendered by several mechanisms, including systemic resistance. Researchers concluded that whether these suppression mechanisms were prevalent for all types of composts needs further assessment.

Literature on vermicompost-mediated suppression mostly suggested that besides the amount of organic matter and nutrient content, microbiological component of vermicompost determines its efficacy in plant protection and other applications. Given that there are no widely acceptable and reliable criteria for assessment of suppression potential of vermicomposts derived from diverse sorts of feedstock, molecular methods are recently being utilized to fulfill that task (Gopalakrishnan et al. 2011; Yasir et al. 2009a, b; Fernández-Gómez et al. 2012). One example of these studies is the work by Fernández-Gómez et al. (2012) who manipulated the denaturing gradient gel electrophoresis (DGGE) and COMPOCHIP (i.e., a microarray targeting typical bacteria of stabilized organic materials and pathogenic bacteria) for investigating the bacterial communities of four different vermicomposts. Statistically assuring data were obtained from DGGE on composition of bacterial communities and corresponding particular chemical features. In addition COMPOCHIP showed differences in the abundance of particular bacterial taxa among the vermicomposts, giving an idea on if the vermicompost harbor inhabitant bacteria, which are potent for disease suppression of a given plant pathogen. Results by Fernández-Gómez et al. (2012) supported the idea by Yasir et al. (2009a, b) and Gopalakrishnan et al. (2011) that detection of *Streptomyces* sp. in a vermicompost renders the vermicompost to exhibit strong antifungal activities against several soilborne pathogenic fungi, including *F. oxysporum* f. sp. *ciceri*. Fernández-Gómez et al. (2012) further showed that the joint use of DGGE and COMPOCHIP could distinguish among different vermicomposts on the basis of their inhabiting bacterial communities that facilitate vermicompost-mediated suppression mechanisms. Therefore, DGGE can be a rapid fingerprinting method, in a single experiment, useful to ascertain the degree of similarity among bacterial communities of vermicomposts produced from wastes of different natures and origins.

15.4.1 Direct Mechanisms: Microbial Suppression

Thermophilic and mesophilic compost products are known to be exceedingly rich in microbial population and diversity, partially fungi, bacteria, and actinomycetes (Hoitink and Grebus 1997; Hoitink et al. 1997; Lazcano et al. 2008; Stone et al. 2004). Traditional thermophilic composts promote only selected microbes, while non-thermophilic vermicomposts are rich sources in microbial diversity and activity that are derived from a wide variety of antagonistic bacteria, acting as effective biocontrol agents in suppression of soilborne phytopathogenic fungi (Chaoui et al. 2002; Scheuerell et al. 2005; Singh et al. 2008; Jack et al. 2011). Manandhar and Yami (2008) compared disease suppression efficacy of aerated and non-aerated composts and vermicompost teas on foot rot disease of rice caused by *F. moniliforme* Sheldon–*Gibberella fujikuroi* in field. Aerated vermicompost tea extract provided the highest level of disease control followed by aerated compost tea, non-aerated compost tea, and non-aerated vermicompost tea in decreasing order.

There are an increasing number of reports, revealing high microbial diversity in vermicomposts, which harbors a wide variety of notorious antagonistic bacteria (Szczech 1999; Chaoui et al. 2002; Singh et al. 2008; Simsek Ersahin 2007; Pathma and Sakthivel 2012). For example, the bacterial strain *Chitinophaga vermicomposti*, isolated from paper mill and dairy sludge vermicompost, was antagonistic to many soilborne fungi, including *Colletotrichum coccodes*, *Pythium ultimum*, and *Fusarium moniliforme* (Yasir et al. 2010). In another study by Gopalakrishnan et al. (2011), five selected strains of actinomycetes, isolated from an herbal vermicompost, effectively controlled Fusarium wilt disease caused by *F. oxysporum* f. sp. *ciceri* (FOC) in chickpea under green house and field conditions. Gopalakrishnan et al. (2011) stated that two of the five antagonistic actinomycetes, producing cellulase and protease, would act as effective biocontrol agents on cellulose and protein cell wall-bearing pathogens, such as *Phytophthora* and *Pythium* spp. The researchers also underlined that the broad range of antifungal activity of the five antagonistic actinomycetes demonstrates the multiple mechanisms of suppression effect (antibiosis, HCN, siderophore, IAA, and cell wall-degrading enzymes) which may involve more than one antifungal metabolite production. They concluded that the indigenous microbial populations harbored by vermicomposts have the potential for biological control of *Fusarium* diseases.

Earthworm castings are rich in nutrients and calcium humate, which is a binding agent that reduces desiccation of individual castings and also favors the proliferation of beneficial microbes, such as *Trichoderma* sp. (Tiunov and Scheu 2000), *Pseudomonas* spp. (Schmidt et al. 1997), and mycorrhizal spores (Doube et al. 1995). Earthworm activity increased the communities of Gram-negative bacteria (Elmer 2009). Vermicompost-associated chitinolytic bacterial communities, viz., *Nocardioides oleivorans*, several species of *Streptomyces*, and *Staphylococcus epidermidis*, showed inhibitory effects against plant phytopathogens such as *R. solani*, *Colletotrichum coccodes*, *P. ultimum*, *P. capsici*, and *F. moniliforme* (Yasir et al. 2009a, b, 2010).

Some research reports on vermicomposts exhibited the presence of bacteria, such as *Bradyrhizobium japonicum*, which improved nodulation on soybean roots, and pseudomonads and actinomycetes, which suppressed *F. oxysporum* f. sp. *asparagi* and *F. proliferatum* in asparagus, *Verticillium dahlia* in eggplant, and *F. oxysporum* f. sp. *lycopersici* race 1 in tomato (Elmer 2009). In addition, diverse fluorescent pseudomonads, free-living N_2 fixers, *Azospirillum* spp., *Azotobacter* spp., autotrophic *Nitrosomonas* spp., *Nitrobacter* spp., ammonifying bacteria, and phosphate solubilizers were identified in a vermicompost produced from coconut leaves by *Eudrilus* spp. (Gopal et al. 2009). Those studies exhibited the presence of indigenous beneficial and antagonistic bacteria with high diversity, nourished fundamentally by the parental organic waste used and the earthworm species involved in vermicompost production (Fernandez-Gomez et al. 2012).

Simsek-Ersahin (2007) evaluated the disease suppression efficacy of mature (9 months old) solid vermicompost, produced from apple scabs, potato, and tree bark (*Salix* spp.), on root and stem rot of cucumber (*Cucumis sativus* L.) caused by *F. oxysporum* f. sp. *cucumerinum*. After vermicompost-amended substrates (0, 10, 20, and 30 % vermicompost mixed with pot substrate) were prepared, the substrates were inoculated with the pathogen, and then 2-day-old cucumber seedlings were transferred into the pots. *Fusarium* disease symptoms on the roots of cucumber seedlings were observed throughout 2 months. At the second week after the inoculation, disease symptoms were distinct on the primary roots of all cucumber seedlings from all treatments, including 20 and 30 % vermicompost-amended pots (Fig. 15.1). Four weeks after the inoculation, previously explicit symptoms of *Fusarium* infection on primary and even lateral roots of the seedlings in 20 and 30 % vermicompost-amended pots were eliminated (Fig. 15.3). The symptoms caused by *Fusarium* in seedlings from the pots amended with 20 and 30 % vermicompost were completely ameliorated, while seedlings from the pots with 10 % and 0 % vermicompost (control) were either extremely retarded or died (Figs. 15.2 and 15.3). Aqueous vermicompost exhibited strong antagonistic effect against the fungal growth at in vitro conditions as well (Fig. 15.4). Molecular identification of 16S ribosomal RNA (rRNA) gene sequence analysis of the bacterial strain, isolated from the aqueous vermicompost, was defined as *Lysinibacillus fusiformis*.

Fig. 15.1 The roots of 2-week-old cucumber seedlings grown in pots amended with vermicompost ratios as 30, 20, 10, and 0 % (from *left* to *right*). All inoculated with *F. oxysporum* f. sp. *cucumerinum* at 2-day-old state

Fig. 15.2 Four-week-old cucumber seedlings grown in pots amended with vermicompost ratios as 30, 20, 10, and 0 % (from *left* to *right*). All four inoculated with *F. oxysporum* f. sp. *cucumerinum* at 2-day-old state. The first pot at the far *right* is the negative control for the pathogen with no vermicompost amendment

The antagonistic activity in vermicompost-mediated suppression is primarily attributed to the nature of their autochthonous microorganisms with plant growth-promoting and biocontrol traits. Members of genus *Pseudomonas*, *Bacillus*, *Streptomyces*, *Azospirillum*, *Azoarcus*, *Azotobacter*, *Burkholderia*, *Cyanobacteria*, *Herbaspirillum*, and *Chryseobacterium* are known for their potential for plant growth-promoting and plant disease suppression (Gandhi et al. 2009; Pathma et al. 2010, 2011). Pathma and Sakthivel (2013) isolated 193 bacteria originated from straw and goat manure vermicompost and taxonomically defined them as bacteria with high antagonistic and biofertilizing potential. Predominant genera of the 193 bacteria were *Bacillus* (57 %), *Pseudomonas* (15 %), and *Microbacterium* (12 %); the remaining genera were comprised of *Acinetobacter* (5 %), *Chryseobacterium* (3 %), *Arthrobacter*, *Pseudoxanthomonas*, *Stenotrophomonas*, *Paenibacillus*, *Rhodococcus*, *Enterobacter*, *Rheinheimera*, and *Cellulomonas*. Of the 193 bacteria, 49 % showed antagonistic potential against phytopathogenic

Fig. 15.3 The roots of 4-week-old cucumber seedlings grown in pots amended with vermicompost ratios as 0 % (negative control for pathogen), 0, 10, 20, and 30 % (from *left* to *right*). All four, except the one at the far *left*, inoculated with *F. oxysporum* f. sp. *cucumerinum* at 2-day-old state. *Fusarium* infection was entirely alleviated on the seedlings from pots with 20 and 30 % vermicomposts

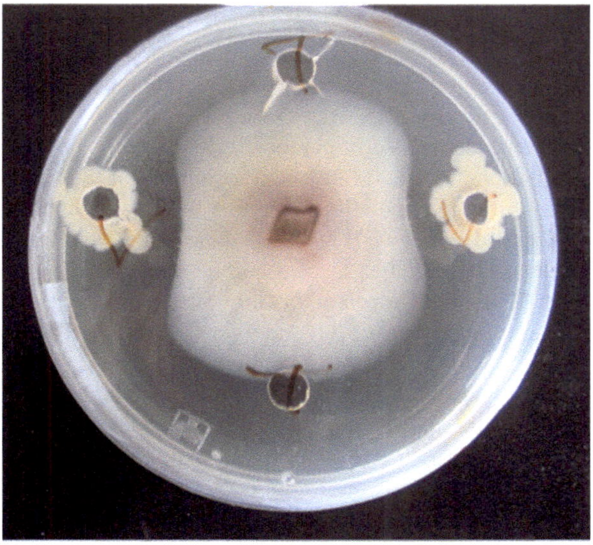

Fig. 15.4 Antagonism of water extract of the vermicompost on *F. oxysporum* f. sp. *cucumerinum* in a malt agar medium. Unsterilized vermicompost water extract was injected into the *right* and *left* holes and sterile water into the *top* and *bottom* holes, and then a mycelium disk with *F. oxysporum* f. sp. *cucumerinum* was placed in the center of the plate

fungi, e.g., the strains of *Pseudomonas* against *F. oxysporum*. Additional functional characterization of the bacteria revealed that particularly strains of both *Pseudomonas* and spore-forming *Bacillus* bacteria exhibited broad-spectral plant growth-promoting traits and antagonistic potential. These results by Pathma and Sakthivel (2012, 2013) supported those previously found by Simsek-Ersahin (2007) that the genus *Bacillus* have potential for suppression of the genus *Fusarium*.

Research on vermicompost-mediated suppression indicated that soilborne disease suppression is exclusively related to diversity and richness of bacterial community and activity in the vermicompost (Chaoui et al. 2002; Scheuerell et al. 2005). Recently, studies manipulating molecular techniques to identify that microbial variety and richness responsible for plant growth promotion and disease suppression traits of vermicomposts have been reported (Yasir et al. 2009a, b, 2010; Pathma and Sakthivel 2012, 2013; Gopalakrishnan et al. 2010, 2011, 2012). For example, Yasir et al. (2009a) assessed whether earthworm activity influences the variety of resident bacterial community sludge mixtures throughout vermicomposting process and the suppression potential of the end product, vermicompost, on spore germination of *F. moniliforme* in comparison with sludge. They also examined whether chitinase gene diversity is involved in the suppression of soilborne plant pathogenic fungi, by using culture-dependent and independent methods. Analysis of chitinolytic isolates and chitinase gene diversity revealed that vermicomposting process enriched chitinolytic bacterial communities, for example, Actinobacteria, which exclusively inhibited plant fungal pathogens.

Recently, Barocio-Ceja et al. (2013) have evaluated in vitro activities of *Trichoderma* sp. and *Aspergillus* sp., isolated from chicken-manure vermicompost, on two important species of tomato wilt pathogens (i.e., *Fusarium* and *Rhizoctonia*). Tomato pathogens, isolated from wilted tomato plants, were identified as *F. oxysporum*, *F. solani*, *F. subglutinans*, and *Rhizoctonia* sp., using molecular analysis. Of 11 isolates from chicken-manure vermicompost, only *Trichoderma* sp. and *Aspergillus* sp. exclusively inhibited growth of *F. oxysporum* and moderately *Rhizoctonia* sp. Barocio-Ceja et al. (2013) stated that soil applications of chicken-manure vermicompost can significantly increase the soil microbe populations (6.5×10^4–1.8×10^5 conidia/g), thus diversifying the microbiota and promoting the populations of these antagonists against phytopathogenic fungi.

Castillo et al. (2013) designed a study to elucidate the interactions among biochemical parameters in relation to microbial dynamics during and 2 months after (2 months) vermicomposting of two different types of lignocellulosic organic wastes: wet olive cake (O) and vine shoots (W). To do that, they analyzed chemical changes with respect to chemical properties as well as biochemical functions (dehydrogenase, b-glucosidase, acid phosphatase, urease, and ortho-diphenol oxidase). Furthermore, they assayed the microbial community changes in the bacterial and fungal structure by determination of biomass of the total abundance of bacteria and fungi as well as microbial taxa (Alpha-, Beta-, and Gammaproteobacteria and Actinobacteria), using taxa-specific real-time PCR assay. Multivariate correlation analysis between microbial structure and abundance and enzyme activities revealed significant correlations between b-glucosidase activity and bacterial and fungal structure. In the vermicomposting period of O and W, a decline was found in bacteria (94 and 77 %), fungi (93 and 94 %), and Gammaproteobacteria (56 and 71 %), but an increase in Betaproteobacteria and Actinobacteria (62–79 %). Alphaproteobacteria increased only in O (26 %). Despite the difference in content of initial lignocellulosic wastes, the mature vermicomposts contain similar microbial communities and biochemical parameters in relation to the indigenous microbial community.

15.4.2 Indirect Mechanisms

Induced systemic resistance (ISR; or systemic acquired resistance, SAR) is "a state of enhanced defensive capacity developed by a plant when appropriately stimulated" (Bakker et al. 2003; van Loon et al. 1998). ISR can provide protection against viral, fungal, and bacterial plant pathogens and root, vascular, and foliar diseases of plants. A variety of soil and rhizosphere bacterial and fungal isolates have been reported to turn on ISR in plants (van Loon et al. 1998). Microbial metabolites such as salicylic acid, siderophores, antibiotics, and lipopolysaccharides have been implicated in microbially mediated ISR (Bakker et al. 2003).

Previously, induced resistance has been implicated in the genus *Fusarium* in that nonpathogenic *F. oxysporum* soil isolates induced systemic resistance in watermelon against Fusarium wilt in some suppressive soil systems and in greenhouse systems (Larkin et al. 1996; Keener et al. 2000). Compost-mediated induced resistance to Fusarium wilt of tomato resulted in reduced fungal colonization of the tomato roots due to an increase in physical barriers, such as callose-enriched, multilayered wall appositions and osmiophilic deposits to fungal penetration (Pharand et al. 2002). Inoculation of composts with *Pythium* also induced the SAR in tomato, leading to increased level of suppression and protection by formation of physical barriers (Pharand et al. 2002). There are reports stated that compost products, applied as soil amendments, control several important soilborne diseases such as wilts caused by *Fusarium* (Reuveni et al. 2002). A recent study by Singh et al. (2012) reported a significant increase in vermicompost-mediated suppression in disease incidence of root rot/wilt, a complex disease of *C. forskohlii* involving *F. chlamydosporum* and *Ralstonia solanacearum*, in organic field conditions.

15.5 Conclusion

Considerable research, conducted on solid and aqueous vermicompost products as plant growth substrates and plant disease-suppressive agents, has demonstrated a great potential for both plant growth promotion and plant protection. However, research findings exhibited a large variability and inconsistency. In addition, an adequate understanding of vermicompost-mediated suppression mechanisms of plant diseases caused by *Fusarium* requires more comprehensive research to better identify pathogen- and host-specific components of suppression mechanisms. There are several difficulties to overcome, such as obtaining a better understanding of vermicompost microbial population variability through vermicomposting and maturation process and interactions between the pathogen and indigenous microbial community. Since suppression effect is lost after sterilization or pasteurization, suppression activity is attributed to a diverse microbial community of vermicomposts. Therefore, advanced understanding of microbial composition and structure and function of the antagonistic populations in vermicompost could reduce variability and increase predictability of disease control efficacy. Developing methodology for assessment of vermicompost-mediated suppression for a single vermicompost batch is one of the major tasks needed to be executed by the researchers.

At present, there is information on thermophilic compost-mediated *Fusarium* suppression that suggests involvement of different mechanisms, including induced resistance, observed in suppression of *F. oxysporum*: *melonis*, *basillici*, *radicis-lycopersici*, and *radicis-cucumerinum*. High genetic flexibility of the genus *Fusarium* requires case-by-case, host- and formae speciales-specific studies for a better understanding of vermicompost-mediated suppression mechanisms of diseases caused by the genus *Fusarium*. Recently, research on vermicompost microbiology has extensively been manipulated by non-cultivating methods based on molecular characteristics, targeting information-bearing macromolecules (RNA, DNA, or lipids) using PCR-based molecular methods (DGGE, clone library, and micro-array). There are a few pioneering attempts to analyze phylogenetic relationships in vermicompost microbial community, as well as reveal the microbial community composition, and structure to functions, using metagenomics and metatranscriptomics.

Biological control of plant diseases by vermicompost products provides an environmentally friendly agricultural practice and employs many benefits as well. There are many obstacles in obtaining information on vermicompost-mediated soilborne disease suppression, specifically those caused by *Fusarium*. The overarching challenges are (1) research that will lead to a better understanding of interactions between indigenous antagonistic microbial populations, involved in vermicompost-mediated suppression, pathogen, and plant in rhizosphere zone, (2) developing reliable quality disease control tools, (3) and integrating research findings into commercial vegetable production.

References

Agrios GN (2005) Plant pathology, 5th edn. Elsevier Academic Press, London, 922 p

Anonymous (2014a) Available at http://www.broadinstitute.org

Anonymous (2014b) Dr. Elaine Ingham of SFI. Available at http://www.soilfoodweb.com

Arancon NQ, Edwards EA, Yardim E, Lee S (2002) Management of plant parasitic nematode population by vermicomposts. Proc Brighton Crop Prot Conf Pests Dis 2(8B–2):705–710

Arancon NQ, Edwards CA, Bierman P (2006) Influences of vermicomposts on field strawberries: Part 2. Effects on soil microbiological and chemical properties. Bioresour Technol 97:831–840

Bakker PAHM, Ran LX, Pieterse CMJ, van Loon LC (2003) Understanding the involvement of rhizobacteria-mediated induction of systemic resistance in biocontrol of plant diseases. Can J Plant Pathol 25:5–9

Barocio-Ceja NB, Ceja-Torres LF, Morales-García JL, Silva-Rojas HV, Flores-Magallón R, Ochoa-Estrada S (2013) In vitro bio-control of tomato pathogens using antagonists isolated from chicken-manure vermicompost. Phyton 82(1):15–22

Booth C (1971) The genus *Fusarium*. Commonwealth Mycological Institute, Kew Surrey, 237 p

Carvalho SIM, Otero M, Duarte AC, Santos EBH (2008) Effects of solar radiation on the fluorescence properties and molecular weight of fulvic acids from pulp mill effluents. Chemosphere 71:1539–1546

Castillo JM, Romero E, Nogales R (2013) Dynamics of microbial communities related to biochemical parameters during vermicomposting and maturation of agro-industrial lignocelluloses wastes. Bioresour Technol 146:345–354

Chaoui H, Edwards CA, Brickner A, Lee S, Arancon NQ (2002) Suppression of the plant diseases, *Pythium* (damping-off), *Rhizoctonia* (root rot) and *Verticillium* (wilt) by vermicomposts. Proc Brighton Crop Prot Conf Pests Dis 3:711–716

Chen W, Hoitink HAJ, Schmitthenner AF, Tuovinen OH (1988) The role of microbial activity in suppression of damping-off caused by *Pythium ultimum*. Phytopathology 78:314–322

Cook RJ, Baker KF (1983) The nature and practice of biological control of plant pathogens. APS Press, St. Paul

Doube BM, Stephens PM, Davorena CW, Ryderb MH (1994) Interactions between earthworms, beneficial soil microorganisms and root pathogens. Appl Soil Ecol 1:3–10

Doube BM, Ryder MH, Davoren CW, Meyer T (1995) Earthworms: a down under delivery system service for biocontrol agents of root disease. Acta Zool Fennica 196:219–223

Edwards CA (1995) Commercial and environmental potential of vermicomposting: a historical overview. BioCycle 36(6):62–63

Edwards CA (1998) The use of earthworms in processing organic wastes into plant growth media and animal feed protein. In: Edwards CA (ed) Earthworm ecology. CRC press, Boca Raton, FL, pp 327–354

Edwards CA, Arancon N (2004) Vermicompost suppress plant pests and disease attacks. In: Rednova News. http://www.rednova.com/display

Edwards CA, Burrows I (1988) The potential of earthworm composts as plant growth media. In: Edwards CA, Neuhauser EF (eds) Earthworm environmental and waste management. SPB Academic Publishers, The Hague, pp 211–220

Edwards CA, Arancon NQ, Greytak S (2006) Effects of vermicompost teas on plant growth and diseases. Biocycle 2006:28–31

Elmer WH (2009) Influence of earthworm activity on soil microbes and soilborne diseases of vegetables. Plant Dis 93:175–179

Fernández-Gómez MJ, Nogales R, Insam H, Romero E, Goberna M (2012) Use of DGGE and COMPOCHIP for investigating bacterial communities of various vermicomposts produced from different wastes under dissimilar conditions. Sci Total Environ 414:664–671

Fokkema NJ (1993) Opportunities and problems of control of foliar pathogens with microorganisms. Pestic Sci 37:411–416

Fravel D, Olivain C, Alabouvette C (2003) *Fusarium oxysporum* and its biocontrol. New Phytol 157(3):493–502

Gandhi PM, Narayanan BK, Naik RP, Sakthivel N (2009) Characterization of *Chryseobacterium aquaticum* strain PUPC1 producing a novel antifungal protease from rice rhizosphere soil. J Microbiol Biotechnol 19:99–107

Gold MV (2007) Sustainable agriculture: definitions and terms. Special reference briefs series no. SRB 94-5. National Agricultural Library, ARS, USDA. Available at www.nal.usda.gov/afsic/pubs/terms/srb9902.shtml

Goldman LR (2008) Encyclopedia of public health: fungicides. Cited on 6 June 2008. Available at http://www.answers.com/topic/fungicide

Gopal M, Gupta A, Sunil E, Thomas VG (2009) Amplification of plant beneficial microbial communities during conversion of coconut leaf substrate to vermicompost by *Eudrilus* sp. Curr Microbiol 59:15–20

Gopalakrishnan S, Kannan IGK, Alekhya G, Humayun P, Meesala SV, Kanala D (2010) Efficacy of *Jatropha*, *Annona* and *Parthenium* biowash on *Sclerotium rolfsii*, *Fusarium oxysporum* f. sp. *ciceri* and *Macrophomina phaseolina*, pathogens of chickpea and sorghum. Afr J Biotechnol 9(47):8048–8057

Gopalakrishnan S, Pande S, Sharma M, Humayun P, Kiran BK, Sandeep D, Vidya MS, Deepthi K, Rupela O (2011) Evaluation of actinomycete isolates obtained from herbal vermicompost for the biological control of fusarium wilt of chickpea. Crop Prot 30:1070–1078

Gopalakrishnan S, Humayun P, Vadlamudi S, Vijayabharathi R, Bhimineni RK, Rupela O (2012) Plant growth-promoting traits of *Streptomyces* with biocontrol potential isolated from herbal vermicompost. Biocontrol Sci Technol 22:1199–1210

Grantina-Ievina L, Malisevs A, Nikolajeva V (2014) Antifungal activity of water extracts of vermicompost. Environ Exp Biol 12:71–72

Hoitink HAJ, Boehm MJ (1999) Biocontrol within the context of soil microbial communities: a substrate-dependent phenomenon. Annu Rev Phytopathol 37:427–446

Hoitink HA, Grebus ME (1997) Composts and control of plant diseases. In: Hayes MHB, Wilson WS (eds) Humic substances, peats and sludges: health and environmental aspects. Royal Society of Chemistry, Cambridge, pp 359–366

Hoitink HAJ, Stone AG, Han DY (1997) Suppression of plant diseases by compost. Hortic Sci 32: 184–187

Hoitink HAJ, Krause MS, Han DY, Ceuster TJJ (1998) Soil organic matter quality and induced resistance of plants to root rots and foliar diseases. Int Plant Propagators Soc Proc 48:361–367

ILO (1996) Wage workers in agriculture: condition of employment and work; sectoral activities programme. Report for discussion at the tripartite meeting on improving conditions of employment and work of agricultural wage workers in the context of economic restructuring. International Labor Organization, Geneva

Jack ALH, Rangarajan SA, Culman T, Sooksa-Nguan Thies JE (2011) Choice of organic amendments in tomato transplants has lasting effects on bacterial rhizosphere communities and crop performance in the field. Appl Soil Ecol 48(1):94–101

Kanangara T, Utkhede RS, Paul JW, Punja ZK (2000) Effects of mesophilic and thermophilic composts on suppression of Fusarium root and stem rot of greenhouse cucumber. Can J Microbiol 46:1021–1029

Kai H, Ueda T, Sakaguchi M (1990) Antimicrobial activity of bark–compost extracts. Soil Biol Biochem 22:983–986

Keener HM, Dicks WA, Hoitink HAJ (2000) Composting and beneficial utilization of composted by-product materials. In: Bartels JM, Dicks WA (eds) Land application on municipal agricultural, industrial and municipal by-products. Book series, vol 6. Soil Science Society of America, Madison, pp 315–343

Kistler HC, Rep M (2010) Comparative genomics reveals mobile pathogenicity chromosomes in *Fusarium*. Nature 464:367–373

Kumar V, Garkoti A, Tripathi HS (2013) Management of vascular wilt of lentil through biocontrol agents and organic amendments in tarai area of Uttarakhand State. Bioscan 8(2):575–577

Larkin RP, Hopkins DL, Martin FN (1996) Suppression of Fusarium wilt of watermelon by nonpathogenic *Fusarium oxysporum* and other microorganisms recovered from a disease-suppressive soil. Phytopathology 86:812–819

Lazcano C, Gomez-Brandon M, Dominguez J (2008) Comparison of the effectiveness of composting and vermicomposting for the biological stabilization of cattle manure. Chemosphere 72:1013–1019

Lockwood JL (1990) Relation of energy stress to behaviour of soilborne plant pathogens and to disease development. In: Hornby D (ed) Biological control of soilborne plant pathogens. CAB International, Wallingford, pp 197–214

Lodha S (1995) Soil solarization, summer irrigation, and amendments for the control of *Fusarium oxysporum* f. sp. *cumini* and *Macrophomina phaseolina* in arid soils. Crop Prot 14:215–219

Ma LJ, van der Does HC, Borkovich KA, Coleman JJ, Daboussi MJ, Di Pietro A, Dufresne M, Freitag M, Grabherr M, Henrissat B, Houterman PM, Kang S, Shim WB, Woloshuk C, Xie X, Xu JR, Antoniw J, Baker SE, Bluhm BH, Breakspear A, Brown DW, Butchko RA, Chapman S, Coulson R, Coutinho PM, Danchin EG, Diener A, Gale LR, Gardiner DM, Goff S, Hammond-Kosack KE, Hilburn K, Hua-Van A, Jonkers W, Kazan K, Kodira CD, Koehrsen M, Kumar L,

Lee YH, Li L, Manners JM, Miranda-Saavedra D, Mukherjee M, Park G, Park J, Park SY, Proctor RH, Regev A, Ruiz-Roldan MC, Sain D, Sakthikumar S, Sykes S, Schwartz DC, Turgeon BG, Wapinski I, Yoder O, Young S, Zeng Q, Zhou S, Galagan J, Cuomo CA, Kistler HC, Rep M (2010) Comparative genomics reveals mobile pathogenicity chromosomes in *Fusarium*. Nature 464(7287):367–373

Malathi S, Mohan S (2013) Induction of defence-related enzymes in onion by combined application of fungal and bacterial biocontrol agents against *Fusarium oxysporum* f. sp. *cepae*. Arch Phytopathol Plant Prot 46:243–252

Manandhar T, Yami KD (2008) Biological control of foot rot disease of rice using fermented products of compost and vermicompost. Sci World 6(6):52–57

Moody SA, Piearce TG, Dighton J (1996) Fate of some fungal spores associated with wheat straw decomposition on passage through the guts of *Lumbricus terrestris* and *Aporrectodea longa*. Soil Biol Biochem 28:533–537

Nakamura Y (1996) Interactions between earthworms and microorganisms in biological control of plant root pathogens. Farming Jpn 30(6):37–43

Nakasone AK, Bettiol W, de Souza RM (1999) The effect of water extracts of organic matter on plant pathogens. Summa Phytopathol 25:330–335

Nelson PE, Toussoun TA, Marasas WFO (1983) *Fusarium* species. An illustrated manual for identification. The Pennsylvania State University Press, University Park, 193 p

Oritsejafor JJ, Adeniji MO (1990) Influence of host and non-host rhizospheres and organic amendments on survival of *Fusarium oxysporum* f. sp. *elaeidis*. Mycol Res 94:57–63

Orlikowski LB (1999) Vermicompost extract in the control of some soil borne pathogens. Int Symp Crop Prot 64:405–410

Oyarzun PJ, Postma J, Luttikholt AJG, Hoogland AE (1994) Biological control of foot and root rot in pea caused by *Fusarium solani* with nonpathogenic *Fusarium oxysporum* isolates. Can J Bot 72:843–852

Park D (1958) The saprophytic status of *Fusarium oxysporum* Schl. causing vascular wilt of oil palm. Ann Bot 22:19–35

Pathma J, Sakthivel N (2012) Microbial diversity of vermicompost bacteria that exhibit useful agricultural traits and waste management potential. SpringerPlus 1:26

Pathma J, Sakthivel N (2013) Molecular and functional characterization of bacteria isolated from straw and goat manure based vermicompost. Appl Soil Ecol 70:33–47

Pathma J, Kamaraj KR, Sakthivel N (2010) Mechanisms of fluorescent pseudomonads that mediate biological control of phytopathogens and plant growth promotion of crop plants. In: Maheshwari DK (ed) Bacteria in agrobiology: plant growth. Springer, Berlin, pp 77–105

Pathma J, Rahul GR, Kamaraj KR, Subashri R, Sakthivel N (2011) Secondary metabolite production by bacterial antagonists. J Biol Control 25:165–181

Pereira JCR et al (1996) Control of Sclerotium cepivorum by the use of vermicompost, solarization, Trichoderma harzianum and Bacillus subtilis. Summa Phytopathol 22:228–234

Pharand B, Carisse O, Benhamou N (2002) Cytological aspects of compost-mediated induced resistance against Fusarium crown and root rot in tomato. Phytopathology 92:424–438

Pimentel D (1997) Techniques for reducing pesticide use. Economic and environmental benefit. Wiley, Chichester, pp 51–78

Prabha ML (2009) Waste management by vermitechnology. Indian J Environ Prot 29(9):795–800

Reddy SA, Bagyaraj DJ, Kale RD (2012) Management of tomato bacterial spot caused by *Xanthomonas campestris* using vermicompost. J Biopest 5(1):10–13

Reuveni R, Raviv M, Krasnovsky A, Freiman L, Medina S, Bar A, Orion D (2002) Compost induces protection against *Fusarium oxysporum* in sweet basil. Crop Prot 21:583–587

Rodriguez JA, Zavaleta E, Sanchez P, Gonzalez H (2000) The effect of vermicomposts on plant nutrition, yield and incidence of root and crown rot of gerbera (*Gerbera jamesonii* H. Bolus). Fitopatologia 35:66–79

Sahni S, Sarma BK, Singh KP (2008) Management of *Sclerotium rolfsii* with integration of non-conventional chemicals, vermicompost and *Pseudomonas syringae*. World J Microbiol Biotechnol 24(4):517–522

Scheuerell S, Mahaffee W (2002) Compost tea: principles and prospects for plant disease control. Compost Sci Util 10(4):313–338

Scheuerell SJ, Sullivan DM, Mahaffee WF (2005) Suppression of seedling damping-off caused by *Pythium ultimum*, *P. irregulare*, and *R. solani* in container media amended with a diverse range of Pacific Northwest compost sources. Phytopathology 95:306–315

Schmidt O, Doubre BM, Ryder MH, Killman K (1997) Population dynamics of *Pseudomonas corrugata* 2140R LUX8 in earthworm food and in earthworm cast. Soil Biol Biochem 29: 523–528

Serra-Wittling C, Houot S, Alabouvette C (1996) Increased soil suppressiveness to Fusarium wilt of flax after addition of municipal solid waste compost. Soil Biol Biochem 28:1207–1214

Simsek Ersahin Y, Haktanir K, Yanar Y (2009) Vermicompost suppresses Rhizoctonia solani Kuhn in cucumber seedlings. J Plant Dis Prot 9:15–17

Simsek-Ersahin Y (2007) Determination of suppressive effect of vermicast and its humic substances on root and stem rot of cucumber (*Cucumis sativus* L.) caused by *Rhizoctonia solani* (Kuhn) and *Fusarium oxysporum* f. sp. *cucumerinum*. PhD thesis, GOP University, Tokat

Simsek-Ersahin Y (2011) The use of vermicompost products to control plant diseases and pests. In: Karaka A (ed) Biology of earthworms. Springer, Berlin, pp 191–214

Singh UP, Maurya S, Singh DP (2003) Antifungal activity and induced resistance in pea by aqueous extract of vermicompost and for control of powdery mildew of pea and balsam. J Plant Dis Prot 110:544–553

Singh R, Sharma RR, Kumar S, Gupta RK, Patil RT (2008) Vermicompost substitution influences growth, physiological disorders, fruit yield and quality of strawberry (*Fragaria xananassa* Duch.). Bioresour Technol 99:8507–8511

Singh R, Soni SK, Awasthi A, Kaira A (2012) Evaluation of vermicompost doses for management of root-rot disease complex in *Coleus forskohlii* under organic field conditions. Aust Plant Pathol 41(4):397–403

Stone AG, Scheuerell SJ, Darby HM (2004) Suppression of soilborne diseases in field agricultural systems: organic matter management, cover cropping and other cultural practices. In: Magdoff F, Weil R (eds) Soil organic matter in sustainable agriculture. CRC Press, Boca Raton, FL, pp 131–177

Szczech M (1999) Suppressiveness of vermicompost against fusarium wilt of tomato. J Phytopathol 147:158–161

Szczech M, Smolinska U (2001) Comparison of suppressiveness of vermicompost produced from animal manures and sewage sludge against *Phytophthora nicotianae* Breda de Haan var. *Nicotianae*. J Phytopathol 149:77–82

Szczech M, Rondomanski W, Brzeski MW, Smolinska U, Kotowski JF (1993) Suppressive effect of a commercial earthworm compost on some root infecting pathogens of cabbage and tomato. Biol Agric Hortic 10:47–52

Tiunov AV, Scheu S (2000) Microfungal communities in soil litter and casts of *Lumbricus terrestris* (Lumbricidae): a laboratory experiment. Appl Soil Ecol 14:17–26

Toyota K, Kimura M (1993) Colonization of chlamydospores of *Fusarium oxysporum* f. sp. *raphani* by soil bacteria and their effects on germination. Soil Biol Biochem 25:193–197

Toyota K, Ritz K, Young IM (1996) Microbiological factors affecting the colonisation of soil aggregates by *Fusarium oxysporum* f. sp. *raphani*. Soil Biol Biochem 28:1513–1521

Trillas-Gay MI, Hoitink HAJ, Madden LV (1987) Nature of suppression of fusarium wilt of radish in a container medium amended with composted hardwood bark. Plant Dis 70:1023–1027

van Loon LC, Bakker PAHM, Pieterse CMJ (1998) Systemic resistance induced by rhizosphere bacteria. Annu Rev Phytopathol 36:453–483

Weltzien HC (1989) Some effects of composted organic materials on plant health. Agric Ecosyst Environ 27:439–446

Yasir M, Aslam Z, Kim SW, Lee SW, Jeon CO, Chung YR (2009a) Bacterial community composition and chitinase gene diversity of vermicompost with antifungal activity. Bioresour Technol 100:4396–4403

Yasir M, Aslam Z, Song GC, Jeon CO, Chung YR (2009b) *Eiseniicola composti* gen. nov., sp. nov., with antifungal activity against plant pathogenic fungi. Int J Syst Evol Microbiol 60:268

Yasir M, Aslam Z, Song GC, Bibi F, Jeon CO, Chung YR (2010) *Chitinophaga vermicomposti* sp. nov. with antifungal activity, isolated from vermicompost. Int J Syst Evol Microbiol 60:268

Yogev A, Raviv M, Hadar Y, Cohen R, Katan J (2006) Plant waste-based composts suppressive to diseases caused by pathogenic *Fusarium oxysporum*. Eur J Plant Pathol 116:267–278

Zaller JG (2006) Foliar spraying of vermicompost extracts: effects on fruit quality and indications of late-blight suppression of field-grown tomatoes. Biol Agric Hortic 24:165–180

Chapter 16
Impact of Organic Amendments on the Suppression of Fusarium Wilt

Christel Baum, Bettina Eichler-Löbermann, and Katarzyna Hrynkiewicz

16.1 Introduction

Fusarium wilts are important diseases of diverse horticultural and agricultural crops and lead to significant yield losses. They are caused by pathogenic *formae speciales* (f. sp.) of the soilborne fungus *Fusarium oxysporum*. *F. oxysporum* is found worldwide and comprises not only pathogenic but also saprophytic nonpathogenic strains. The host-specialist pathotypes (f. sp.) may cause vascular wilt or damping-off. Since pathogenic f. sp. of *F. oxysporum* in general are facultative parasitic fungi, their regulation by crop rotation is ineffective. Also fungicides and host resistance often do not give adequate and sustainable control (Weller et al. 2002). However a positive aspect is, that *Fusarium* spp. have diverse microbial antagonists, which can contribute to their regulation (Toyota et al. 1994).

Soil suppressiveness of Fusarium wilt can be based on suppression (1) of pathogenic f. sp. of *F. oxysporum* only or (2) of *Fusarium* spp. strains in general (Scher and Baker 1980). The specific suppressiveness can limit Fusarium wilt

C. Baum (✉)
Department of Agricultural and Environmental Sciences, Soil Science, University of Rostock, Justus-von-Liebig-Weg 6, 18059 Rostock, Germany
e-mail: christel.baum@uni-rostock.de

B. Eichler-Löbermann
Department of Agricultural and Environmental Sciences, Agronomy and Crop Science, University of Rostock, Justus-von-Liebig-Weg 6, 18059 Rostock, Germany
e-mail: bettina.eichler@uni-rostock.de

K. Hrynkiewicz
Department of Microbiology, Faculty of Biology and Environmental Protection, N. Copernicus University, Lwowska 1, Torun, Poland
e-mail: hrynk@umk.pl

M.K. Meghvansi, A. Varma (eds.), *Organic Amendments and Soil Suppressiveness in Plant Disease Management*, Soil Biology 46,
DOI 10.1007/978-3-319-23075-7_16

caused by pathogenic *F. oxysporum* and is not effective against diseases caused by nonvascular *Fusarium* species including *F. roseum* and *F. solani* or other soilborne pathogens (Alabouvette 1986, 1999).

However, the single introduction of microorganisms isolated from suppressive soils into conducive soils can fail to develop *Fusarium*-suppressive conditions by the lack of reproduction of the microbial community structure and interactions which occur in natural suppressive soils. Therefore, an application of organic amendments, which introduce the substrate in combination with advantageous microbial populations, might be favorable.

In the present chapter an overview on the causal agents of suppression of Fusarium wilt is given, and the quality of different organic amendments for this reason was evaluated. It is aimed to facilitate a selection and optimization of the use of organic amendments in the arable management by reviewing the actual state of knowledge.

16.2 Causes of the Suppression of Fusarium Wilt by Organic Amendments

The causes of the suppression of Fusarium wilt by organic amendments are usually combinations of biotic and abiotic impacts on the pathogen. The application of organic amendments leads to microbial and chemical/biochemical changes in the soils and substrates, which involve properties which are known to control the growth and/or pathogenicity of *F. oxysporum* and can thereby cause suppression of Fusarium wilt (Fig. 16.1).

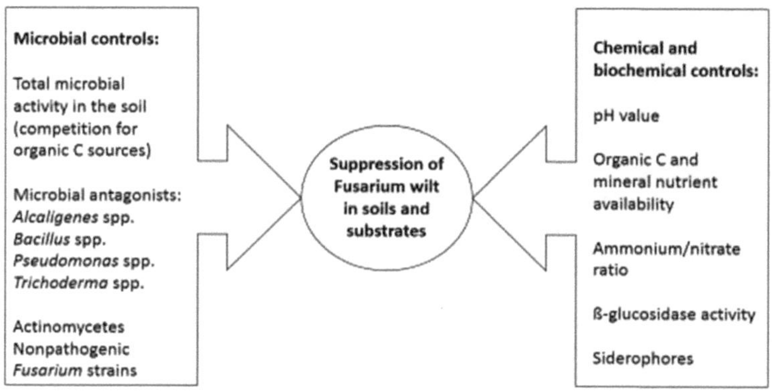

Fig. 16.1 Controls of Fusarium wilt mediated by organic amendments

16.2.1 Microbial Controls of Fusarium *Wilt*

A general increased microbial activity after application of organic amendments leads to pathogen suppression (Ha and Huang 2007; Escuadra and Amemiya 2008), although *F. oxysporum* in general can be isolated even more frequently in these conditions, e.g., after treatments with farmyard manure (Joffe 1963).

Diverse microbial antagonistic taxa contribute with their specific mechanisms to the pathogen control. Among these microbial antagonists of *F. oxysporum* are bacterial and fungal taxa, which were described to contribute to Fusarium wilt suppressiveness like *Alcaligenes* and *Pseudomonas* spp. (e.g., Sayyed and Patel 2011), *Bacillus* spp., and *Trichoderma* spp. Furthermore, actinomycetes, like *Streptomyces lydicus* (Himmelstein et al. 2014) and nonpathogenic *F. oxysporum*, were described to contribute to the suppression (Weller et al. 2002). Specific mechanisms of the antagonists include mycoparasitism, antibiosis, and nutrient or space competition. The competition for organic C sources, which were released from the plants, is a fundamental aspect, since it controls the germination of plant pathogenic fungi in the rhizosphere. Furthermore, the inhibition of pectinases and other enzymes that are essential for the pathogen growth and infection of the host plant can be involved and may include development of induced resistance and/or growth promotion of the host plant in general (Harman et al. 2004; Verma et al. 2007).

Bacteria involved in the nitrogen cycling control Fusarium wilt severity by their impact on the ammonium/nitrate ratio. The diversity of ammonia-oxidizing bacteria was negatively correlated with the Fusarium wilt severity (Senechkin et al. 2014).

16.2.2 Chemical and Biochemical Controls of Fusarium *Wilt*

After addition of fresh organic amendments to soil, ammonium concentration and pH increase temporarily, followed by increased nitrification and a decline in ammonium and pH (Zelenev et al. 2005). Between the chemical properties, the pH value was revealed to be a leading control of the soil suppressiveness. The substrate pH and furthermore the ß-glucosidase activity in the substrate explained more than 91 % of the severity of Fusarium wilt on tomato in experiments of Borrero et al. (2004). Fang et al. (2012) underlined the significant impact of the soil pH and found strongest suppression of Fusarium wilt in the soil pH of 6.7, with significant advantages compared to acidic soils (pH 5.2 or 5.8).

High nitrogen concentrations in general and high ammonium-to-nitrate ratios increase Fusarium wilt incidence and severity (Jones et al. 1993). This means that the impact of organic amendments on the C/N and on the ammonium/nitrate ratios in soil can depend on these properties (Borrero et al. 2012; Huang et al. 2012).

Siderophores, released by soil microorganisms, are high-affinity iron-chelating chemical compounds, which were described to be causal elicitors of *Fusarium* suppression (Sayyed and Patel 2011). This leads back to the fact that low iron availability reduces the growth and sporulation of *F. oxysporum* and was described to be associated with suppression of Fusarium wilt of tomato (Borrero et al. 2004).

16.3 The Impact of the Source and Quality of Organic Amendments

In general, the significance of organic amendments for the control of Fusarium wilt is based on the initial status of the soil and the amount and quality of the organic amendment. Diverse qualities of organic matter were tested on their impact on Fusarium wilt severity.

The significance of different organic amendments as controls of the suppressiveness of soils was indicated in an in vitro experiment by Toyota et al. (1995) and reviewed for several soilborne pathogens in soils and substrates by Bonanomi et al. (2007). In more than half of the observed cases of a total of more than 150 studies, organic amendments contributed to suppression of *Fusarium* spp. The tested organic amendments had decreasing suppressive effects on *Fusarium* spp. in the following order: compost (74 %) > crop residues (56 %) > waste (46 % of the observed cases with highly suppressive or at least suppressive effects). However, peat application even leads often (58 % of the observed cases) to conducive effects and never to induction of suppression of *Fusarium* spp. (Bonanomi et al. 2007).

The significance of different sources and qualities of organic matter to support the suppressiveness of soils against Fusarium wilt will be described in the following subsections by dividing the organic amendments in three groups: plant residues, amendments of animal waste, and composts and complex organic amendments.

16.3.1 The Impact of Plant Residues

In general effects of plant residues on soilborne fungi vary significantly. They can be suppressive (45 % of the observed cases), conducive (28 % of the observed cases), or neutral by reviewing of results of about 2400 experimental case study (Bonanomi et al. 2007). The effect of plant residues on Fusarium wilt is based on a general growth effect, which can include even fungal growth promotion, and on the plant-specific effect on the spore germination. Pathogenic *F. oxysporum* can survive on plant residues over long periods (10 years and more), once it is established in a field (Zhou and Everts 2004). Thereby plant residues can promote the spreading of the pathogen after return of the host plant. However, the impact on the later germination ability of *F. oxysporum* spores varies plant genotype specific (Elmer and Lacy 1987). The germination-lysis mechanism (proposed by Chinn and Ledingham 1961) seems to be the basis of increased suppression of Fusarium wilt after application of such organic amendments in soil, which are suppressive to Fusarium wilt in general (Toyota et al. 1995). These authors described the effects of organic amendments and alterations of environmental conditions on the inoculum potential of *F. oxysporum* f. sp. *raphani* strain PEG-4, estimated from its population dynamics and spore germinability. They tested soils which were suppressive and soils which were conducive to Fusarium wilt of radish. In this investigation it was found that suppressive soil possessed a greater degree of fungistasis than soil which was conducive to *Fusarium*. Rice straw and fresh radish residue brought about suppressive effects on the germination of spores of the tested *Fusarium* strain PEG-4 in both soils along with their decomposition. The autecology of the *F. oxysporum* strain PEG-4 was quite different in suppressive and conducive soil and affected by the presence or absence of organic amendments (Toyota et al. 1995).

In comparisons of legumes and other plant species (e.g., grasses), residues of legumes usually were the most effective plant species to suppress Fusarium wilt. This might be explained by their low C/N ratio, which results in a fast, extensive breakdown of foliage and a significant stimulation of the soil microbial activity (Himmelstein et al. 2014).

In field experiments with watermelon, four different fall-planted cover crops (*Vicia villosa*, *Trifolium incarnatum*, *Secale cereale*, *Brassica juncea*) that were tilled in the spring as green manures and bare ground were evaluated on their impact on Fusarium wilt severity caused by *F. oxysporum* f. sp. *niveum* and measured in the yield and quality of watermelon fruits (Himmelstein et al. 2014). In this investigation *V. villosa* and *T. incarnatum* were able to reduce Fusarium wilt of watermelon.

Also in watermelon production systems soil amendment with hairy vetch (*Vicia villos* a Roth) at 0.25 or 0.50 % (w/w) resulted in 54–69 % decreased wilt incidence by *F. oxysporum* f. sp. *niveum* (Zhou and Everts 2004). In greenhouse experiments by these authors, soil amendment with hairy vetch (5 %, w/w) reduced significantly the population density of the pathogen, which was attributed primarily to increased levels of fungicidal ammonia produced during decomposition. This effect was missing in microplot and field experiments with this treatment, most probably caused by strong temperature differences. Incorporation of hairy vetch into mulched soil was indicated to be a supplement to cultivar resistance for management of Fusarium wilt of watermelon.

16.3.2 The Impact of Animal Waste

Animal wastes (e.g., slurry and dung, shell powder) have been tested on their impact on diverse pathotypes of *F. oxysporum*. Slurry and dung were preferably used after composting and shell powder was mainly added in complex organic amendments (Senechkin et al. 2014; see Sect. 16.3.3).

Shrimps and crap shell powder was tested on its impact on Fusarium wilt caused by *F. oxysporum* f. sp. *tracheiphilum* on asparagus bean (*Vigna sesquipedalis*) by

Ha and Huang (2007). They found that amendments of 1 % (w/w) in pathogen-infested soil were the most effective in reducing population densities of this pathogen and reduced the disease severity by 56 %. The combination of shrimps and crap shell powder (0.5 %, w/w) with two tested *Bacillus* spp. strains was more effective than the organic amendments or the bacterial inoculation alone. Also in general, organic amendments with animal waste were used rather in combination with other organic matters and/or after composting (e.g., Escuadra and Amemiya 2008).

16.3.3 The Impact of Composts and Complex Organic Amendments

Compost is organic matter that has been decomposed, which caused increased stability against further microbial decomposition and is a key ingredient in horticulture and organic farming. Its quality and effects are significantly controlled by the quality of the basic raw material and the duration of composting. Composts are between the most suppressive organic materials with more than 50 % of cases showing effective disease control of several soilborne pathogens (Bonanomi et al. 2007). Composts were also used in combination with microbial inoculants or animal wastes. The effects of the combined amendments were sometimes stronger as compost alone (Pharand et al. 2002; Escuadra and Amemiya 2008).

The potential of compost based on pulp and paper mill residues for the control of crown and root rot of greenhouse-grown tomato caused by *F. oxysporum* f. sp. *radicis-lycopersici* was ultrastructurally investigated by Pharand et al. (2002). In this investigation one of the most prominent facets of compost-mediated induced resistance concerned the formation of physical barriers at sites of attempted fungal penetration. These structures, likely laid down to prevent pathogen ingress toward the vascular elements, included callose-enriched wall appositions and osmiophilic deposits around the sites of potential pathogen ingress. A substantial increase in the extent and magnitude of the cellular changes induced by compost was observed when *Pythium oligandrum* was supplied to the potting substrate as microbial agent. This finding corroborates the current concept that amendment of composts with specific antagonists may be a valuable option for amplifying their beneficial properties in terms of plant disease suppression (Pharand et al. 2002).

Complex organic amendments using different bacterial strains were tested by de Boer et al. (2003). They combined specific strains of antagonistic bacteria, using multiple traits antagonizing the pathogen to achieve a higher level of protection. The tested strain *Pseudomonas putida* WCS358 suppressed Fusarium wilt of radish by effectively competing for iron through the production of its pseudobactin siderophore. The strain *P. putida* RE8 induced systemic resistance against Fusarium wilt. When WCS358 and RE8 were mixed through soil together, disease suppression was significantly enhanced to approximately 50 % as compared to the 30 % reduction for the single-strain treatments. Moreover, when one strain failed to suppress disease in the single application, the combination still resulted in disease control. The authors concluded that the enhanced disease suppression by the combination of *P. putida* strains WCS358 and RE8 was the result of the combination of their different disease-suppressive mechanisms. These demonstrate that combining biocontrol strains can lead to more effective or, at least, more reliable biocontrol of Fusarium wilt of radish.

Ntougias et al. (2008) studied nine composts of wastes and by-products of the olive oil, wine, and *Agaricus* mushroom agro-industries. The composts were mixed with peat at a ratio of 1:3 (w/w) and evaluated on their impacts of diverse pathogens of tomato including *F. oxysporum* f. sp. *radicis-lycopersici*. Suppressiveness of Fusarium wilt by the compost amendments was relatively low and varied widely among compost types (8–95 % decrease in plant disease incidence).

The effect of different composts and complex organic amendments on the suppression of Fusarium wilt of spinach caused by *F. oxysporum* f. sp. *spinaciae* was evaluated in a continuous cropping system in both containers and in microplot field trials by Escuadra and Amemiya (2008). They tested amendments with wheat bran alone, wheat bran and sawdust, coffee grounds, chicken manure, or a mixture of different composts with and without 5 % (w/w) crab shell powder either once (5 %, w/w) or continuously (2.5 %) into the test soils infested with the pathogen. In their container trials, the soil amended with composts became suppressive to disease development on the second and third cropping. The suppressive effect was notable in the soil amended with the mixture of compost with and without crab shell powder. The coffee compost lowered soil pH but became suppressive to the disease after modifying the soil pH. In the field trial using the mixture of the different composts containing 5 % crab shell powder, a combination of 5 % before the first cropping and 2.5 % every second cropping gave stable disease control and promoted plant growth. After compost amendment, the total microbial activity increased and population of the pathogen gradually decreased. These phenomena were especially notable in soils amended with the mixture of different composts. The results indicative of these investigation revealed that diversity in the organic materials promotes higher microbial activity and population in the soil thereby enhancing disease suppressiveness (Escuadra and Amemiya 2008).

In cucumber production Fusarium wilt caused by *F. oxysporum* f. sp. *cucumerinum* is one of the most destructive soilborne diseases without any efficient fungicide available for its control. Therefore, organic amendments using compost of sewage sludge and pig manure were tested in cucumber production by Huang et al. (2012). They formulated plant-growth media with peat mixture and compost in a 4:1 ratio (v/v) and inoculated artificially with *F. oxysporum* conidia (5×10^5 conidia mL^{-1}) by root-dip method. In this investigation Fusarium wilt was effectively suppressed in sludge-compost-amended media, while the disease suppression effect of pig manure compost was limited. Sludge compost was indicated as a potential biocontrol of Fusarium wilt in cucumber production.

Also on strawberry (*Fragaria* × *ananassa*) compost of manure was the most effective organic amendment to suppress Fusarium wilt in the soil (at 5 %, w/w), and this effect was significantly based on the increased soil pH after the application of compost (Fang et al. 2012). These authors stated a great potential for manipulating soil pH, adding soil organic amendments and utilizing crop rotation, not only to successfully manage Fusarium wilt on strawberry but to do so in a sustainable way without current reliance upon chemical fumigants.

In flax production four types of organic amendments (plant-derived fresh compost, steer-derived slurry, slurry plus dung, slurry plus compost and dung) were tested on their ability to promote the suppression of Fusarium wilt caused by *F. oxysporum* f. sp. *lini* (Senechkin et al. 2014). In this investigation complex amendment with slurry compost and dung suppressed flax Fusarium wilt, whereas single amendments with fresh compost even enhanced it. Senechkin et al. (2014) suggested that ammonium-oxidizing bacteria could be useful indicators for suppression of soilborne pathogens.

16.4 Conclusion

Organic amendments from plant and animal origins can significantly contribute to suppress Fusarium wilt of diverse crop plant species; however, differences between different qualities and quantities of organic amendment and different sites or substrates can be assumed. Composts in general and complex organic amendments, including combinations of composts with microbial antagonists of *Fusarium* and/or animal wastes, were promising organic amendments for this reason. The selection of production system-specific optimized organic amendments to suppress Fusarium wilt is essential and can be accelerated by consideration of the presented state of knowledge. For instance, application of peat has general rather conducive than suppressive effects on Fusarium wilt and can be excluded for this reason.

The advantage of compost and complex organic amendments to single microbial inoculations to induce suppression of Fusarium wilt was a combination of chemical and microbial effects on *F. oxysporum*. Indicated environmental controls of the efficiency of organic amendments to suppress Fusarium wilt were especially the temperature, the soil pH, and the nutrient concentrations and the ammonium/nitrate ratio.

Yet, concerted research activities are required to develop fast and efficient selection schemes for case-specific optimized organic amendments for an efficient suppression of Fusarium wilt, considering the production system (e.g., greenhouse vs. field), the initial soil or substrate conditions, and the host plant and pathogen genotype combination.

References

Alabouvette C (1986) Fusarium wilt-suppressive soils from Châteaurenard region: review of a 10-year study. Agronomie 6:273–284

Alabouvette C (1999) Fusarium wilt suppressive soils. Australas Plant Pathol 18:57–64

Bonanomi G, Antignani V, Pane C, Scala F (2007) Suppression of soilborne fungal diseases with organic amendments. J Plant Pathol 89:311–324

Borrero C, Trillas MI, Ordovás J, Tello JC, Avilés M (2004) Predictive factors for the suppression of Fusarium wilt of tomato in plant growth media. Phytopathology 94:1094–1101

Borrero C, Trillas MI, Delgado A, Avilés M (2012) Effect of ammonium/nitrate ratio in nutrient solution on control of Fusarium wilt of tomato by *Trichoderma asperellum* T34. Plant Pathol 61:132–139

Chinn SHF, Ledingham RJ (1961) Mechanisms contributing to the eradication of spores of *Helminthosporium sativum* from amended soil. Can J Bot 39:739–748

de Boer M, Bom P, Kindt F, Keurentjes JJ, van der Sluis I, van Loon LC, Bakker PA (2003) Control of Fusarium wilt of radish by combining *Pseudomonas putida* strains that have different disease-suppressive mechanisms. Phytopathology 93:626–632

Elmer WH, Lacy ML (1987) Effects of crop residues and colonization of plant tissues on propagule survival and soil populations of *Fusarium oxysporum* f. sp. *apii* Race 2. Phytopathology 77:381–387

Escuadra GME, Amemiya Y (2008) Suppression of Fusarium wilt of spinach with compost amendments. J Gen Plant Pathol 74:267–274

Fang X, You MP, Barbetti MJ (2012) Reduced severity and impact of Fusarium wilt on strawberry by manipulation of soil pH, soil organic amendments and crop rotation. Eur J Plant Pathol 134: 619–629

Ha MT, Huang JW (2007) Control of Fusarium wilt of asparagus bean by organic soil amendment and microorganisms. Plant Pathol Bull 16:169–180

Harman GE, Howell CR, Viterbo A, Chet I, Lorito M (2004) *Trichoderma* species-opportunistic, avirulent plant symbionts. Nat Rev Microbiol 2:43–56

Himmelstein JC, Maul JE, Everts KL (2014) Impact of five cover crop green manures and Actinovate on Fusarium wilt of watermelon. Plant Dis 98:965–972

Huang X, Shi D, Sun F, Lu H, Lui J, Wu W (2012) Efficacy of sludge and manure compost amendments against Fusarium wilt of cucumber. Environ Sci Pollut Res 19:3895–3905

Jones JP, Engelhard AW, Woltz SS (1993) Management of Fusarium wilt of vegetables and ornamentals by macro- and microelement nutrition. In: Engelhard WA (ed) Soilborne plant pathogens: management of diseases with macro- and microelements. APS Press, St. Paul, MN, pp 18–32

Joffe AZ (1963) The mycoflora of a continuously cropped soil in Israel, with special reference to effects of manuring and fertilizing. Mycologia 55:271–282

Ntougias S, Papadopoulou KK, Zervakis GI, Kavroulakis N, Ehaliotis C (2008) Suppression of soil-borne pathogens of tomato by composts derived from agro-industrial wastes abundant in Mediterranean regions. Biol Fertil Soils 44:1081–1090

Pharand B, Carisse O, Benhamou N (2002) Cytological aspects of compost-mediated induced resistance against Fusarium crown and root rot in tomato. Phytopathology 92:424–438

Sayyed RZ, Patel PR (2011) Biocontrol potential of siderophore producing heavy metal resistant *Alcaligenes* sp. and *Pseudomonas aeruginosa* RZS3 vis-à-vis organophosphorus fungicide. Indian J Microbiol 51:266–272

Scher FM, Baker R (1980) Mechanism of biological control in a *Fusarium*-suppressive soil. Phytopathology 70:412–417

Senechkin IV, van Overbeek LS, van Bruggen AHC (2014) Greater Fusarium wilt suppression after complex than after simple organic amendments as affected by soil pH, total carbon and ammonia-oxidizing bacteria. Appl Soil Ecol 73:148–155

Toyota K, Kamesaka T, Kimura M (1995) Autecology of *Fusarium oxysporum* f. sp. *raphani* in soils suppressive and conductive to Fusarium-wilt of radish. FEMS Microbiol Ecol 16: 261–268

Toyota K, Yamamoto K, Kimura M (1994) Mechanisms of suppression of *Fusarium oxysporum* f. sp. *raphani* in soils so-called suppressive to Fusarium-wilt of radish. Soil Sci Plant Nutr 40: 373–338

Verma M, Brar SK, Tyagi RD, Surampalli RY, Valero JR (2007) Antagonistic fungi, *Trichoderma* spp.: panoply of biological control. Biochem Eng J 37:1–20

Weller DM, Raaijmakers JM, McSpadden Gardener BB, Thomashow LS (2002) Microbial populations responsible for specific soil suppressiveness to plant pathogens. Annu Rev Phytopathol 40:309–348

Zelenev VV, van Bruggen AHC, Semenov AM (2005) Short-term wavelike dynamics of bacterial populations in response to nutrient input from fresh plant residues. Microb Ecol 49: 83–93

Zhou XG, Everts KL (2004) Suppression of Fusarium wilt of watermelon by soil amendment with hairy vetch. Plant Dis 88:1357–1365

Chapter 17
Role of Soil Amendment with Micronutrients in Suppression of Certain Soilborne Plant Fungal Diseases: A Review

Sazada Siddiqui, Saad A. Alamri, Sulaiman A. Alrumman, Mukesh K. Meghvansi, K.K. Chaudhary, Mona Kilany, and Kamal Prasad

17.1 Introduction

Mineral nutrients are essential elements for normal growth and development of plants and microorganisms (Fig. 17.1). Some of the common mineral nutrients are boron (B), chlorine (Cl), copper (Cu), iron (Fe), manganese (Mn), molybdenum (Mo) and zinc (Zn), which are required by plants in very small amounts. Therefore, they are called as micronutrients. Apart from their role in normal development and growth of the plants, micronutrients are essential factors in protection from adverse environmental conditions and disease control (Agrios 2005; Dordas 2008). The occurrence of micronutrients in the soil has direct implications on the severity of plant disease and thereby plays a key role in controlling, scavenging and detoxification of free oxygen radicals. Nutrients can affect disease resistance or tolerance of plants. However, the severity of plant disease can be affected by several factors

S. Siddiqui (✉) • S.A. Alamri • S.A. Alrumman • M. Kilany
Biology Department, College of Science, King Khalid University, Abha, PO Box 10255 61321, Saudi Arabia
e-mail: guddank@rediffmail.com; kalasaz@yahoo.co.in

M.K. Meghvansi
Ministry of Defence, Defence R&D Organisation, Defence Research Laboratory, Post Bag 2, Tezpur 784001, Assam, India
e-mail: mk_meghvansi@yahoo.co.in

K.K. Chaudhary
Department of Biotechnology, Institute School of Life Sciences, Jaipur National University, Jaipur, Rajasthan, India
e-mail: kamal22055@yahoo.co.in

K. Prasad
Symbiosis Sciences Pvt Ltd., Sector-37, Gurgaon 122001, India
e-mail: kamalprsd27@gmail.com

M.K. Meghvansi, A. Varma (eds.), *Organic Amendments and Soil Suppressiveness in Plant Disease Management*, Soil Biology 46,
DOI 10.1007/978-3-319-23075-7_17

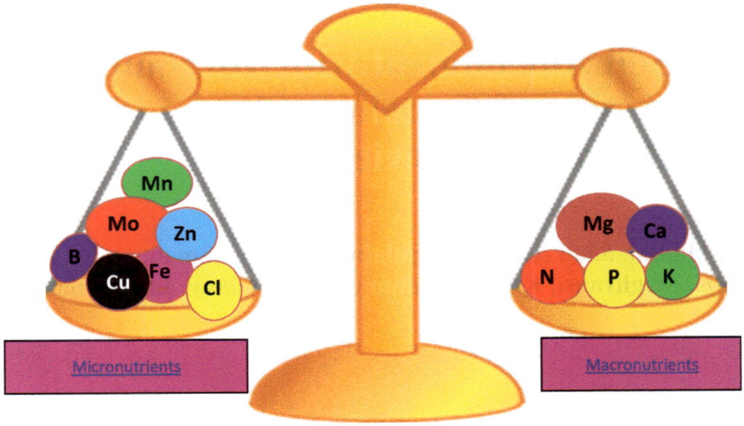

Fig. 17.1 Nutrient balance is important for normal plant growth

such as seeding date, crop rotation, mulching, mineral nutrients, organic amendments, irrigation, pH adjustment, management of the nutrient availability through fertiliser addition and plant disease control strategies (Marschner 1995; Huber and Graham 1999). In recent years, micronutrient management has received considerable attention. However, there is very little information available about whether any specific nutrient can decrease or increase the severity of any specific plant disease with an increase in micronutrient concentrations following soil amendments. It has been shown that high levels of micronutrients in soils significantly prevented pathogenic infection (Graham and Webb 1991; Huber and Graham 1999). The use of micronutrients as fertilisers reduces the severity of many diseases and together with the cultural practices can affect disease control. The micronutrient level in soil can affect plant physiology or pathogens, alone or in concert, which will affect the development of diseases. In addition, pathogenic infection and sporulation can be affected by the uptake of micronutrients (Atkinson and McKinlay 1997; Oborn et al. 2003). Pathogens can directly influence plant physiology, specifically nutrient uptake, assimilation, translocation from root to shoot and immobilisation of nutrients near the rhizosphere zone, which deprives root tissues, while others cause nutrient toxicity or nutrient deficiency by interfering with translocation (Marschner 1995). In addition, significant amounts of micronutrients can be consumed by other organisms for their growth, causing a reduction in the availability of micronutrients for the plants and increasing its susceptibility to pathogenic infections due to nutrient deficiency (Dordas 2008).

Considering the current information available, the role of micronutrients in regulating the soil system and controlling certain plant fungal diseases per se is important for future research. The aim of the present article is to evaluate the role of micronutrients in managing soilborne plant fungal diseases. However, very little literature is available on this topic. Therefore, we need a more thorough understanding on the importance of micronutrients in agro-ecosystem processes, and the mechanisms of suppression of fungal activity is yet to be fully understood due to its complexity. In the present chapter, we attempt to discuss the role of micronutrients, which can lead to a less disease-favourable environment and increase host plant resistance. Here, our main focus is the critical analysis of various factors responsible for the suppression of certain plant fungal diseases due to micronutrients. In addition, we have identified efforts to determine key areas where sincere research efforts are still needed to develop strategies for manipulating micronutrient application in such a way that it could be more efficiently utilised in managing soilborne plant fungal diseases.

17.2 Biology of Soilborne Plant Fungal Diseases

Fungi are considered the most important soilborne pathogens among the four major groups (bacteria, viruses, nematodes and protozoa) of plant pathogens (Agrios 2005). On the basis of morphological and biological characteristics, plant pathogenic fungi are commonly divided into five main taxonomic classes, i.e. Plasmodiophoromycetes, Zygomycetes, Oomycetes, Ascomycetes and Basidiomycetes. Many soilborne fungi persist in soil for long periods, because they produce resilient survival structures like melanised mycelium, chlamydospores, oospores and sclerotia. Only a few groups of bacteria are soilborne, because none of the spore-forming bacteria can survive in soil for long periods. Soilborne pathogens share the soil environment with many other organisms and compete with them for limited resources. In addition, many of the microorganisms in soil are directly or indirectly antagonistic to soilborne pathogens. In the current chapter, we focus on fungi because they are the most important soilborne pathogens causing a number of plant diseases.

Numerous diseases caused by soilborne pathogens are difficult to detect, diagnose and predict. In addition to this, soil ecology is extremely complex, making it a challenge to understand all aspect of diseases caused by soilborne pathogens (Koike et al. 2003). Most of the soilborne plant pathogens decrease the ability of the root to provide the plant with water and nutrients (Huber and Graham 1999). Rot fungi are the most common soilborne fungal pathogens that damage plant tissues below the ground (damping off of seedlings, root and crown rots and seed decay), and vascular wilt initiated by root infection is also mostly reported in the field. A few soilborne pathogens, however, cause foliar diseases with symptoms and damage appearing in above-ground plant parts.

17.3 Biological Control of Soilborne Plant Fungal Diseases

Biocontrol may be defined as any condition or practice where the survival or activity of pathogens is reduced through the living organism used as the biological agent with the result that there is a reduction in the incidence of disease caused by the pathogen (Singh 2002). During the 1970s, biological control was developed as an academic discipline and is now an established method supported in both the public and private sectors (Baker and Cook 1974). It is a potential nonhazardous alternative method for preventing crop loss due to diseases (Chet 1990). In nature, several bio-agents are available and were tested against pathogens during the 1970s and 1980s, with *Gliocladium* and *Trichoderma* gaining high popularity and success. It is now well established that certain bio-agents have tremendous potential and can be exploited successfully in modern agriculture for the control of plant diseases (Mukhopadhyay 1994). Despite that, Van Lenteren (1995) showed that biological control is practised in just 5 % of the estimated 299989.42 ha in greenhouses worldwide. The important factors for adopting biocontrol are predictability, efficacy and cost (Heinz et al. 2004). There are many general micro- and mesofauna predators, such as protists, collembolans, mites, nematodes, annelids and insect larvae whose activities can not only reduce pathogen biomass but also facilitate infection and/or stimulate plant host defences by virtue of their own herbivorous activities. Because plant diseases may be suppressed by the activities of one or more plant-associated microbes, researchers have attempted to characterise the organisms involved in biological control. High soil organic matter supports a large and diverse mass of microbes resulting in decreased ecological niche availability for pathogens. The extent of general suppression will be substantially different depending on the quantity and quality of organic matter present in a soil (Hoitink and Boehm 1999). Manipulation of agricultural systems, through additions of composts, green manures and cover crops, is aimed at improving endogenous levels of general suppression. Few reports regarding the application of microbes as biocontrol agents have negative effects on rhizosphere microbiota (Scherwinski et al. 2006).

The utilisation of organic amendments such as green manure, animal manure, incorporation of crop residues into the soil, peats and composts has been proposed, both for conventional and biological systems of agriculture, to improve soil structure and fertility (Magid et al. 2001; Conklin et al. 2002; Cavigelli and Thien 2003) and decrease the incidence of disease caused by soilborne pathogens (Litterick et al. 2004; Noble and Coventry 2005; Bonanomi et al. 2007). Vermicompost was the most suppressive material, with more than 50 % of cases showing effective disease control. Sahni et al. (2008) studied the collar rot disease caused by *Sclerotium rolfsii* and demonstrated that substituting the soil with different amounts of vermicompost showed a significant reduction in mortality of chickpea compared with the control. In a more recent review, Meghvansi et al. (2011) determined that earthworm populations can suppress soilborne fungal diseases.

17.4 Suppressive Soil

A suppressive soil is one in which disease incidence or severity is at minimum levels, despite the presence of the pathogens and susceptible plant hosts (Baker and Cook 1974). However, non-suppressive (conducive) soil is one in which disease occurs and progresses. Induction of soil suppressiveness to soilborne fungi may provide long-term plant protection. Suppressive soil is maintained by different methods such as addition of organic matter and crop rotation, which improves the presence of microbes that are antagonistic to soilborne pathogens. Farmers have been trying to manipulate soil ecology by the addition of organic matter for a few decades. However, organic matter helps to maintain soil aeration, structure, drainage, moisture holding capacity, nutrient availability and microbial ecology (Davey 1996; Zaccardelli et al. 2013).

Organic amendments such as animal manure, composts, peats and green manure (the incorporation of crop residues into the soil) have been proposed, both for conventional and biological systems of agriculture, to improve soil structure and fertility (Magid et al. 2001; Conklin et al. 2002; Cavigelli and Thien 2003). Organic amendments are useful for controlling diseases caused by pathogens such as *Sclerotium* spp. (Coventry et al. 2005), *Pythium* spp. (McKellar and Nelson 2003; Veeken et al. 2005), *Phytophthora* spp. (Szczech and Smolinska 2001), *Sclerotinia* spp. (Boulter et al. 2002), *Rhizoctonia solani* (Diab et al. 2003) and *Verticillium dahliae* (Lazarovits et al. 1999). Studies have shown that after a few years of reduced organic input, organic matter and soil fertility decreased over the time, and a large number of diseases caused by soilborne plant pathogens spread in agroecosystems (Bailey and Lazarovits 2003). Incorporating organic amendments and managing crop residue type and quantity have a direct impact on plant health and crop productivity. Crop rotations, consisting of wheat, beans or legumes followed by either a fallow period or a green manure, were frequently used in the times of ancient Greece and Rome (Karlen et al. 1994).

Soil is crucial for micronutrient storage such as Br, Mn, Zn, Cu, Fe and Cl, which can reduce the severity of plant disease by increasing disease tolerance and resistance of plants to pathogens. Once a plant is infected by a fungus, its natural defences are triggered and it causes increased production of fungus inhibiting phenolic compounds and flavonoids both at the site of infection and in other parts of the plant. The production and transport of these compounds are controlled in large part by the nutrition of the plant (Lattanzio et al. 2006). Some products of the seafood and livestock industries as well as manures have been used by farmers to maintain productivity in agricultural soil for millennia (Lazarovits 2001). Liu and Baker (1980) showed that successive monocultures of radishes generated soil suppressiveness to *Rhizoctonia solani*, and enhanced *Trichoderma harzianum* propagule density was closely accompanied by soil suppressiveness. Chung et al. (1988) postulated that high propagule density of *Trichoderma* was found to be associated with naturally suppressive Colombian soils than the conductive soils due to acidic pH (5.1), which enhanced the propagation of fungi and *Trichoderma*.

17.5 Micronutrients Suppress Certain Soilborne Plant Fungal Diseases

Plant nutrition affects disease and pathogen resistance mechanisms in two ways: mechanical barriers or cell wall thickening and production of pathogen defence compounds like flavonoids and antioxidants. Micronutrients play an important role in the resistance mechanisms of plants against pathogens, and increases in micronutrient concentrations in soils significantly prevent pathogenic infections (Marschner 1995). Plant damage caused by pathogens can be reduced or controlled using micronutrients, by direct toxicity to the pathogen or by promoting induced system resistance. The use of micronutrients such as B, Cu and Mn can release, through cation exchange, Ca from cell walls. Once released, the Ca ions act together with salicylic acid to activate a systemic acquired resistance (Reuveni et al. 1996, 1997; Reuveni and Reuveni 1998). Micronutrients play an important role in plant metabolism by affecting the phenolics and lignin content and also membrane stability (Graham and Webb 1991).

17.5.1 Boron

Boron (B) is an important micronutrient in reducing the incidence of plant fungal diseases. B provides direct strength and stability for the cell wall and has a beneficial effect on reducing disease severity. In addition, B also contributes to plant resistance and tolerance. B reduces disease susceptibility because it plays an important role in cell wall structure and maintains cell membrane permeability required for metabolism of phenolics or lignin (Blevins and Lukaszewski 1998; Brown et al. 2002; Mustafa and Murat 2013). Plant tissues contain and produce different types of defensive compounds, which hinder the fungal attachment. B plays a main role in the synthesis of these compounds. Borate compounds trigger the enhanced formation of a number of plant defensive chemicals at the site of nitrification. The level of these substances and their fungistatic effect also decrease when the nitrogen supply is too high. It has been shown that B amendment in soil reduces soilborne plant fungal diseases caused by *Fusarium solani* (Mart.) (Sacc.) in bean, *Plasmodiophora brassicae* (Woron.) in crucifers, *Verticillium albo-atrum* (Reinke and Berth) in toma)to and cotton (Graham and Webb 1991) and *Blumeria graminis* in wheat (Marschner 1995). B-deficient plants are susceptible to a wide range of diseases such as an ergot, fusarium wilt, powdery mildew and rust (Graham 1983). It is therefore imperative that future research focuses on the understanding of the exact role of B in management of plant diseases.

17.5.2 Manganese

Manganese (Mn) is a highly effective micronutrient in plant resistance to diseases by affecting cell wall composition and lignin synthesis, and Mn suppresses penetration of pathogens into plant tissue. Mn increases the production of soluble phenols, which play a role in plant resistance mechanisms against fungal pathogens and also inhibits the production of aminopeptidase necessary to supply pathogens with amino acids. It also inhibits pectin methylesterase, which is needed by the pathogen in order for the organism to penetrate plant cell walls (Graham and Webb 1991). Marschner (1995) and Graham and Webb (1991) reported that Mn plays a crucial role in photosynthesis, lignin and phenol biosynthesis and several other functions. Induction of pectin methylesterase, a fungal enzyme that degrades host cell walls, and aminopeptidase, an enzyme that supplies essential amino acids for fungal growth, is inhibited by Mn. Many pathogenic diseases such as take-all, downy mildew, powdery mildew, tan spot and several others can be controlled by Mn fertiliser (Heckman et al. 2003; Simoglou and Dordas 2006). Mn also activates many enzymes that participate in the shikimic acid and phenylpropanoid pathways and also controls lignin and suberin biosynthesis; both of these compounds are important biochemical barriers to fungal pathogen invasion, because they are phenolic polymers resistant to enzymatic degradation (Hammerschmidt and Nicholson 2000; Vidhyasekaran 1997). Lignin and suberin play an important role in wheat resistance and also all diseases caused by *Gaeumannomyces graminis* (Sacc.) (Rovira et al. 1983; Krauss 1999). When Mn occurs in different redox states, it performs different functions. When it is present in healthy tissues such as the Mn^{2+} ion, it accumulates at the sites of pathogen attack in the Mn^{4+} form. Mn might improve host resistance either by alteration of metabolic status or by production of toxic metabolites (Thompson and Huber 2007).

17.5.3 Zinc

Zinc (Zn) appears to be involved in resistance to many diseases. However, the mechanisms of how Zn is involved in disease resistance are unclear. In some cases, it decreased or increased, and in others, it had no effect on plant susceptibility to disease (Grewal et al. 1996). Zn acts as a cofactor for numerous enzymes and also affects membrane stability. It also plays a crucial role in protein and starch synthesis; therefore, a low Zn concentration induces accumulation of amino acids and reduces sugars in plant tissue (Römheld and Marschner 1991; Rice 2007; Duffy 2007). In most of cases, application of Zn in soil reduced disease severity. Zn is important for maintaining defence mechanisms in plants because it participates in superoxide production, which is responsible for a cascade of plant defence pathways against fungal and bacterial pathogens (Doke et al. 1996). Zn deficiency induced NADPH-dependent superoxide radical generation and membrane damage and also free radical damage to critical cell constituents; these free radicals damage membranes, DNA, chlorophyll and protein and finally lead to cell death (Cakmak 2000). Application of Zn to soils reduces root pathogen attack in tomato, including *Fusarium solani*, *Rhizoctonia solani*, *Macrophoma phaseoli* and also *Rhizoctonia* root rots in wheat, chickpea, cowpea and medicago (Kalim et al. 2003). However, low Zn level in soils and leaf tissues was associated with a high incidence of *Phytophthora* pod rot (or black pod) in cocoa in Papua New Guinea (Nelson et al. 2011).

17.5.4 Copper

Copper (Cu) is an essential micronutrient for higher plants as well as for fungi and bacteria. Cu is also very toxic to all plants when present at high levels. However, reported deficiency of Cu decreases lignification in the xylem and has been linked to lodging in cereals, and application of Cu to soil protects grapes and hops from downy mildew caused by *Plasmopara viticola* and *Pseudoperonospora humuli*, respectively (Evans et al. 2007). Cu causes direct toxic effects on pathogens. Cu increases cuticle thickness by lignin formation and acts as barrier for infections. It plays an important role in polyphenol oxidase activity; it produces some phytoalexins and other antipathogenic molecules. Phytoalexins are antimicrobial compounds produced by plants in response to a host-parasite interaction. Some phytoalexins are phenolics (Robinson and Hodges 1981). When a plant becomes infected by a fungus, its natural defences are triggered. The infection causes increased production of fungus inhibiting phenolic compounds and flavonoids, both at the site of infection and in other parts of the plant. The production and transport of these compounds are controlled in large part by the nutrition of the plant (Graham and Webb 1991). Cu deficiency) leads to impaired defensive compound production, accumulation of soluble carbohydrates and reduced wood lignification, all of which contribute to lower disease resistance. Cu is extensively used as a commercial fungicide. Cu deficiency results in impaired synthesis of chemical defence compounds that provide protection against pathogens.

17.5.5 Iron

Iron (Fe) is a necessary micronutrient for plants and animals. However, the role of Fe in disease resistance is not well studied in plants. Therefore, Fe differs from the other micronutrients such as Mn, Cu and B, for which microbes have lower requirements. However, addition of B, Cu and Mn to deficient soils, generally, benefits the host, whereas the effect of Fe application is unclear; it has been shown to have positive or negative effects on the host. A few studies suggested that Fe can reduce or control the disease severity of several diseases such as rust in wheat leaves, smut in wheat and *Colletotrichum musae* in banana (Graham and Webb 1991). Fe has an essential role in plant cells as a cofactor in redox reactions and other functions. Fe is mainly available to plants as its reduced ion Fe^{2+}. Fe stimulates other enzymes involved in the biosynthetic pathway as it is a component of peroxidase but it did not affect lignin synthesis. Synthesis of fungal antibiotics by soil bacteria and siderophore synthesis by rhizosphere microorganisms can be promoted by Fe and it results in lowering Fe level in the soils. The antagonisms for Fe and the production of siderophores are not the only processes that lead to the limitation of the growth of parasitic fungus (Graham and Webb 1991).

17.5.6 Chlorine

Chlorine (Cl) is an essential micronutrient, which is required in very small amounts for plant growth and development. It is thought that Cl might interact with other nutrients such as Mn. A number of diseases such as *Septoria* in wheat, downy mildew in millet, take-all in wheat, northern corn leaf blight, stripe rust in wheat and stalk rot in corn are controlled by Cl (Mann et al. 2004). The mechanisms regarding the effect of Cl on resistance are not clearly understood. It seems that Cl can participate with NO_3 absorption and affects the rhizosphere pH. Thus, it can increase the availability of Mn and suppress nitrification. In) addition, Cl ions can increase Mn for the plant and increase pathogen tolerance and mediate reduction of Mn oxides. It is important to conduct future research on understanding the more precise role of Cl in suppressing soilborne plant fungal diseases.

17.5.7 Molybdenum

Molybdenum (Mo) is an essential trace element and soil is the primary source of Mo. Mo is used by plants as molybdate ions (MoO_4^-). Mo is an essential micronutrient that enables plants to make use of nitrogen. Without molybdenum, plants cannot convert nitrate nitrogen to amino acids and legumes cannot fix atmospheric nitrogen (Rice 2007). It is considered a mobile element, as it moves in both the xylem and phloem conductive tissue of the plant. Palti (1981) reported that Mo reduced ascochyta blight in beans and peas caused by *Ascochyta* spp. and late blight in potato caused by *Phytophthora infestans*. Hahlbrock and Scheel (1989) reported that Mo increases photosynthetic pigments leading to an increase in carbohydrate content. Carbohydrates are the main reservoir for photosynthesis. Polysaccharides of the plant cell wall such as cellulose, hemicellulose and pectin that are barriers against plant pathogens, as well as phenolic compounds, are associated with carbohydrates that play an important role in plant defence. There are very few reports about the potential effects of Mo on plant diseases. It has been reported that the production of a toxin by *Myrothecium roridum*, a pathogen of muskmelon, is reduced by Mo. Also the symptoms of verticillium wilt are reduced when Mo is applied to tomato roots (Kuti et al. 1989). A study showed that the reproduction of *Phytophthora drechsleri* and *Phytophthora cinnamomi* diseases is slightly decreased by Mo. It is not clear that Mo plays any role in protection against diseases within a plant. The deficiency of Mo can reduce nitrate reductase production, which converts nitrates to proteins; therefore, a small amount of Mo is essential to plant health. When Mo is applied to soil, it can inhibit growth of certain soilborne fungi (Mortvedt and John 1991). So, in order to understand the role of Mo in management of plant disease, further research is required.

17.6 Factors Affecting and Improving the Availability of Micronutrients in Soil and Disease Resistance

Various methods have been employed to improve the availability of micronutrients in soil and limit the imbalance of certain elements that can affect growth and disease tolerance. The most common approaches are discussed below (Fig. 17.2).

17.6.1 Fertilisers

Inorganic and organic fertilisers are generally used in maintaining the appropriate soil fertility. Applying organic and inorganic fertilisers is a simple approach, which

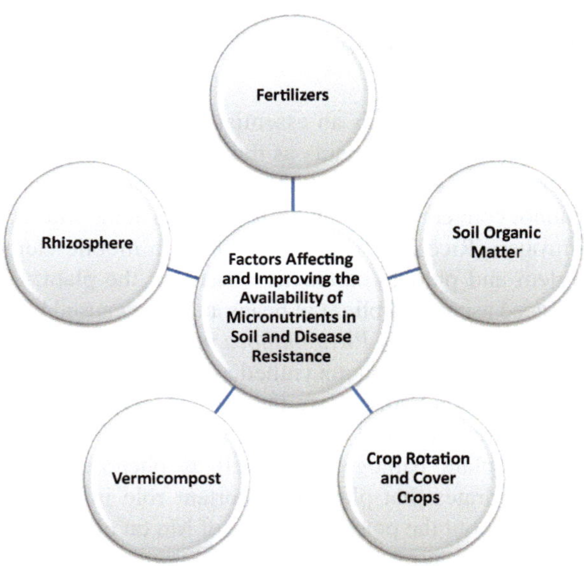

Fig. 17.2 Factors affecting and improving the availability of micronutrients in soil and disease resistance

has been practised for years in overcoming soil fertility constraints (Kazemeini et al. 2010; Abedi et al. 2010). Soil amendments by organic and mineral fertilisers can lead to beneficial interactions between macro- and micronutrients; thus, they provide the optimum need for micronutrient requirements. Fertilisers have been reported to improve crop yield and quality and play a key role in the maintenance of soil productivity (Akande et al. 2006). In addition, the presence of micronutrients and plant uptake can be affected by the availability of macronutrients present in these amendments due to either negative or positive interactions between macro- and micronutrients (Fageria 2001). Few plant diseases are completely controlled by application of fertilisers; for example, botrytis disease can be alleviated by proper application and management of micronutrients. The use of fertilisers produces a more direct means of utilising nutrients to reduce or control the severity of many diseases (Atkinson and McKinlay 1997).

17.6.2 Soil Organic Matter

Soil organic matter (SOM) consists of plant and animal residues at various stages of decomposition. SOM increases soil fertility by providing action exchange sites and acting as a reserve of essential nutrients, especially nitrogen (N), phosphorus (P) and sulphur (S), along with micronutrients. As such, there is a significant correlation between SOM content and soil fertility. SOM is known to affect soil aeration, structure, drainage, moisture holding capacity, nutrient availability and microbial ecology (Davey 1996). SOM plays a key role in promoting the uptake of Fe, Mn, Zn and Cu by higher plants and in the use of micronutrient-enriched organic wastes and naturally occurring metal organic complexes as soil amendments (Bonanomi et al. 2010).

Manures and compost are considered a rich source of N and might reduce soilborne diseases by releasing certain allelochemicals generated during product storage or by subsequent microbial decomposition. The modes of action for disease suppression are elucidated for) a number of diseases including verticillium wilt and common scab in potato (Chakraborty et al. 2011; Chaoui et al. 2003). Stone et al. (2004) reported that fields with organic residue applications such as crop residues, cover crops and organic waste can affect soilborne pathogen and diseases and also affect the availability of nutrients. Addition of sphagnum peat to soil has been shown to suppress diseases caused by *Pythium* spp. Also, addition of different organic amendments has been shown to reduce *Phytophthora* root rot in a number of species (Szczech et al. 1993). A recent study by Pane et al. (2013) has shown that agricultural waste-based composts exhibiting suppressiveness of diseases are caused by the phytopathogenic soilborne fungi *Rhizoctonia solani* and *Sclerotinia minor*.

17.6.3 Crop Rotation and Cover Crops

Crop rotation is the practice of growing a sequence of different crops on the same field. The idea that crop rotation improves overall agricultural productivity is not new; crop rotation was practised in China during the Han dynasty (ca. 206 B.C. to A.D. 220) to improve productivity (MacRae and Mehuys 1985). Long-term experiments showed that crop rotation together with other fertility management practices is fundamental to long-term agricultural productivity and sustainability (Reid et al. 2001; Stone et al. 2004). Crop rotation is considered the most effective disease control strategy because plant pathogen propagules have a finite lifetime in soils, and rotation with non-host crops limits their food supply. Crop rotation can increase N levels and can also affect the availability of other nutrients, which can then affect disease severity (Reid et al. 2001). Rotation is the most powerful and effective practice to control bean diseases, and it remains one of the most important disease management strategies available in many cropping systems (Hall and Nasser 1996).

One of the primary uses of cover crops is to increase soil fertility and affect plant health. They are used to manage a range of soil macronutrients and micronutrients. Mustards belonging to the family Brassicaceae have been widely shown to suppress fungal disease populations through the release of naturally occurring toxic chemicals during the degradation of glucosinolate compounds in their plant cell tissues (Lazzeri and Mancini 2001). The trace element Mn is affected by crop rotation; it was found that crop rotation with lupin increased the availability of Mn (Graham and Webb 1991). Micronutrients such as P, Zn and Mn availability in the soil increase by adding green manure to soil, which can also affect disease tolerance (Huber and Graham 1999). Most of the green manure species that are used can fix nitrogen with N-fixing bacteria and can increase soil N levels (Cherr et al. 2006). This can have a significant effect on disease development.

17.6.4 Vermicompost

Vermicompost is a nutrient-rich, microbiologically active organic amendment that results from the interactions between earthworms and microorganisms during the breakdown of organic matter. It is characterised by high porosity and high water-holding capacity, in which most nutrients are present in forms that are readily taken up by plants (Domfnguez 2004). Vermicompost constitutes an excellent source of plant macro- and micronutrients. Although some of these nutrients are present in inorganic forms and are readily available to plants, most are released gradually through mineralisation of organic matter, thus constituting a slow-release fertiliser that supplies the plant with a gradual and constant source of nutrients (Chaoui et al. 2003). However, in contrast to chemical fertilisers, the amount of nutrients provided may vary greatly, depending on the original feedstock, processing time and maturity of the vermicompost (Campitelli and Ceppi 2008).

17.6.5 Rhizosphere

Root exudates may alter the chemical environment of the root either directly by interaction with element soil constitutes or indirectly by their influence on the microbial community. Plant roots are known to exude a variety of compounds to alter the availability of nutrients in their environment. Root exudates play a fundamental role in the mineral nutrition of plants. They either contain signals that act as regulators of microbial growth and function or possess molecules that directly control rhizosphere processes, which enhance nutrient uptake and assimilation (Dakora and Phillips 2002). N is a main component of protein and DNA in cells. It combines with Mg and forms a main constituent of chlorophyll and takes part in photosynthesis (Soetan et al. 2010). In addition, certain concentrations of P, S, Ca, Mg, Fe and Cu stimulate the production of isoflavonoids in plants, and these molecules function as signals to mutualistic soil microbes and/or phytoalexins against infecting pathogens (Dakora and Muofhe 1996).

17.7 Conclusion

With an extensive literature search, it can be concluded that the addition of micronutrients or application of fertilisers has significant effects on controlling soilborne plant fungal diseases. Micronutrients play a vital role in gene expression; biosynthesis of proteins, nucleic acids and growth substances; and metabolism of carbohydrates and lipids through their involvement in various plant enzyme systems and other physiologically active molecules (Rangel 2003). Disease resistance is genetically controlled but mediated through physiological and biochemical processes, interrelated with the nutritional status of the plant or pathogen.

It has been confirmed that the micronutrient activity in the soil creates a favourable environment for the growth of plant beneficial microbes and suppresses the growth of pathogenic microbes. Therefore, by improving genetic efficiency of the plant and modification of the plant environment, it is possible to expect improved agricultural production. In sustainable agriculture practices, balanced nutrition is an essential component of any integrative crop protection programme, because in most cases it is a more cost-effective and also environmentally friendly approach to control plant disease. Micronutrients can reduce disease to an acceptable level or at least to a level at which further control by other cultural practices or conventional organic biocides are more successful and less expensive. Extensive research is required in order to understand the mechanisms by which micronutrients can reduce disease severity and cause alterations in disease tolerance and plant metabolism. This may help in understanding the association between any specific micronutrient(s) and the susceptibility of the plant to a particular disease.

References

Abedi T, Alemzadeh A, Kazemeini SA (2010) Effect of organic and inorganic fertilizers on grain yield and protein banding pattern of wheat. Aust J Crop Sci 4:384–389

Agrios NG (2005) Plant pathology, 5th edn. Elsevier, Amsterdam, p 635

Akande MO, Oluwatoyinbo FI, Kayode CO, Olowokere FI (2006) Response of maize (*Zea mays*) and Okra (*Abelmoschus esculentus*) intercrop relayed with cowpea (*Vigna unguiculata*) to different levels cow dung amended phosphate rock. World J Agric Sci 2:119–122

Atkinson D, McKinlay RG (1997) Crop protection and its integration within sustainable farming systems. Agric Ecosyst Environ 64:87–93

Bailey KL, Lazarovits G (2003) Suppressing soil-borne diseases with residue management and organic amendments. Soil Tillage Res 72:169–180

Baker KF, Cook RJ (1974) Biological control of plant pathogens. W.H. Freeman and Company, San Francisco, p 433

Blevins DG, Lukaszewski KM (1998) Boron in plant structure and function. Annu Rev Plant Physiol 49:481–500

Bonanomi G, Antignani V, Pane C, Scala F (2007) Suppression of soil-borne fungal diseases with organic amendments. J Plant Pathol 89(3):311–324

Bonanomi G, Antignani V, Capodilupo M, Scala F (2010) Identifying the characteristics of organic soil amendments that suppress soil-borne plant diseases. Soil Biol Biochem 42:136–144

Boulter JI, Boland GJ, Trevors JT (2002) Evaluation of composts for suppression of dollar spot (*Sclerotinia homoeocarpa*) of turf grass. Plant Dis 86:405–410

Brown PH, Brellaloui N, Wimmer MA, Bassil ES, Ruiz J, Hu H, Pfeffer H, Dannel F, Römheld V (2002) Boron in plant biology. Plant Biol 4:205–223

Cakmak I (2000) Possible roles of zinc in protecting plant cells from damage by reactive oxygen species. New Phytol 146:185–205

Campitelli P, Ceppi S (2008) Effects of composting technologies on the chemical and physico-chemical properties of humic acids. Geoderma 144:325–333

Cavigelli MA, Thien SJ (2003) Phosphorus bioavailability following incorporation of green manure crops. Soil Sci Soc Am J 67:1186–1194

Chakraborty A, Chakrabarti K, Chakraborty A, Ghosh S (2011) Effect of long-term fertilizers and manure application on microbial biomass and microbial activity of a tropical agricultural soil. Biol Fertil Soils 47:227–233

Chaoui HI, Zibilske LM, Ohno T (2003) Effects of earthworm casts and compost on soil microbial activity and plant nutrient availability. Soil Biol Biochem 35:295–302

Cherr CM, Scholberg JMS, McSorley R (2006) Green manure approaches to crop production: a synthesis. Agron J 98:302–319

Chet I (1990) In: Hornby D (ed) Biological control of soil-borne plant pathogens. CAB International, Wallingford, pp 274–277

Chung YR, Hoitink HAJ, Dick WA, Herr LJ (1988) Effects of organic matter decomposition level and cellulose amendment on the inoculum potential of *Rhizoctonia solani* in hardwood bark media. Phytopathology 78:836–840

Conklin AE, Erich MS, Liebman M, Lambert D, Gallandt ER, Halteman WA (2002) Effects of red clover (*Trifolium pratense*) green manure and compost soil amendments on wild mustard (*Brassica kaber*) growth and incidence of disease. Plant Soil 238:245–256

Coventry E, Noble R, Mead A, Whipps JM (2005) Suppression of *Allium* white rot (*Sclerotium cepivorum*) in different soils using vegetable wastes. Eur J Plant Pathol 111:101–112

Dakora FD, Muofhe ML (1996) Diverse functions of isoflavonoids in legumes transcend anti-microbial definitions of phytoalexins. Physiol Mol Plant Pathol 49:1–20

Dakora FD, Phillips DA (2002) Root exudates as mediators of mineral acquisition in low-nutrient environments. Plant Soil 245:35–47

Davey CB (1996) Nursery soil management organic amendments. In: Landis TD, Douth DB (Tech. Coordinators) National proceedings, forest and conservation nursery associations. General technical Report PNW-GTR-389. U.S. Department of Agriculture Forest Service, Pacific Northwest Research Station, pp 6–18

Diab HG, Hu S, Benson DM (2003) Suppression of *Rhizoctonia solani* on impatiens by enhanced microbial activity in composted swine waste-amended potting mixes. Phytopathology 93:1115–1123

Doke N, Miura Y, Sanchez LM, Park HJ, Noritake T, Yoshioka H, Kawakita K (1996) The oxidative burst protects plants against pathogen attack: mechanism and role as an emergency signal for plant bio-defence. A review. Gene 179:45–51

Domfnguez J (2004) State of the art and new perspectives on vermicomposting research. In: Edwards CA (ed) Earthworm ecology, 2nd edn. CRC Press LLC, Boca Raton, pp 401–424

Dordas C (2008) Role of nutrients in controlling plant diseases in sustainable agriculture: a review. Agron Sustain Dev 28:33–46

Duffy B (2007) Zinc and plant disease. In: Datnoff LE, Wade HE, Huber DM (eds) Mineral nutrition and plant disease. American Phytopathological Society, St. Paul, MN, pp 155–178

Evans I, Solberg E, Huber DM (2007) Copper and plant disease. In: Datnoff LE, Wade HE, Huber DM (eds) Mineral nutrition and plant disease. American Phytopathological Society, St. Paul, MN, pp 177–188

Fageria VD (2001) Nutrient interactions in crop plants. J Plant Nutr 24:1269–1290

Graham DR (1983) Effects of nutrient stress on susceptibility of plants to disease with particular reference to the trace elements. Adv Bot Res 10:221–276

Graham DR, Webb MJ (1991) Micronutrients and disease resistance and tolerance in plants. In: Mortvedt JJ, Cox F, Shuman LM, Welch RM (eds) Micronutrients in agriculture, 2nd edn. Soil Science Society of America Inc., Madison, WI, pp 329–370

Grewal HS, Graham RD, Rengel Z (1996) Genotypic variation in zinc efficiency and resistance to crown rot disease (*Fusarium graminearum* Schw. Group 1) in wheat. Plant Soil 186:219–226

Hahlbrock K, Scheel D (1989) Physiology and molecular biology of phenylpropanoid metabolism. Annu Rev Plant Physiol Plant Mol Biol 40:347–369

Hall R, Nasser LCB (1996) Practice and precept in cultural management of bean diseases. Can J Plant Pathol 18:176–185

Hammerschmidt R, Nicholson RL (2000) A survey of plant defense responses to pathogens. In: Agrawal AA, Tuzun S, Bent E (eds) Induced plant defenses against pathogens and herbivores. APS, Minneapolis, MN, p 390

Heckman JR, Clarke BB, Murphy JA (2003) Optimizing manganese fertilization for the suppression of take-all patches disease on creeping bent grass. Crop Sci 43:1395–1398

Heinz KM, Driesche RV, Parrella MP (eds) (2004) Biocontrol in protected culture. Ball Publishing, Batavia, IL, p 552

Hoitink HAJ, Boehm MJ (1999) Biocontrol within the context of soil microbial communities: a substrate dependent phenomenon. Annu Rev Phytopathol 37:427–446

Huber DM, Graham RD (1999) The role of nutrition in crop resistance and tolerance to disease. In: Rengel Z (ed) Mineral nutrition of crops fundamental mechanisms and implications. Food Product Press, New York, pp 205–226

Kalim S, Luthera YR, Gandhi SK (2003) Cowpea root rot severity and the metabolic changes in relation to manganese application. J Phytopathol 151:92–97

Karlen DL, Varvel GE, Bullock DG, Cruse RM (1994) Crop rotations for the 21st century. In: Sparks DL (ed) Advances in agronomy, vol 53. Academic Press, London, pp 1–45

Kazemeini SA, Hamzehzarghani H, Edalat M (2010) The impact of nitrogen and organic matter on winter canola seed yield and yield components. Aust J Crop Sci 4:335–342

Koike ST, Subbarao KV, Davis RM, Turini TA (2003) Vegetable diseases caused by soil borne pathogens. Publication 8099, Division of Agriculture and Natural Resources, University of California. http://www.anrcatalog.ucdavis.edu

Krauss A (1999) Balanced nutrition and biotic stress, IFA Agricultural conference on managing plant nutrition, 29 June–2 July 1999. Barcelona, Spain

Kuti JO, Bean GA, Macky WA, Ng TJ (1989) Influence of muskmelon cell wall polysaccharides on roridin E production by a pathogenic strain of *Myrothecium roridum*. Mycopathologia 108:139–144

Lattanzio V, Lattanzio VMT, Cardinali A (2006) Role of phenolics in the resistance mechanisms of plants against fungal pathogens and insects. In: Imperato F (ed) Phytochemistry: advances in research. Research Signpost, Trivandrum, pp 23–67. ISBN 81-308-0034-9

Lazarovits G (2001) Management of soil-borne plant pathogens with organic amendments: a disease control strategy salvaged from the past. Can J Plant Pathol 23:1–7

Lazarovits G, Conn KL, Potter J (1999) Reduction of potato scab, verticillium wilt, and nematodes by soy meal and meat and bone meal in two Ontario potato fields. Can J Plant Pathol 21:345–353

Lazzeri L, Mancini M (2001) Allelopathic effect of glucosinolate containing plant green manure on *Pythium* sp. and total fungal population in soil. Hortic Sci 36:1283–1289

Litterick AM, Harrier L, Wallace P, Watson CA, Wood M (2004) The role of un composted materials, composts, manures, and compost extracts in reducing pest and disease incidence and severity in sustainable temperate agricultural and horticultural crop production—a review. Crit Rev Plant Sci 23:453–479

Liu S, Baker R (1980) Mechanism of biological control in soil suppressive to *Rhizoctonia solani*. Phytopathology 70:404–412

MacRae RJ, Mehuys GR (1985) The effect of green manuring on the physical properties of temperate area soils. Adv Soil Sci 3:71–94

Magid J, Henriksen O, Thorup-Kristensen K, Mueller T (2001) Disproportionately high N-mineralisation rates from green manures at low temperatures: implications for modelling and management in cool temperate agro-ecosystems. Plant Soil 228:73–82

Mann RL, Kettlewell PS, Jenkinson P (2004) Effect of foliar applied potassium chloride on septoria leaf blotch of winter wheat. Plant Pathol 53:653–659

Marschner H (1995) Mineral nutrition of higher plants, 2nd edn. Academic, London, p 889

McKellar ME, Nelson EB (2003) Compost-induced suppression of *Pythium* damping-off is mediated by fatty-acid metabolizing seed-colonizing microbial communities. Appl Environ Microbiol 69:452–460

Meghvansi MK, Singh L, Srivastava RB, Varma A (2011) Assessing the role of earthworms in biocontrol. In: Karaca A (ed) Biology of earthworms, vol 24, Soil biology. Springer, Berlin, pp 173–189

Mortvedt JJ, John J (1991) Micronutrients in agriculture, 2nd edn. Soil Science Society of America, Madison, WI, 760 pp. http://trove.nla.gov.au/version/23937086

Mukhopadhyay AN (1994) Bio-control of soil borne fungal plant pathogens-current status, future prospects and potential limitation. Indian Phytopathol 47:119–126

Mustafa K, Murat SA (2013) Biological effects of boron. In: Whitacre DM (ed) Reviews of environmental contamination and toxicology, vol 225. Springer Science+Business Media, New York, pp 57–75

Nelson PN, Webb MJ, Berthelsen S, Curry G, Yinil D, Fidelis C (2011) Nutritional status of cocoa in Papua new guinea, vol 76. ACIAR technical reports, Australian Centre for International Agricultural Research, Canberra

Noble R, Coventry E (2005) Suppression of soil-borne plant diseases with composts: a review. Biocontrol Sci Technol 15:3–20

Oborn I, Edwards AC, Witter E, Oenema O, Ivarsson K, Withers PJA, Nilsson SI, Richert Stinzing A (2003) Element balances as a tool for sustainable nutrient management: a critical appraisal of their merits and limitations within an agronomic and environmental context. Eur J Agron 20:211–225

Palti J (1981) Cultural practices and infectious crop diseases. Advanced series in agricultural sciences, vol 9. Springer, Berlin, pp 73–189

Pane C, Piccolo A, Spaccini R, Celano G, Villecco D, Zaccardelli M (2013) Agricultural waste-based composts exhibiting suppressivity to diseases caused by the phytopathogenic soil-borne fungi *Rhizoctonia solani* and *Sclerotinia minor*. Appl Soil Ecol 65:43–51

Rangel Z (2003) Heavy metals as essentials nutrients. In: Prasad MNV, Hagemeyer J (eds) Heavy metal stress in plants: molecules to ecosystems. Springer, Berlin, pp 271–294

Reid LM, Zhu X, Ma BL (2001) Crop rotation and nitrogen effects on maize susceptibility to gibberella (*Fusarium graminearum*) ear rot. Plant Soil 237:1–14

Reuveni R, Reuveni M (1998) Foliar-fertilizer therapy—a concept in integrated pest management. Crop Prot 17:111–118

Reuveni M, Agapov V, Reuveni R (1996) Controlling powdery mildew caused by *Sphaerotheca fuliginea* in cucumber by foliar sprays of phosphate and potassium salts. Crop Prot 15:49–53

Reuveni M, Agapov V, Reuveni R (1997) A foliar spray of micronutrient solutions induces local and systemic protection against powdery mildew (*Sphaerotheca fuliginea*) in cucumber plants. Eur J Plant Pathol 103:581–588

Rice RW (2007) The physiological role of minerals in plants. In: Datnoff LE, Wade HE, Huber DM (eds) Mineral nutrition and plant disease. American Phytopathological Society, St. Paul, MN, pp 9–30

Robinson PW, Hodges CF (1981) Nitrogen-induced changes in the sugars and amino acids of sequentially senescing leaves of *Poa pratensis* and pathogenesis by *Drechslera sorokinian*. J Phytopathol 101:348–361

Römheld V, Marschner H (1991) Function of micronutrients in plants. In: Mortvedt JJ, Cox FR, Shuman LM, Welch RM (eds) Micronutrients in agriculture. Soil Science Society of America, Inc., Madison, WI, pp 297–328

Rovira AD, Graham RD, Ascher JS (1983) Reduction in infection of wheat roots by *Gaeumanomyces graminis* var. *Tritici* with application of manganese to soil. In: Parker CA, Rovira AD, Moore KJ, Wong PTW, Kollmorgen JF (eds) Ecology and management of soil-borne plant pathogens. APS, Minneapolis, MN

Sahni S, Sarma BK, Singha DP, Singha HB, Singha KP (2008) Vermicompost enhances performance of plant growth-promoting rhizobacteria in *Cicer arietinum* rhizosphere against *Sclerotium rolfsii*. Crop Prot 27:369–376

Scherwinski K, Wolf A, Berg G (2006) Assessing the risk of biological control agents on the indigenous microbial communities: *Serratia plymuthica* HRO-C48 and *Streptomyces* sp. HRO-71 as model bacteria. BioControl 52:87–112

Simoglou K, Dordas C (2006) Effect of foliar applied boron, manganese and zinc on tan spot in winter durum wheat. Crop Prot 25:657–663

Singh RS (2002) Introduction to principles of plant pathology, 4th edn. Oxford and IBH Publishing, New Delhi

Soetan KO, Olaiya CO, Oyewole OE (2010) The importance of mineral elements for humans, domestic animals and plants: a review. Afr J Food Sci 4:200–222

Stone AG, Scheuerell SJ, Darby HM (2004) Suppression of soil borne diseases in field agricultural systems: organic matter management, cover cropping, and other cultural practices. In: Magdoff F, Ray RR (eds) Soil organic matter in sustainable agriculture. CRC Press, New York, pp 131–177

Szczech M, Smolinska U (2001) Comparison of suppressiveness of vermicomposts produced from animal manures and sewage sludge against *Phytophthora nicotianae* Breda de Haar var. *nicotianae*. J Phytopathol 149:77–82

Szczech MM, Rondomanski W, Brzeski MW, Smolinska U, Kotowski JF (1993) Suppressive effect of a commercial earthworm compost on some root infecting pathogens of cabbage and tomato. Biol Agric Hortic 10:47–52

Thompson IA, Huber DM (2007) Manganese and plant disease. In: Datnoff LE, Wade HE, Huber DM (eds) Mineral nutrition and plant disease. American Phytopathological Society, St. Paul, MN, pp 139–154

Van Lenteren JC (1995) Integrated pest management in protected crops. In: Dent D (ed) Integrated pest management. Chapman and Hall, London, pp 311–343
Veeken AHM, Blok WJ, Curci F, Coenen GCM, Temorshuizen AJ, Hamelers HVM (2005) Improving quality of composted bio waste to enhance disease suppressiveness of compost-amended, peat based potting mixes. Soil Biol Biochem 37:2131–2140
Vidhyasekaran P (1997) Fungal pathogenesis in plants and crops: molecular biology and host defense mechanisms. Dekker, New York, p 568
Zaccardelli M, De Nicola F, Villecco D, Scotti R (2013) The development and suppressive activity of soil microbial communities under compost amendment. J Soil Sci Plant Nutr 13(3):730–742

Chapter 18
Impact of Green Manure and Vermicompost on Soil Suppressiveness, Soil Microbial Populations, and Plant Growth in Conditions of Organic Agriculture of Northern Temperate Climate

L. Grantina-Ievina, V. Nikolajeva, N. Rostoks, I. Skrabule, L. Zarina, A. Pogulis, and G. Ievinsh

18.1 Introduction

Several aspects of agricultural management regime, such as crop rotation, tillage frequency, compost or manure type, application of pesticides and synthetic fertilizers, and water regime, are key determinants of microbial community structure in the soil. Vegetation is also an important factor since plants are providing soil

L. Grantina-Ievina
Faculty of Biology, Department of Microbiology and Biotechnology, University of Latvia, 4 Kronvalda blvd., LV-1586 Riga, Latvia

Latvian Plant Protection Centre, Struktoru 14a, LV-1039 Riga, Latvia
e-mail: lelde.grantina-ievina@lu.lv

V. Nikolajeva (✉) • N. Rostoks
Faculty of Biology, Department of Microbiology and Biotechnology, University of Latvia, 4 Kronvalda blvd., LV-1586 Riga, Latvia
e-mail: vizma.nikolajeva@lu.lv; nils.rostoks@lu.lv

I. Skrabule • L. Zarina
State Priekuli Plant Breeding Institute, 1a Zinatnes Str., Priekuli distr., Priekuli LV-4126, Latvia
e-mail: ilze.skrabule@priekuliselekcija.lv; livija.zarina@priekuliselekcija.lv

A. Pogulis
BALTORGPOTATO, Project "Baltic Organic Potato for the World Markets", Alojas novads, Latvia
e-mail: aivars.pogulis@inbox.lv

G. Ievinsh
Faculty of Biology, Department of Plant Physiology, University of Latvia, 4 Kronvalda blvd., LV-1586 Riga, Latvia

Latvian Plant Protection Centre, Struktoru 14a, LV-1039 Riga, Latvia
e-mail: gederts.ievins@lu.lv; gederts.ievins@laapc.lv

M.K. Meghvansi, A. Varma (eds.), *Organic Amendments and Soil Suppressiveness in Plant Disease Management*, Soil Biology 46,
DOI 10.1007/978-3-319-23075-7_18

microorganisms with specific carbon sources (Garbeva et al. 2004), but, on the other hand, microbial products can influence the decomposition of organic matter in the soil (Lützow et al. 2006). Several investigations show long-term positive influence of organic farming practices on soil quality and microbiological activity in comparison with conventional farming, due to regular crop rotation, and the absence of synthetic fertilizers and pesticides (Shannon et al. 2002). Fertilizing the soil rather than the plant is an organic farmer's goal to assure sufficient nutrient mineralization (Fliessbach and Mader 2000). In the meta-analysis of several investigations about the impact of organic agriculture on soil organisms, it was concluded that soil fungal populations mostly respond positively to organic management, but effects on microbial biomass and activity have been contradictory (Bengtsson et al. 2005).

The objective of this study was to provide an analysis of the impact of organic amendments, i.e., green manure and vermicompost, on the soil microorganisms and plant growth and health in conditions of organic agriculture of Northern temperate climate.

Some case studies dealing with green manure or vermicompost amendments in organic agriculture are discussed giving deeper analyses of the vermicompost impact on plant growth. The first case study is about the impact of green manure on soil microbial populations and soil suppressiveness against such pathogens as late blight, potato scab, and black scurf of potato in organic agriculture. The second case study is about the use of vermicompost in organic starch potato cultivation. Vermicompost produced from composted grass and starchless potato pulp was amended in the field experiment in two growing seasons. The development and severity of the late blight were assessed, as well as the impact on several groups of soil microorganisms. During the growing season, the plant response to the vermicompost amendments was monitored in the terms of photosynthetic activity and leaf chlorophyll content. The possible acting mechanisms of the vermicompost on plant growth are also discussed.

18.2 Green Manure

18.2.1 Impact of Green Manure on the Soil Biochemical and Microbiological Properties and Plant Parameters

Truu et al. (2008) studied a set of microbiological and biochemical properties of soil to assess the influence of agricultural practices on the three most widespread soil types (calcaric regosols, calcaric cambisols, and stagnic luvisols) in the fields of horticultural farms throughout Estonia. Investigation showed that soils managed according to organic farming principles were generally characterized by elevated microbiological parameter (microbial activity and biomass) values, but at the same time the variation of those parameters among soils from these fields was also highest. Researchers offer an opinion that the reason for such large deviations may be the different durations of organic management practice as well as differences in management history among fields, such as different amount and types of organic fertilizers (green or brown manure) applied and differences in crop rotation. Truu et al. (2008) also found that legume-based (mainly clover) crop rotation increased soil respiration and microbial biomass.

In an investigation in the semiarid Canadian prairie comparing annual legumes as green manure (green fallow) with tilled fallow–wheat and continuing wheat cultivation, it was estimated that after 6 years of these management practices, significant improvements were detected in several microbiological characteristics such as colony counts of aerobic bacteria and filamentous fungi. Four green manure crops, black lentil (*Lens culinaris* Medikus), tangier flat pea (*Lathyrus tingitanus* L.), chickling vetch (*Lathyrus sativus* L.), and feed pea (*Pisum sativum* L.), were used. This investigation also proved that the microbiological attributes of the soil are sensitive and responsive to the beneficial influence of the particular cropping systems (Biederbeck et al. 2005).

It is reported that organic farming with various cover crops and green manure in combination with animal manure in the long term results in higher biodiversity of soil organisms. The diversity of bacterial functional communities has been recorded to be higher in soils from organic farms, while species diversity was similar (Liu et al. 2007). Higher abundance and diversity of actinomycetes, important decomposers of organic material, is reported in organic tomato fields (manured with leguminous green manures and/or organic soil amendments) than conventional ones in Mediterranean climate (Drinkwater et al. 1995). The ratios of Gram-positive to Gram-negative bacteria and of bacteria to fungi have been reported to be higher in the fields with organic treatments (plant residues and straw incorporated into the soil) than in the conventional treatments (Marschner et al. 2003).

In an investigation in Maine (USA), it has been observed that green manure (rapeseed) has increased the total population of cultivable bacteria, mainly Gram-negative bacteria in organic farming system and Gram-positive bacteria in conventional farming system (Bernard et al. 2012).

Edesi et al. (2013) studied the influence of organic cultivation with green manure and cattle manure, organic cultivation with green manure, and conventional cultivation with green manure, cattle manure, mineral fertilizers and pesticides on soil microbial activity, and plate count microorganisms in podzoluvisol in Estonia. They found that the total number of bacteria was not different under various management regimes. All soil samples were examined for molds, yeasts, mesophilic spore-forming bacteria, *Fusarium* spp., actinomycetes, azotobacteria, cellulose decomposers, and denitrifying and nitrifying bacteria. In this investigation, the abundance of abovementioned groups of microorganisms did not differ significantly among treatments with exception of nitrifying bacteria. The amount of nitrifying bacteria was higher in both organic and conventional systems treated with cattle manure than in organic cultivation system treated only with green manure. Researchers conclude that although the green manuring is considered to be an important management practice in organic farming to maintain and increase soil microbial activity and the abundance of microbes in different microbial populations, it is important to use also other organic fertilizers such as animal manure in addition to green manure (Edesi et al. 2013).

Cover crops have traditionally been used to reduce soil erosion and build soil quality, but more recently cover crops are being used as an effective tool in organic weed management. Wortman et al. (2013) demonstrated that weeds may alter soil microbial community structure as a means of increasing competitive success in arable soils. However, the relationship between weeds and soil microbial communities requires further investigations.

Tein et al. (2014) investigated how different farming systems influence tuber yields and quality of potato as well as how potato cultivation within a crop rotation under different farming systems affects soil quality. Experiments were carried out on stagnic luvisol in Estonia. In this study, potato was part of a five crop rotation experiment in which red clover (*Trifolium pratense* L.), winter wheat (*Triticum aestivum* L.), peas (*Pisum sativum* L.), potato, and barley (*Hordeum vulgare* L.) followed each other simultaneously on a same field. In the first organic farming system, catch crops were used to provide organic green manure. In the second organic system, a fully composted cattle manure at a rate of 40 tons/ha was also added as a fertilizer. It was estimated that the first system significantly decreased the average soil potassium (K) concentration after potato cultivation. The second system significantly increased the average soil organic carbon (C) and phosphorus (P) concentrations after potato cultivation. The fresh tuber yield differences between both systems were found to be nonsignificant. There were no significant differences among both systems in average tuber K, calcium (Ca), dry matter, and starch concentrations.

Green manure can be incorporated in the soil as a fresh plant material or processed. Direct incorporation of red clover-derived slurry and compost (both with equal nitrogen (N) and C in comparison to fresh red clover) in the leek field in Sweden resulted in the immediate increase in the abundance of bacteria and fungi (estimated according to fatty acid analysis). Mulching with fresh red clover sustained a higher bacterial and fungal biomass until the end of the cropping season and stimulated arbuscular mycorrhizal fungi (estimated as amount of neutral lipid fatty acid 16:1ω5) at the end of the cropping season (Elfstrand et al. 2007). Although in another investigation in Sweden various N amendments were used for 53 years, it was found out that soil fungal populations did not differ among treatments, including the treatment with green manure (fodder crops) every second year (Börjesson et al. 2012). The protease, acid phosphatase, and arylsulphatase activities were highest in the direct incorporation treatment, whereas enzyme activity in treatments with processed red clover was never higher than in the control treatments. There were no differences in leek harvest yield, but the N, P, and sulfur (S) concentrations in the leek crop at harvest increased in response to higher amounts of slurry and compost amendment. The authors concluded that direct incorporation of a red clover ley before planting of the leek was most effective for enhancing and sustaining a high microbial biomass and high rates of enzyme activity in the soil in comparison to other treatments: mulching with fresh red clover, incorporation of biogas slurry from fermented red clover and composted red clover (Elfstrand et al. 2007).

Arlauskiene et al. (2013) presented the analysis of application of grass biomass in organic manure production using innovative technologies, i.e., after additional mulching. Field experiments with different methods of perennial grasses (festulolium, red clover, and lucerne) aboveground biomass removed from the field mulching four times during the period of vegetation and mixed—first cut removed from the field, second, and third—mulching for green manure) were carried out in Lithuania on an endocalcari–endohypogleyic cambisol. As a result, the mulch of grasses was partially mineralized. Late in the autumn, $N_{inorganic}$ content in soil increased the least after application of the aboveground mass of grasses in a combined manner. It was concluded that it is purposeful to apply the aboveground mass of perennials in a combined manner from the environmental approach because the mulch of perennials affects the soil $N_{inorganic}$ content in spring more than in the autumn.

Olesen et al. (2009) studied the influence of green manure on the yield of winter cereal in organic arable farming on three different soil types varying from coarse sand to sandy loam in Denmark. All cuttings of the grass–clover were left on the soil as the mulch. Catch crops did not significantly affect grain yield and total aboveground biomass but reduced grain N concentration for 0.4–0.5 N kg^{-1} dry matter. The authors are of the opinion that the slower mineralization of the organic matter in the incorporated grass–clover probably increased late season N uptake, thereby primarily affecting grain protein content. The dry matter biomass in catch crops was considerably smaller than the weed biomass. The dominating leaf diseases for winter wheat were Septoria, mildew, and stripe rust. The dominating leaf diseases on winter rye were rye leaf rust and scald. There was no significant relationship between disease severity and grain yield, when yield was corrected for effects of year and N input. The results obtained by Olesen et al. (2009) showed that N in grass–clover green manure crops can be an important source of N for winter cereals on soils with good N retention, but they should be avoided on sandy soils with high rates of N leaching.

Results provided by Doltra and Olesen (2013) indicate that in Nordic climates, legume-based catch crops can contribute to the ecological intensification of spring cereals, not only by reducing the nitrate leaching and increasing N retention but also by improving yields.

However, investigations about soil fungal communities do not clearly indicate that they are always positively influenced by organic agriculture practices. In an investigation in southern Germany, it was determined by the cultivation-independent approach that fungal populations were almost entirely uninfluenced by the farming management practices, whereas active population, investigated by the isolation of hyphae using a soil-washing technique, showed a clear response to farming management practices (Hagn et al. 2003). The propagule number of *Trichoderma* has been shown to be higher in soils from conventional farms that used animal manure with synthetic fertilizers in comparison with organic farms using animal manure and deep litter (Elmholt and Labouriau 2005), but it depended on the year of analyses. It is assumed that *Trichoderma* spp. are less affected by a soil disturbance (after the use of pesticides) than other soil fungi and are able to quickly colonize niches left by other organisms in conventional fields with monoculture (Liu et al. 2007). In an investigation in Denmark, it was determined that there were no significant differences of amount of cultivable filamentous fungi and yeasts among organically cultivated fields and fields with synthetic fertilizer and/or animal manure. There were differences only in the abundance of particular genera, i.e., *Penicillium* spp. and *Gliocladium roseum* were more represented under organic than conventional farming (Elmholt and Labouriau 2005).

In microcosm studies with various types of manure, including green manure (grass–clover), it was detected that fresh grass–clover amendment to the soil increased several times the easy degradable organic carbon content, microbial biomass, and significant changes in microbial diversity measures compared to the raw cattle slurry and the two anaerobically digested materials (cattle slurry/maize, cattle slurry/grass–clover). At the same time, the increased microbial biomass depleted the soil for mineral nitrogen (Johansen et al. 2013). Soil microbial parameters alone do not give broad understanding about the soil quality. For agricultural purposes, it is important to reduce the level of soilborne fungal and bacterial pathogens.

Two classical types of soil suppressiveness to soilborne plant pathogens are known (Weller et al. 2002). General suppression owes its activity to the total microbial biomass and is not transferable between soils. Specific suppression owes its activity to the effects of select groups of microorganisms and is transferable. Take-all decline results from the building of fluorescent *Pseudomonas* spp. that produce the antifungal metabolite 2,4-diacetylphloroglucinol. Producers of this metabolite may have a broader role in disease-suppressive soils worldwide (Weller et al. 2002).

Disease-suppressive properties of the soil depend on various factors: soil texture, structure, pH, Ca content, agricultural practices (crop rotation, tillage, fertilizers, and organic amendments), and soil biota (microbial activity or soil respiration, microbial community diversity and composition, population size of particular microbial groups like actinomycetes) (Postma et al. 2008). The soil can act as a reservoir of the inoculum of pathogenic fungi, for example, oospores of late blight *Phytophthora infestans* can survive in the soil in the absence of the host for several years (Drenth et al. 1995). In order to estimate the impact of agricultural practices, it is important to evaluate both soil microbial parameters and disease-suppressive capacity of the soil.

Brassica crops used in crop rotations and as green manures have been associated with reductions in soilborne pests and pathogens. These reductions have been attributed to the production of volatile sulfur compounds through a process known as biofumigation and to changes in soil microbial community structure (Larkin and Griffin 2007). It is reported that green manure from white mustard (*Sinapis alba*), oriental mustard (*Brassica juncea*), and a sorghum–sudangrass hybrid in Newport (USA) reduced the verticillium wilt in the subsequent potato crop. The mustard mixture reduced also other diseases—black scurf and common scab (Larkin et al. 2011a) and, in other investigation, also the rhizoctonia stem canker of potato (Larkin et al. 2011b). Green manure from rye and vetch reduced the incidence of southern blight of tomatoes caused by *Sclerotium rolfsii* (Bulluck III and Ristaino 2002). Many vegetables, primarily the family *Brassicaceae*, are rich in glucosinolates (beta-thioglucoside-*N*-hydroxysulfonates), the precursors of isothiocyanates, and/or their breakdown products known for their fungicidal, nematocidal, and allelopathic properties (Fahey et al. 2001). Lord et al. (2011) assessed the effects of brassica green manures on pale potato cyst nematode *Globodera pallida*. Three *Brassica juncea* lines containing high concentrations of 2-propenyl glucosinolate were the most effective, causing over 95 % mortality of encysted eggs of *G. pallida* in the polyethylene-covered soil. The toxic effects of green manures were greater in the polyethylene-covered than in open soil. In this research, toxicity in the soil correlated with the concentration of isothiocyanate-producing glucosinolate but not total glucosinolate in green manures. However, disease reductions are not always associated with higher glucosinolate-producing crops and have been also observed with non-*Brassica* crops (barley and ryegrass), indicating other mechanisms and interactions are important, particularly for control of *Rhizoctonia solani* (Larkin and Griffin 2007).

18.2.2 Case Study

Only a small part of soil fungi (17 %) and bacteria (0.1–1 %) (Bridge and Spooner 2001; Torsvik et al. 1996; Val-Moraes et al. 2013) are cultivable, and therefore, currently two approaches are used to analyze soil microbial communities, i.e., conventional plating of cultivable microorganisms and DNA-based analyses that are independent of cultivation. Amplified rRNA gene restriction analysis (ARDRA) gives genetic fingerprinting of communities, populations, or phylogenetic groups. In soil microbiology, this method is used to determine the diversity within phylogenetic or functional groups of microorganisms (Lynch et al. 2004). Several studies have shown that quantitative PCR can be used successfully to determine the abundance of specific groups of microorganisms in the soil. An important genus of soil fungi analyzed with this method is *Trichoderma* that is known for its antagonistic activities against plant pathogens (Cordier et al. 2006).

The objective of our study (Grantina et al. 2011) was to conduct complex investigation of microbial attributes in the soil of three organic and four conventional agriculture fields in order to estimate the impact of 6-year-long organic agriculture practices in Northern temperate zone conditions and to compare the characteristics of microbial populations with crop plant health and pathogen suppression. For the characterization of soil bacteria, only classical microbiological methods that analyze cultivable bacteria were used, but soil fungal populations were assessed using both classical and molecular biology methods targeting also those organisms that are uncultivable under laboratory conditions. The hypothesis was that 6 years of organic agriculture practices after long-term conventional agriculture can result in some improvements in the conditions of soil microbial populations and/or plant health and pathogen suppression. Three fields of organic agriculture and four fields of conventional agriculture were examined at the State Priekuli Plant Breeding Institute. Fields of organic agriculture were treated with this type of management for 6 years. The crop rotation in organic fields was as follows: spring crops with clover undersown, clover, winter crops, potatoes, and crucifers (*Brassicaceae*) for green manure and spring crops. The green manure was incorporated in each field every 6 years. In other years, the amelioration of the soil was achieved by cultivating the clover (symbiotic nitrogen fixation), as well as with turning the plant residues into the soil. Similar to organic fields, in the conventional fields, winter crops were grown before potatoes. In all analyzed fields, there was sod-podzolic soil. Soil pH and soil moisture contents were similar in organic and conventional agriculture fields.

Soil samples were taken in the fields in June and in August 2008 and 2009. Nine subsamples were collected on transect of each field at a sampling depth of 10–15 cm (three subsamples in each corner of the field and three subsamples in the middle of the field, 100 g each). The subsamples were pooled together to create three larger samples for every field. Altogether, 84 soil samples were analyzed. The information about the time of outbreak and severity of late blight (*Phytophthora infestans*), potato scab (*Streptomyces scabies*), and black scurf of potato (*Rhizoctonia solani*) was recorded each growing season. On average, the total number of bacteria was significantly higher in organic agriculture fields in comparison with conventional fields. The increase of bacterial colony-forming units (CFU) was on average approximately 70 %. There was a trend that at the end of summer 2008, the number of Actinobacteria in all fields decreased (except one organic field with green manure and cover crops in this year), but in 2009 the number of Actinobacteria increased in all fields; however, these changes were not statistically significant. Overall, the total number of Actinobacteria was significantly higher in organic agriculture fields—on average almost four times if results of both years are combined. The total number of yeasts and maltose-utilizing bacteria was fluctuating during the analyzed period, and on average it was higher in samples of 2009 and also in organic agriculture fields in general in comparison to conventional fields—on average by 190 % (statistically not significant).

The ratio of bacteria to fungi differed significantly in particular sampling times. On average, the ratio of bacteria to fungi was significantly higher in the conventional fields (498 vs. 312). A common trend was observed that the total number of cultivable filamentous fungi (CFF) increased in 2009 in all fields with the exception of conventional barley field. It is still unclear, why the total number of CFF increased significantly in the second year in almost all fields, since none of the factors included in the regression models explained this shift. In spite of the fact that one conventional field received fungicides (mancozeb and others) several times during the second summer, the total number of CFF was increased 9.5 times at the end of August 2009 in comparison with the previous level. Data about dominating CFF genera showed that especially the number of CFU of *Mucor* spp. and sterile mycelia increased in 2009, while members of other genera remained unchanged. It contradicts other investigations that found that the application of such fungicide as mancozeb in amount of 10 mg kg^{-1} in soil decreased the amount of fungi for at least 3 months (Doneche et al. 1983), although the concentration of mancozeb applied on the abovementioned conventional field was significantly lower. In general, the total number of CFF was significantly higher in organic fields. The increase of CFF numbers in organic agriculture fields was on average approximately by 110 %.

Changes in the abundance of dominant fungal genera (*Trichoderma*, *Mucor*, *Mortierella*, *Penicillium*, and *Verticillium*) and sterile mycelia (not sporulating after 10 days of incubation) were evaluated in the two-year period. Similar to the investigations of Liu et al. (2007), in our investigation there were no statistically significant differences in the propagule numbers of *Trichoderma* genus among fields of organic and conventional agriculture. The most abundant genus was *Penicillium*—on average 37.8 ± 14.4 % of all fungi, while other genera were represented by 5–10 % of all CFF, and sterile mycelia covered 33.0 ± 10.1 %. In organic fields, only propagule numbers of *Penicillium* and *Verticillium* were significantly higher than in conventional fields. Higher numbers of *Penicillium* have been recorded in organic fields amended with animal manure and deep litter in the work of Elmholt and Labouriau (2005). Other genera were similarly abundant in both groups of fields.

Consequently, in our investigation we found that colony counts of all groups of cultivable microorganisms (total bacterial count, Actinobacteria, yeasts and maltose-utilizing bacteria, and CFF) were significantly higher in organic agriculture fields after a 6-year-long period of organic agriculture practices than in continued conventional fields. This is in line with the results of Biederbeck et al. (2005) in the semiarid Canadian prairie after the period of 6 years. Similarly, two times higher bacterial numbers under low-input (integrated) agriculture in comparison to high-input agriculture have been recorded in an investigation in the Netherlands (Bloem et al. 1992). There were no statistically significant differences among fields of organic and conventional agriculture for the results obtained by molecular methods, although the mean Shannon diversity index of fungal population was higher in the organic fields in comparison to the conventional agriculture fields (2.56 vs. 2.43). Similar to our study, no significant differences were detected between the two agricultural regimes (organic farms with ecological or biodynamical practices and conventional farms) regarding the number of phylotypes per field and Shannon diversity indices of arbuscular mycorrhizal fungi in onion fields in the Netherlands using molecular methods (Galván et al. 2009).

Quantitative PCR indicated an increase in the amount of *Trichoderma* spp. DNA in 2009, especially in August. However, there were no statistically significant differences among fields of organic and conventional agriculture, although the mean values of this parameter were higher in organic fields, i.e., 9.23 ng g^{-1} dry soil vs. 7.17 ng g^{-1} dry soil. In 2008, the first damage of the late blight (*Phytophthora infestans*) in organic fields was observed 7–10 days earlier than in conventional fields. Late blight significantly destroyed foliage (30–100 %) in organic field 10–14 days before it reached such level in conventional fields. In

2009, the first spots of the disease on potato leaves were observed at the same time on both environments, but significant foliage damages (5–100 %) were assessed after 10 days in organic field and only after 24 days in conventional field. The application of fungicide delayed the late blight development in conventional field and saved crop vegetation for longer time. The late blight development was faster in 2008 than in 2009 due to more favorable weather conditions (more rainfall during August) in 2008. The precipitation in August 2009 was approximately two times less than in two previous years. The prevalence of potato scab caused by *Streptomyces scabies* and black scurf of potato caused by *Rhizoctonia solani* was similar in the fields of both agricultural practices. Consequently, in contrast to the soil microbiological indicators that showed improvement after 6 years of organic agricultural practices in comparison to the conventional agricultural fields, the plant health, in terms of plant disease suppression, had not been improved. Controversial results about the capacity of low tillage and organic agriculture systems to reduce the disease levels, for example, of common root rot of cereals caused by *Cochliobolus sativus*, verticillium wilt, and common scab of potato, have been obtained in previous investigations (Bailey and Lazarovits 2003). Fungal activity measured as fungal biomass has been proved to correlate with *R. solani* suppression in soil (Postma et al. 2008). Our investigation showed that the increase in the number of CFF did not result in the disease suppression, possibly because a 6-year organic management period was too short to reduce the plant pathogen levels in the soil, and crop rotation had gone through the whole cycle only once.

18.3 Vermicompost

18.3.1 Impact of Vermicompost on Plant Growth

The use of vermicompost in agriculture is increasing. Among beneficial effects of vermicompost in agriculture, it is usually generally stated that vermicompost application leads to the improvement of soil's physical properties, including porosity, water retention capacity, etc. (Ferreras et al. 2006). However, in short-term studies in controlled conditions, soil mechanical properties are of less importance in comparison to field experiments. Therefore, potential beneficial effect from vermicompost application could be more easily related to changes in the chemical composition of substrate, e.g., mineral nutrients and plant hormonelike substances. Within the present review, instead of analyzing agronomic properties, we will focus on direct and indirect physiological effects of vermicompost on plants.

An overview of possible direct or indirect physiological effects of vermicompost on plants is given in Table 18.1. Due to a different degree of mineralization and variation in mineral nutrient content in feeding material, it is evident that the beneficial effect of vermicompost needs to be analyzed at least at two levels of soil mineral nutrient availability. In conditions of low mineral supply, plant growth

Table 18.1 Possible direct and indirect physiological effects of vermicompost constituents on plants

Constituent	Concentration or level	Possible benefits	Possible negative consequences
Minerals	Relatively low, variable, and unbalanced in respect to particular elements	Directly used for needs of mineral nutrition, increase plant growth and development	Do not meet optimum needs at low level of application. Certain elements can be at toxic level
Organic matter	Relatively high	Indirect benefit from improving soil properties, long-term effect from acting as nutrients for microorganisms	Decrease in plant availability of certain minerals
Biologically active substances	Highly variable, usually high	Promote plant growth, improve uptake of minerals, induce resistance against pests and diseases	Positive effect will be seen only at optimum level of mineral supply. Include growth inhibitory substances
Microorganisms	Highly variable, usually high	Promote availability of mineral nutrients through mineralization and solubilization. Release biologically active substances	Can contain potentially harmful microorganisms

and development will be promoted due to the increasing doses of plant-available mineral nutrients with the application of vermicompost. Consequently, any amount of vermicompost in relatively poor soil will benefit plant growth. This is especially important in organic agriculture, where organically derived fertilizers with a relatively high degree of mineralization are a valuable choice for increasing plant productivity. However, it is necessary to note that a special care needs to be taken to balance mineral nutrient content in feeding material for earthworms to better address plant needs for essential elements. Usually, vermicomposts are relatively rich in Ca, Mg, Zn, and B and deficient in N, S, Fe, Mn, Cu, and Mo, while P and K can reach extremely high levels (Karlsons et al. 2015). In addition, Na and Cl concentration can be high, especially, if composted livestock manure has been used as a feed for earthworms.

In conditions of optimal soil mineral nutrient availability, high doses of vermicompost might even lead to toxicity of some elements. Consequently, a direct beneficial effect of vermicompost application can be related to (1) high content of hormonelike substances promoting plant growth and development and (2) protection against pests and pathogens. Irrespective of original soil mineral nutrient content, high organic matter and occurrence of microorganisms in vermicompost will promote renovation of soil fertility.

While plant hormonelike activity in compost and vermicompost preparations is a well-known phenomenon (Krishnamoorthy and Vajranabhaiah 1986; Tomati et al. 1988), no attempts have been made to quantify this effect of plants. Recently, we used two different approaches to assess plant growth-affecting activity of organic fertilizers (Ievinsh 2011; Grantina-Ievina et al. 2013, 2014a; Karlsons et al. 2015). The first approach includes measuring an effect of water extract from fertilizers on seed germination and growth of etiolated vegetable seedlings. Four vegetable crop species with a relatively wide range of physiological responses against vermicompost application were selected for the test including beetroot (*Beta vulgaris* L.), Swedish turnip (*Brassica napus* var. *napobrassica* L.), carrot (*Daucus carota* L.), and tomato (*Lycopersicon esculentum* L.). Seed samples were imbibed in water or vermicompost extract at various concentrations and germinated in darkness in the presence or absence of the respective test solution. After 6 days, the hypocotyl height and radicle length of the seedlings were measured, and a degree of stimulation vs. inhibition was calculated. Possible effect of soluble mineral nutrients on plant growth was eliminated by using a second control with mineral nutrient solution at concentration identical to that in vermicompost extract. The method revealed significant differences in plant growth-affecting activity between different organic waste-derived compost and vermicompost samples (Grantina-Ievina et al. 2013). In particular, the highest growth-promoting activity was found for cow manure vermicompost stored wet for 1 year at 4 °C, while storage of the same preparation dry for 1 year at room temperature significantly decreased growth-promoting activity and increased growth-inhibiting activity. Also, plant growth-promoting activity significantly increased when composted sewage sludge were vermicomposted for a short or further for a relatively long period of time.

The second approach allowed to eliminate possible mineral nutrient effects during plant cultivation studies in controlled conditions with organic fertilizer as a substrate amendment (Grantina-Ievina et al. 2014a; Karlsons et al. 2015). The experimental setup allowed to discriminate whether changes in plant growth and development resulted from plant growth-affecting activity or were related to changes in mineral nutrient supply. This was achieved by using two types of control, e.g., pure quartz sand and quartz sand with optimum level of mineral nutrients added. Treatment with increasing doses of organic fertilizers was performed both in the case of pure sand and mineral-enriched sand. It was shown that even 10 % substrate substitution treatment with vermicompost at optimum mineral nutrient conditions resulted in 90 and 98 % increase of fresh and dry mass of winter rye (*Secale cereale* L.) plants (Karlsons et al. 2015). Moreover, further increase of substrate substitution rate with vermicompost (30 and 50 %) resulted in a near-linear concentration-dependent increase in both fresh and dry mass accumulations of rye plants. In consequence, it was concluded that in conditions of optimal soil mineral nutrient availability, a beneficial effect of vermicompost application results mainly from plant growth-promoting activity, while in nutrient-poor soils increase in plant-available minerals due to vermicompost treatment is the most important aspect.

18.3.2 Microbiological Quality of Vermicompost

The wide variety of organic waste (plant residues, animal manure, activated sludge from wastewater treatment plants, etc.) available as feedstock in vermicomposting represents a rich source of microbial diversity. It is reported that vermicompost can significantly increase the amount of plant growth-promoting (free-living nitrogen fixers, nitrifying bacteria, phosphate solubilizers, silicate solubilizers, and fluorescent pseudomonads) and plant disease-protective microorganisms, such as *Trichoderma* spp. fungi in comparison to the initial substrate (coconut leaves with cow manure) used for vermicomposting (Gopal et al. 2009). The application of vermicompost has been used in an investigation in India to increase the level of potentially favorable soil microorganisms such as nitrogen fixers and mycorrhizal fungi (Kale et al. 1992). It has been shown in previous studies that the addition of pig manure and food waste vermicompost significantly increased the microbial activity in commercial substrates (Atiyeh et al. 2000, 2001). Based on molecular analysis, it was found that microbial diversity and species composition of vermicomposts, prepared from mixed organic materials, mainly green plant parts, cattle manure, and agricultural plant waste, were similar to those of vermicompost extracts produced from them. For example, the saprophytic bacteria, *Sphingobacterium* and *Actinomyces*, and ammonium-oxidizing bacteria, *Nitrosovibrio* and *Nitrosospira*, were found in both vermicompost and subsequent extracts (Fritz et al. 2012).

Evidently, vermicompost-associated microorganisms can affect humans during processing; therefore, vermicompost handling needs to be conducted similarly as in conventional composting (Deportes et al. 1995). For example, in a study in Italy of fungal populations of vermicompost produced from 70 % dung (from cows, poultry, and various zoo animals) and 30 % plant debris from various sources, it was found that the fungal populations were dominated by two species: *Pseudallescheria boydii* and *Aspergillus fumigatus* (Anastasi et al. 2005). Both species are potential human and animal pathogens and have been found also in vermicompost samples produced in Latvia from various substrates—cow manure, cow manure with tree leaves, sewage sludge and starchless potato pulp, and composted grass (Grantina-Ievina et al. 2013).

It has been shown that the level of artificially inoculated potentially harmful microorganisms such as *Escherichia coli*, *Enterococcus* spp., and *Salmonella* spp. is significantly reduced due to the activity of earthworms already after 6 days of vermicomposting biosolids from municipal plants (Eastman et al. 2001). Selective reduction of pathogenic bacteria was observed during the vermicomposting of cow manure: the level of fecal enterococci, fecal coliforms, and *Escherichia coli* was reduced, but the level of *Clostridium*, total coliforms, and enterobacteria remained unchanged (Aira et al. 2011). The indicators of fecal contamination such as bacteria *E. coli* and enterococci have been detected in composted sewage sludge and in two consecutive immature vermicompost samples, but in mature vermicompost only *E. coli* was present (Grantina-Ievina et al. 2013). There is also some evidence that the level of potentially pathogenic fungi may remain unchanged during vermicomposting (Beffa et al. 1998) or even increases (Grantina-Ievina et al. 2013). Nevertheless, it has been demonstrated in several investigations that water extracts from vermicompost possess antifungal activity. For example, it is reported that aqueous extracts of air-dried vermicompost inhibited spore germination of several fungi from *Alternaria*, *Curvularia*, and *Helminthosporium* genera and the development of powdery mildews on balsam and pea in India (Singh et al. 2003). In another study, water extracts of vermicompost that was produced from paper sludge and dairy sludge inhibited spore germination of *Fusarium moniliforme* in vitro, but spore germination of such plant pathogens as *Rhizoctonia solani*, *Colletotrichum coccodes*, *Pythium ultimum*, and *Phytophthora capsici* was not reduced (Yasir et al. 2009). Water extracts from vermicomposts produced from cow manure, cow manure with tree leaves, sewage sludge and starchless potato pulp, and composted grass have shown antifungal activity in vitro against fungi from genera *Pseudeurotium*, *Beauveria*, *Nectria*, and *Fusarium* (Grantina-Ievina et al. 2014b).

Much research has been conducted with general bacterial populations, and it is known that particular production conditions (feedstock, time and method of vermicomposting) result in similar species composition of bacterial populations of vermicompost samples if the same earthworm species is used. For example, the average similarity coefficient among various products was nearly 80 % when estimated by comparable methods (Fernández-Gómez et al. 2012).

18.3.3 Case Study: The Impact of Vermicompost on Soil Microorganisms and Potato Yield

The second case study is about the use of vermicompost in organic starch potato cultivation. In the first growing season (2012), the vermicompost produced from composted grass and starchless potato pulp was amended in the amount of 0, 4, 6, 8, 10, and 12 tons/ha in field experiment. The development and severity of the late blight caused by *Phytophthora infestans* were assessed. It was estimated that vermicompost amendments did not reduce the potato late blight infection as it was expected, but in contrary, it was significantly increased (Table 18.2). It can be explained by observed encouraged growth of potato foliage that resulted in more favorable conditions and microclimate for the development of potato late blight infection. The impact of plant density to the potato late blight infection has been described (Hospers-Brands et al. 2008). Nevertheless, the vermicompost increased the potato yield. For example, application of 12 tons/ha of the vermicompost increased potato and starch yields by 15 % and 10 %, respectively, in the first growing season (unpublished data). In the second year of field experiments (2013), granulated form of vermicompost from starchless potato pulp and composted grass was used in the amount of 0, 1, 2, and 3 tons/ha. The largest amount of the granules

Table 18.2 Incidence of potato late blight pathogen *Phytophthora infestans* Deb. infection (%)

	Time of assessment		
Amount of vermicompost (tons/ha)	24 July 2012	31 July 2012	09 August 2012
0	0	7.9	33.4
4	0	9.9	33.8
6	0	10.9	43.8
8	0	12.1	52.8
10	0	13.1	63.8
12	0	12.8	68.1

increased the potato and starch yields by 15–30 % depending on the field (unpublished data). During the growing season, the plant response to the vermicompost amendments was monitored in the terms of photosynthetic activity and leaf chlorophyll content, and in particular measurement times, significant changes of these parameters were detected.

The impact of the vermicompost on several groups of soil microorganisms (total bacterial population, number of Actinobacteria, and filamentous fungi) was assessed. It was concluded that vermicompost amendments did not significantly change the abundance of these microorganisms, while the species spectrum of filamentous fungi was altered. For example, the application of 1 tons/ha significantly increased the amount of plant growth-promoting filamentous fungi, such as *Mortierella* and *Trichoderma* spp. (unpublished data).

18.4 Conclusions

It is expected that organic farming with the application of green manure or vermicompost would result in high biodiversity of soil organisms and plant growth promotion. On average, significantly higher numbers of all groups of analyzed cultivable microorganisms were observed in organic agriculture fields in comparison to conventional fields, e.g., total bacterial population had increased by 70 %, Actinobacteria by 290 %, and cultivable filamentous fungi by 110 %. Results obtained by molecular methods regarding fungal diversity did not show such an increase.

In contrast to the soil microbiological indicators, controversial results about plant health, in terms of disease suppressiveness, have been obtained. Our studies raise particular concerns about the vermicompost. Definitely, the unique nature of organic amendments in each case must be taken into account. Further studies are needed to explain the impact of green manure and vermicompost on the plant health.

References

Aira M, Gómez-Brandón M, González-Porto P, Domínguez J (2011) Selective reduction of the pathogenic load of cow manure in an industrial-scale continuous-feeding vermireactor. Bioresour Technol 102:9633–9637

Anastasi A, Varese GC, Marchisio VF (2005) Isolation and identification of fungal communities in compost and vermicompost. Mycologia 97:33–44

Arlauskiene A, Nemeiksiene D, Slepetiene A (2013) Innovative green manure technologies on organic farming. In: Proceedings of 12th international scientific conference "Engineering for rural development", Jelgava, 23–24 May 2013

Atiyeh RM, Subler S, Edwards CA, Bachman G, Metzger JD, Shuster W (2000) Effects of vermicomposts and composts on plant growth in horticultural container media and soil. Pedobiologia 44:579–590

Atiyeh RM, Edwards CA, Subler S, Metzger JD (2001) Pig manure vermicompost as a component of a horticultural bedding plant medium: effects on physicochemical properties and plant growth. Bioresour Technol 78:11–20

Bailey KL, Lazarovits G (2003) Suppressing soil-borne diseases with residue management and organic amendments. Soil Tillage Res 72:169–180

Beffa T, Staib F, Fischer LJ, Lyon PF, Gumowski P, Marfenina OE, Dunoyer-Geindre S, Georgen F, Roch-Susuki R, Gallaz L, Latgé JP (1998) Mycological control and surveillance of biological waste and compost. Med Mycol 36:137–145

Bengtsson J, Ahnström J, Weibull AC (2005) The effects of organic agriculture on biodiversity and abundance: a meta-analysis. J Appl Ecol 42:261–269

Bernard E, Larkin RP, Tavantzis S, Erich MS, Alyokhin A, Sewell G, Lannan A, Gross SD (2012) Compost, rapeseed rotation, and biocontrol agents significantly impact soil microbial communities in organic and conventional potato production systems. Appl Soil Ecol 52:29–41

Biederbeck VO, Zentner RP, Campbell CA (2005) Soil microbial populations and activities as influenced by legume green fallow in a semiarid climate. Soil Biol Biochem 37:1775–1784

Bloem J, de Ruiter PC, Koopman GJ, Lebbink G, Brussaard L (1992) Microbial numbers and activity in dried and rewetted arable soil under integrated and conventional management. Soil Biol Biochem 24:655–665

Börjesson G, Menichetti L, Kirchmann H, Kätterer T (2012) Soil microbial community structure affected by 53 years of nitrogen fertilisation and different organic amendments. Biol Fertil Soils 48:245–257

Bridge P, Spooner B (2001) Soil fungi: diversity and detection. Plant Soil 232:147–154

Bulluck LR III, Ristaino JB (2002) Effect of synthetic and organic soil fertility amendments on southern blight, soil microbial communities, and yield of processing tomatoes. Phytopathology 92:181–189

Cordier C, Edel-Hermann V, Martin-Laurent F, Blal B, Steinberg C, Alabouvette C (2006) SCAR-based real time PCR to identify a biocontrol strain (T1) of *Trichoderma atroviride* and study its population dynamics in soils. J Microbiol Methods 68:60–68

Deportes I, Benoit-Guyod JL, Zmirou DE, Bouvier MC (1995) Hazard to man and environment posed by the use of urban waste compost: a review. Sci Total Environ 172:197–222

Doltra J, Olesen JE (2013) The role of catch crops in the intensification of spring cereals in organic farming under Nordic climate. Eur J Agron 44:98–108

Doneche B, Seguin G, Ribereau-Gayon P (1983) Mancozeb effect on soil microorganisms and its degradation in soils. Soil Sci 135:361–366

Drenth A, Janssen EM, Govers F (1995) Formation and survival of oospores of *Phytophthora infestans* under natural conditions. Plant Pathol 44:86–94

Drinkwater LE, Letourneau DK, Wokneh F, van Bruggen AHC, Shennan C (1995) Fundamental differences between conventional and organic tomato agroecosystems in California. Ecol Appl 5:1098–1112

Eastman BR, Kane PN, Edwards CA, Trytek L, Gunadi B, Sterne AL, Mobley JR (2001) The effectiveness of vermiculture in human pathogen reduction for USEPA biosolids stabilization. Compost Sci Util 9:38–49

Edesi L, Jarvan M, Lauringson E, Akk E, Tamm K (2013) The effect of solid cattle manure on soil microbial activity and on plate count microorganisms in organic and conventional farming systems. Int J Curr Microbiol Appl Sci 2:476–488

Elfstrand S, Bath B, Martensson A (2007) Influence of various forms of green manure amendment on soil microbial community composition, enzyme activity and nutrient levels in leek. Appl Soil Ecol 36:70–82

Elmholt S, Labouriau R (2005) Fungi in Danish soils under organic and conventional farming. Agric Ecosyst Environ 107:65–73

Fahey JW, Zalcmann AT, Talalay P (2001) The chemical diversity and distribution of glucosinolates and isothiocyanates among plants. Phytochemistry 56:5–51

Fernández-Gómez MJ, Nogales R, Insam H, Romero E, Goberna M (2012) Use of DGGE and COMPOCHIP for investigating bacterial communities of various vermicomposts produced from different wastes under dissimilar conditions. Sci Total Environ 414:664–671

Ferreras L, Gomez E, Toresani S, Firpo I, Rotondo R (2006) Effect of organic amendments on some physical, chemical and biological properties in a horticultural soil. Bioresour Technol 97:635–640

Fliessbach A, Mader P (2000) Microbial biomass and size-density fractions differ between soils of organic and conventional agricultural system. Soil Biol Biochem 32:757–768

Fritz JI, Franke-Whittle IH, Haindl S, Insam H, Braun R (2012) Microbiological community analysis of vermicompost tea and its influence on the growth of vegetables and cereals. Can J Microbiol 58:836–847

Galván GA, Parádi I, Burger K, Baar J, Kuyper TW, Scholten OE, Kik C (2009) Molecular diversity of arbuscular mycorrhizal fungi in onion roots from organic and conventional farming systems in the Netherlands. Mycorrhiza 19:317–328

Garbeva P, van Veen JA, van Elsas JD (2004) Microbial diversity in soil: selection of microbial populations by plant and soil type and implications for disease suppressiveness. Annu Rev Phytopathol 42:243–270

Gopal M, Gupta A, Sunil E, Thomas GV (2009) Amplification of plant beneficial microbial communities during conversion of coconut leaf substrate to vermicompost by *Eudrilus* sp. Curr Microbiol 59:15–20

Grantina L, Kenigsvalde K, Eze D, Petrina Z, Skrabule I, Rostoks N, Nikolajeva V (2011) Impact of six-year-long organic cropping on soil microorganisms and crop disease suppressiveness. Zemdirbyste-Agriculture 98:399–408

Grantina-Ievina L, Andersone U, Berkolde-Pīre D, Nikolajeva V, Ievinsh G (2013) Critical tests for determination of microbiological quality and biological activity in commercial vermicompost samples of different origins. Appl Microbiol Biotechnol 97:10541–10554

Grantina-Ievina L, Karlsons A, Andersone-Ozola U, Ievinsh G (2014a) Effect of freshwater sapropel on plants in respect to growth-affecting activity and cultivable microorganism content. Zemdirbyste-Agriculture 101:355–366

Grantina-Ievina L, Malisevs A, Nikolajeva V (2014b) Antifungal activity of water extracts of vermicompost. Abstract of the 72nd scientific conference of the University of Latvia. Environ Exp Biol 12:71–72

Hagn A, Pritsch K, Schloter M, Munch JC (2003) Fungal diversity in agricultural soil under different farming management systems, with special reference to biocontrol strains of *Trichoderma* spp. Biol Fertil Soils 38:236–244

Hospers-Brands AJTM, Ghorbani R, Bremer E, Bain R, Litterick A, Halder F, Leifert C, Wilcockson SJ (2008) Effects of presprouting, planting date, plant population and configuration on late blight and yield of organic potato crops grown with different cultivars. Potato Res 51:131–150

Ievinsh G (2011) Vermicompost treatment differentially affects seed germination, seedling growth and physiological status of vegetable crop species. Plant Growth Regul 65:169–181

Johansen A, Carter MS, Jensen ES, Hauggard-Nielsen H, Ambus P (2013) Effects of digestate from anaerobically digested cattle slurry and plant materials on soil microbial community and emission of CO_2 and N_2O. Appl Soil Ecol 63:36–44

Kale RD, Mallesh BC, Kubra B, Bagyaraj DJ (1992) Influence of vermicompost application on the available macronutrients and selected microbial populations in a paddy field. Soil Biol Biochem 24:1317–1320

Karlsons A, Osvalde A, Andersone-Ozola U, Ievinsh G (2015) Vermicompost from municipal sewage sludge affects growth and mineral nutrition of winter rye (*Secale cereale*) plants. J Plant Nutr (in press)

Krishnamoorthy RV, Vajranabhaiah SN (1986) Biological activity of earthworm casts: an assessment of plant growth promotor levels in the casts. Proc Indian Acad Sci Anim Sci 95:341–351

Larkin RP, Griffin TS (2007) Control of soilborne diseases using *Brassica* green manures. Crop Prot 26:1067–1077

Larkin RP, Honeycutt CW, Olanya OM (2011a) Management of *Verticillium* wilt of potato with disease-suppressive green manures and as affected by previous cropping history. Plant Dis 95:568–576

Larkin RP, Honeycutt CW, Griffin TS, Olanya OM, Halloran JM, He Z (2011b) Effects of different potato cropping system approaches and water management on soilborne diseases and soil microbial communities. Phytopathology 101:58–67

Liu B, Tu C, Hu S, Gumpertz M, Ristaino JB (2007) Effect of organic, sustainable, and conventional management strategies in grower fields on soil physical, chemical, and biological factors and the incidence of Southern blight. Appl Soil Ecol 37:202–214

Lord JS, Lazzeri L, Atkinson HJ, Urwin PE (2011) Biofumigation for control of pale potato cyst nematodes: activity of brassica leaf extracts and green manures on *Globodera pallida in vitro* and in soil. J Agric Food Chem 59:7882–7890

Lützow MV, Kögel-Knabner I, Ekschmitt K, Matzner E, Guggenberger G, Marschner B, Flessa H (2006) Stabilization of organic matter in temperate soils: mechanisms and their relevance under different soil conditions – a review. Eur J Soil Sci 57:426–445

Lynch JM, Benedetti A, Insam H, Nuti MP, Smalla K, Torsvik V, Nannipieri P (2004) Microbial diversity in soil: ecological theories, the contribution of molecular techniques and the impact of transgenic plants and transgenic microorganisms. Biol Fertil Soils 40:363–385

Marschner P, Kandeler E, Marschner B (2003) Structure and function of the soil microbial community in a long-term fertilizer experiment. Soil Biol Biochem 35:453–461

Olesen JE, Askegaard M, Rasmussen IA (2009) Winter cereal yields as affected by animal manure and green manure in organic arable farming. Eur J Agron 30:119–128

Postma J, Schilder MT, Bloem J, van Leeuwen-Haagsma WK (2008) Soil suppressiveness and functional diversity of the soil microflora in organic farming systems. Soil Biol Biochem 40:2394–2405

Shannon D, Sen AM, Johnson DB (2002) A comparative study of the microbiology of soils managed under organic and conventional regimes. Soil Use Manag 18:274–283

Singh UP, Maruya S, Singh DP (2003) Antifungal activity and induced resistance in pea by aqueous extract of vermicompost and for control of powdery mildew of pea and balsam. J Plant Dis Prot 110:544–553

Tein B, Kauer K, Eremeev V, Luik A, Selge A, Loit E (2014) Farming system affect potato (*Solanum tuberosum* L.) tuber and soil quality. Field Crop Res 156:1–11

Tomati U, Grappelli A, Galli E (1988) The hormone-like effect of earthworm casts on plant growth. Biol Fertil Soils 5:288–289

Torsvik V, Sorheim R, Goksoyr J (1996) Total bacterial diversity in soil and sediment communities – a review. J Ind Microbiol Biotechnol 17:170–178

Truu M, Truu J, Ivask M (2008) Soil microbiological and biochemical properties for assessing the effect of agricultural management practices in Estonian cultivated soils. Eur J Soil Biol 44:231–237

Val-Moraes SP, Pedrinho EAN, Lemos EGM, Carareto-Alves LM (2013) Molecular identification of fungal communities in a soil cultivated with vegetables and soil suppressiveness to *Rhizoctonia solani*. Appl Environ Soil Sci, Article ID 268768, 7 p. doi:10.1155/2013/268768

Weller DM, Raaijmakers JM, Gardener BB, Thomashow LS (2002) Microbial populations responsible for specific soil suppressiveness to plant pathogens. Annu Rev Phytopathol 40:309–348

Wortman SE, Drijber RA, Francis CA, Lindquist JL (2013) Arable weeds, cover crops, and tillage drive soil microbial community composition in organic cropping systems. Appl Soil Ecol 72:232–241

Yasir M, Aslam Z, Kim SW, Lee SW, Jeon CO, Chung YR (2009) Bacterial community composition and chitinase gene diversity of vermicompost with antifungal activity. Bioresour Technol 100:4396–4403

Chapter 19
The Impact of Silicon Amendment on Suppression of Bacterial Wilt Caused by *Ralstonia solanacearum* in Solanaceous Crops

Henok Kurabachew

19.1 Introduction

Bacterial wilt caused by *Ralstonia solanacearum* is one of the most destructive bacterial diseases of bacterial origin in the world (Hayward 1995; Yabuuchi et al. 1995). *R. solanacearum* is a Gram-negative, strictly aerobic rod bacterium (0.5–0.7 × 1.5–2.0 μm in diameter) classified in the subdivision of the *Proteobacteria* (Kersters et al. 1996). The species *R. solanacearum* is severe in tropical, subtropical, and some relatively warm temperate regions of the world where the environmental condition is optimal for the pathogen (Hayward 1991). Recently, the geographical spectrum has extended to more temperate countries in Europe and North America as a result of dissemination of strains adapted to cooler environmental conditions (Genin and Boucher 2004). The host range of the bacterium is exceptionally wide, and many economically important crops as well as many weed hosts have been recognized (Hayward 1991). It is a major constraint in the production of several important crops particularly *Solanaceae* crops such as tomato, potato, tobacco, eggplant, and tobacco (French and Sequeira 1970). Generally, *R. solanacearum* has an extended host range that includes over 450 host species in 54 botanical families (Wicker et al. 2007).

Ralstonia solanacearum is a highly heterogeneous bacterial species, based on host range the species divided into five races (Buddenhagen et al. 1962; He et al. 1983; Pegg and Moffett 1971) and into six biovars according to the ability of species to metabolize three sugar alcohols and three disaccharides (Hayward 1964, 1991, 1994; He et al. 1983). Both classifications lack an exact concordance

H. Kurabachew (✉)
School of Nutrition, Food Science and Technology, College of Agriculture, Hawassa University, P.O. Box 05, Hawassa, Ethiopia
e-mail: henok.k5@gmail.com

M.K. Meghvansi, A. Varma (eds.), *Organic Amendments and Soil Suppressiveness in Plant Disease Management*, Soil Biology 46,
DOI 10.1007/978-3-319-23075-7_19

with the genetic background of the complex members. Therefore, molecular-based assessment of the genetic diversity of *R. solanacearum* employing restriction fragment length polymorphism analysis resulted in two clusters of strains as divisions 1 Asiaticum and 2 Americanum (Cook et al. 1989; Cook and Sequeira 1994). More recently, a phylogenetically meaningful classification scheme was developed based on DNA sequence analysis (Fegan and Prior 2005, 2006). This scheme divides the complex species into four phylotypes that broadly reflect the ancestral relationships and geographical origins of the strains. Accordingly, phylotype I, II, III, and IV strains originated in Asia, America, Africa, and Indonesia, respectively. The phylotypes are further subdivided into sequevars based on the sequence of the endoglucanase (*egl*) gene (Fegan and Prior 2005, 2006). This phylotyping scheme proposed by Fegan and Prior (2005) is consistent with the former phenotypic and molecular typing schemes and adds valuable information about the geographical origin and in some cases the pathogenicity of strains.

Symptoms of *R. solanacearum* include leaf yellowing, wilting, and necrosis as well as vascular browning (Swanson et al. 2005). Typically, stem and tuber cross-sections ooze whitish bacterial exudates (Genin and Boucher 2002). The bacterium scurvies in infected plants, volunteer crops, susceptible weed hosts, and infested soil. Its dissemination is mainly through the use of infected plants, latently infected planting material, and contaminated irrigation water (Hayward 1991, 1994).

19.2 Management of Bacterial Wilt

Control of *R. solanacearum* is difficult due to its wide host range and its survival capacity in various environments. Unlikely a single strategy cannot separately control bacterial wilt in epidemic regions (Saddler 2005). However, losses can be reduced by following integrated disease management and application of multiple-control measures (Denny 2006). The control measures such as plant breeding, field sanitation, crop rotation, and biological control have only limited success (Ciampi-Panno et al. 1989). Also the use of pesticide is limited, so no commercial pesticide is available against the pathogen other than chemical fumigants (Wang and Lin 2005). Although disease resistance is an important component of integrated disease management, it is generally agreed that breeding for resistance is not completely effective, producing only modest gains and often lacking stability and/or durability (Hayward 1991; Boucher et al. 1992). The stability of resistant varieties highly affected by pathogen strains, temperature, soil moisture, and presence of root-knot nematodes (Wang and Lin 2005). Alternatively, enhancing the host resistance against the pathogen can be an effective control strategy. Recently, Si amendment has significantly reduced bacterial wilt incidence and enhanced the host resistance in tomato. The enhanced resistance was attributed to an induced resistance (Diogo and Wydra 2007; Kurabachew and Wydra 2014).

19.3 Induced Disease Resistance in Plants

Plants are subjected to numerous infections with pathogens and parasites. The only way to face the infection is activating resistance mechanisms against the pathogenic agent. "Resistance is the ability of an organism to exclude or overcome, completely or in some degree, the effect of a pathogen or other damaging factor" (Agrios 1997). During evolution plants have developed sophisticated defensive strategies to perceive pathogen attack and to translate this perception into an appropriate adaptive response. In response to microbial attack, plants activate a complex series of responses that lead to the local and systemic induction of a broad spectrum of antimicrobial defenses (Hammond-Kosack and Jones 1996).

Resistance in plants to pathogen is a natural phenomenon that is often observed as hypersensitive response (HR), a necrotic lesion that surrounds the site of infection and limits the spread of the pathogen (Van Loon et al. 1998). Local infection by a necrotizing pathogen leads to a HR, and the enhanced state of resistance extends systemically into the uninfected plant parts. This long-lasting and broad-spectrum induced disease resistance is referred to as systemic acquired resistance (SAR) (Ross 1961; Durrant and Dong 2004). The induction of SAR is accompanied by local and systemic accumulation of endogenous levels of the plant hormone salicylic acid (SA), followed by the coordinate activation of a specific set of pathogenesis-related (PR) genes, many of which encode PR proteins with antimicrobial activity (Van Loon et al. 2006b).

Systemic resistance against plant pathogens can also be induced by plant growth-promoting rhizobacteria (PGPR) known as induced systemic resistance (ISR) (Van Loon and Glick 2004). ISR is mediated through jasmonic acid (JA) in concert with the ethylene (ET) pathway. Such systemic resistance triggered by beneficial microorganisms confers a broad-spectrum resistance that is effective against different types of plant pathogens such as viruses, bacteria, and even insect herbivores (Van Wees et al. 2008).

Application of SAR and ISR in pest management seems promising. Unlike traditional pesticides, synthetic elicitors and PGPR strains provide a way to control disease without applying additional selective pressure on pathogen populations, as they generally do not exhibit any direct antimicrobial activity. In addition, the inducers of SAR and ISR seem to be friendly to the environment relative to the current pesticides. Therefore, SAR and ISR are attractive approaches for managing crop pests in a sustainable manner within the scope of a conventional agriculture system. Although induced resistance has benefits, like all technologies, there may be undesirable costs that need to be considered. A consistent problem from several field studies using benzothiadiazole (BTH) or 2,6 dichloroisonicotinic and its methyl ester (INA) has been the reduction of crop yield (Louws et al. 2001; Romero et al. 2001), but this reduction is not significant (Iriti and Faoro 2003).

19.4 Role of Silicon in Plant Biology

Silicon (Si) is the second most abundant element in the lithosphere following oxygen and comprises approximately 28 % of the earth crust. The element and its role in plant did not seize much attention for decades (Epstein 1994). Si is found in nature in the form of silica, SiO_2, and aluminum, iron, or calcium silicates. The simplest source of monosilicic acid is quartz, which reacts with water to form silicic acid. The roots of plants interplay with the soil minerals and play a major role in the solubilization of Si, and hence, Si in its uncharged form, the silicic acid (H_4SiO_4), is provided in the soil solution for absorption. Actual concentrations in the soil solution vary widely in space and time, depending on the particular soil minerals present and many other factors, both abiotic and biotic. However, the range of concentrations 0.1–0.6 mM may be considered as a normal range (Dahlgren 1993; Epstein 1999; Dakora and Nelwamondo 2003).

Si accumulation in plants varies greatly due to the differences in ability to uptake Si. Plants are classified into three groups regarding Si uptake. The Si accumulators are defined as plants which contain higher than 1.0 % Si and show a Si/Ca mol ratio higher than 1, the intermediate plants contain 0.5–1.0 % Si or even higher but show a Si/Ca mol ratio less than 1, while Si non-accumulators contain less than 0.5 % Si. The uptake mode is active for the first group, passive for the second, and rejective for the third group. The most popular examples representing these groups are rice, cucumber, and tomato which are Si accumulator, intermediate accumulator, and Si non-accumulator, respectively (Ma et al. 2001; Mitani and Ma 2005).

Si is a multifunctional element that significantly influences plant growth resulting in greater yields, e.g., in rice, or increases the sugar content, e.g., in sugarcane (Savant et al. 1999; Seebold et al. 2000). It enhances soil fertility; improves soil physical properties; increases photosynthesis; improves the efficiency of water use; regulates evapotranspiration; alleviates abiotic and biotic stresses; increases tolerance to metal toxicity such as Fe, Mn, and Cd; reduces frost damage; and improves disease and pest resistance (Dakora and Nelwamondo 2003; Gao et al. 2004; Ma 2004; Liang et al. 2005b).

19.5 Role of Silicon Amendment in Plant Resistance Induction

Silicon alleviates biotic stresses and increases the resistance of plants to pathogens. Several studies have suggested that Si activates plant defense mechanisms, yet the exact nature of the interaction between the element and biochemical pathways leading to resistance remains unclear (Fauteux et al. 2005). Silicon amendment showed not only increased resistance toward fungal and bacterial diseases but also toward insects, such as a reduced preference, longevity, and production of nymphs of the green aphids *Schizaphis graminum* on wheat (Basagli et al. 2003).

Si induces plant defense only in response to infection with pathogens, in order to invest energetic costs only in infected plants (Chérif et al. 1994; Schneider and Ullrich 1994). Si pre-sensitizes the cellular metabolism of the plant, so after exposure to pathogen or biological stress, these pre-sensitized or "primed" plants are able to respond quicker, and with higher level of resistance capacity than non-primed plants, and thus cope better with the challenge. Ample evidence showed that Si alone has apparently no effect on the metabolism of plants growing in a controlled unstressed environment (Cai et al. 2009). Plants expressing SAR, ISR, or BABA-IR exhibit a faster and stronger activation of specific defense responses after they have been infected by a pathogen. This capacity for augmented defense expression is called priming (Conrath et al. 2002; Van Hulten et al. 2006). The priming phenomenon has been demonstrated in different plant species against biotic and abiotic stress (Conrath et al. 2002). Thus, priming is likely a common property of the plant's immune system (Van Hulten et al. 2006).

Disease resistance induced by Si has been observed in many plant species including rice, cucumber, and wheat. Si enhances rice (Si accumulator) resistance to many diseases such as blast, sheath blight, brown spot leaf scab, and stem rot (Datnoff et al. 1997; Rodrigues et al. 2003; Fauteux et al. 2005; Cai et al. 2008). Si also increases plant resistance to powdery mildew in wheat, barley, cucumber, and *Arabidopsis* (Fauteux et al. 2005, 2006; Ma and Yamaji 2006). Recently, Si has been shown to induce resistance in tomato against bacterial wilt caused by *R. solanacearum* (Dannon and Wydra 2004; Diogo and Wydra 2007; Kurabachew and Wydra 2014).

19.5.1 Mode of Action of Silicon-Induced Resistance

Plants, being sessile, have evolved a battery of defense response genes to protect themselves against biotic and abiotic stress. Defense in plant can be constitutive or induced. Induced plant defenses are regulated by highly interconnected signaling networks in which the plant hormones such as jasmonic acid (JA), ethylene (ET), and salicylic acid (SA) play a central role (Asselbergh et al. 2008; Pozo et al. 2004; Van Loon et al. 2006a).

In induced resistance, the defense capacity of the plant can be enhanced biologically by beneficial rhizobacteria and mycorrhizal fungi or chemically by exogenous application of low doses of SA, its functional analogue benzothiadiazole (BTH), acibenzolar-S-methyl (ASM), JA or ß-aminobutyric acid (BABA), or silicon (Conrath et al. 2006; Dannon and Wydra 2004; Fauteux et al. 2005; Frost et al. 2008). Silicon is known to induce systemic acquired resistance (SAR) and modulate the defense response of the plant by participating in signal transduction through accumulation of salicylic acid, which leads to the enhancement of host resistance (Fauteux et al. 2005). The onset of SAR is associated with increased levels of salicylic acid (SA) and is characterized by the coordinate activation of a specific set of pathogenesis-related (PR) genes, many of which encode PR proteins with antimicrobial activity (Van Loon et al. 2006b).

19.5.2 Biochemical Mode of Action of Silicon-Induced Resistance

Plants develop an enhanced resistance against further pathogen attack when infected with necrotizing pathogens, which is referred to as systemic acquired resistance (SAR) (Conrath 2006). Silicon induces defense responses similar to SAR. Different studies showed that Si treatment increased the activity of the common protective enzymes, i.e., peroxidase (PO), polyphenol oxidase (PPO), and phenylalanine ammonia lyase (PAL) in stem of tomato (Kurabachew and Wydra 2014), leaves of rice (Cai et al. 2008), wheat (Yang et al. 2003), and cucumber (Liang et al. 2005a). These enzymes played a pivotal role in regulating the production and accumulation of antimicrobial compounds such as phenolic metabolism product (lignin), phytoalexins, and pathogenesis-related proteins in plants. Si application can induce the production of antifungal compounds after the penetration of pathogens (Liang et al. 2005a; Rémus-Borel et al. 2005). Furthermore, Si treatment resulted in the increase of flavonoid phytoalexin in cucumber plants infected by powdery mildew (*Podosphaera xanthii*) (Fawe et al. 1998).

19.5.3 Molecular Mode of Action of Silicon-Induced Resistance

Debatably, Si has been suggested to be a SAR inducer. A difference between known SAR inducers and Si is the loss of activity when Si donation is interrupted, as a result of its deposition in the cell wall which leads to its inactivation as SAR inducer. Therefore, Fauteux et al. (2005) suggested that Si acted as a signal in triggering defense responses. Additionally, it is speculated that Si modulates the defense response of the plant by its involvement in signal transduction. If Si is involved in the signaling events leading to the enhancement of the host resistance, it should also influence the systemic signals. The signals are transmitted to the cell nucleus, where the signal is translated into expression of the defense-related genes, through the activation of specific kinase/phosphatase cascades. In other words, the gene expression is modulated by activating defense-regulating transcription factors or deactivating inhibitors of defense response (Fauteux et al. 2005).

Si is known to bind to hydroxyl groups and may thus affect protein activity or conformation. The mode of action of Si in signal transduction may also derive from interactions with phosphorus. Thus, it was suggested that Si could act as an activator of strategic signaling proteins interacting with several key components of plant stress signaling systems ultimately leading to induced resistance against pathogens. On the other hand, metals play a crucial role for many enzymes. Excess of toxic metal concentrations may lead to enzymatic dysfunctions. Si was mentioned above to extenuate the toxic effect of such metals. Thus, Si may lead to improvement of the enzymatic catalysis. However, to affirm whether Si enhances plant defenses indirectly by sequestering toxic metals or directly by modulating signal transduction and subsequent gene expression, more detailed analysis at the molecular level is required (Fauteux et al. 2005).

Si acted as a signal in triggering plant defense mechanisms similar to SAR (Fauteux et al. 2005; Cai et al. 2009). If Si is involved in the signaling events leading to the enhancement of the host resistance, it should also influence the systemic signals. The signals are transmitted to the cell nucleus, where the signal is translated into expression of the defense-related genes, through the activation of specific kinase/phosphatase cascades. In other words, the gene expression is modulated by activating defense-regulating transcription factors or deactivating inhibitors of defense response (Fauteux et al. 2005). Si can also bind to hydroxyl groups of proteins strategically involved in signal transduction, or it can interfere with cationic cofactors of enzymes influencing pathogenesis-related events. Therefore, Si interacts with several key components of plant stress signaling systems leading to induced resistance.

19.5.3.1 Gene Expression During Silicon-Induced Resistance

Gene expression profiling using microarrays has been recognized as a powerful approach to obtain an overall view on gene expression and physiological processes involved in response to a particular stimulus (Maleck et al. 2000; Schenk et al. 2000). Transcriptome analysis of tomato stem after challenge inoculation with the bacterial pathogen *R. solanacearum* strain ToUdk2 (race1,phylotype 1) revealed amplified expression patterns defense genes, indicating that the plants were primed by silicon to respond more rapidly and/or more strongly to pathogen attack (Kurabachew et al. 2013). In this setup, the silicon-mediated upregulated defense-related genes were pathogenesis-related protein1 precursor (PR-1); endo-1,3-beta glucanase-like protein; basic endochitinase; disease resistance protein (NBS-LRR class); hevein-related protein precursor (PR-4); pathogenesis-related protein; glycoside hydrolase family 19 (basic endochitinase); leucine-rich repeat protein; defensin; disease resistance protein; cytochrome P450; germin-like, putative cytochrome P450; and peroxidase (Kurabachew et al. 2013). Additionally a variety of transcription factors and signal transduction elements such as myb family transcription factor, homeodomain protein containing "homeobox" domain signature, Zip transcription factor ATB2, putative WRKY-type DNA binding protein, zinc finger protein putative, WRKY transcription factor 3 and mitogen-activated protein kinase, transmembrane protein, leucine-rich repeat protein family, receptor-related serine/threonine kinase, tyrosine phosphatase, phosphatidylinositol-4-

phosphate 5-kinase, MAP3K-like protein kinase, protein phosphatase 2C (PP2C), and NADPH oxidase were upregulated (Kurabachew et al. 2013).

Inoculation of *R. solanacearum* in tomato primed with silicon triggered changes in the expression of defense response genes. Most of the upregulated defense-related genes and transcripts belong to the salicylic acid-dependent pathway that leads to induction of systemic acquired resistance (SAR). SAR is induced after local infection of the plant by the pathogen or elicitor accompanied by an increase in the level of endogenous salicylic acid (SA) and subsequent PR protein expression (Ross 1961; Durrant and Dong 2004).

In microarray analysis, upregulation of PR-1 protein, a marker for SAR, was found. PR proteins function either directly on the pathogen through production of antimicrobial substances or indirectly by creating physical barriers to the pathogen infection process or by upstream intrinsic PR signaling (Jiang et al. 2009). Furthermore, pathogenesis-related (PR) proteins such as endo-1,4-beta-glucanase, basic endochitinase, and glucan endo-1,3-beta-glucosidase are known to disrupt the cell wall of fungal/bacterial pathogens (Datta and Muthukrishnan 1999). All these genes participate in the induction of systemic resistance in the plant. Furthermore, results indicated induction of SAR against the vascular pathogen by silicon application which was also depicted by reduction of bacterial wilt severity and incidence in the *ad planta* experiment. This indicated the pivotal role of silicon in resistance induction in tomato against the pathogen. In another silicon-induced gene expression profiling in tomato against tomato *Ralstonia solanacearum*, Ghareeb et al. (2011) reported upregulation of jasmonic acid/ethylene marker genes JERF3, TSRF1, and ACCO, oxidative stress markers FD-I and POD, and basal defense marker AGP-1g.

For analysis of gene expression profiles in molecular plant microbe interactions, the use of an internal control or housekeeping gene with high expression stability under the experimental conditions is needed as a prerequisite for accurate relative quantification of gene expression. In recent study Ghareeb et al. (2011) conclude the expression stability of two housekeeping genes: phosphoglycerate kinase genes (PGK) and α-tubulin (TUB) in silicon-primed and *R. solanacearum*-inoculated tomato plants. However, the expression stability of actin (ACT) severely varied, in particular at the early phase after inoculation with the pathogen, suggesting the possibility of disabling the cytoskeleton that mediates resistance. However, application of silicon resulted in more expression stability of the three housekeeping genes, showing alleviation of the biotic stress imposed by the pathogen.

19.6 Conclusions

Silicon is a bioactive element associated with beneficial effects on mechanical and physiological properties of plants. Silicon alleviates abiotic and biotic stresses and increases the resistance of plant pathogens. The element possesses a unique biochemical property that may explain its bioactivity as a regulator of plant defense mechanisms. Silicon can act as a modulator influencing the timing and extent of plant defense responses upon infection pathogens. It may also interact with several key components of plant stress signaling systems leading to induced resistance.

Different biochemical and molecular studies have indicated that silicon activates plant defense mechanisms; however, the exact nature of the interaction between the element and biochemical pathways leading to resistance still remains unclear. Silicon triggered the regulation of different defense-related genes involved in signal transduction and transcription factors that increase plant resistance toward bacterial wilt providing a higher protective role against the pathogen. This strengthens the hypothesis that silicon alleviates and induces resistance after pathogen inoculation triggering the expression of a variety of defense-related genes. Furthermore, the phenotypic and biochemical investigation on tomato, which is a non-silicon accumulator plant, supports the idea that silicon-related protection is based on induction of systemic resistance rather than on the formation of a mechanical barrier. Therefore, based on different research and literature analysis, silicon can be part of an integrated disease management package against bacterial wilt.

References

Agrios GN (1997) Plant pathology, 4th edn. Academic Press Inc., San Diego, CA

Asselbergh B, De Vleesschauwer D, Höfte M (2008) Global switches and fine-tuning-ABA modulates plant pathogen defense. Mol Plant Microbe Interact 21:709–719

Basagli MAB, Moraes JC, Carvalho GA, Ecole CC, Gonçalves-Gervásio RCR (2003) Effect of sodium silicate application on the resistance of wheat plants to the green-aphids *Schizaphis graminum* (Rond.) (*Hemiptera*: *Aphididae*). Neotrop Entomol 32:659–663

Boucher CA, Gough CL, Arlat M (1992) Molecular genetics of pathogenicity determinants of *Pseudomonas solanacearum* with special emphases on hrp genes. Annu Rev Phytopathol 30:443–461

Buddenhagen IW, Sequeira L, Kelman A (1962) Designation of races in *Pseudomonas solanacearum*. Phytopathology 52:726

Cai KZ, Gao D, Luo SM, Zeng RS, Yang JY, Zhu XY (2008) Physiological and cytological mechanisms of silicon induced resistance in rice against blast disease. Physiol Plant 134: 324–333

Cai K, Gao D, Chen J, Luo S (2009) Probing the mechanisms of silicon-mediated pathogen resistance. Plant Signal Behav 4:1–3

Chérif M, Menzies JG, Ehret DL, Bogdanoff C, Bélanger RR (1994) Yield of cucumber infected with *Pythium aphanidermatum* when grown with soluble silicon. Hortic Sci 29:896–897

Ciampi-Panno L, Fernandez C, Bustamante P, Andrade N, Ojeda S, Conteras A (1989) Biological control of bacterial wilt of potatoes caused by *Pseudomonas solanacearum*. Am Potato J 66:315–332

Conrath U (2006) Systemic acquired resistance. Plant Signal Behav 4:179–184

Conrath U, Pieterse CMJ, Mauch-Mani B (2002) Priming in plant-pathogen interactions. Trends Plant Sci 7:210–216

Conrath U, Beckers GJM, Flors V, García-Agustín P, Jakab G, Mauch F et al (2006) Priming: getting ready for battle. Mol Plant Microbe Interact 19:1062–1071

Cook D, Sequeira L (1994) Strain differentiation of *Pseudomonas solanacearum* by molecular genetic methods. In: Hayward AC, Hartman GL (eds) Bacterial wilt: the disease and its causative agent, *Pseudomonas solanacearum*. CAB International, Wallingford, pp 77–94

Cook D, Barlow E, Sequeira L (1989) Genetic diversity of *Pseudomonas solanacearum* : detection of restriction fragment length polymorphisms with DNA probes that specify virulence and the hypersensitive response. Mol Plant Microbe Interact 2:113–121

Dahlgren RA (1993) Comparison of soil solution extraction procedures: effect on solute chemistry. Commun Soil Sci Plant Anal 24:1783–1794

Dakora FD, Nelwamondo A (2003) Silicon nutrition promotes root growth and tissue mechanical strength in symbiotic cowpea. Funct Plant Biol 30:947–953

Dannon EA, Wydra K (2004) Interaction between silicon amendment, bacterial wilt development and phenotype of *Ralstonia solanacearum* in tomato genotypes. Physiol Mol Plant Pathol 64: 233–243

Datnoff LE, Deren CW, Snyder GH (1997) Silicon fertilization for disease management of rice in Florida. Crop Prot 16:525–531

Datta K, Muthukrishnan S (1999) Pathogenesis-related proteins in plants, 1st edn. CRC Press, Boca Raton

Denny TP (2006) Plant pathogenic *Ralstonia* species. In: Gnanamanickam SS (ed) Plant associated bacteria. Springer, Dordrecht, pp 573–644

Diogo VCR, Wydra K (2007) Silicon-induced basal resistance in tomato against *Ralstonia solanacearum* is related to modification of pectic cell wall polysaccharide structure. Physiol Mol Plant Pathol 70:120–129

Durrant WE, Dong X (2004) Systemic acquired resistance. Annu Rev Phytopathol 42:185–209

Epstein E (1994) The anomaly of silicon in plant biology. Proc Natl Acad Sci U S A 91:11–17

Epstein E (1999) Silicon. Annu Rev Plant Physiol Plant Mol Biol 50:641–664

Fauteux F, Remus-Borel W, Menzies JG, Bélanger RR (2005) Silicon and plant disease resistance against pathogenic fungi. FEMS Microbiol Lett 249:1–6

Fauteux F, Chain F, Belzile F, Menzies J, Bélanger RR (2006) The protective role silicon in the *Arabidopsis*-powdery mildew pathosystem. Proc Natl Acad Sci U S A 103:17554–17559

Fawe A, Abou-Zaid M, Menzies JG, Bélanger RR (1998) Silicon-mediated accumulation of flavonoid phytoalexins in cucumber. Phytopathology 88:396–401

Fegan M, Prior P (2005) How complex is the "*Ralstonia solanacearum* species complex". In: Allen C, Prior P, Hayward AC (eds) Bacterial wilt disease and the *Ralstonia solanacearum* species complex. APS Press, Madison, WI, pp 449–462

Fegan M, Prior P (2006) Diverse members of the *Ralstonia solanacearum* species complex cause bacterial wilts of banana. Aust Plant Pathol 35:93–101

French ER, Sequeira L (1970) Strains of *Pseudomonas solanacearum* from Central and South America: a comparative study. Phytopathology 60:506–512

Frost CJ, Mescher MC, Carlson JE, De Moraes CM (2008) Plant defense priming against herbivores: getting ready for a different battle. Plant Physiol 146:818–824

Gao X, Zou C, Wang L, Zhang F (2004) Silicon improves water use efficiency in maize plants. J Plant Nutr 27:1457–1470

Genin S, Boucher C (2002) *Ralstonia solanacearum*: secrets of major pathogen unveiled by analysis of its genome. Mol Plant Pathol 3(3):111–118

Genin S, Boucher C (2004) Lessons learned from the genome analysis of *Ralstonia solanacearum*. Annu Rev Phytopathol 42:107–134

Ghareeb H, Bozsó Z, Ott PG, Repenning C, Stahl F, Wydra K (2011) Transcriptome of silicon-induced resistance against *Ralstonia solanacearum* in the silicon non accumulator tomato implicates priming effect. Physiol Mol Plant Pathol 75:83–89

Hammond-Kosack KE, Jones JDG (1996) Resistance gene-dependent plant defense responses. Plant Cell 8:1773–1791

Hayward AC (1964) Characterization of *Pseudomonas solanacearum*. J Appl Bacteriol 27: 265–277

Hayward AC (1991) Biology and epidemiology of bacterial wilt caused by *Pseudomonas solanacearum*. Annu Rev Phytopathol 29:67–87

Hayward AC (1994) The hosts of *Pseudomonas solanacearum*. In: Hayward AC, Hartman GL (eds) Bacterial wilt: the disease and its causative agent *Pseudomonas solanacearum*. CAB International, Oxford, pp 9–24

Hayward AC (1995) *Pseudomonas solanacearum* pathogens and host specificity in plant disease: histopathological, biochemical, genetic and molecular bases. In: Singh US, Singh RP, Kohmoto K (eds) Prokaryotes. Elsevier Science, Inc., Tarry Town, NY, pp 139–151

He LY, Sequiera L, Kelman A (1983) Characteristics of strains of *Pseudomonas solanacearum* from China. Plant Dis 67:1357–1361

Iriti M, Faoro F (2003) Does benzothiadiazole-induced resistance increase fitness cost in bean? J Plant Pathol 85:265–270

Jiang F, Zheng X, Chen J (2009) Microarray analysis of gene expression profile induced by the biocontrol yeast *Cryptococcus laurentii* in cherry tomato fruit. Gene 430:12–16

Kersters K, Ludwig W, Vancanneyt M, De Vos P, Gillis M, Schleifer KH (1996) Recent changes in the classification of pseudomonas: an overview. Syst Appl Microbiol 19:465–477

Kurabachew H, Wydra K (2014) Induction of systemic resistance and defense related enzymes after elicitation of resistance by rhizobacteria and silicon application against *Ralstonia solanacearum* in tomato (*Solanum lycopersicum*). Crop Prot 57:1–7

Kurabachew H, Stahl F, Wydra K (2013) Global gene expression of rhizobacteria-silicon mediated induced systemic resistance in tomato (*Solanum lycopersicum*) against *Ralstonia solanacearum*. Physiol Mol Plant Pathol 84:44–52

Liang YC, Sun WC, Si J, Römheld V (2005a) Effects of foliar- and root-applied silicon on the enhancement of induced resistance to powdery mildew in *Cucumis sativus*. Plant Pathol 54: 678–685

Liang Y, Wong JWC, Wei L (2005b) Silicon-mediated enhancement of cadmium tolerance in maize (*Zea mays* L.) grown in cadmium contaminated soil. Chemosphere 58:475–483

Louws FJ, Wilson M, Campbell HL, Cuppels DA, Jones JB, Shoemaker PB, Sahin F, Miller SA (2001) Field control of bacterial spot and bacterial speck of tomato using a plant activator. Plant Dis 85:481–488

Ma JF (2004) Role of silicon in enhancing the resistance of plants to biotic and abiotic stresses. Soil Sci Plant Nutr 50:11–18

Ma JF, Yamaji N (2006) Silicon uptake and accumulation in higher plants. Trends Plant Sci 11: 392–397

Ma JF, Miyake Y, Takahashi E (2001) Silicon as a beneficial element for crop plants. In: Datnoff LE, Snyder GH, Korndorfer GH (eds) Silicon in agriculture. Elsevier Science, Amsterdam, pp 17–39

Maleck K, Levine A, Eulgem T, Morgan A, Schmid J, Lawton KA (2000) The transcriptome of Arabidopsis thaliana during systemic acquired resistance. Nat Genet 26:403–410

Mitani N, Ma JF (2005) Uptake system of silicon in different plant species. J Exp Bot 56: 1255–1261

Pegg K, Moffett M (1971) Host range of the ginger strain of *Pseudomonas solanacearum* in Queensland. Aust J Exp Agric Anim Husb 11:696–698

Pozo MJ, Van Loon LC, Pieterse CMJ (2004) Jasmonates-signals in plant-microbe interactions. J Plant Growth Regul 23:211–222

Rémus-Borel W, Menzies JG, Bélanger RR (2005) Silicon induces antifungal compounds in powdery mildew-infected wheat. Physiol Mol Plant Pathol 66:108–115

Rodrigues FA, Vale FXR, Korndorfer GH, Prabhu AS, Datnoff LE, Oliveira AMA, Zambolim L (2003) Influence of silicon on sheath blight of rice in Brazil. Crop Prot 22:23–29

Romero AM, Kousik CS, Ritchie DF (2001) Resistance to bacterial spot in bell pepper induced by acibenzolar-S-methyl. Plant Dis 85:189–194

Ross AF (1961) Systemic acquired resistance induced by localized virus infections in plants. Virology 14:340–358

Saddler GS (2005) Management of bacterial wilt disease. In: Allen C, Prior P, Hayward AC (eds) Bacterial wilt disease and *Ralstonia solanacearum* species complex. The American Phytopathological Society, St. Paul, MN, pp 121–132
Savant NK, Korndörfer GH, Datnoff LE, Snyder GH (1999) Silicon nutrition and sugarcane production: a review. J Plant Nutr 22:1853–1903
Schenk PM, Kazan K, Wilson I, Anderson JP, Richmond T, Somerville SC (2000) Coordinated plant defense responses in Arabidopsis revealed by microarray analysis. Proc Natl Acad Sci U S A 97:11655–11660
Schneider S, Ullrich WR (1994) Differential induction of resistance and enhanced enzyme activities in cucumber and tobacco caused by treatment with various abiotic and biotic inducers. Physiol Mol Plant Pathol 45:291–304
Seebold KW, Kucharek TA, Datnoff LE, Correa-Victoria FJ, Kucharek TA, Snyder GH (2000) Effect of silicon rate and host resistance on blast, scald, and yield of upland rice. Plant Dis 84: 871–876
Swanson JK, Yao J, Tans-Kersten J, Allen C (2005) Behaviour of *Ralstonia solanacearum* race 3 biovar 2 during latent and active infection of geranium. Phytopathology 95:136–143
Van Hulten M, Pelser M, Van Loon LC, Pieterse CMJ, Ton J (2006) Costs and benefits of priming for defense in Arabidopsis. Proc Natl Acad Sci U S A 113:5602–5607
Van Loon LC, Glick GR (2004) Increased plant fitness by rhizobacteria. In: Sandermann H (ed) Molecular ecotoxicology of plants. Springer, Berlin, pp 177–205
Van Loon LC, Bakker PAHM, Pieterse CMJ (1998) Systemic resistance induced by rhizosphere bacteria. Annu Rev Phytopathol 36:453–483
Van Loon LC, Geraats BPJ, Linthorst HJM (2006a) Ethylene as a modulator of disease resistance in plants. Trends Plant Sci 11:184–191
Van Loon LC, Rep M, Pieterse CMJ (2006b) Significance of inducible defense related proteins in infected plants. Annu Rev Phytopathol 44:135–162
Van Wees SCM, Van der Ent S, Pieterse CMJ (2008) Plant immune responses triggered by beneficial microbes. Curr Opin Plant Biol 11:443–448
Wang JF, Lin CH (2005) Colonizing capacity of *Ralstonia solanacearum* tomato strains differing in aggressiveness on tomato and weeds. In: Allen C, Prior P, Hayward AC (eds) Bacterial wilt disease and the *Ralstonia solanacearum* species complex. The American Phytopathological Society, St. Paul, MN, pp 73–79
Wicker E, Grassart L, Coranson-Beaudu R, Mian D, Guilbaud C, Fegan M (2007) *Ralstonia solanacearum* strains from Martinique (French West Indies) exhibiting a new pathogenic potential. Appl Environ Microbiol 71:6790–6801
Yabuuchi E, Kosako Y, Yano I, Hotta H, Nishiuchi Y (1995) Transfer of two *Burkholderia* and an *Alcaligenes* species to *Ralstonia* gen. Nov.: proposal of *Ralstonia pickettii* (Ralston, Palleroni and Doudoroff 1973) comb. Nov., *Ralstonia solanacearum* (Smith 1896) comb. Nov. and *Ralstonia eutropha* (Davis 1969) comb. Nov. Microbiol Immunol 39:897–904
Yang YF, Liang YC, Lou YS, Sun WC (2003) Influences of silicon on peroxidase, superoxide dismutase activity and lignin content in leaves of wheat *Tritium aestivum* L. and its relation to resistance to powdery mildew. Sci Agric Sin 36:813–817

Chapter 20
Suppression of Soilborne Plant Pathogens by Cruciferous Residues

Ritu Mawar and Satish Lodha

20.1 Introduction

In most agricultural ecosystems, occurrence of soilborne plant pathogens is a major limiting factor in the production of marketable yields. They are also more recalcitrant to management and control compared to pathogens that attack the aboveground portions of the plant (Bruehl 1987). Due to limitation of suitable lands, crops are frequently or even continuously planted on the same piece of land, leading to rapid buildup of host-specific pest population confounding the problems. The inoculum density of soilborne plant pathogens increases with increased years of cultivation of susceptible crops and the inoculum density is directly proportional to the disease intensity in the field.

In severe cases, total devastation forces aggrieved farmers to either abandon the land or shift to less susceptible but often less profitable crops. Knowing the quantity of inocula in the soil and their potential for damage constitute a challenge for both farmers and soil biologists who seek to avoid or minimize the damage by applying effective and practical measures to manage the pathogens and suppress the induced diseases. The major challenge in the control of soilborne plant pathogens is to bring the control agents to all desired sites in the soil. It is also equally important to avoid undesirable effects on nontarget biotic and abiotic components. These issues are relevant to any soil disinfestation method. A host of management strategies are

R. Mawar (✉)
Division of Plant Improvement, Propagation and Pest Management, Central Arid Zone Research Institute, Jodhpur 342003, Rajasthan, India
e-mail: mawar_ritu@rediffmail.com

S. Lodha
Division of Plant Improvement, Propagation and Pest Management, Central Arid Zone Research Institute, Jodhpur 342003, Rajasthan, India
e-mail: satish_lodha@rediffmail.com

M.K. Meghvansi, A. Varma (eds.), *Organic Amendments and Soil Suppressiveness in Plant Disease Management*, Soil Biology 46,
DOI 10.1007/978-3-319-23075-7_20

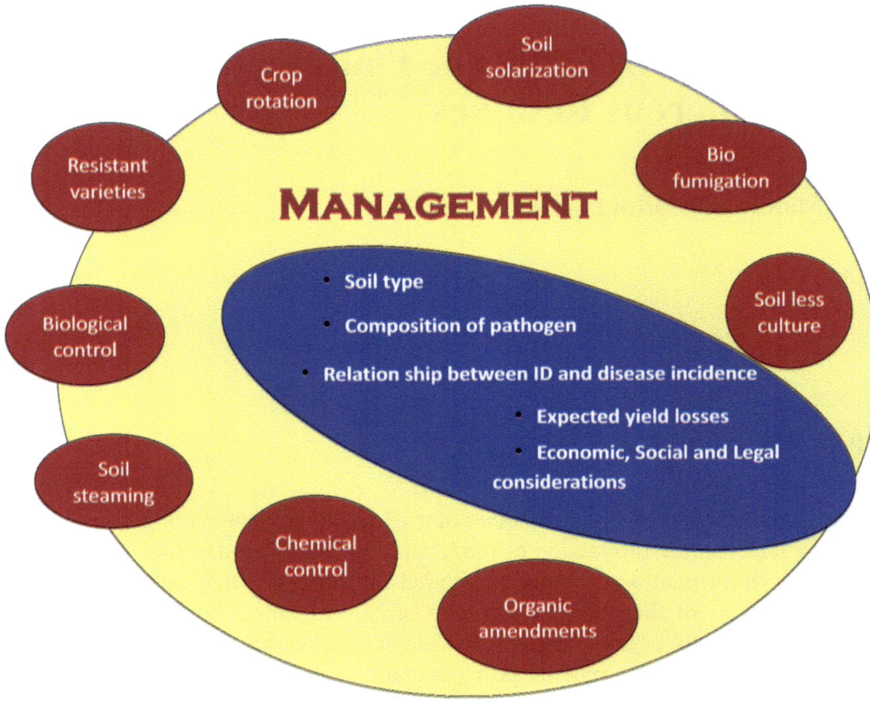

Fig. 20.1 Various strategies to manage soilborne plant pathogens

advocated to reduce or eliminate inoculum density of soilborne plant pathogens but their use depends on consideration of many factors (Fig. 20.1).

Broad-spectrum pesticides have been used for a long time to control soilborne plant pathogens. An example is metam sodium or sodium *N*-methyl dithiocarbamate, which has been used since the 1950s to control pathogenic soilborne organisms. Metam sodium in contact with water generates the compound methyl isothiocyanate, which is effective against nematodes, fungi, pathogens, insects and weeds. However, since 2005 this compound has been designed a class 1 ozone-depleting substance under the Montreal Protocol. Due to restrictions on the use of chemical pesticides, many producers are seeking biological alternatives.

Among management strategies, use of organic amendments as crop residues, composts or manures has found to be of wider acceptance and practical relevance in most of the agricultural production systems. The incorporation of plant residues in soil as green manure or at the end of crop growth has been a common practice for years. Higher plants contain and release an enormous variety of biologically active compounds, some of which have been exploited as potential pesticides.

20.2 Biofumigation: Use of Crucifers

There was a growing interest for use of bioactive plant materials for high-value crops. Glucosinolates (GSLs) are present in various quantities in many dicotyledonous plants. Enzymatic hydrolysis of GSLs in the presence of enzyme myrosinase results in the production of various sulphur compounds, some of which possess antimicrobial activity (Duncan 1991). The Cruciferae are among the plant families with high content of GSLs in their tissues. Biofumigation is a term used to describe the suppression of soilborne pests and pathogens by *Brassica* rotation or green manure crops (Angus et al. 1994; Kirkegaard et al. 1993). These are also characterized by a high content of other sulphur-containing compounds. Antifungal volatile compounds such as allyl isothiocyanates have been found in leaf extracts of various *Brassica* species (Mayton et al. 1996; Sang et al. 1984). There are about 20 different types of GSLs commonly found in *Brassicas* which vary in their structure depending on the type of organic side chain (aliphatic, aromatic or indolyl) on the molecule. The profile, concentration and distribution of these GSLs vary within and between *Brassica* species and in different plant tissues, and consequently the concentration and type of biocidal hydrolysis products that evolved also vary (Mithen 1992). Among the major hydrolysis products, isothiocyanates (ITCs) are generally considered the most toxic; however, individual ITCs also vary in their toxicity to different organisms (Brown and Morra 1997). For example, ITCs derived from aromatic GSLs have been found to be 40 times more toxic to eggs of black vine weevil (*Otiorhynchus sulcatus* F.) than the aliphatic moiety (Borek et al. 1994). The range in GSL profiles, the differential toxicity of the ITCs that evolved to different pests of plants and the wide range of phonological and morphological diversity of *Brassicas* provide scope to select or breed *Brassicas* with enhanced biofumigation potential for particular target organisms. Kirkegaard and Sarwar (1998) investigated the potentials to enhance biofumigation by considering the variation in GSL production in the roots and shoots of 76 entries from 13 *Brassica* and related weed species in Australia. The types of GSLs present in the tissues varied considerably between species but were consistent within species. By contrast, the concentration of individual and total GSLs in both root and shoot tissues varied four- to tenfolds both between and within all species. Shoots contained predominantly aliphatic GSLs, while aromatic GSLs, particularly 2-phenylethyl GSL, were dominant in the roots of all entries. The variation in the biomass, GSL profiles and concentrations in both roots and shoots provide significant scope to select or develop *Brassicas* with enhanced biofumigation potential.

Lewis and Papavizas (1970) measured volatile compounds produced a week following incorporation of a wide variety of different crucifer tissues into soil and in no case detected GSL hydrolysis products. Only low-molecular weight non-GSL-derived volatile S compounds including dimethyl-disulphide, dimethyl-sulphide and methanethiol were found, none of which were produced from non-crucifer tissues. In a semi-quantitative study, Gamliel and Stapleton (1993) detected low amounts of ITCs in volatiles collected from soil amended with cabbage residues, again finding large quantities of dimethyl-disulphide and methanethiol. Non-GSL-derived S compounds, including dimethyl-sulphide, are themselves toxic to a broad range of organisms including fungi, bacteria and invertebrates. The generation of toxic compounds from decomposing organic amendments increases with increased temperature (Gamliel and Stapleton 1993). Possible mechanisms for the enhanced generation of volatile compounds with increased soil temperature include: (1) increase of the vapour pressure of compounds present in the liquid or solid soil fractions, resulting in greater release to the soil atmosphere; (2) changes in soil chemical and physical properties; and (3) heat-induced breakdown of more complex compounds and release of polar molecules from clay particles.

Bending and Lincoln (1999) compared concentrations of GSL hydrolysis products and other non-GSL derive toxic volatile S compounds, during decomposition of leaf tissues of *B. juncea* in sandy–loam and clay–loam soils. The tissues were shown to be rich in 2-propenylglucosinolate, which is hydrolyzed to 2-propenyl-ITC on tissue damage. Patterns of formation of the compounds differed in two soils, with smaller amounts of all compounds detected in the clay–loam, in which microbial respiration was higher. It was suggested that the biofumigant properties of crucifer tissues represent the combined effect of the low quantities of highly toxic ITC and large quantities of mildly toxic non-GSL-derived volatile S-containing compounds produced during decomposition.

Morra and Kirkegarrd (2002) conducted experiments to determine the concentration and pattern of ITCs released from GSLs in *Brassicaceous* residues like rapeseed and Indian mustard. A flush in ITCs occurred immediately after tissue incorporation into soil because cell membranes were broken during plough down. Freezing caused extensive cell membrane disruption and thus permitted greater contact between GSLs and myrosinase. The flush in ITC from frozen tissue correspondingly was much more dramatic. This study indicates that soilborne pest suppression is likely to be improved by choosing a high GSL-containing variety of rapeseed or mustard and providing adequate moisture to increase ITC release and soil retention. However, the greater improvements in the use of *Brassica* biofumigants to control soilborne plant pests will be achieved by focusing on methods to increase cell disruption thereby maximizing GSL hydrolysis and ITC release.

20.3 Persistence

The question concerning the persistence of biological effects of amending soil with *Brassica* tissues on soilborne pathogens had to date only dealt with the kinetics of disappearance of ITC (Gardiner et al. 1999; Gimsing and Kirkegaard 2006). Brown and Morra (1997) reviewed the factors contributing to biofumigation efficacy and suggested that, as the lifetime of GSL products in the environment was shown to be short, "a short residence time places limits on achieving effective control and may contribute to the variability observed in the suppression of soil borne plant pests". These studies provided the first insight into understanding some of the mechanisms, which might be involved in the persistence of control, but although the disappearance of ITC in soil is rapid, no firm conclusion can be drawn concerning the noxious action of residues after the period of ITC detection has passed.

Persistence of control of primary infections caused by *Rhizoctonia solani* and *G. graminis* var. *tritici*, following the incorporation of above-ground parts (AP), below-ground parts (BP) or both (AP+BP) of *B. juncea* into soil, was studied by Motisi et al. (2009). Control was quantified by measuring disease incidence in bioassays where inoculum was introduced at different dates after the incorporation of plant residues. All types of residues showed an unexpected long-term persistence that lasted at least 13 days, while the predominant GSLs contained in AP (20.9 μmol sinigrin g^{-1} dry matter) and BP (2.3 μmol gluconasturtiin g^{-1} dry matter) were hydrolyzed in less than 3 days. Temporal trends in the efficacy of the residues behaved mostly in a quadratic manner, suggesting that the noxious effect of residues may be attributable to the release of ITCs during the first days following incorporation but that other mechanisms are most likely to contribute to lasting persistence. Persistence of action of *B. juncea* residues may be caused due to persistence of unhydrolysed GSL in soil detected 5–8 days after residue incorporation (Gimsing and Kirkegaard 2006). As myrosinase activity can be detected in soils with no recent history of cultivation of GSL-containing plants (Gimsing et al. 2006), GSL can potentially be hydrolysed by extracellular microbial myrosinase several days after residue incorporation (Al-Turki and Dick 2003). Several non-GSL-derived volatile S-containing compounds, such as sulphides and thiols, are formed by microbial degradation of *Brassica* residues in soil (Bending and Lincoln 1999), which are known to be toxic to a range of organisms and are likely to contribute to biofumigation by acting in association with ITC. Mazzola et al. (2007) demonstrated that long-term control of *R. solani* AG-5 by *B. juncea* seed meal amendment was attributed to increased populations of *Streptomyces* spp. antagonistic to pathogen (Cohen et al. 2005). Across all treatments, AP and AP+BP suppressed *R. solani* by 54 and 63 %, respectively, and *G. graminis* var. *tritici* by 40 and 40 %, respectively, compared with controls. While BP did not cause any additional detectable effect when combined with AP, they had a significant effect when incorporated alone, suggesting the existence of a complex interaction between these two types of residues. Hence, many other phenomena are likely to contribute to the persistent effect of *Brassica* residues on the infectivity of soil inoculum. The exact cause of this phenomenon is unknown, but it suggests that environmental conditions determine diverse and complex interactions between above- and below-ground residues of *B. juncea* and disease suppression by *Brassica* amendments does not derive solely from ITC or other GSL-related compounds, but from other chemical or biological changes in the soil microbial profile that can influence disease expression. Furthermore, certain epidemiological factors (inoculum survival, disease development and expression) must be taken into account at the field scale as they are likely to restrict the benefits of biofumigation to specific seasonal conditions (Kirkegaard et al. 2000).

20.4 Fungal Pathogen Management

Research has been conducted world over on the significant effect of incorporated crucifer tissues on activity and control of many soilborne plant pathogens (Table 20.1). Climatic, edaphic and biotic factors have all been reported to influence the GSL concentration in *Brassica* tissues (Rosa et al. 1997). Environmental factors such as day length and temperature also influence the phenology and biomass production of *Brassicas* (Nanda et al. 1996). As a result, the total production of GSL on a ground area basis (the product of GSL concentrations × biomass) and therefore biofumigation potential will be significantly influenced by growing conditions.

Use of broccoli as rotation crop or as residues for the control of *Verticillium* wilt of cauliflower has been extensively studied in California, USA. Amendment of soil with broccoli residues resulted in significant reductions in the numbers of *V. dahlia* microsclerotia in soil and incidence of wilt in the following cauliflower crop (Subbarao et al. 1999). Although broccoli and cauliflower are related with the

Table 20.1 Important soilborne plant pathogens managed by cruciferous residues

Pathogen	Disease	Crop	Reference
Aphanomyces euteiches f. sp. *pisi*	Root rot	Pea	Muehlchen and Parke (1990), Smolinska et al. (1997)
Fusarium oxysporum f. sp. *conglutinans*	Cabbage yellows	Cabbage	Ramirez-Villapudua and Munnecke (1987, 1988)
F. o. f. sp. *cumini*	Wilt	Cumin	Israel et al. (2005), Mawar and Lodha (2002)
F. o. f. sp. *spinacia*	Wilt	Spinach	Mowlick et al. (2013)
F. o. f. sp. *niveum*	Wilt	Watermelon	Njoroge et al. (2008)
Gaeumannomyces graminis var. *tritici*	Take all	Wheat	Motisi et al. (2009)
Meloidogyne chitwoodi	Root knot	Potato	Mojtahedi et al. (1993)
Pythium ultimum	Damping off	Tomato/pepper	Handiseni et al. (2012)
Phytophthora capsici, *P. parasitica*	–	Pepper	Guerrero et al. (2010)
Rhizoctonia solani	Hypocotyle rot/crown rot	Bell pepper/ snap bean/sugar beet	Hansen and Keinath (2013), Manning and Crossan (1969), Motisi et al. (2009)
Thielaviopsis basicola	Root rot	Sesame	Adams (1971)
Verticillium dahliae	Wilt	Eggplant/ cauliflower	Garibaldi et al. (2010), Subbarao et al. (1999)
Didymella bryoniae	Gummy stem blight	Watermelon	Keinath (1996)

genus *Brassicas*, they differ markedly in their response to *V. dahliae*. Subsequently, mechanisms of broccoli-mediated wilt reduction were studied. The reduction in colony density of pathogen was presumably caused by the reduction in the number of microsclerotia and subsequently by decreased root colonization potential of surviving microsclerotia under the influence of broccoli residue (Shetty et al. 2000). Colonization of the root cortex by *V. dahaliae* alone does not always lead to the disease because not all successful root infections result in colonization of the vascular tissue. The zone near the root apex consists of tissues in their early stages of growth and maturation; it is particularly vulnerable to vascular invasion. The increased root exudation near the root apex and the zone of root elongation, compared with that of the older tissue, is likely to be conducive for *V. dahaliae* activity, which initially colonizes near the root tip. All the colonies are initiated behind the zone of elongation. The latter suggests that, for vascular infection, other conducive factors in addition to colonization of the root cortical surface are needed.

Studies on rotation-crop residue amendment suggest a biological mode of action: sustained suppression of soilborne pathogens results from the activation of biological components that are already present in the soil (Stapleton and Duncan 1998; Subbarao et al. 1999). It is plausible that the microbial population changes resulting from broccoli residue decomposition also lead to greatly increased competition among root colonizers. Increased microbial activity following broccoli amendment and the resulting competition for colonization of root cortical surface may also limit infection foci for *V. dahaliae*. Mature broccoli residues are rich in lignin, and the enzymes involved in lignin biodegradation can also degrade fungal melanin (Butler and Day 1998). Melanin is known to protect the fungus from various abiotic and biotic stresses and the microsclerotium of *V. dahaliae* is a melanized structure; therefore, it can be hypothesized that biodegradation of broccoli residues may also affect *V. dahaliae* microsclerotia. Xiao et al. (1998) suggested that this disease can be managed by developing a rotation scheme that includes broccoli as cash crop and then incorporating the residues into the soil. This rotation scheme also fits in current cropping systems and can be easily adapted by growers. The required length of rotation of susceptible crops with broccoli may depend on the initial level of soil infestation and the relative susceptibility of the crop. Such a crop rotation having wheat–mustard–cumin has been suggested for Indian arid region, where mustard fits well in current cropping system. This rotation scheme also saves irrigation water as mustard and cumin are less water-requiring crops.

Gummy stem blight (*Didymella bryoniae*) is the most destructive foliar disease of watermelon and other cucurbits in the USA. A minimum 2-year rotation away from cucurbits is recommended to reduce soilborne inoculum of the pathogen. But most growers are unwilling to employ rotations longer than 1 year due to profitability. Therefore, additional management strategies are needed for gummy stem blight control. Three cropping sequences, watermelon–cabbage–soil solarization–watermelon, watermelon–wheat–soybean–watermelon and 3-year watermelon, were evaluated (Keinath 1996). Cabbage–soil solarization and the wheat–soybean double crop reduced area under the disease progress curve for gummy stem blight.

Cabbage followed by soil solarization increased the weight and number of marketable-sized and total healthy fruits compared with the non-solarized treatments.

In Italy, efficacy of a biofumigant green manure of *B. juncea* selection ISC120 used in combination with grafting and soil mulching was investigated in an eggplant production system in naturally infested soil with *V. dahaliae* (Garibaldi et al. 2010). In a second set of trials, effectiveness of soil application of a patented formulation (Lazzeri et al. 2008) of *B. carinata* biofumigant defatted seed meals combined or not with a simulation of soil solarization against Fusarium wilt of lettuce and of basil was studied. Severe infection of *Verticillium* wilt was recorded in non-grafted eggplants on both solarized, biofumigated soil and in the plots where grafting and biofumigation were combined. Combination of biofumigation and grafting onto *Solanum torvum* improved only partial resistance of the root stock. In second set, defatted seed meals alone at 2 and 4 g l^{-1} showed a partial but significant effect. The combination of defatted seed meals and soil solarization provided the best results against both *F. o.* f. sp. *lactucae* and *F. o.* f. sp. *basilica*. These results demonstrated that combining biofumigation will reduce polyethylene mulching period, increasing at the same time its efficacy.

In Spain, disinfest efficacy of biosolarization with *B. carinata* pellets at 300 g m^{-2} alone or mixed with fresh sheep manure in different dates of application has been evaluated against *Phytophthora* spp. and *M. incognita* (Guerrero et al. 2010). When biosolarization was carried out in August, the survival of *P. capsici* oospores was as low as that obtained with methyl bromide and the incidence of *M. incognita* was similar to methyl bromide. When biosolarization was initiated in October, disinfest efficacy decreased, since the incidence of *Meloidogyne* and the survival of *P. capsici* inoculum increased using either pellets alone or mixed with manure.

In India, farmers were using paste of white mustard in warding off diseases and pests, particularly those attacking roots in ancient times. Many studies have reported use of mustard oil cake in control of soilborne pest and diseases. In Indian arid region, solar irradiations, high soil temperature and cruciferous residues are amply available during crop-free period (April–June). In the laboratory, significant reduction in the population of *M. phaseolina* occurred in the mustard oil-cake amended soil, where complete reduction in population of *Fusarium oxysporum* f. sp. *cumini* (*Foc*-cumin wilt pathogen) was also achieved within 30 days (Sharma et al. 1995). The effect was mainly attributed to the release of toxic volatiles such as mercaptan, methyl sulphide and isothiocyanate (Gamliel and Stapleton 1993). The population of bacteria and actinomycetes increased considerably in amended soils. Over 90 % of the total actinomycetes were antagonistic to *M. phaseolina*, but populations of actinomycetes antagonistic to *Foc* were less as compared to *M. phaseolina*. Thus, apart from the toxic effects of oil cake, increased population of antagonists might have also contributed in reducing the population of both test pathogens.

Efficacy of mustard oil-cake (4 tons ha^{-1}) and cauliflower leaf residues (5 tons ha^{-1}) combined with summer irrigation and/or solarization was ascertained on the population of *M. phaseolina* (Lodha et al. 1997). One summer irrigation of the dry non-amended plots caused 40 % reduction in *M. phaseolina* counts at 0–30 cm depth. Amending soil with cruciferous residues augmented the efficiency of irrigation by eliminating a sizeable proportion of *M. phaseolina* in non-solarized plots. Soil solarization of amended and irrigated plots elevated soil temperature by 4–6 °C compared to non-solarized plots. Combined effect of moisture, amendments and temperature completely eliminated viable propagules of *M. phaseolina*, irrespective of soil depth. In yet another study (1998–2000), combined effects of *Brassica* amendments (mustard oil cake or pod straw -2.5 tons ha^{-1}) and summer irrigation on survival of *M. phaseolina* and *Foc* in soil and charcoal rot intensity on cluster bean (July–October) and wilt of cumin (November–March) were studied in the same field. Both the amendments were significantly superior in reducing incidence of both the diseases (Mawar and Lodha 2002). Of the residues, mustard oil cake was significantly more effective than pod straw with a 34 % greater reduction in wilt incidence.

Effectiveness of different doses of *Brassica* amendments in reducing viable propagules of *M. phaseolina* was ascertained in the field (Lodha and Sharma 2002). In amended pits, soil temperature remained 0.5–3.0 °C (unshaded) and 0.5–1.5 °C (shaded) higher than non-amended pits. *Brassica* amendments significantly reduced *M. phaseolina* under both the environments. Mustard oil cake was significantly better where complete reduction in viable propagules was achieved at 0.9 % (2 tons ha^{-1}) compared to 0.22 % (5 tons ha^{-1}) with mustard pod residues. Under shade, magnitude of reduction in *M. phaseolina* propagules was low but significant improvement in the reduction was estimated with increased concentration of amendments. Efficacy of mustard oil cake even at low concentration could be attributed to the presence of 7–8 % oil that releases more quantity of volatiles at high temperature besides having 5 % nitrogen. A significant improvement in lytic bacterial density was estimated in amended compared to dry and irrigated non-amended pits. The possible role of increased lytic bacterial density in reducing *M. phaseolina* counts cannot be excluded as these bacteria are capable of lysing fungal mycelium of soilborne pathogens (Mitchell and Alexander 1963). As a result, effective doses of *Brassica* amendment for the control of *M. phaseolina* in hot arid region have been worked out (Lodha and Sharma 2002).

Since mustard oil cake was considered expensive, a need was felt to improve the efficiency of mustard pod residue by integrating it with other easily available, cost-effective practical management strategies. Therefore, effects of soil solarization, residue incorporation, summer irrigation and biocontrol agents alone or in combination on survival of *M. phaseolina* and *Foc* were ascertained (Israel et al. 2005). Combining amendments and soil solarization elevated the soil temperatures by 0.5–5 °C and 2.5–13.0 °C compared to non-amended solarized and non-solarized plots, respectively. These treatment combinations significantly reduced *M. phaseolina* and *Foc* propagules compared to control. Of these, combining mustard pod residues with soil solarization almost eliminated viable propagules of both the pathogens at 0–30 cm soil depth. However, a combination of mustard pod residues and oil cake (2.5 + 0.5 tons ha^{-1}) also caused pronounced reduction in pathogenic propagules,

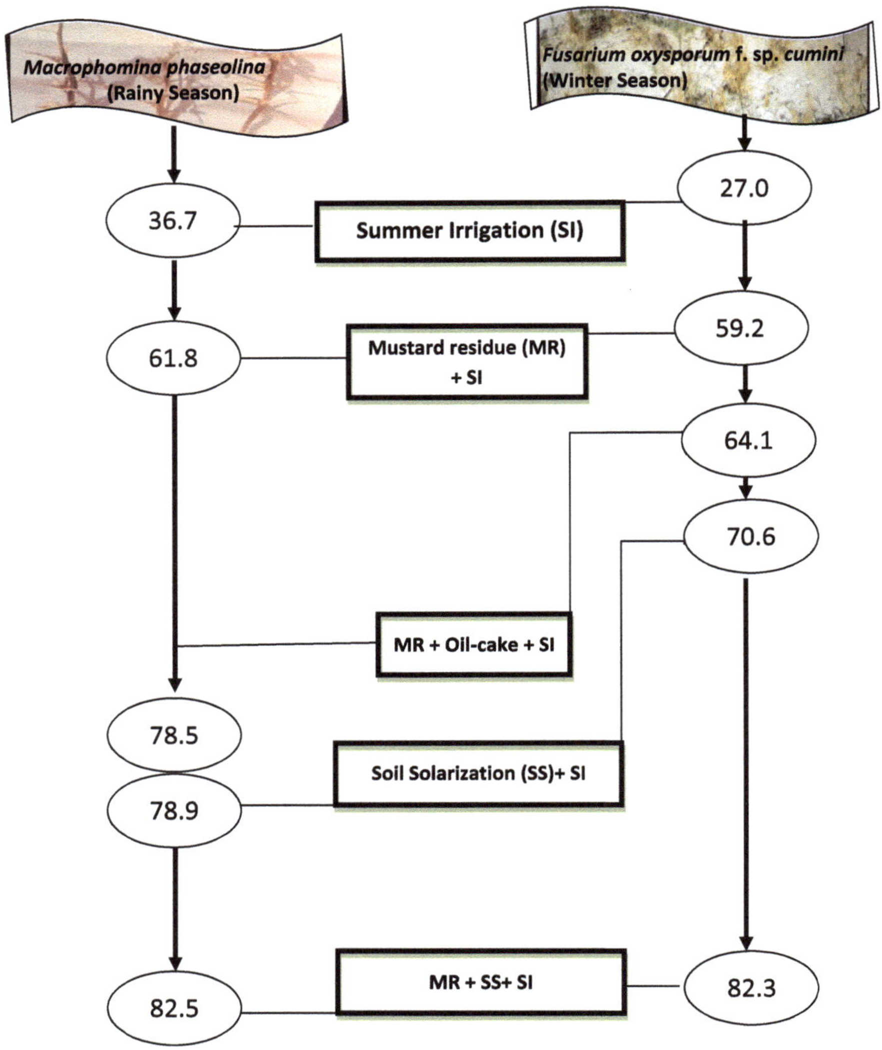

Fig. 20.2 Per cent improvement in reduction in viable propagules of *Macrophomina phaseolina* and *Fusarium oxysporum* f. sp. *cumini* by soil solarization/*Brassicas*

which was equal to that recorded in non-amended solarized plots (Fig. 20.2). When the effect of surviving propagules of *M. phaseolina* and *Foc* on incidence of charcoal rot on cluster bean and wilt on cumin was studied in subsequent rainy and winter seasons, respectively, significant reductions in both diseases were recorded in residue and biocontrol amended plots with or without soil solarization compared to non-amended control. The least plant mortality was observed in amended solarized plots. However, the disease indices in the plots having a combination of mustard residues and oil-cake amendment were equal to that achieved in the treatment having solarization. These results suggest that in hot arid regions, the use of *Brassica* residues can be a practical and feasible substitute for polyethylene mulching in managing soilborne plant pathogens and induced diseases.

20.5 Nematode Management

Cruciferous residues were also found to reduce nematode population in the soil. Several in vitro and in vivo trials have shown a wide biocidal activity of GSL degradation productions (GLDPs) on several nematode species (Mojtahedi et al. 1993; Potter et al. 1998). Lazzerri et al. (2004) evaluated in vitro the biocidal activity of 11 GSLs and their degradation products on second-stage juveniles of the root-knot nematode *Meloidogyne incognita* expressed by the nematicidal (LS_{50}) and immobilization effects after 24 and 48 h. None of the intact glucosinolates had any biological effect, but after myrosinase addition, their hydrolysis products (essentially ITCs) resulted in highly different biocidal activities. Among the hydrolysis products of the tested GSLs, 2-phenylethyl, benzyl, 4-methylthiobutyl and prop-2-enyl isothiocyanate showed the stronger activity, with an LD_{50} at concentration of 11, 15, 21 and 34 μM, respectively. The results seem to be an important starting point for studying the possibility of restricting *M. incognita* infestation by the use of plants selected for GSL content of their roots. These plants could be used as biocidal catch crops, supposing that when *Meloidogyne* J2s penetrate the roots, several cellular lesions where contact between root GSLs and MYR occurs can be determined; so there is production in situ of the corresponding GLDPs characterized from a clear nematicidal activity. In this way, nematode lives in a medium poisoned by GLDPs, and their development should stop a few days after root penetration. Therefore, the nematode does not produce any progeny, with a consequent decrease of the soil infestation level.

Roubtsova et al. (2007) determined the direct localized and indirect volatile effects of amending soil with broccoli tissue on root-knot nematode populations. *M. incognita* infested soil in 50-cm-long tubes was amended with broccoli tissue, which was mixed throughout the tube or concentrated in a 10-cm layer. After 3 weeks at 28 °C, *M. incognita* populations in the amended tubes were 57–80 % less than in non-amended tubes. Mixing broccoli throughout the tubes reduced *M. incognita* more than concentrating broccoli in a 10-cm layer. Amending a 10-cm layer with tissue reduced *M. incognita* in the non-amended layers of those tubes by 31–71 %, probably due to a nematicidal effect of released volatiles. However, the localized direct effect was much stronger than the indirect effect of volatiles suggesting that residues should be distributed uniformly in the soil profile for better pathogen control.

20.6 Combining Weakening Effects with *Brassicas*

Exposing the infested soil to dry summer heat (sublethal temperature) weakens the propagules and often renders them more vulnerable to other management strategies (Freeman and Katan 1988; Lifshitz et al. 1983). Any effective amendment combined with prolonged duration of sublethal temperatures may require less energy, time and quantity for improving the control. This information can be used to work out appropriate time of application of cost-effective concentration of *Brassica* amendments to improve or augment control of soilborne plant pathogens.

In controlled environment conditions, effects of amending soil with fresh and dried residues of *B. nigra*, *B. oleracea* var. *chinensis*, *B. oleracea* var. *italiensis*, *B. oleracea* var. *capitata*, *B. oleracea* var. *compacta* and *Raphanus sativus* and of a sublethal soil heating regime (38 °C/27 °C night) on survival and activity of *M. incognita*, *S. rolfsii* and *Pythium ultimum* were studied by Stapleton and Duncan (1998). The addition of the various cruciferous amendments to soil without heating resulted in significantly reduced tomato root galling (38–100 %) by *M. incognita* or reduced recovery of active fungal pathogens (0–100 %) after 7 days of incubation. When cruciferous soil amendments were combined with the sublethal regime, nematode galling was reduced by 95–100 % and recovery of active fungi was reduced by 85–100 %. However, no differences were found between fresh or dried cruciferous residues.

In Indian arid region, sublethal heating (45–55 °C) of *M. phaseolina*-infested dry soil reduced the viable propagules by only 12.8 % in a period of 90 days (Lodha et al. 2003). One summer irrigation without sublethal heating caused 33.9 % reduction in pathogenic propagules, which improved to 43.3 % when it was combined with 60 days of sublethal heating. The addition of the *Brassica* amendments to moist soil significantly reduced (60.4–71.6 %) counts of *Macrophomina*, but reduction improved (89.4–96.1 %) when sublethal heating was combined with amendment. Mustard oil cake (0.18 w/w) was found to be the most effective with 96 % reduction, but a 94 % inoculum reduction by mustard pod straw (0.36 % w/w) was also achieved at 0–30 soil depth. These results suggest that combining sublethal heating and *Brassica* amendments with one summer irrigation can improve pathogen control. In the next phase, effect of varying intensities of sublethal heating was ascertained on efficiency of *Brassica* amendments in reducing viable population of *M. phaseolina* and *Foc*. After 30 days of dry summer exposure of pathogen-infested soil, incorporation of mustard residues and oil cake (0.18 and 0.04 % w/w) and then application of irrigation significantly reduced viable counts of *M. phaseolina* by 75.3–81.3 % and those of *Foc* by 93.9 % at 0–15 and 16–30 cm depths (Mawar and Lodha 2009). Increased duration (60 days) of summer exposure improved the reductions in *M. phaseolina* by 83.6–90.4 % and in *Foc* by 78.2–94.8 % at the same soil depths. Significantly low levels of reduction in pathogenic propagules of *Macrophomina* (63.9–71.4 %) and *Foc* (48.0–57.2 %) under shade compared to unshaded conditions indicated that mild heating did not cause discernible weakening effect.

Effects of four intensities of sublethal heating on the efficiency of readily available on-farm wastes as soil amendments in controlling *Foc* were ascertained. Significant improvement in reduction of *Foc* propagules was achieved with the increased duration and intensity of heat (Israel et al. 2010). In 2000, under shade conditions (heat level 4), 31.8–65.9 % reduction in *Foc* propagules was estimated in all the amendments at 0–30 cm soil depth. Soil brought from laboratory and exposed to bright sunlight (heat level 3) and then application of amendments and irrigation improved this reduction by 75.7–86.5 % with maximum being in *Verbesina* residue-amended soil. Reduction in *Foc* propagules to the tune of 76.6–88.3 % was achieved when infested soil was continuously exposed to dry heat for 56 days (heat level 2) leading to improved efficiency of amendments by 0.9–13.5 % compared to heat level 3. After 56 days of exposure, elevation of soil temperature by polyethylene mulching for 20 days to amended soil (heat level 1) augmented this reduction by 80.2–95.5 %. In the second season, combining a small dose (0.04 %) of onion, *Verbesina* or mustard oil cake with mustard residues (0.18 %) improved the reduction in *Foc* propagules at all the heat levels compared to alone application of onion and *Verbesina* residues (0.18 %). Among these, maximum reduction (94.9–100 %) in *Foc* propagules at 1–3 heat levels was achieved when *Verbesina* residues were supplemented with mustard residues. Combination of amendments also improved the reduction in viable *Foc* propagules at lower soil depth. The results demonstrated that interactive effects of sublethal heating, achieved by prolonged exposure of pathogenic propagules to natural solar heat in dry sandy soil, *Brassica* residues and summer irrigation, improved the reduction of viable *M. phaseolina* and *Foc* propagules in a hot arid environment. However, the magnitude of reduction varied with the type of amendment, level of heat and pathogen involved.

Soil moisture in the form of irrigation affected the sensitivity of sclerotia and chlamydospores to a heat treatment (Lodha et al. 1997). Greater reduction in *M. phaseolina* than *Foc* propagules with irrigation was a result of increased microbial antagonism against *Macrophomina*. More than one mechanism might have operated concurrently or in a sequence in eliminating viable propagules of *M. phaseolina and Foc* from amended soil. Sublethal heating in dry soil for 90 days (April 1–June 30) exerted a weakening effect on the surviving propagules, which depends on temperature level, exposure time and the environment into which the preheated propagules are introduced. However, a certain threshold of heating has to be reached to obtain a detectable weakening effect (Freeman and Katan 1988). An improvement of 10–14 % reduction in viable *Macrophomina* and *Foc* propagules was evident by merely increasing exposure time. Decomposition of cruciferous residues in moist soil at high temperature subsequent to the pronounced weakening effect enhanced the action of sulphur-containing toxic volatiles and microbial antagonism. In the final soil samples, populations of bacteria and actinomycetes were invariably greater in amended than in non-amended pits. The presence of residual soil moisture in amended pits further encouraged microbial antagonism against remaining weakened sclerotia and chlamydospores particularly by bacteria at lower soil depth. Increased bacterial colonization of heat-treated sclerotia of

S. rolfsii has also been reported by Lifshitz et al. (1983). Disruption in activity of enzyme(s) involved in melanin production in case of *M. phaseolina* can be yet another possible factor for increased susceptibility to microbial antagonism. In the case of *Fusarium*, weakening was expressed in delayed spore germination and germ tube growth, reduction in viable florescent staining and enhanced decline in viability of propagules (Freeman and Katan 1988). The cumulative effect of these factors resulted in the ultimate reduction in viable propagules of both the pathogens. These results suggested a new approach to improve control of soilborne plant pathogens by combining prior weakening, effective cruciferous residues and one summer irrigation.

20.7 Factors Determining Effectiveness

Many factors may influence success of using *Brassica* amendments in managing soilborne plant pathogens and associated diseases. A scientific and clear understanding is required for their use; otherwise inconsistent results are obtained even at experimental stage.

20.7.1 Choice of Crop Residue and Variety

Different species and varieties contain varying amounts of bioactive chemicals like GSL content. *Brassicaceae* species should be selected based on the amount of GSL, the type of resulting ITC that will be produced as well as the amount of biomass they are capable of producing (Matthiessen and Kirkegaard 2006). Indian mustard, canola and broccoli are considered most superior as break crops. Indian mustard cv. Pacific Gold was reported to have the highest above-ground biomass (5.7 kg m^{-2}) and GSL content of seven species tested (Antonious et al. 2009). Another cover crop, *R. sativus* (oilseed radish), has the potential to produce approximately 10 kg m^{-2} biomass (Sundermeier 2008). The variations in GSLs are also evident in root and shoot tissues. However, their use in pest management depends on the availability of residues in a farming system as a component of rotation. The *Brassica* cover crops are usually planted in late summer (August) or early fall and incorporated in spring before planting mustard in the USA.

20.7.2 Composition of Pathogen Complex

Fungicidal concentration of ITCs is also known to differ by an order of magnitude for different fungal species (Brown and Morra 1997). Some studies have investigated the toxicity of pure ITCs in the headspace experiments where the volatility of the compound may influence its activity, while others have used ITCs dissolved in the growing media. Sarwar et al. (1998) investigated in vitro toxicity of ITCs against different pathogens and found variation in fungal response.

20.7.3 Time of Application

Incorporation and application of irrigation is the most unique and critical phase in improving the efficiency of crucifer residues in controlling a particular pathogen. Ambient temperature and corresponding soil temperature influence the release of GSLs and the hydrolysis products. Gamliel and Stapleton (1993) analyzed profiles of volatiles in headspace and reported that the concentration of volatiles, which increased with an increase in soil temperature, was higher in heated, amended soils than in non-heated amended soil. Therefore, it is generally recommended that cruciferous residues should be incorporated during warmer months to get greater release of volatiles. In addition, growth stage at incorporation is also important to consider for the success of biofumigation. GSL concentration was highest at the bud-raised growth stage prior to flowering and higher in spring- versus fall-seeded *Brassicas* (Sarwar and Kirkegaard 1998).

20.7.4 Amount and Size of Residue Tissues

It has been observed in many studies that large size residues caused sharp reduction in beneficial microbes in soil. Amending soil with milled plant material or placing crop debris on the soil surface and roto tilling is the most effective way. Incorporating relatively large fragments may lead to uneven distribution of the amendment in the soil profile.

20.7.5 Fresh or Dry

Studies have shown that incorporation of fresh residues as soil amendment is better than dried residues in suppression of pathogens. Reduction in wilt incidence and population of microsclerotia of *Verticillium* was higher in the plots amended with fresh broccoli than those amended with dried residues. Therefore, efforts should be made to use fresh residues or the quantity of dried residues should be accordingly adjusted.

20.7.6 Environment

Total GSL concentration is known to be influenced by climatic (temperature, day length, radiation, water stress), edaphic (soil type, nutrients) and biotic factors (pest and diseases). Shorter days, lower radiation and cooler temperatures accompanied by frost induce lower levels of GSLs in vegetative material. In addition, both insect attack and water stress can increase total GSLs.

20.8 Limitations

It is true that there are many positive attributes in a new innovation, particularly that which deals with the management of biological entities. However, there are certain shortcomings also for its application in a wider perspective. It is better to get disappointment at the level of research rather than at the level of farmers. Therefore, negative aspects should also be considered before any management strategy is recommended to growers.

20.8.1 Effects on Beneficial Microbes

Many studies have shown that beneficial microbes like mycorrhiza, population of fluorescent *Pseudomonas*, antagonistic actinomycetes and fungi are reduced when a particular plant residue or higher quantity of residue is incorporated. Therefore, recommendation of a species or quantity of residue incorporation should be made only after scientific investigations. By contrast, variations in soil community compositions were observed when clone library analysis based on 16S rRNA gene sequences was done to determine relationship between the bacterial compositions in the *B. juncea* amended soil and suppression of the disease (Mowlick et al. 2013). Results revealed that members of *Firmicutes* mainly from the class *Clostridia* dominated in treated soils. These changes in the soil condition might affect the population of soil pathogens and bring about the suppression of disease incidence.

20.8.2 Pathogen Specific

Pathogens also differ in their sensitivity to ITCs. *Gaeumannomyces* is the most sensitive; *Rhizoctonia* and *Fusarium* are intermediate, while *Bipolaris* and *Pythium* are least sensitive. Such specific sensitivity requires clear knowledge of selection of pathogen as well as residue for getting maximum benefit in terms of control.

Biofumigation treatments did not reduce populations of *Pythium* spp. or *S. rolfsii* compared to plots covered with virtually impermeable film (CVIF) and did not reduce plant mortality of pepper (Hansen and Keinath 2013). However, pepper yields were highest in biofumigation treatments compared to CVIF. Njoroge et al. (2008) and Collins et al. (2006) also found that the incorporation of *Brassica* tissue did not significantly reduce, and in some cases increased, soil population densities of *Pythium* spp.

20.8.3 Phytotoxicity

In some studies, plants grown after residue incorporation developed phytotoxicity either as reduced germination or bronzing of leaves at seedling stage. This is possible if volatiles are not released properly during decomposition either due to inadequate soil moisture or large size of residue particles. Residual volatile compounds in the soil, in turn, lead to phytotoxicity in the next crop, particularly those having small size seeds. In a study, oilseed radish, oriental mustard and yellow mustard green manure reduced direct-seeded muskmelon stand count as well as transplant survival (Ackroyd and Ngouajio 2011). Oilseed radish had the greatest effect with 0 % muskmelon stand. These results suggest that species and tissue-dependent toxicity of the cover crops as well as differential susceptibility of the cucurbit crops be tested. Therefore, a plant-back period no longer than 8 days used in this study should be observed after cover crop incorporation before cucurbit seeding or transplanting.

20.9 Conclusion

In developing feasible alternatives to chemical soil fumigants, it is essential that they provide reliable, predictable and relatively rapid reductions of pathogen/pest inoculum. Addition of bioactive soil amendments may fulfil partly or fully these requirements. Their exists great potential in the use of cruciferous plant residues for suppression of soilborne plant pathogens as an alternative to hazardous chemical means. This approach of pathogen control is renewable and biodegradable and has no impact on CO_2 level and relatively low toxicity in terms of long-term use. It is expected that this management strategy may often result in partial control due to many factors listed above, but it can be safely integrated with other management approaches like use of biocontrol agents, partial host resistance, sound crop rotation, low doses of chemicals, etc. In many countries, such approach holds great promise because cruciferous plant species are a common component of most of the cropping systems, which ensures regular availability of residues. Incidentally, ample availability of solar irradiations during crop-free periods allows quick and effective decomposition of residues with greater release of biotoxic volatiles.

However, there is a need to investigate in detail about the feasibility of using crucifers for control of economically important pathogen(s), comparative evaluation with other available management strategies, time and amount of application, effects on beneficial microbes, weed suppression, compatibility with biocontrol agents, soil fertility improvement and effect of crop rotation with crucifers on succeeding crops. After generation of scientific data, large-scale field demonstrations will be useful for fine-tuning of this practical cultural control for different agricultural zones. At present growers in many developing countries are dependent on cultural control measures for the partial reduction of soilborne plant pathogens. Use of cruciferous residues or making crucifers as a part of rotation will not only provide reasonable control of these pathogens but may also improve population of antagonists in soil, which will induce soil suppressiveness.

References

Ackroyd VJ, Ngouajio M (2011) Brassicaceae cover crops affect seed germination and seedling establishment in cucurbit crops. HortTechnology 21(5):525–532

Adams PB (1971) Effect of soil temperature and soil amendments on *Thielaviopsis* root rot of sesame. Phytopathology 61:93–97

Al-Turki AI, Dick WA (2003) Myrosinase activity in soil. Soil Sci Soc Am J 67:139–145

Angus JF, Gardner PA, Kirkegaard JA, Deshmarchelier JM (1994) Biofumigation: isothiocyanates released from *Brassica* roots inhibit growth of the take-all fungus. Plant Soil 162:107–112

Antonious GF, Bomford M, Vincelli P (2009) Screening *brassica* species for glucosinolate content. J Environ Sci Health B 44:311–316

Bending GD, Lincoln SD (1999) Characterization of volatile sulphur-containing compounds produced during decomposition of *Brassica juncea* tissues in soil. Soil Biol Biochem 31:695–703

Borek V, Morra MJ, Brown PD, McCaffrey JP (1994) Allelochemicals produced during sinigrin decomposition in soil. J Agric Food Chem 42:1030–1034

Brown PD, Morra MJ (1997) Control of soil borne plant pests using glucosinolate containing plants. Adv Agron 61:167–231

Bruehl GW (1987) Soilborne plant pathogens. Macmillan, New York, 368 p

Butler MJ, Day AW (1998) Fungal melanins: a review. Can J Microbiol 44:1115–1136

Cohen MF, Yamasaki H, Mazzola M (2005) *Brassica napus* seed meal soil amendment modifies microbial community structure, nitric oxide production and incidence of rhizoctonia root rot. Soil Biol Biochem 37:1215–1227

Collins HP, Alva A, Boydston RA, Cochran RL, Hamm PB, McGuire A, Riga E (2006) Soil microbial, fungal, and nematode responses to soil fumigation and cover crops under potato production. Biol Fertil Soils 42:247–257

Duncan A (1991) Glucosinolates. In: Mello JP, Duffs CM, Duffs JH (eds) Toxic substances in crop plants. Royal Society of Chemistry, Cambridge, pp 127–147

Freeman S, Katan J (1988) Weakening effect on propagules of *Fusarium* by sub-lethal heating. Phytopathology 78:1656–1661

Gamliel A, Stapleton JJ (1993) Characterization of antifungal volatile compounds evolved from solarized soil amended with cabbage residues. Phytopathology 83:899–905

Gardiner JB, Morra MJ, Eberlein CV, Brown PD, Borek V (1999) Allelochemicals released in soil following incorporation of rapeseed (*Brassica napus*) green manures. J Agric Food Chem 47:3837–3842

Garibaldi A, Gilardi G, Clematis F, Gullino ML, Lazzeri L, Malaguti L (2010) Effect of green Brassica manure and Brassica defatted seed meals in combination with grafting and soil solarization against Verticillium wilt of eggplant and Fusarium wilt of lettuce and basil. Acta Hortic 883:295–302

Gimsing AL, Kirkegaard JA (2006) Glucosinolate and isothiocyanate concentration in soil following incorporation of *Brassica* biofumigants. Soil Biol Biochem 38:2255–2264

Gimsing AL, Sorensen JC, Tovgaard L, Jorgensen AMF, Hansen HCB (2006) Degradation kinetics of glucosinolates in soil. Environ Toxicol Chem 25:2038–2044

Guerrero MM, Ros C, Lacasa CM, Martinez MA, Martinez V, Lacasa A, Fernanadez P, Nunez-Zofio M, Larregla S, Diez-Rojo MA, Bello A (2010) Effect of biosolarization using pellets of Brassica carinata on soil-borne pathogens in protected pepper crops. Acta Hortic 883:337–344

Handiseni M, Brown J, Zemetra R, Mazzola M (2012) Use of Brassicaceous seed meals to improve seedling emergence of tomato and pepper in *Pythium ultimum* infested soils. Arch Phytopathol Plant Protect 45:1204–1209

Hansen ZR, Keinath AP (2013) Increased pepper yields following incorporation of biofumigation cover crops and the effects on soilborne pathogen populations and pepper diseases. Appl Soil Ecol 63:67–77

Israel S, Mawar R, Lodha S (2005) Soil solarization, amendments and bio-control agents for the control of *Macrophomina phaseolina* and *Fusarium oxysporum* f. sp. *cumini* in aridisols. Ann Appl Biol 146:481–491

Israel S, Mawar R, Lodha S (2010) Combining sub-lethal heating and on-farm wastes: effects on *Fusarium oxysporum* f. sp. *cumini* causing wilt on cumin. Phytoparasitica 39:73–82

Keinath AP (1996) Soil amendment with cabbage residue and crop rotation to reduce gummy stem blight and increased growth and yield of watermelon. Plant Dis 80:564–570

Kirkegaard JA, Sarwar M (1998) Biofumigation potentials of *Brassicas.* I. Variation in glucosinolate profiles of diverse field grown *brassicas.* Plant Soil 201:71–89

Kirkegaard JA, Gardner PA, Desmarcheleir JM, Angus JF (1993) Biofumigation-using *Brassica* species to control pest and diseases in horticulture and agriculture. In: Wrotten V, Mailer RJ (eds) 9th Australian research assembly on *Brassicas.* Agricultural Research Institute, Wagga Wagga, pp 77–82

Kirkegaard JA, Sarwar M, Wong PTW, Mead A, Howe G, Newell M (2000) Field studies on the biofumigation of take-all by *Brassica* break crops. Aust J Agric Res 51:445–456

Lazzeri L, Leoni O, Manici LM, Palmieri S, Patalano G (2008) Brevetto Europeo EP1530421B1, Use of seed flour as soil pesticide. Register of European patents

Lazzerri L, Curto G, Leoni O, Dallavalle E (2004) Effects of glucosinolates and their enzymatic hydrolysis products via myrosinase on the root- knot nematode *Meloidogyne incognita* (Kofoid et White) Chitw. J Agric Food Chem 52(22):6703–6707

Lewis JA, Papavizas GC (1970) Evolution of volatile sulfur containing compounds from decomposition of crucifers in soil. Soil Biol Biochem 2:239–246

Lifshitz R, Tabachnik M, Katan J, Chet I (1983) The effect of sub-lethal heating on Sclerotia of *Sclerotium rolfsii.* Can J Microbiol 29:1607–1610

Lodha S, Sharma SK (2002) Effect of natural heating and different concentration of *Brassica* amendments on survival of *Macrophomina phaseolina.* Indian Phytopathol 55:303–305

Lodha S, Sharma SK, Aggarwal RK (1997) Natural and solar heating of irrigated soil amended with cruciferous residue and improved control of *Macrophomina phaseolina.* Plant Pathol 46:186–190

Lodha S, Sharma SK, Mathur BK, Aggarwal RK (2003) Integrating sub-lethal heating with Brassica amendments and summer irrigation for control of *Macrophomina phaseolina.* Plant Soil 256:423–430

Manning WJ, Crossan DF (1969) Field and greenhouse studies on the effects of plant amendments on rhizoctonia hypocotyls rot of snapbean. Plant Dis Report 53:227–231

Matthiessen JN, Kirkegaard JA (2006) Biofumigation and enhanced biodegradation: opportunity and challenge in soilborne pest and disease management. Crit Rev Plant Sci 25:235–265

Mawar R, Lodha S (2002) *Brassica* amendments and summer irrigation for the control of *Macrophomina phaseolina* and *Fusarium oxysporum* f. sp. *cumini* in hot arid regions. Phytopathol Mediterr 41:45–54

Mawar R, Lodha S (2009) Prior weakening of *Macrophomina phaseolina* and *Fusarium propagules* for enhancing efficiency of *Brassica* amendments. Crop Prot 28:812–817

Mayton HS, Olivier C, Vaughan SF, Loria R (1996) Correlation of fungicidal activity of Brassica species with allylisothiocyanate production in macerated leaf tissue. Phytopathology 86:267–271

Mazzola M, Brown J, Izzo AD, Cohen MF (2007) Mechanism of action and efficacy of seed meal-induced pathogen suppression differ in a Brassicaceae species and time-dependent manner. Phytopathology 97:454–460

Mitchell R, Alexander M (1963) Lysis of soil fungi by bacteria. Can J Microbiol 9:169–171

Mithen R (1992) Leaf glucosinolate profiles and their relationship with pest and disease resistance in oilseed rape. Euphytica 63:71–80

Mojtahedi H, Santo GS, Wilson JH, Hang AN (1993) Managing *Meloidogyne chitwoodi* on potato with rape seed as green manure. Plant Dis 77:42–46

Morra MJ, Kirkegarrd JA (2002) Isothiocyanate release from soil- incorporated *Brassica* tissues. Soil Biol Biochem 34:1683–1690

Motisi N, Montfort F, Dore T, Romillac N, Lucas P (2009) Duration of control of two soil borne pathogens following incorporation of above- and below-ground residues of *Brassica juncea* into soil. Plant Pathol 58:470–478

Mowlick S, Yasukawa H, Inoue T, Takehara T, Kaku N, Ueki K, Ueki A (2013) Suppression of spinach wilt disease by biological soil disinfestation incorporated with *Brassica juncea* plants in association with changes in soil bacterial communities. Crop Prot 54:185–193

Muehlchen AM, Parke JL (1990) Evaluation of crucifer green manure for controlling *Aphanomyces* root rot of peas. Plant Dis 74:651–654

Nanda R, Bhargava SC, Tomar DPS, Rowson HM (1996) Phenological development of *Brassica campestris*, *B. juncea*, *B. napus* and *B. carinata* grown in controlled environments and from 14 sowing dates in the field. Field Crops Res 46:93–103

Njoroge SMC, Riley MB, Keinath AP (2008) Effect of incorporation of *Brassica* spp. residues on population densities of soil borne microorganisms and on damping-off and *Fusarium* wilt of watermelon. Plant Dis 92:287–294

Potter MJ, Davies JK, Rathjen A (1998) Suppressive impact of glucosinolates in Brassica vegetative tissues on root lesion nematode *Pratylenchus neglectus*. J Chem Ecol 24:67–80

Ramirez-Villapudua J, Munnecke DE (1987) Control of cabbage yellows (*Fusarium oxysporum* f. sp. *conglutinans*) by solar heating of field soil amended with dry cabbage residues. Plant Dis 71:217–221

Ramirez-Villapudua J, Munnecke DE (1988) Effects of solar heating and soil amendments of cruciferous residues on *Fusarium oxysporum* f. sp. *conglutinans* and other micro-organisms. Phytopathology 78:289–295

Rosa EAS, Heaney RK, Fenwick GR (1997) Glucosinolates in crop plants. Hortic Rev 19:99–215

Roubtsova T, Lopez-Perez JA, Edward S, Pleog A (2007) Effect of broccoli (*Brassica oleracea*) tissue incorporation at different depths in a soil column on *Meloidogyne incognita*. J Nematol 39:111–117

Sang JP, Minchinton PK, Johnstone PK, Truscott JW (1984) Glucosinolate profiles in the seed, root and leaf tissue of cabbage, mustard, rapeseed, radish and swede. Can J Plant Sci 64:77–93

Sarwar M, Kirkegaard JA (1998) Biofumigation potential of brassicas II. Effect of environment and ontogeny on glucosinolate production and implications for screening. Plant Soil 201:91–101

Sarwar M, Kirkegaard JA, Wong PTW, Desmarchelier JM (1998) Biofumigation potentials of brassicas-III. In vitro toxicity of isothiocyanates to soil-borne fungal pathogens. Plant Soil 201:103–112

Sharma SK, Aggarwal RK, Lodha S (1995) Population changes of *Macrophomina phaseolina* and *Fusarium oxysporum* f. sp. *cumini* in the oil cake and crop residue amended sandy soils. Appl Soil Ecol 2:281–284

Shetty KG, Subbarao KV, Huisman OC, Hubbard JC (2000) Mechanism of broccoli-mediated *Verticillium* wilt reduction in cauliflower. Phytopathology 90:305–310

Smolinska U, Knudsen GR, Morra MJ, Borek V (1997) Inhibition of *Aphanomyces euteiches* f. sp. *pisi* by volatiles produced by hydrolysis of *Brassica napus* seed meal. Plant Dis 81:288–292

Stapleton JJ, Duncan RA (1998) Soil disinfestation with cruciferous amendments and sub lethal heating: effects on *Meloidogyne incognita*, *Sclerotium rolfsii* and *Pythium ultimum*. Plant Pathol 47:737–742

Subbarao KV, Hubbard JC, Koike ST (1999) Evaluation of broccoli residue incorporation into field soil for *Verticillium* wilt control in cauliflower. Plant Dis 83:124–129

Sundermeier A (2008) Oilseed radish cover crop. Ohio State University Extension SAG-5-08

Xiao CL, Subbarao KV, Schulbach KF, Koike ST (1998) Effects of crop rotation and irrigation on *Verticillium dahliae* microsclerotia in soil and wilt in cauliflower. Phytopathology 88:1046–1055

Part IV
Combinatorial Approaches in Plant Disease Management

Chapter 21
Biodisinfestation with Organic Amendments for Soil Fatigue and Soil-Borne Pathogens Control in Protected Pepper Crops

Santiago Larregla, María del Mar Guerrero, Sorkunde Mendarte, and Alfredo Lacasa

21.1 Soil Phytopathological Problems and Soil Fatigue in Protected Pepper Crops

Phytophthora root rot is a destructive disease for pepper plants (*Capsicum annuum* L.) worldwide (Wang et al. 2014). The mortality of pepper plants ranges between 30 and 40 % and in severe cases, even 100 % (Liu et al. 2008). As a consequence of the high plant mortality, relevant economic losses have been reported in pepper crops from Spain, not only in the Mediterranean region (Tello and Lacasa 1997) but also in areas characterised by a humid temperate climate, such as the Basque Country (Northern Spain) (Larregla 2003). The main causal agents of this disease in greenhouse pepper crops are the oomycetes *Phytophthora capsici* and *P. cryptogea* in Northern Spain and *P. capsici* and *P. parasitica* in South-eastern Spain. In the last region, the nematode *Meloidogyne incognita* is also a recurring and persistent problem that causes substantial crop damages (Tello and Lacasa 1997; Bello et al. 2004). In South-eastern Spain (Murcia and Alicante provinces) pepper occupies more than 90 % of the area dedicated to greenhouse crops and has been a monoculture for the last 20 years (Lacasa and Guirao 1997). The normal

S. Larregla (✉)
Plant Protection, NEIKER-Tecnalia, C/Berreaga 1, E-48160 Derio, Bizkaia, Spain
e-mail: slarregla@neiker.eus

M. del Mar Guerrero • A. Lacasa
Biotechnology and Crop Protection, IMIDA, C/Mayor, s/n, E-30150 La Alberca, Murcia, Spain
e-mail: mariam.guerrero@carm.es; alfredo.lacasa@carm.es

S. Mendarte
Conservation of Natural Resources, NEIKER-Tecnalia, C/Berreaga 1, E-48160 Derio, Bizkaia, Spain
e-mail: smendarte@neiker.eus

M.K. Meghvansi, A. Varma (eds.), *Organic Amendments and Soil Suppressiveness in Plant Disease Management*, Soil Biology 46,
DOI 10.1007/978-3-319-23075-7_21

crop cycle lasts 9–10 months (November–December to September–October). Strategies recommended for management of phytophthora root rot involve integrated approaches that focus on cultural practices: reduced soil moisture, reduction of pathogen propagule in soil, utilisation of cultivars with resistance to the disease and the judicious fungicide applications (Ristaino and Johnston 1999). Until the year 2005, methyl bromide (MB) was used to disinfect soils to control both pathogens (Gilreath and Santos 2004) and to lessen the effects of fatigue caused by repeated monocultures (Martínez et al. 2011a). Since 2005, MB has been replaced by a mixture of 1,3-dichloropropene and chloropicrin, but these will also be banned by European legislation in the near future. The increasing demand for ecological foods produced by sustainable agricultural practices must be added to this forthcoming ban, meaning that non-chemical methods will have to be developed for controlling soil-borne plant pathogens and plant parasitic nematodes adapted for use in intensive horticulture.

In recent years, numerous alternatives for chemical disinfection have been studied, and of these, those based on organic amendments alone or in combination with solarisation seem to be the most promising (Guerrero et al. 2013) in intensive protected horticultural crops.

In this book chapter, we aim to review both the mechanisms involved in disease suppression and the organic amendment management strategies for the control of protected pepper crops soil-borne diseases and soil fatigue. Several disease suppression mechanisms following the addition of organic matter such as (1) the release of compounds that are toxic to the pathogens, (2) the stimulation of non-pathogenic microorganisms that inhibit or kill the pathogens and (3) the improvement of soil physical, chemical and biological properties will be explained.

21.2 Mechanisms Involved in Disease Suppression by Soil Organic Amendments

Several mechanisms have been identified as contributing to disease suppression following the addition of organic matter. These namely include the stimulation of non-pathogenic microorganisms that inhibit or kill the pathogens through competition (Lockwood 1988) or parasitism (Hoitink and Boehm 1999), the release of compounds that are toxic for the pathogens (Bailey and Lazarovits 2003) and the stimulation of the host plant's disease defence system (Zhang et al. 1996, 1998). Other indirect mechanisms that explain the ability of organic amendments to increase soil suppressiveness are: improvement of nutrition and vigour in the host plants and improvement of physicochemical and biological properties of soil. Some organic amendments are thought to work primarily by altering the structure of the microbial communities in the soil or by changing the physical and chemical properties of the soil.

21.2.1 Production of Biocidal Compounds

The application of animal manure leads to the generation of ammonia (NH_3), which is the mechanism most often implicated in killing soil pathogens (Tenuta and Lazarovits 2002), although other lethal molecules such as nitrous acid and volatile fatty acids (VFA) have also been reported (Tenuta et al. 2002; Conn et al. 2005). Tenuta and Lazarovits (2002) summarised that NH_3 is thought to kill cells by disrupting membranes, eliminating proton gradients across membranes, through the assimilation of NH_3 into glutamine and the exhaustion of the chemical energy of cells removing cytosolic NH_3. Accumulation of NH_3 and VFA derived from manure application have been described as mechanisms capable to kill soil pathogens (Conn et al. 2005), which is largely influenced by several factors that include moisture content, pH, soil organic matter content and quality, soil texture and buffering capacity and nitrification rate soil buffering capacity. These authors concluded that high VFA toxicity was achieved in acid soils (pH about 5.0) while high NH_3 toxicity was related with alkaline soils (pH about 7.5).

The application of animal manure followed by soil plastic covering during spring period reduced inoculum survival of the fungal pathogen *P. capsici* and disease incidence in a greenhouse pepper crop in Northern Spain (Arriaga et al. 2011). Northern Spain is an area characterised by a temperate climate with annual mean temperature of 12 °C, a maximum mean temperature in summer of 25 °C and rainfall of 1200 mm per year.

Ammonia volatilisation, among other volatile compounds, and the increase in soil suppressiveness contributed to minimise *P. capsici* inoculum survival rate and subsequent greenhouse crop disease incidence, respectively. The use of fresh manure favoured NH_3 volatilisation as organic nitrogen (N_{org}) mineralisation was higher than in semicomposted manure, with more stable N_{org} content. Mean NH_3 concentration increased with fresh sheep manure and dry chicken litter (SCM) during biodisinfestation process compared with semicomposted mixture of horse manure and chicken litter (HCM) (Fig. 21.1).

NH_3 concentration increased significantly after manure amendment with respect to control plots (C) and also differed between SCM and HCM manure. Ammonia concentration from C plots averaged 3.9 mg NH_3 m^{-3}, while SCM averaged 14.8 mg NH_3 m^{-3} and 9.1 mg NH_3 m^{-3} in HCM. The highest NH_3 concentrations were reached at the beginning of the experiment in SCM and HCM treatments and decreased 45.0 % after 35 days of soil biodisinfestation (Fig. 21.1). The reduction of NH_3 concentrations could be related to adsorption of NH_3 or the increasing anaerobic conditions during manure decomposing process (Kirchmann and Witter 1989). The high water condensation observed on the plastic inner surface, which would trap volatilised NH_3 (Kroodsma et al. 1993), and the overall anaerobic conditions under plastic sheets might have reduced NH_3 accumulation (Kirchmann and Witter 1989).

P. capsici inoculum survival rate in infected plant residues was significantly different among treatments. The application of fresh SCM under plastic sheets

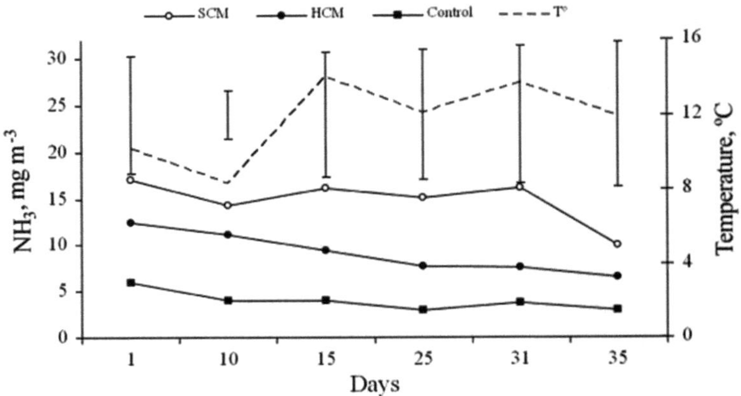

Fig. 21.1 Evolution of NH_3 concentration under the plastic sheets from non-amended (control), fresh manure (SCM) and semicomposted manure (HCM) amended plots during soil biodisinfestation starting on March 14, 2008, showing variation of air temperature throughout the biodisinfestation period. The *vertical bars* indicate least significance difference at 0.05 between treatments. Soil was tarped with 50-μm-thick (two million) transparent low density polyethylene plastic film. The greenhouse field experiment was located in Derio (Biscay) (Northern Spain). Reprinted from Journal of Crop Protection, 30(4), H. Arriaga et al. (2011), Gaseous emissions from soil biodisinfestation by animal manure on a greenhouse pepper crop, 412–419, Copyright (2015), with permission from Elsevier

reduced *Phytophthora* inoculum survival in relation to HCM and S treatments (Table 21.1) ($P < 0.05$). Biodisinfestation by SCM manure reduced by 50 % inoculum survival rate compared with C plots (61.1 %), while *Phytophthora* inoculum survived in 75.0 % and 94.4 % of plant residues in HCM and S treatments, respectively, which was significantly higher than survival reported in C plots. The higher NH_3 concentration in SCM contributed to reduce the inoculum survival rate of *P. capsici* (30.6 % and 75.0 % in SCM and HCM treatments, respectively). Inoculum survival rate was not reduced in solarised non-amended plots (94.4 %) as soil temperature at 15 cm depth did not exceed 33 °C under plastic sheets in S, SCM and HCM treatments (a temperature known to be insufficient to inactivate resistant propagules of *P. capsici* since this pathogen normally shows an optimum temperature range at 24–33 °C) (Erwin and Ribeiro 1996; Etxeberria et al. 2011). Additionally, warm soil temperatures and water condensation detected during the biodisinfestation process might have favoured the conditions for *Phytophthora* inoculum survival in S and HCM plots when compared with C plots. The lower inoculum survival rate in C plots was related to the higher water evaporation and the subsequent lower soil volumetric water content in these uncovered soils.

Higher *P. capsici* inoculum inactivation observed in SCM was attributed to the effect of toxic volatile compounds generated from the decomposition of organic amendments. Soil pH of our experiment averaged 6.9 in SCM and HCM treatments, which might suggest that NH_3 contributed significantly to the reduction of *Phytophthora* inoculum survival in SCM. Moreover, Oka et al. (2007) summarised

Table 21.1 Infected plant rate (inoculum survival), disease incidence and crop yield in non-treated control (C), solarized (S), fresh sheep manure and dry chicken litter (SCM) and semicomposted mixture of horse manure and chicken litter (HCM) biodisinfestated plots

Treatment	Infected plant residues (%)	Disease incidence (%)	Crop yield (kg m^{-2})
C	61.1 ab	40.7 a	3.0 b
S	94.4 a	42.6 a	3.4 b
SCM	30.6 b	2.8 b	4.6 a
HCM	75.0 a	8.3 b	4.3 a

Soil was tarped with 50-μm-thick (two million) transparent low density polyethylene plastic film. The greenhouse field experiment was located in Derio (Biscay) (Northern Spain). Reprinted from Journal of Crop Protection, 30(4), H. Arriaga et al. (2011), Gaseous emissions from soil biodisinfestation by animal manure on a greenhouse pepper crop, 412–419, Copyright (2015), with permission from Elsevier

For each variable, values followed by the same letter are not significantly different according to Fisher's protected LSD test ($P < 0.05$). Mean values ($n = 3$)

that soil amendments with low C/N ratios have been reported to have fungicidal activity mainly through the release of NH_3. In our experiment, the amount of manure amended exceeded those rates applied by Oka et al. (2007), which would support that NH_3 was the main factor controlling inoculum survival of Arriaga et al. (2011).

21.2.2 *Increase in Microbial Activity*

Phytophthora disease incidence decreased significantly in biodisinfestated SCM and HCM plots compared with C and S treatments (Table 21.1). Plant disease incidence was reduced by 90 % in SCM and HCM plots in relation to inoculum survival rate observed in plant residues 4 months before. Of note, disease was only reduced by 33 % and 54 % in C and S plots, respectively. The application of high amounts of organic amendments contributes to the suppressive capacity of soils through enhanced activity and growth of edaphic microorganisms, which may play an important role in reducing disease incidence by an antagonistic mechanism (Hoitink and Boehm 1999). The significant reduction of disease incidence compared with the high inoculum survival rate could explain this phenomenon in SCM and HCM plots. Several authors have also reported the success of organic matter applications in the control of *Phytophthora* spp., suggesting that the competition for nutrients and antibiosis are the main mechanisms involved in *Phytophthora* spp. suppressiveness (Leoni and Ghini 2006).

21.2.3 Improvement of Plant Nutrition and Vigour

Soil organic matter management affects not only soil biological properties but also soil chemical and physical properties and plant nutrient status. All of them improve plant health and vigour (Stone et al. 2004) and thus may help the plants to overcome pathogen infection. The increase in soil-borne disease suppression by organic amendments may also be attributed to other effects such as increase in plant nutritional status and vigour (Hoitink et al. 1997).

Pepper fruit yield increased with manure amendment in SCM and HCM (Table 21.1), as the application of organic amendments improves soil quality, increasing the amount of plant-available nutrients and, in consequence, crop yield (Liu et al. 2008).

Highest values of crop vigour (plant height) were observed in plots amended with animal manures, and differences increased during crop development (Fig. 21.2) (Núñez-Zofío et al. 2010; Arriaga et al. 2011).

21.2.4 Improvement of Soil Physical and Chemical Properties

Detected differences in plant nutrient status have been generally found between nonamended and amended soils. This could be due to the improved nutrient content, water holding capacity and soil structure imparted to the soil by the amendments (Vallad et al. 2003).

The effects of repeated biodisinfestation with different organic amendments after three consecutive crop seasons improved soil physical properties through a reduction in soil bulk density and an increase in soil water infiltration (Table 21.2) (Núñez-Zofío et al. 2012). This management strategy provided an effective control of phytophthora root rot in protected pepper crops. Improvements in soil water properties that prevent water flooding are known to facilitate soil-borne pathogen control, mainly in the case of oomycetes (Liu et al. 2008).

In general terms, biodisinfestation with non-composted and semicomposted manures increased the values of all soil chemical properties, except for pH (Table 21.3). Besides, non-composted manure was the only treatment that significantly increased P_2O_5, Cl^-, K^+ and Zn^{2+} contents. Significantly, higher values of Cu^{2+} content were found only in semicomposted manure-biodisinfected soils. However, no significant differences were observed between Brassica-treated plots and control soils (Table 21.3) (Núñez-Zofío et al. 2012).

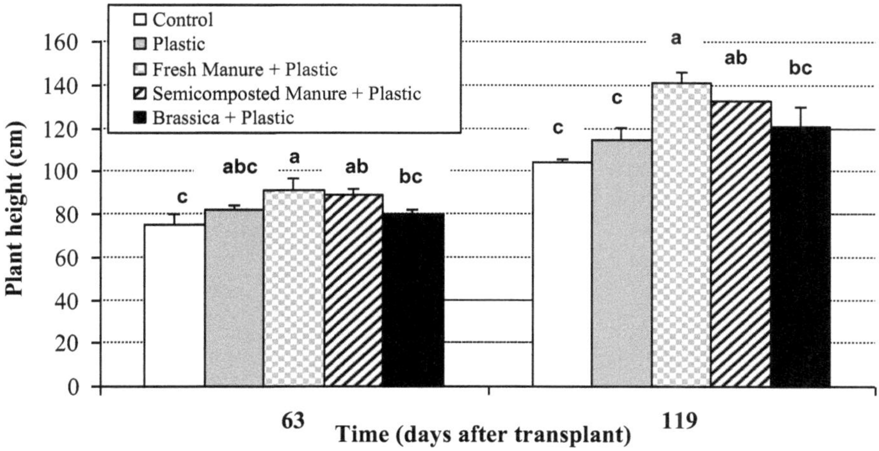

Fig. 21.2 Effect of treatments on pepper plant height, measured at the middle (63 days) and at the end (119 days) of crop development. Error bars represent standard error of the mean ($n = 3$) from three replicate plots. Different *letters* indicate significant differences ($P < 0.05$) (least significance difference = 9.71 and 17.72 cm at 63 and 119 days after transplant, respectively) according to Fisher's protected LSD test. Reprinted from Acta Horticulturae (ISHS), http://www.actahort.org/books/883/883_44.htm, 883, Núñez-Zofío et al. (2010), Application of organic amendments followed by plastic mulching for the control of *Phytophthora* root rot of pepper in Northern Spain, 353–360, with permission from International Society for Horticultural Science

Table 21.2 Effect of biodisinfestation treatments on soil physical properties

	Bulk density (g cm^{-3})			
Treatments[a]	0–10 cm	10–20 cm	20–30 cm	Infiltration (cm)
Control	1.35 a	1.32 a	1.38 a	42.81 c
Plastic-Mulched	1.28 ab	1.31 a	1.34 a	87.17 b
Non-composted	1.20 bc	1.21 ab	1.19 b	173.36 a
Semicomposted	1.13 c	1.17 b	1.27 ab	139.39 a
Brassica	1.28 ab	1.28 ab	1.35 a	182.98 a

Reprinted from Spanish Journal of Agricultural Research, 10(3), Núñez-Zofío et al. (2012), Repeated biodisinfection controls the incidence of *Phytophthora* root and crown rot of pepper while improving soil quality, 794–805, Open Access Journal from the Spanish National Institute for Agricultural and Food Research and Technology

[a]Control: untreated soil, Plastic-Mulched: non-amended plastic-mulched soil, Non-composted: non-composted manure amended soil + plastic-mulched, Semicomposted: semicomposted manure amended soil + plastic-mulched, Brassica: *B. carinata* dehydrated pellets + *S. alba* fresh green manure amended soil + plastic-mulched. For each variable and depth, values followed by the same letter are not significantly different according to Waller-Duncan's *K*-ratio *t* test ($P < 0.05$). Mean values ($n = 6$)

Table 21.3 Effect of biodisinfestation treatments on soil chemical properties

	Treatments[a]				
Variable	Control	Plastic-Mulched	Non-composted	Semicomposted	Brassica
OM[b] (%)	4.83 b	5.32 b	7.28 a	6.81 a	5.23 b
C_{org} (%)	2.80 b	3.09 b	4.22 a	3.95 a	3.03 b
N (%)	0.21 b	0.21 b	0.30 a	0.28 a	0.18 b
P_2O_5 (mg kg^{-1})	109.8 c	108.2 c	288.8 a	247.0 b	126.9 c
Cl^- (meq l^{-1})	0.94 c	0.91 c	9.17 a	3.18 b	1.61 bc
pH	6.87	6.71	7.10	6.89	6.95
EC[c] (dS m^{-1})	1.72 c	2.05 bc	3.84 a	3.09 ab	2.12 bc
K^+ (meq kg^{-1})	0.03 c	0.05 c	0.32 a	0.18 b	0.07 c
Ca^{2+} (meq kg^{-1})	1.18 c	1.42 bc	1.91 a	1.67 ab	1.37 bc
Mg^{2+} (meq kg^{-1})	0.17 b	0.20 b	0.31 a	0.29 a	0.20 b
Na^+ (meq kg^{-1})	0.09 b	0.08 b	0.16 a	0.13 ab	0.10 ab
CEC[d] (meq kg^{-1})	1.33 b	1.56 b	2.40 a	2.05 a	1.55 b
Cu^{2+} (mg kg^{-1})	1.65 b	1.58 b	1.69 b	2.85 a	1.50 b
Zn^{2+} (mg kg^{-1})	5.43 bc	5.86 bc	12.79 a	8.44 b	4.84 c

Reprinted from Spanish Journal of Agricultural Research, 10(3), Núñez-Zofío et al. (2012). Repeated biodisinfection controls the incidence of *Phytophthora* root and crown rot of pepper while improving soil quality, 794–805, Open Access Journal from the Spanish National Institute for Agricultural and Food Research and Technology

[a]Treatments: see Table 21.2. All values are expressed on a dry soil weight basis

[b]OM: organic matter content

[c]EC: electrical conductivity

[d]CEC: cation exchange capacity

For each variable, values followed by the same letter are not significantly different according to Waller-Duncan's *K*-ratio *t* test ($P < 0.05$). Mean values ($n = 3$)

21.2.5 Improvement of Soil Biological Properties

Biodisinfected soils with non-composted and semicomposted manure showed significantly higher values of all enzyme activities when compared with control non-amended soils, whereas *Brassica* treatment significantly increased the values of dehydrogenase, β-glucosidase and acid phosphatase activity compared with control non-amended treatments (Table 21.4). Apart from the organic matter input, this increase of enzyme activities could also be attributed, at least partly, to a stimulatory rhizosphere effect caused by a *Sinapis alba* cover crop during the winter season. This rhizosphere effect could also be responsible for the higher values of microbial population densities (total bacteria, actinomycetes and *Pseudomonas* spp.) detected in *Brassica*-amended soil. No significant differences were found between control and plastic-mulched plots (Table 21.4). Significant positive correlations were obtained between organic matter content and the following

Table 21.4 Effect of biodisinfestation treatments on soil biological properties

	Treatments[a]				
Variable	Control	Plastic-Mulched	Non-composted	Semicomposted	Brassica
FDA[b] (mg F kg^{-1} h^{-1})	77.7 bc	64.0 c	129.8 a	114.7 a	93.5 b
Dehydrogenase (mg INTF kg^{-1} h^{-1})	4.8 b	4.8 b	10.5 a	9.7 a	9.0 a
Urease (mg N-$NH_4^+$$kg^{-1}$ h^{-1})	23.8 b	17.2 b	56.8 a	46.9 a	26.1 b
β-glucosidase (mg NP kg^{-1} h^{-1})	39.3 c	39.1 c	64.5 a	64.6 a	51.2 b
Alkaline phosphatase (mg NP kg^{-1} h^{-1})	245.1 cd	210.3 d	456.5 a	393.7 b	280.2 c
Acid phosphatase (mg NP kg^{-1} h^{-1})	318.3 b	290.1 b	416.5 a	396.1 a	398.6 a
N_{min} (mg N-NH_4^+ kg^{-1})	20.3 c	17.7 c	29.1 ab	28.8 b	33.4 a
WSOC (mg C_{org} kg^{-1})	67.2 c	64.3 c	99.7 a	82.5 b	65.2 c
C_{mic} (mg C kg^{-1})	277.8 c	241.5 c	463.8 a	401.5 b	296.7 c

Reprinted from Spanish Journal of Agricultural Research, 10(3), Núñez-Zofío et al. (2012). Repeated biodisinfection controls the incidence of *Phytophthora* root and crown rot of pepper while improving soil quality, 794–805, Open Access Journal from the Spanish National Institute for Agricultural and Food Research and Technology

[a]Treatments: see Table 21.2. All values are expressed on a dry soil weight basis

[b]*FDA* fluorescein diacetate hydrolysis, *WSOC* water soluble organic carbon, N_{min} potentially mineralisable nitrogen, C_{mic} microbial biomass carbon. For each variable, values followed by the same letter are not significantly different according to Waller-Duncan's *K*-ratio *t* test ($P < 0.05$). Mean values ($n = 6$)

enzyme activities: FDA, dehydrogenase, urease, alkaline phosphatase and acid phosphatase.

Biodisinfested soils had significantly higher values of potentially mineralisable nitrogen than control and plastic-mulched soils, but higher values of water soluble organic carbon and microbial biomass carbon were obtained in manure-biodisinfected soils. Highest values were found in non-composted manure (Table 21.4). The higher values of enzyme activities obtained in non-composted and semicomposted manure-amended soils were concomitant with higher values of microbial biomass carbon, indicating that the higher levels of microbial activity were in this case due to an increase in microbial biomass. It was also observed both in non-composted and semicomposted manure-amended soils, an increase in potentially mineralisable nitrogen and water soluble organic carbon (indicators of biologically active N and C, respectively) (Núñez-Zofío et al. 2012).

Mandal et al. (2007) also found that the incorporation of farmyard manure increased enzyme activities in response to an increase in microbial biomass carbon and an improvement in the soil nutrient status.

The incorporation of fresh manure to the soil must be carried out with caution due to a possible increase in salt content which can alter soil properties and affect crop production and disease control (Litterick et al. 2004; Moral et al. 2009).

Núñez-Zofío et al. (2012) reported that significantly highest values of P_2O_5, K^+ and Cl^- were observed in non-composted soils when compared with semicomposted manure-amended soils. Although salinisation can negatively affect soil microbial properties (Rietz and Haynes 2003), Núñez-Zofío et al. (2012) found no differences in soil microbial properties between non-composted and semicomposted soils. In any event, composting provides a more stabilised product, thereby reducing the risk of soil salinisation, leaching and phytotoxicity (Moral et al. 2009).

21.3 Management of Soil-Borne Diseases with Organic Amendments in Protected Pepper Crops

Several reports from farmers show that plant diseases and the need for chemical control measures are reduced over time when practices that improve soil health are used. Research shows that soil management and microbial diversity are key factors in suppressing plant diseases. Organic amendments are quite diverse, including various types of organic materials such as animal manures, food processing wastes, crop residues and sewage sludge. Materials can be composted or uncomposted. The type of organic matter added to the soil could be of use as substrates, and their quality and quantity determine the types of organisms (both pathogens and natural occurring antagonists) that can profit the nutrients (Stone et al. 2004). It is well documented the use of specific organic amendments with suppressive effects against pathogens such as fungi, nematodes and bacteria. These amendments are primarily used for the control of diseases that these pathogens produce (Stone et al. 2004; Bonanomi et al. 2007, 2010; Oka 2010) although also are secondarily used to control soil fatigue caused by microorganisms that take advantage of plant weakness or subclinic pathogens (Manici et al. 2003; Martínez et al. 2011a, b; Mazzola and Manici 2012; Weerakoon et al. 2012; Guerrero 2013). Soil solarisation is an approach for soil disinfestation which uses passive solar heating of soil with plastic sheeting, usually transparent polyethylene (Stapleton 2000). The resulting soil temperature increase leads to decreased populations of pathogens. Soil solarisation and the application of organic amendments on soil have been described as a valid alternative to the use of chemical fumigants to reduce *Phytophthora* from pepper crops (Ristaino and Johnston 1999). In the technique known as Anaerobic Soil Disinfestation (also termed Biological Soil Disinfestation/Soil reductive sterilisation/Reductive Soil Disinfestation), organic amendments are applied in conjunction with soil tarping with an impermeable film for inducing anaerobiosis in order to generate toxic compounds (Blok et al. 2000; Momma 2008; Butler et al. 2012a, b). The advantage of anaerobic soil disinfestation when compared with soil solarisation is that the method does not require high solar radiation so it can be applied in cloudy areas or periods of low sunlight and, thus, a growing season is not lost (Baysal-Gurel et al. 2012).

21.3.1 *Biofumigation*

The term biofumigation was originally coined for that part of the suppressive effects of *Brassica* species on noxious soil-borne organisms (Kirkegaard et al. 1993) that arose quite specifically through liberation of isothiocyanates from hydrolysis of the glucosinolates that characterise the Brassicaceae (Kirkegaard and Matthiessen 2004). Since being coined, the initial term biofumigation has broadened its initial meaning and currently encompasses any beneficial effect arising from green manure or rotation crops and even composts (Matthiessen and Kirkegard 2006).

The use of Brassicaceae in crop rotations or as green manure amendment in biofumigation treatments (Stapleton and Bañuelos 2009) has proven to reduce the incidence of some soil-borne pathogens and plant parasites, including nematodes (Smolinska et al. 2003; Larkin and Griffin 2007), through their release of isothiocyanates. Improved pathogen and weed control has been achieved by using amendments obtained from by-products produced during the extraction of oil from *Brassica carinata* and *Sinapis alba* seeds (Palmieri 2005; Sachi et al. 2005; Lazzeri and Manici 2000; Cohen et al. 2005; Lazzeri et al. 2010). *Brassica carinata* (BP) pellets or *B. carinata* (BP) + fresh sheep manure (M) were evaluated for biodisinfestation treatments which began on two different dates (August and October), and the results were compared with MB-disinfested and untreated controls in greenhouse pepper crops in South-eastern Spain (Guerrero et al. 2013). During the third year, the gall index for BP was lower than that obtained for BP + M, and it was also lower in August than in October (Table 21.5).

The commercial crop of pepper fruit obtained in August biodisinfestations was similar or higher than the one obtained with MB, but higher than in October biodisinfestation treatments (Table 21.6). The yield of the October biodisinfestation treatments was higher than that of the untreated. In August of all the studied years, the accumulated exposure times were greater than the thresholds required to kill *M. incognita* populations at 15 cm soil depth. The incidence of the nematode did not correspond to the reduction achieved during solarisation and seemed to increase during the crop cycle.

21.3.2 *Biosolarisation*

The approach of combining soil solarisation together with the application of organic matter has been defined as biosolarisation (Ros et al. 2008) or biodisinfestation (de la Fuente et al. 2009). In this book chapter, the term biodisinfestation will be used in a more general sense than biosolarisation. Biodisinfestation will be applied for the combined use of an organic amendment and soil plastic tarping without implying soil heating (solarisation). Combining soil solarisation with the amendment of fresh organic residues elevates soil temperature by an additional

Table 21.5 Incidence of *Meloidogyne incognita* (galling index and % of galled plants) in a protected pepper crop greenhouse experiment located in South-eastern Spain

	First crop season		Second crop season		Third crop season	
Treatment	GI	% of galled plants	GI	% of galled plants	GI	% of galled plants
Untreated	5.7 d	100.0 b	6.3 cd	100.0 b	7.3 e	100.0 b
MB	0.1 a	6.6 a	0.2 a	6.6 a	1.5 a	43.3 a
BP+M August	4.1 b	100.0 b	3.7 b	93.3 b	3.5 c	96.7 b
BP August	3.8 b	93.3 b	4.3 b	86.6 b	2.6 b	86.7 b
BP+M October	4.6 b	93.3 b	6.8 d	100.0 b	6.9 e	100.0 b
BP October	4.1 b	100.0 b	5.4 c	100.0 b	5.4 d	100.0 b

Reprinted from Spanish Journal of Agricultural Research, 11(2), Guerrero et al. (2013), Evaluation of repeated biodisinfestation using *Brassica carinata* pellets to control *Meloidogyne incognita* in protected pepper crops, 485–493, Open Access Journal from the Spanish National Institute for Agricultural and Food Research and Technology

Mean values ($n = 30$). For each variable, values followed by the same letter are not significantly different according to Fisher's protected LSD Test ($P < 0.05$)

GI galling index, *MB* Methyl bromide-treated plots to 30 g m^{-2}, *BP+M* biodisinfestation with *Brassica carinata* pellets + fresh sheep manure, *BP* biodisinfestation with *Brassica carinata* pellets

Table 21.6 Pepper crop yield (kg m^{-2}) in a protected pepper crop greenhouse experiment located in south-eastern Spain

Crop yield (kg m^{-2})	First season	Second season	Third season
Untreated	9.8 c	10.3 b	9.7 d
MB	11.1 b	11.9 a	12.0 b
BP + M August	12.7 a	11.6 a	12.1 ab
BP August	12.6 a	12.2 a	12.7 a
BP + M October	10.9 b	11.8 a	11.7 c
BP October	10.9 b	12.1 a	11.1 c

Reprinted from Spanish Journal of Agricultural Research, 11(2), Guerrero et al. (2013), Evaluation of repeated biodisinfestation using *Brassica carinata* pellets to control *Meloidogyne incognita* in protected pepper crops, 485–493, Open Access Journal from the Spanish National Institute for Agricultural and Food Research and Technology

MB Methyl bromide-treated plots to 30 g m^{-2}, *BP+M* biodisinfestation with *Brassica carinata* pellets + fresh sheep manure, *BP* biodisinfestation with *Brassica carinata* pellets. For each variable, values followed by the same letter are not significantly different according to Fisher's protected LSD Test ($P < 0.05$)

1–3 °C, in addition to the generation of toxic volatile compounds (Biofumigation) which enhance the vulnerability of soil pathogens (Gamliel and Stapleton 1993; Gamliel et al. 2000; Bello et al. 2004; Stapleton and Bañuelos 2009). Several mechanisms are involved: (1) accumulation of toxic volatile compounds generated during organic matter decomposition; (2) creation of anaerobic conditions in the soil and (3) increase in soil suppressiveness due to high levels of microbial activity

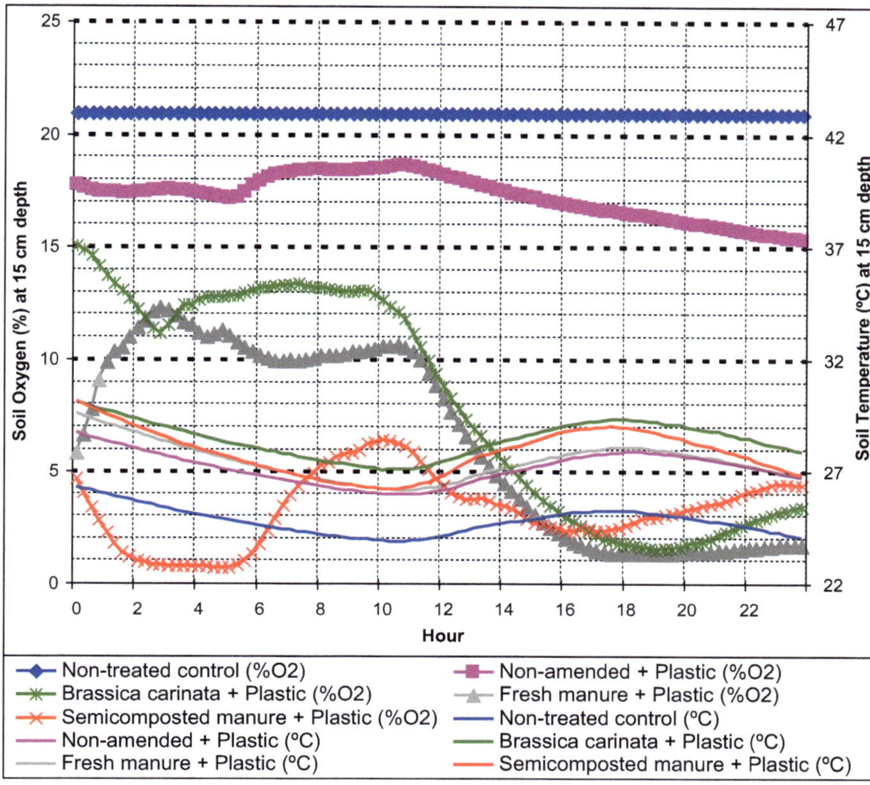

Fig. 21.3 Hourly temperatures (°C) and oxygen volumetric content (%) continuously recorded at 15 cm soil depth during biodisinfestation treatments with different organic amendments on August 7, 2009. The greenhouse field experiment was located in Derio (Biscay) (Northern Spain). Soil was tarped with 50-µm-thick (two million) transparent low density polyethylene plastic film from August 6 to September 22, 2009 (Larregla S; unpublished data)

(Gamliel et al. 2000). In these processes, the effects of anaerobiosis and temperature (Fig. 21.3) are added to the effect of gases and bioactive compounds released during organic matter decomposition (Lazarovits 2001; Tenuta and Lazarovits 2002; Arriaga et al. 2011).

Biosolarisation (BS) using fresh sheep manure (M) as amendment in August provided similar results to those obtained using MB for *Phytophthora* spp. control of protected pepper crops in South-eastern Spain (Guerrero et al. 2004a). Production increased when the application was repeated more than 2 years (Guerrero et al. 2004b, 2006; Cándido et al. 2005), but it seemed to have little effect when applied after the beginning of September (Guerrero et al. 2010).

21.4 The Problem of Soil Fatigue in Greenhouse Pepper Monocultures and its Control with Organic Amendments

Soil fatigue has been defined as: "the reduced development of certain crops when cultivated two or more times in the same soils" (Scotto-La Massese 1983; Bouhot 1983). Soil fatigue can be caused by one or a combination of several factors of physical, chemical or biological nature. Biotic component of soil fatigue usually includes certain microorganisms that take advantage of plant weakness or subclinic pathogens that tend to accumulate in soil with crop repetition (Manici et al. 2003; Martínez et al. 2011a, b; Mazzola and Manici 2012; Weerakoon et al. 2012; Guerrero 2013; Guerrero et al. 2014). These microorganisms have been isolated from plants showing vegetative depression, when they were repeatedly cultivated in the same soil, but did not produce disease when inoculated, nor reproduced the symptoms of depression, so that they were considered "weakness or subclinical pathogens" by Katan and Vanacher (1990).

In greenhouse pepper monocultures, soil fatigue appears in soils without primary pathogens (*Phytophthora capsici* or *P. parasitica, Meloidogyne incognita*) after the second year of crop repetition (Guerrero 2013). The depressive effect on plant development and the loss of production are related to the proliferation of species of *Fusarium* (Martínez et al. 2009, 2011a). Soil fatigue's specific depressive effect on the pepper plots is mitigated by soil disinfection (Guerrero 2013) and the reduction of the population densities of *Fusarium solani, F. oxysporum* and *F. equiseti* (Martínez et al. 2009). Soil disinfection through the use of chemical disinfectants (methyl bromide or chloropicrin) have less durable effects than when an organic amendment is used (fresh sheep manure + poultry manure), either alone (biofumigation) or when the soil is covered with plastic (biosolarisation) (Martínez et al. 2011b) (Table 21.7).

When biosolarisation is repeated, its effectiveness against soil fatigue increases (Martínez et al. 2009), either by providing direct action against fungal microbiota and/or increasing plant health through the improvement of soil chemical and physical characteristics (Fernández et al. 2005). Increase in macro- and micronutrients, increase in water infiltration capacity and decrease in apparent density and compaction are among the improvements in soil characteristics that may be mentioned. The use of biosolarisation combined with organic amendments in pepper greenhouses influences the soil physical characteristics, specifically in relation to the control of *Phytophthora capsici* or *P. parasitica*. These fungal pathogens are found in greater numbers in compact clay soils than in well-ventilated soils with adequate drainage. Even so, the control of the disease (root rot) can also be attributed to the effects of temperature, the released gases in the amendments bio-decomposition (Guerrero et al. 2010; Lacasa et al. 2010) and the suppressiveness connected with bacterial microorganisms (Núñez-Zofío et al. 2011).

Table 21.7 Changes on soil *Fusarium* spp. communities (colony-forming units per gramme of soil) isolated at first and in the last soil sampling of two growing seasons at two pepper greenhouses in Southeast of Spain

	Greenhouse CH				Greenhouse E			
	Year 1		Year 2		Year 1		Year 2	
Soil treatments	First	Last	First	Last	First	Last	First	Last
Control	1652.0[a] b	558.7 a	25.7 a	1097.5 b	1635.1 ns	1874.5 ns	463.0 ns	427.5 ns
Methyl bromide	110.4 a	481.4 b	23.9 a	1356.3 b	187.6 a	1587.2 b	131.5 a	716.0 b
Biofumigation	1481.5 ns	2047.2 ns	3157.2 ns	3608.6 ns	54.6 a	204.7 b	[b]	
Biosolarisation	45.7 ns	52.1 ns	105.5 a	1483.5 b	0.0 a	9.5 b	2.1 a	432.5 b

Reprinted from Bulletin IOBC/WPRS (International Organisation for Biological Control/West Palaeartic Regional Section), https://www.iobc-wprs.org, 71, Martínez et al. (2011b), Long-term effects of the application of organic amendments on soil fungal communities in pepper crops, 81–84, with permission from International Organisation for Biological Control

[a]Values are the means of three replicates. Mean values at first and in the last of the same growing season with the same letter do not differ significantly, according to Fisher's protected LSD test ($P < 0.05$). The $\sqrt{(x+0.5)}$ transformation was performed to normalise data; *ns* not differ significantly

[b]There were not biofumigation assays

21.5 Conclusions

Biosolarisation provides an effective and stable strategy for soil-borne pathogens control and the mitigation of soil fatigue in protected pepper monocultures.

Biosolarisation reiteration improves soil chemical, physical and biological properties with a subsequent increase in *Phytophthora* control effectiveness and crop yield.

However, field studies to establish types and rates of organic amendments should be carried out in different horticultural pathosystems in order to optimise pest, soil and crop responses when organic amendment incorporation is combined with soil plastic tarping at moderate soil temperatures. In-depth knowledge of several mechanisms that are contributing to control of soil-borne pathogens is needed. Future research should focus on the complexity of relationships among microbial communities for the establishment of soil management strategies towards a sustainable plant disease control.

Acknowledgements This research was financially supported by the National Institute for Agricultural and Food Research and Technology (INIA) of the Spanish Ministry of Science and Innovation (projects RTA-2008-00058-C03 and RTA 2011-00005-C03) and by the Department of Environment, Territorial Planning, Agriculture and Fisheries of the Basque Government (projects CIPASAPI and BIOSOL).

References

Arriaga H, Núñez-Zofío M, Larregla S, Merino P (2011) Gaseous emissions from soil biodisinfestation by animal manure on a greenhouse pepper crop. Crop Prot 30:412–419

Bailey KL, Lazarovits G (2003) Suppressing soilborne diseases with residues management and organic amendments. Soil Tillage Res 72:169–180

Baysal-Gurel F, Gardener BM, Miller SA (2012) Soilborne disease management in organic vegetable production. Department of Plant Pathology, The Ohio State University, published online 1 Aug 2012. http://www.extension.org/pages/64951/soilborne-disease-management-in-organic-vegetable-production. Verified 17 Feb 2014

Bello A, López-Pérez JA, García-Álvarez A, Arcos SC, Ros C, Guerrero MM, Guirao P, Lacasa A (2004) Biofumigación con solarización para el control de nematodos en cultivo de pimiento. In: Lacasa A, Guerrero MM, Oncina M, Mora JA (eds) Desinfección de suelos en invernaderos de pimiento. Publicaciones de la Consejería de Agricultura, Agua y Medio Ambiente, Región de Murcia, pp 105–129

Blok WJ, Lamers JG, Termorshuizen AJ, Bollen GJ (2000) Control of soilborne plant pathogens by incorporating fresh organic amendments followed by tarping. Phytopathology 90:253–259

Bonanomi G, Antignani V, Pane C, Scala F (2007) Suppression of soilborne fungal diseases with organic amendments. J Plant Pathol 89:311–324

Bonanomi G, Antignani V, Capodilupo M, Scala F (2010) Identifying the characteristics of organic soil amendments that suppress soilborne plant diseases. Soil Biol Biochem 42:136–144

Bouhot D (1983) Étude de la fatigue des sols dans les aspergerais et les pépinières d'asperge. In: INRA (ed) La fatigue des sols. Diagnostic de la fertilité dans les systèmes culturaux. INRA, Paris, pp 61–64

Butler DM, Rosskopf EN, Kokalis N, Burelle N, Albano JP, Muramoto J, Shennan C (2012a) Exploring warm-season cover crops as carbon sources for anaerobic soil disinfestation (ASD). Plant Soil 335:149–165

Butler DM, Kokalis-Burelle N, Muramoto J, Shennan C, McCollum TG, Rosskopf EN (2012b) Impact of anaerobic soil disinfestation combined with soil solarization on plant–parasitic nematodes and introduced inoculum of soilborne plant pathogens in raised-bed vegetable production. Crop Prot 39:33–40

Candido V, Miccolis V, Basile M, D'Addabbo T, Gatta G (2005) Soil solarization for the control of *Meloidogyne javanica* on eggplant in Southern Italy. Acta Hortic 698:195–199

Cohen MF, Yamasaki H, Mazzola M (2005) *Brassica napus* seed meal soil amendment modifies microbial community structure, nitric oxide production and incidence of *Rhizoctonia* root rot. Soil Biol Biochem 37:1215–1227

Conn KL, Tenuta M, Lazarovits G (2005) Liquid swine manure can kill *Verticillium dahliae* microsclerotia in soil by volatile fatty acid, nitrous acid, and ammonia toxicity. Phytopathology 95:28–35

de la Fuente E, Soria AC, Díez-Rojo MA, Piedra Buena A, García-Álvarez A, Almendros G, Bello A (2009) Solid-phase micro-extraction (SPME) in the early detection of potentially active volatile compounds from organic wastes used for the management of soil-borne pathogens. J Environ Sci Health A 44:1004–1010

Erwin DC, Ribeiro OK (1996) Phytophthora diseases worldwide. American Phytopathological Society, St. Paul, MN

Etxeberria A, Mendarte S, Larregla S (2011) Thermal inactivation of *Phytophthora capsici* oospores. Rev Iberoam Micol 28(2):83–90

Fernández P, Guerrero MM, Martínez MA, Ros C, Lacasa A, Bello A (2005) Effects of biofumigation plus solarization on soil fertility. In: Industrial crops and rural development. Proceedings of annual meeting of the association for the advancement of industrial crops, Murcia, 17–21 Sept 2005, pp 229–236

Gamliel A, Stapleton JJ (1993) Characterization of antifungal volatile compounds evolved from solarized soil amended with cabbage residues. Phytopathology 38:899–905

Gamliel A, Austerweil M, Kritzman G (2000) Non-chemical approach to soilborne pest management – organic amendments. Crop Prot 19:847–853

Gilreath JP, Santos BM (2004) Methyl bromide alternatives for weed and soilborne disease management in tomato (*Lycopersicon esculentum*). Crop Prot 23:1193–1198

Guerrero MM (2013) Biofumigación y desinfección de suelos de invernadero para cultivos de pimiento y la fatiga del suelo. PhD thesis, ETSIA, Universidad Politécnica de Cartagena, Murcia, 215 pp

Guerrero MM, Lacasa A, Ros C, Bello A, Martínez MC, Torres J, Fernández P (2004a) Efecto de la biofumigación con solarización sobre los hongos del suelo y la producción: fechas de desinfección y enmiendas. In: Lacasa A, Guerrero MM, Oncina M, Mora JA (eds) Desinfección de suelos en invernaderos de pimiento. Publicaciones de la Consejería de Agricultura, Agua y Medio Ambiente, Región de Murcia. pp 208–238

Guerrero MM, Ros C, Martínez MA, Barceló N, Martínez MC, Guirao P, Bello A, Contreras J, Lacasa A (2004b) Estabilidad en la eficacia desinfectante de la biofumigación con solarización en cultivos de pimiento. Acta Hortic 42:20–24

Guerrero MM, Ros C, Martínez MA, Martínez MC, Bello A, Lacasa A (2006) Biofumigation vs biofumigation plus solarization to control *Meloidogyne incognita* in sweet pepper. IOBC-WPRS Bulletin 29:313–318. In: Castañé C, Sánchez JA (eds) Proceedings of the meeting at Murcia, 14–18 May 2006. ISBN 92-9067-187-2. Working group "Integrated Control in Protected Crops, Mediterranean Climate". International organisation of biological control of noxious animals and plants – west palaearctic regional section. 379 pp. http://www.iobc-wprs.org/pub/bulletins/index.html

Guerrero MM, Ros C, Lacasa C, Martínez V, Lacasa A, Fernández P, Martínez MA, Núñez M, Larregla S, Díez-Rojo MA, Bello A (2010) Effect of biosolarization using pellets of *Brassica carinata* on soilborne pathogens in protected pepper crops. Acta Hortic 883:337–344

Guerrero MM, Lacasa-Martínez CM, Hernández-Piñera A, Martínez-Alarcon V, Lacasa-Plasencia A (2013) Evaluation of repeated biodisinfestation using *Brassica carinata* pellets to control *Meloidogyne incognita* in protected pepper crops. Span J Agric Res 11(2):485–493

Guerrero MM, Guirao P, Martínez MC, Tello J, Lacasa A (2014) Soil fatigue and its specificity towards pepper plants in greenhouses. Span J Agric Res 12:644–652

Hoitink HAJ, Boehm MJ (1999) Biocontrol within the context of soil microbial communities: a substrate-dependent phenomenon. Annu Rev Phytopathol 37:427–446

Hoitink HAJ, Stone AG, Han DY (1997) Suppression of plant diseases by composts. HortSci 32:184–187

Katan J, Vanacher A (1990) Soil and crop health following soil disinfestation. In: Giraud M, Faure J (eds) Fatigue de sol. Méthode de diagnostic en verger et en fraiseraie. Infos, Paris, pp 21–26

Kirchmann H, Witter E (1989) Ammonia volatilization during aerobic and anaerobic manure decomposition. Plant Soil 115:35–41

Kirkegaard JA, Matthiessen JN (2004) Developing and refining the biofumigation concept. Agroindustria 3:233–239

Kirkegaard JA, Gardner PA, Desmarchelier JM, Angus JF (1993) Biofumigation: using Brassica species to control pests and diseases in horticulture and agriculture. In: Wratten N, Mailer R (eds) Proceedings of the 9th Australian Research Assembly on Brassicas. NSW Agric, Wagga Wagga, pp 77–82

Kroodsma W, Huis in't Veld JWH, Scholtens R (1993) Ammonia emission and its reduction from cubicle houses by flushing. Livest Prod Sci 35:293–302

Lacasa A, Guirao P (1997) Investigaciones actuales sobre alternativas al uso del bromuro de metilo en pimiento en invernaderos del campo de Cartagena. In: López A, Mora JA (eds) Posibilidades de alternativas viables al bromuro de metilo en el pimiento de invernadero. Publicaciones de la Consejería de Agricultura, Agua y Medio Ambiente, Región de Murcia, Jornadas 11, pp 21–36

Lacasa CM, Guerrero MM, Ros C, Martínez V, Lacasa A, Fernández P, Martínez MA, Núñez M, Larregla S, Díez-Rojo MA, Bello A (2010) Efficacy of biosolarization with sugar beet vinasses for soil disinfestation in pepper greenhouses. Acta Hortic 883:345–352

Larkin RP, Griffin TS (2007) Control of soilborne potato diseases using Brassica green manures. Crop Prot 26:1067–1077

Larregla S (2003) Etiología y epidemiología de la "Tristeza" del pimiento en Bizkaia: su control. PhD thesis, Universidad del País Vasco, Bizkaia, 756 pp

Lazarovits G (2001) Management of soilborne plant pathogens with organic soil amendments: a disease control strategy salvaged from the past. Can J Plant Pathol 23:1–7

Lazzeri L, Manici LM (2000) The glucosinolate-myrosinase system: a natural and practical tool for biofumigation. Acta Hortic 532:89–95

Lazzeri L et al (2010) Use of seed flour as soil pesticide. Patent number US7749549B2

Leoni C, Ghini R (2006) Sewage sludge effect on management of *Phytophthora nicotianae* in citrus. Crop Prot 25:10–22

Litterick AM, Harrier L, Wallace P, Watson CA, Wood M (2004) The role of uncomposted materials, composts, manures, and compost extracts in reducing pest and disease incidence and severity in sustainable temperate agricultural and horticultural crop production – a review. Crit Rev Plant Sci 23:453–479

Liu B, Gumpertz ML, Hu S, Ristaino JB (2008) Effect of prior tillage and soil fertility amendments on dispersal of *Phytophthora capsici* and infection of pepper. Eur J Plant Pathol 120:273–287

Lockwood JL (1988) Evolution of concepts associated with soilborne plant pathogens. Annu Rev Phytopathol 26:93–121

Mandal A, Patra AK, Singh D, Swarup A, Masto RE (2007) Effect of long-term application of manure and fertilizer on biological and biochemical activities in soil during crop development stages. Bioresour Technol 98:3585–3592

Manici LM, Ciavatta C, Kelderer M, Erschbaumer G (2003) Replant problems in South Tyrol: role of fungal pathogens and microbial population in conventional and organic apple orchards. Plant Soil 256:315–324

Martínez MA, Lacasa A, Tello J (2009) Ecología de la microbiota fúngica de los suelos de los invernaderos de pimiento y su interés agronómico. Ministerio de Medio Ambiente, Madrid, 374 pp

Martínez MA, Martínez MC, Bielza P, Tello J, Lacasa A (2011a) Effect of biofumigation with manure amendments and repeated biosolarization on *Fusarium* density in pepper crops. J Ind Microbiol Biotechnol 38:3–11

Martínez MA, Martínez MC, Torres J, Tello J, Lacasa A (2011b) Long-term effects of the application of organic amendments on soil fungal communities in pepper crops. Bull IOBC/WPRS 71:81–84

Matthiessen JN, Kirkegard JA (2006) Biofumigation and enahanced biodegradation: opportunity and challenge in soilborne pest and disease management. Crit Rev Plant Sci 25:235–265

Mazzola M, Manici LM (2012) Apple replant disease: role of microbial ecology in cause and control. Annu Rev Phytopathol 50:45–65

Momma N (2008) Biological soil disinfestation (BSD) of soilborne pathogens and its possible mechanisms. Jpn Agric Res Q 42:7–12

Moral R, Paredes C, Bustamante MA, Marhuenda-Egea F, Bernal MP (2009) Utilisation of manure composts by high-value crops: safety and environmental challenges. Bioresour Technol 100:5454–5460

Núñez-Zofío M, Garbisu C, Larregla S (2010) Application of organic amendments followed by plastic mulching for the control of *Phytophthora* root rot of pepper in northern Spain. Acta Hortic 883:353–360

Núñez-Zofío M, Larregla S, Garbisu C (2011) Application of organic amendments followed by soil plastic mulching reduces the incidence of *Phytophthora capsici* in pepper crops under temperate climate. Crop Prot 30:1563–1572

Núñez-Zofío M, Larregla S, Garbisu C (2012) Repeated biodisinfection controls the incidence of *Phytophthora* root and crown rot of pepper while improving soil quality. Span J Agric Res 10 (3):794–805

Oka Y (2010) Mechanisms of nematode suppression by organic soil amendments: a review. Appl Soil Ecol 44:101–115

Oka Y, Shapira N, Fine P (2007) Control of root-knot nematodes in organic farming systems by organic amendments and soil solarization. Crop Prot 26:1556–1565

Palmieri S (2005) Biofumigation: a new technology based on the use of sulphur-containing metabolites in Brassica plants. In: Proceedings of annual meeting of the association for the advancement of industrial crops, Murcia, 17–21 Sept 2005, pp 201–213

Rietz DN, Haynes RJ (2003) Effects of irrigation-induced salinity and sodicity on soil microbial activity. Soil Biol Biochem 35:845–854

Ristaino JB, Johnston SA (1999) Ecologically based approaches to management of *Phytophthora* blight on bell pepper. Plant Dis 83:1080–1089

Ros M, García C, Hernández MT, Lacasa A, Fernández P, Pascual JA (2008) Effects of biosolarization as methyl bromide alternative for *Meloidogyne incognita* control on quality of soil under pepper. Biol Fertil Soils 45:37–44

Sachi S, Odorizzi S, Lazzeri L, Marciano P (2005) Effect of *Brassica carinata* seed meal treatment on the *Trichoderma harzianum* T39-*Sclerotinia* species interaction. Acta Hortic 698:289–292

Scotto La Massesse C (1983) Mesures susceptibles d'apprécier et de limiter les effets de la fatigue des sols dans le cas de reconstitution des vergers. In: INRA (ed) La fatigue des sols. Diagnostic de la fertilité dans les systèmes culturaux. INRA, Paris, pp 99–111

Smolinska U, Morra MJ, Kanudsen GR, James R (2003) Isothiocyanate produced by Brassicaceae species as inhibitors of *Fusarium oxysporum*. Plant Dis 87:407–412

Stapleton JJ (2000) Soil solarization in various agricultural production systems. Crop Prot 19:837–841

Stapleton JJ, Bañuelos GS (2009) Biomass crops can be used for biological disinfestation and remediation of soils and water. Calif Agric 63:41–46

Stone AG, Scheuerell SJ, Darby HM (2004) Suppression of soilborne diseases in field agricultural systems: organic matter management, cover cropping, and other cultural practices. In: Magdoff F, Weil RR (eds) Soil organic matter in sustainable agriculture. CRC Press LLC, Boca Raton, FL, pp 166–223

Tello JC, Lacasa A (1997) Problemática fitosanitaria del suelo en el cultivo del pimiento en el campo de Cartagena. In: López A, Mora JA (eds) Posibilidades de Alternativas Viables al Bromuro de Metilo en el Pimiento de Invernadero. Publicaciones de la Consejería de Agricultura, Agua y Medio Ambiente, Región de Murcia, Jornadas 11, pp 11–17

Tenuta M, Lazarovits G (2002) Ammonia and nitrous acid from nitrogenous amendments kill the microsclerotia of *Verticillium dahliae*. Phytopathology 92:255–264

Tenuta M, Conn KL, Lazarovits G (2002) Volatile fatty acids in liquid swine manure can kill microsclerotia of Verticillium dahliae. Phytopathology 92:548–552

Vallad GE, Cooperband LR, Goodman RM (2003) Plant foliar disease suppression mediated by composted forms of paper mill residuals exhibits molecular features of induced resistance. Physiol Mol Plant Pathol 63:65–77

Wang Q, Ma Y, Wang G, Gu Z, Sun D, An X, Chang Z (2014) Integration of biofumigation with antagonistic microorganism can control Phytophthora blight of pepper plants by regulating soil bacterial community structure. Eur J Soil Biol 61:58–67

Weerakoon DMN, Reardon CL, Paulitz TC, Izzo AD, Mazzola M (2012) Long-term suppression of *Pythium abappressorium* induced by *Brassica juncea* seed meal amendment is biologically mediated. Soil Biol Biochem 51:44–52

Zhang W, Dick WA, Hoitink HAJ (1996) Compost-induced systemic acquired resistance in cucumber pythium root rot and anthracnose. Phytopathology 86:1066–1070

Zhang W, Han DY, Dick WA, Davis KR, Hoitink HAJ (1998) Compost and compost water extract-induced systemic acquired resistance in cucumber and Arabidopsis. Phytopathology 88:450–455

Chapter 22
Combining Biocontrol Agents and Organics Amendments to Manage Soil-Borne Phytopathogens

David Ruano-Rosa and Jesús Mercado-Blanco

22.1 Introduction

A huge amount of agrochemicals are currently used to ensure the health of our crops. Thus, world sales of fungicides reached US$9.91 billion in 2010 and have increased annually by 6.5 % since 1999 (Hirooka and Ishii 2013). In 2013, FAO and WHO published the maximum tolerated levels for residues of 57 different fungicides used in agriculture worldwide (Codex Alimentarius database 2013, www.codexalimentarius.net). The increasing use/misuse of chemicals poses serious collateral problems such as environmental pollution (Ongley 1996), development of pathogen/pest resistance (Sparks 2013; Tupe et al. 2014), residual toxicity towards (micro)organisms (Yoom et al. 2013), and loss of biodiversity (Ghorbani et al. 2008). For example, the emergence of resistant strains of diverse phytopathogens to widely used, chemically based biocides is an increasing problem arising in many areas after the continuous use of these products (Brent and Hollomon 2007). The Fungicide Resistance Action Committee (2013, www.frac.info) periodically reviews the list of resistant plant pathogenic microorganisms, and the number increases after each report release. Indeed, five new pathogen resistances were documented and registered only in 2013. Development of pathogen resistance does not only affect crop production but also human health in two ways: (1) directly, since increasing biocide dosages means more residues potentially enhancing the risk for human (and animal) health and (2) indirectly, because resistance can also be acquired by opportunistic human pathogens (Lelièvre et al. 2013). Moreover, agrochemical treatments are mostly nonspecific and do not only affect target

D. Ruano-Rosa • J. Mercado-Blanco (✉)
Laboratory of Plant-Microorganism Interactions, Department of Crop Protection, Institute for Sustainable Agriculture, Agencia Estatal Consejo Superior de Investigaciones Científicas (CSIC), c/Alameda del Obispo s/n, 14004 Córdoba, Spain
e-mail: ruanodavid@gmail.com; jesus.mercado@ias.csic.es

M.K. Meghvansi, A. Varma (eds.), *Organic Amendments and Soil Suppressiveness in Plant Disease Management*, Soil Biology 46,
DOI 10.1007/978-3-319-23075-7_22

pathogens but also other microorganisms which are potentially beneficial to soil/plant health (Ranganathswamy et al. 2013). While problems related to the abuse/misuse of chemically based biocides are evident and perceived by consumers as highly concerning because of their side effects, many crop diseases are currently difficult, if not impossible, to manage without the use of chemicals. Therefore, an urgent need to develop and implement novel plant disease control strategies is highly demanded. Furthermore, these strategies are claimed to fit synonymous concepts such as "eco-friendly," "environmentally friendly," "nature friendly," or even "green," which can be applied at any stage from production to commercialization of a given crop. All these terms have the same meaning, i.e., "not harmful to the environment and to humans." Strategies based on this concept are thus considered healthier and safer than the traditional disease/pest control measures by means of chemical inputs. Nevertheless, according to sustainable agriculture criteria, the interdependence between economic and environmental aspects should not be forgotten. Thus, to attain sustainability a complete ban of chemical inputs is not always possible without compromising the viability of many farms devoted to specific crops in defined geographical areas. In order to achieve this primary goal, research on disease control management must therefore be focused on strategies aiming to avoid, or at least to greatly reduce, the high dependence on chemical inputs by implementing integrated disease management (IDM) frameworks (see, for instance, López-Escudero and Mercado-Blanco 2011). These approaches consist in the combined use of all available countermeasures effective against a given crop disease. The phytopathological challenge thus consists in that the increasing utilization of nonchemical strategies to control plant diseases and pests (i.e., lower dependence of pesticides, fungicides, and soil volatile disinfectants) should affect neither the production of food nor the economic viability of the farming business (Hamblin 1995). Profits derived from these strategies are not only economic and environmental but they also constitute the best approach to confront emerging pathogen(s) resistance(s) derived from the continuous use of fungicides (Brent and Hollomon 2007).

The aim of the present chapter is to provide a brief overview on research efforts devoted to the use of biological control agents (BCAs) and organic amendments (OAs) against soil-borne diseases within IDM strategies. More specifically, we will focus on the ad hoc combination of BCAs and OAs. Furthermore, we have tried to discuss aspects such as how these approaches may influence soil microbial communities or the suitability of using OAs as carriers to develop more stable and effective formulations of BCAs. Finally, even though literature about the combined application of soil amendments and BCAs against soil-borne diseases is abundant, information regarding its implementation in woody plants is very scant. Therefore, we will also discuss whether this control approach is feasible in tree crops and forestry under field conditions. But first, we will briefly present a few general concepts that the reader will find closely associated along the text.

22.2 Biological Control Agents

Biological control (biocontrol) emerges as one of the most promising alternatives to chemical control. Biocontrol can be defined as "the reduction of a phytopathogen inoculum amount, or its ability to cause disease, by means of the activities of one or more [micro]organisms (except human being)" (Cook and Baker 1983), or "the use of natural or modified organisms, genes, or gene products to reduce the effects of pests and diseases" (Cook 1988). Besides this main aim, implementation of biocontrol measures can lead to an increase in the number, diversity, and activity of nonpathogenic microbial communities originally present in soils and that can antagonize deleterious microorganisms. Without any doubt, biocontrol tools are environmentally friendly and can be implemented in combination with additional chemical, physical, and/or agronomical measures within IDM frameworks (López-Escudero and Mercado-Blanco 2011). Biological control can be used either as preventive or palliative strategy. Concerning plant diseases, biocontrol mainly relies on the artificial introduction of microbial antagonists, the so-called BCAs, to the targeted pathosystem. Nevertheless, biocontrol can also be based on strategies aiming to the modification of the microbial communities present in a particular agro-ecosystem, and/or their activities, by implementing specific agricultural practices. This can be achieved, for instance, by using suppressive soils (see, for instance, Mazzola 2002) or OAs (see below). The effective utilization of BCAs should be based on a profound knowledge of the mechanisms involved in biocontrol (i.e., competition, antibiosis, mycoparasitism, induction of defense responses, etc.), and on how the BCA performance can be affected by the broad range of (a) biotic factors which are dynamically interacting in any given pathosystem. Among BCAs, the species belonging to the genus *Trichoderma* are one of the most widely used microorganisms as biofungicides (Zaidi and Singh 2013). Characteristics like cosmopolitan distribution, adaptability to different soils, direct antagonism against plant pathogens (through mechanisms such as mycoparasitism, production of a large number of secondary metabolites, and/or competition), plant growth promotion, induction of systemic resistance, enhanced tolerance to abiotic stresses, compost colonization, and decomposition of organic matter (Zaidi and Singh 2013) make these fungi as one of the microorganisms best studied (and utilized) not only as BCA but also as biofertilizers (Woo et al. 2014). *Trichoderma* spp. isolates have thus been used to control pathogens from roots to leaves, either in herbaceous or woody plants (Zaidi and Singh 2013). Besides *Trichoderma*, many beneficial bacteria have been also studied as BCAs, the most frequent genera being *Agrobacterium* (e.g., Kawaguchi and Inoue 2012), *Bacillus* (e.g., Ruano-Rosa et al. 2014), *Pseudomonas* (e.g., Mercado-Blanco and Bakker 2007), and *Streptomyces* (e.g., Weiland 2014). Their biocontrol mechanisms can be antibiosis, competition for (micro) nutrients, colonization for specific sites needed for the pathogen to infect the plant, and/or induction of resistance by activating host plant defense responses (Narayanasami 2013). Many examples in which biocontrol bacteria have been successfully applied are available. However, this topic falls out the scope of this chapter and has been reviewed extensively elsewhere (see, for instance, Compant et al. 2013; Suárez-Estrella et al. 2013). Besides these two groups of microorganisms, mycorrhizal fungi (e.g., Ismail et al. 2013), nonpathogenic fungi (e.g., Abeysinghe 2009), or hypovirulent isolates of mycoviruses (Milgroom and Cortesi 2004) have also been studied and used as BCAs.

22.3 Organic Amendments Specified

The aim of this chapter is not to perform a comprehensive review of all materials considered as OA. We particularly aim to review cases in which such substrates have been used in combination with BCAs (see below). FAO defines Soil Amendment as "those materials that are applied to the soil to correct a major constraint other than low nutrient content" (Food and Agriculture Organization of the United Nations 2010a). The Soil Science Society of America defines OA as "any material such as lime, gypsum, sawdust, compost, animal manures, crop residue, or synthetic soil conditioners that is worked into the soil or applied on the surface to enhance plant growth. Amendments may contain important fertilizer elements, but the term commonly refers to added materials other than those used primarily as fertilizers" (Soil Science Glossary Terms Committee 2008). Organic amendments are used with the objective to improve the physical properties of soil, either directly or by activating living (micro) organisms present in the soil. They include organic materials, sometimes considered as waste, with a highly diverse composition and from a wide range of animal and vegetal origins (Food and Agriculture Organization of the United Nations 2010a). Sphagnum peat, wood chips, grass clippings, straw, compost, manure, biosolids, sawdust, and wood ash are considered, among others, OA (Davis and Whiting 2014). Amendments like charcoal or biochar, a solid carbon-rich product from biomass pyrolysis, will not be considered in this chapter. However, it is worth mentioning that these soil amendments are applied not only as fertilizers but also against foliar and soil-borne diseases. On the effect of biochar application on crop productivity and disease suppression, interested readers can consult, for instance, Atkinson et al. (2010) or Jaiswal et al. (2014).

Organic amendments have been used in many ways in agriculture, mainly as non-synthetic fertilizers. The use of OA contributes to reduce agrochemical inputs, thereby minimizing residues originated from farming activity (Trillas et al. 2006). One of the most interesting and promising applications of OAs relies on their ability to lessen the deleterious effects of pathogen attacks to acceptable thresholds (Boulter et al. 2002). There are many examples describing the successful use of OAs to control pathogens (including bacteria, fungi, and nematodes) (Bailey and Lazarovits 2003), to reduce their incidence (e.g., Borrego-Benjumea et al. 2014), or to isolate OA-residing microorganisms that may be applied against phytopathogens because of their proven antagonistic activity (e.g., Kavroulakis et al. 2010). Concerning the use of OA in plant disease control, Agrios (2005) includes soil amendment within biological control methods since they can stimulate antagonistic microbiota to pathogens present in soil, have an organic origin, and usually harbor beneficial microorganisms. Others, however, consider this approach within the category of farming practices control measures or even as category on its own: soil amendment control (Deepak 2011). Considering these premises (stimulation of soil microbiota, content of beneficial microorganisms, etc.) we consider the use of OA as a biological control strategy.

The effectiveness and consistency of OA in disease suppression are influenced, among other factors, by the target pathosystem and by the own variability (i.e., original sources, chemical characteristics, etc.) of the OA. Indeed, the number of pathosystems is huge and modifications/changes in the composition and characteristics of any given OA can enormously vary as well. Mechanisms of disease suppression displayed by OA can also be diverse. Furthermore, increase of disease incidence after the use of an amendment has been occasionally reported (Noble 2011). Therefore, finding the right application strategy for any OA needs of an in-depth knowledge of (1) the pathosystem, (2) the characteristics of the OA, (3) the environmental (biotic and abiotic) factors present in the site of application, and (4) how multitrophic interactions taking place in this site can be influenced by the addition of the OA, which usually carries a diverse microbiota as well. It has thus been shown that results obtained after OAs application can be highly variable and inconsistent. For instance, household waste-based compost batches usually present lack of uniformity. It is therefore of utmost importance to develop protocols to guarantee reproducible disease suppression results upon application of these amendments (Giotis et al. 2009).

Finally, it is also crucial to pay attention to the original source from which materials employed as OAs are derived since they might even contribute to pathogen spread. Indeed, it has been demonstrated that fresh manure from sheep previously fed in a cotton field affected with *Verticillium dahliae* Kleb., contained and transmitted pathogen propagules (microsclerotia) thereby contributing to the increase of the pathogen population in soil (López-Escudero and Blanco-López 1999).

22.4 Soil-Borne Pathogens: The Specific Target of OA and BCA in Disease Management Strategies

Soils contain a huge amount of organisms, many of them with the capacity to cause diseases in plants, viz., viruses, phytoplasmas, nematodes, protozoa, parasitic phanerogams, fungi, and bacteria. Fungi and oomycetes are likely the most important groups of soil-borne pathogens because of their number, diversity, and crop production losses produced by their attacks (García-Jiménez et al. 2010). For example, some 40 soil-borne pathogens cause important diseases in potato (*Solanum tuberosum* L.) tubers, the fourth main food crop in the world (Fiers et al. 2012). Numerous contributing factors help to understand why soil-borne pathogens are serious biotic constraints for many plants and why their efficient control is so difficult. For instance, many of them are able to produce resistance structures (i.e., microsclerotia, chlamydospores, oospores, etc.) enabling their endurance in soils under adverse situations during prolonged periods of time until favorable conditions allow germination. This is the case of microsclerotia produced by *V. dahliae*, the causal agent of verticillium wilts in many plants (Pegg and Brady 2002). Consequently, plausible management strategies to control these diseases, including biocontrol, should aim to eradicate microsclerotia or to avoid their germination (Antonopoulos et al. 2008). The potential use and efficacy of soil amendments to control *Verticillium* spp., including their effects on microsclerotia viability, have been thoroughly reviewed by Goicoechea (2009). Similarly, *Phytophthora* spp. can develop oospores, thick-walled sexual spores enabling this oomycete to survive under unfavorable conditions (e.g., drought, presence of microbial antagonists, etc.). Furthermore, many species of *Phytophthora* can develop other resistance structures like chlamydospores (Jung et al. 2013). Indeed, control strategies aimed to control these pathogens must take into account the possibility they produce resistance structures.

22.5 Effects of Introduced Inputs on the Microbial Soil Communities

Soils are the reservoir of a huge microbial biodiversity compared with other ecosystems. The use of culture-independent and metagenomics approaches is revealing a much wider diversity in soil microbial communities than that uncovered by traditional culture-dependent methods (Daniel 2005). These communities are not static and their composition, abundance, and activity, as well as the multitrophic interactions established among their constituents, can be affected by a number of (a) biotic factors along time and space. For instance, microbial diversity can be influenced by different stresses (e.g., nutrients shortage, environmental factors, or pH) and man-induced perturbations (e.g., soil management practices) (Decaëns 2010). Management practices like irrigation, tillage, cropping, and fertilizer and pesticide application are considered among the most influential factors affecting the composition of the rhizosphere microbiome (Prashar et al. 2014). Therefore, any (a) biotic input introduced into soils will result in short- and/or long-term changes of the microbial community structure. Since soil microbiota, either deleterious or beneficial, is crucial for plant fitness, potential alterations of its structure and functioning due to introduced inputs (chemical such as fungicides or fertilizers, or biological like OA or microorganisms) must be seriously considered to avoid unexpected side effects for the target crop.

Chemical inputs can affect both the composition and the structure of the soil-inhabiting microbial populations. Moreover, their effects can be different depending on the microbial group. Jacobsen and Hjelmsø (2014) point out that changes in microbial diversity vary according to the type of pesticide used. They provide a comprehensive list of agrochemicals (herbicides, soil fumigants, fungicides, insecticides) with variable effects on the bacterial community composition. For instance, it has been reported that copper decreases acidobacteria abundance, or that methyl bromide increases that of Gram positive bacteria (Jacobsen and Hjelmsø 2014, and references therein).

Introduction of BCAs into soils, either directly (e.g., by application of microbial antagonists formulations) or indirectly (e.g., as part of the microbiota present in OAs), has also a potential impact on indigenous soil microbial communities (Fig. 22.1). A given BCA bioformulation usually consists of a high cell/propagule density of the beneficial microorganism to ensure effective colonization of the plant rhizosphere (Trabelsi and Mhamdi 2013). This strategy provokes, at least transiently, a perturbation of the ecological equilibrium present in soil communities because the "new comer" and the indigenous microbiota must now compete for nutrients and space, which are usually scarce. In this scenario, mechanisms such as antibiosis or production of siderophores (Varma and Chincholkar 2007) deployed by the BCA can play an important role to efficiently displace native microorganisms. Similarly, the latter can use their own weapons to confront the invasion of the artificially introduced BCAs. The soil, and particularly the rhizosphere, becomes a battlefield where a multiplicity of trophic interactions takes place to (re)shape the structure of microbial communities (Raaijmakers et al. 2009). Trabelsi and Mhamdi (2013) compile an extensive number of research works and analyze how introduction of BCAs affects microbial communities. They also stressed the importance of the technique used to study the influence that artificial microbial inoculations have

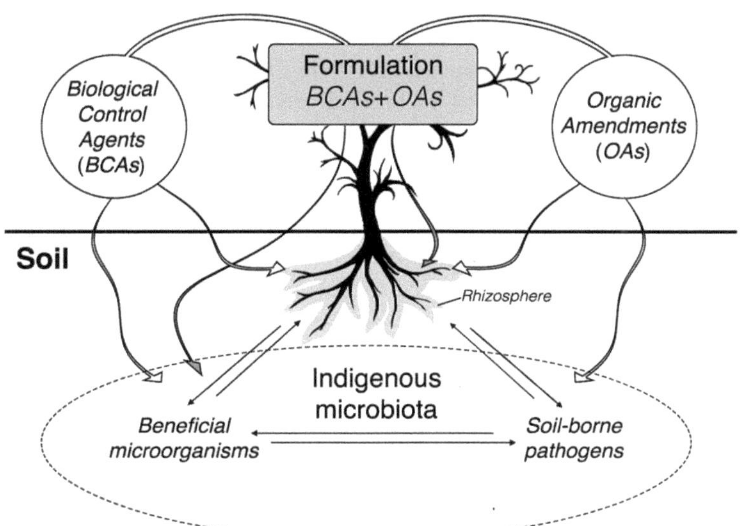

Fig. 22.1 Effects that the introduction of biological control agents (BCAs) and/or organic amendments (OAs) have in soil microbial communities networks (see main text for details)

in soils. For instance, the true impact of BCA introductions may vary depending on whether fatty acid methyl esters or terminal restriction fragment length polymorphism methodologies are used. They also conclude that the effects on plant growth and health are not necessarily a direct consequence of the introduced BCA, but they can be related to induction or repression of the resident microbial populations upon BCA inoculation. Therefore, synergistic and/or antagonistic interactions can take place after BCA inoculation, and they may endure for short and/or long periods of time.

Soil amendments, particularly OA, have the capability to modify soil characteristics such as concentration of nutrients (e.g., P, K, Fe), pH, NO_3 content, organic material, and structure. Since these traits are decisively shaping the structure of the soil-resident microbiota, there is no doubt that OA addition into soil will eventually affect microbial communities and their activity (Fig. 22.1). For instance, Yao et al. (2006) reported the influence that compost treatment had over soil microbial composition in apple (*Malus domestica*) orchards. Overall, they found differences in bacterial and fungi soil activities (measured as soil respiration) and community composition between non-treated and compost-treated soils. In their experiments, soil treated with compost showed the highest respiration rate and cumulative CO_2 production after 10 months, although these parameters eventually decreased and reached normal levels. Similarly, Giotis et al. (2009) observed that the incorporation of organic matter increased soil microbial activity and/or the number of microbial antagonists. Doan et al. (2014) also demonstrated that the nature of OAs has important consequences on soil microbial abundance and diversity. Finally, Gu et al. (2009) studied how long-term chemical fertilization (N-, P-, and K-based fertilizer) and farmyard manure affected soil microbial biomass (expressed as mg kg^{-1} of N and C) and diversity of bacterial communities in paddy soils. They observed that OA resulted in highest soil microbial biomass and diversity of bacterial communities. Moreover, combining OA with N, P, or K, increased microbial biomass and enhanced bacterial diversity compared to those observed with chemical fertilizers alone. The interested reader can consult many works on this particular subject (e.g., Liu et al. 2009; Zhang et al. 2012).

Modification of soil microbial communities and their implication in disease control has also been reported when different control measures are combined. Thus, effective control of Verticillium wilt of cotton due to changes in the fungal structure of rhizosphere soil (reducing fungal diversity) was observed after long-term (three growing seasons) greenhouse pot experiments when a combination of a bioorganic fertilizer (amino acid fertilizer from rapeseed meal fermentation), pig manure compost, and *Bacillus subtilis* was used (Luo et al. 2010). Larkin (2008) combined an aerated compost tea amendment, microorganisms (*B. subtilis*, *Trichoderma virens*, and *T. harzianum*), and even crop rotation to analyze how these inputs altered microbial populations and their activity in the soil. Results showed that different combinations of these treatments not only modified the soil microbial community characteristics but also reduced soil-borne diseases (stem canker and black scurf, caused by *Rhizoctonia solani*, and common scab, caused by *Streptomyces scabiei*) in potato. These authors support the idea that using a combination of treatments within an integrated soil management strategy yields better outcomes than the application of single management approaches. Related to this, Zhao et al. (2011) also observed that application of different formulations such as BIO I (pig manure compost, canola cake fermentation material, *Penicillium* sp., and *Aspergillus* sp.) significantly altered the soil microbial community structure, thereby suppressing Fusarium wilt of melon (*Cucumis melo* L.) effectively. In summary, evidence that inputs like BCAs and OAs can modify microbial community structures and that these changes can persist for a long time is available. However, the actual contribution of each component still remains to be unraveled.

22.6 Use of OAs in Integrated Disease Management Frameworks

Once we have introduced BCA and OA, tools that can be used on their own to control soil-borne diseases, we will now focus our attention on examples showing the potential that the ad hoc combination of BCA and OA has to effectively confront soil phytopathogens. Actually, this strategy has not yet been sufficiently explored, but promising results can be expected within IDM frameworks. It seems to be a general opinion among researchers that the effective control of a disease by means of a single BCA is difficult to achieve. Some authors have thus proposed alternatives such as the use of better adapted microorganisms, e.g., those from the same ecological niche where they will be applied (Ruano-Rosa and López-Herrera 2009), or the combination of BCAs (Xu et al. 2011), especially when they display complementary modes of action against the target pathogen. Examples of the successful use of combinations of BCAs, either fungus–fungus (Abo-Elyousr et al. 2009; Ruano-Rosa and López-Herrera 2009) or fungus–bacterium (Roberts et al. 2005; Ruano-Rosa et al. 2014), are available. Nevertheless, the limited efficacy observed for many available BCAs encourages the search for alternative and sustainable disease control approaches (Boukaew et al. 2013) which usually intend the combination of different control methods fitting IDM framework criteria.

Even though it falls out of the scope of this chapter, we would like to briefly mention that OAs can also be applied in combination with disease control strategies such as crop rotation (Larkin 2008) or soil solarization (Melero-Vara et al. 2011). For instance, soil solarization effects can be improved and/or enhanced by the addition of OAs because of the decomposition of organic matter increases heat generation and production of volatile compounds toxic for pathogenic (and beneficial) soil microbiota (Pokharel 2011). Interested readers can find excellent examples in the literature on the combination of these approaches, even including BCAs, to improve soil-borne pathogen control (e.g., Israel et al. 2005; Porras et al. 2007; Joshi et al. 2009; Melero-Vara et al. 2011; Domínguez et al. 2014). As mentioned in the previous section, implementation of these control measures (alone or in combination) can also greatly alter soil-resident microbial communities, including beneficial microorganisms that can be important for the health and fitness of the target crop (Israel et al. 2005; Porras et al. 2007; Larkin 2008).

22.6.1 Organic Carriers as Physical Support to Deliver BCAs

Selection of beneficial microorganisms that could be applied to a crop either as BCAs, biofertilizers, or for bioremediation purposes, is an arduous process that needs to take into account many factors (e.g., pathogen antagonism range, compatibility between BCAs, stress tolerance, plant growth promoting ability, environmental and human health risk assessment, etc.). A detailed evaluation and proper knowledge of beneficial traits displayed by the selected microbe will greatly determine its potential success when introduced into target agro-ecosystems. After this long process, the production, formulation, storage, and effective application of the selected microorganism usually represent additional bottlenecks prior to the implementation of a successful biocontrol strategy (Alabouvette and Steinberg 2006). Antagonistic microorganisms must therefore be formulated and applied in a way enabling the successful colonization and endurance in the targeted ecological niche (soil, rhizosphere, etc.) (El-Hassan and Gowen 2006; Nakkeeran et al. 2006). This has been recently well documented by Bashan et al. (2014), who comprehensively reviewed recent advances in plant growth promoting bacteria (PGPB) inoculant technology. In our opinion, most of the considerations addressed by these authors for PGPB could be also applied to microorganisms aimed to be used in biological control. In fact, microbe-mediated biocontrol is an indirect way to promote plant growth (Hayat et al. 2010). According to Bashan et al. (2014), two main factors contribute to the success of a PGPB-based formulation: (1) the own capabilities of the bacteria and (2) the technology used to deliver it. For instance, the introduction of any PGPB (or BCA) lacking an appropriate support (carrier) may lead to a rapid decline of its population level after inoculation. This means that its biocontrol potential might not be deployed regardless how powerful the beneficial traits have been previously demonstrated. Moreover, since native soil microbial communities are often better adapted than inoculated (artificially introduced) microorganisms, some advantages should be given to the inoculum once it is formulated.

We use the term "carrier" as any type of physical support, either organic or inorganic employed to develop a suitable formulation to be effectively applied in a given agro-ecosystem. A large number of carriers can be found as part of a bioformulation. Regarding inorganic carriers talc, kaolin, clay, perlite, or vermiculite among others (e.g., El-Hassan and Gowen 2006) and more recently microencapsulation (Kim et al. 2012) are being widely used. Peats and composts are among the most commonly used organic carriers. However, many others are available, even combinations of several of them. The abundance of organic carriers is reflected by the extensive bibliography available on this topic (see Table 22.1 for some examples).

Table 22.1 Examples of studies where organic amendments (OAs) were combined with biological control agents (BCAs) against soil-borne diseases

Organic amendment	Biological control agent	Disease/host (Pathogen)	Reference[a]
Wheat bran, peat moss	*Tricoderma harzianum*	Allium white-rot (*Sclerotium cepivorum*)	Avila et al. (2006)[a]
Vermicompost, neem cake	*T. harzianum*	Brinjal Fusarium wilt (*Fusarium solani* f. sp. *melongenae*)	Bhadauria et al. (2012)
Vineyard pruning wastes	*T. harzianum*	Fusarium wilt (*Fusarium oxysporum* f. sp. *melonis*)	Blaya et al. (2013)[a]
Pig manure compost, canola cake	*Bacillus subtilis*	Cucumber Fusarium wilt (*F. oxysporum* f. sp. *cucumerinum*)	Cao et al. (2011)[a]
Fresh chicken manure	*Trichoderma asperellum*, *Trichoderma atroviride*	Strawberry charcoal rot (*Macrophomina phaseolina*)	Domínguez et al. (2014)[b]
Sawdust, potato processing wastes, and rice straw	*T. harzianum*, *Penicillium oxalicum*, *Chaetomium globosum*	Legumes Fusarium wilt (*F. oxysporum*)	Haggag and Saber (2000)[a]
Cow dung	*T. harzianum*	Foot rot of lentil (*F. oxysporum* and *Sclerotium rolfsii*)	Hannan et al. (2012)
Amino acid fertilizer (from rapeseed meal fermentation)	*Bacillus pumilus*	Cucumber Damping-off disease (*Rhizoctonia solani*)	Huang et al. (2012)[a]
Farm yard manure, compost, poultry manure, press mud, vermicompost, and neem cake	*Pseudomonas fluorescens*	Tomato damping-off (*Pythium aphanidermatum*)	Jayaraj et al. (2007)
Farm yard manure, and poultry manure	*Trichoderma viride*	Tomato damping-off (*Pythium* spp., *R. solani*, *Phytophthora* spp., *Fusarium* spp.)	Joshi et al. (2009)
Amino acid fertilizer (from rapeseed meal fermentation), pig manure compost	*B. subtilis*	Cotton Verticillium wilt (*Verticillium dahliae*)	Lang et al. (2012)[a]
Neem cake and Farm yard manure	*T. viride*, *P. fluorescens*, *B. subtilis*	Physic nut collar and root rot (*Lasiodiplodia theobromae*)	Latha et al. (2011)
Pig manure compost/ microbe-hydrolyzed rapeseed cake	*Brevibacillus brevis*, *Streptomyces rochei*	Tobacco bacterial wilt (*Ralstonia solanacearum*)	Liu et al. (2013)[a]
Compost	*Pisolithus tinctorius*, *Scleroderma verrucosum*	Oak decline (*P. cinnamomi*)	Moreira et al. (2007)

(continued)

Table 22.1 (continued)

Organic amendment	Biological control agent	Disease/host (Pathogen)	Reference[a]
Mustard oil cake	*P. fluorescens*, *Glomus sinuosum*, *Gigaspora albida*	French bean root rot (*R. solani*)	Neeraj and Singh (2011)
Compost from agricultural waste (from cork, grape and olive marc, and spent mushroom)	*T. asperellum*	Cucumber (*R. solani*)	Trillas et al. (2006)
Olive mill wastes	*Bacillus amyloliquefaciens*, *Burkholderia cepacia*	Olive Verticillium wilt (*V. dahliae*)	Vitullo et al. (2013)
Pig manure compost, canola cake	*B. amyloliquefaciens*	Banana Fusarium wilt (*F. oxysporum* f. sp. *cubense*)	Wang et al. (2013)[a]
Pig manure, rice straw	*B. amyloliquefaciens*	Tomato Bacterial wilt (*R. solanacearum*)	Wei et al. (2011)[a]
Pig manure compost, canola cake	*Paenybacillus polymyxa*, *T. harzianum*	Watermelon Fusarium wilt (*F. oxysporum* f. sp. *nevium*)	Wu et al. (2009)
Compost (pig manure, rice straw, residues from medicine, alcohol, and vinegar production)	*T. harzianum*	Cucumber Fusarium wilt (*F. oxysporum* f. sp. *cucumerinum*)	Yang et al. (2011)[a]
Commercial organic fertilizer (pig manure compost, canola cake)	*P. polymyxa*, *B. subtilis*, *Penicillium* sp., *Aspergillus* sp.	Melon Fusarium wilt (*F. oxysporum* f. sp. *melonis*)	Zhao et al. (2011)[a]

[a]Studies in which the OA was used as a carrier of the BCA
[b]Additional control treatment was used in combination with OA+BCA

The development of carriers based on organic matter emerges as an excellent alternative for a more effective application of disease control treatments based on OA plus BCA (OA+BCA) combinations. Indeed, the own nature of this type of carriers provide an adequate nutrient reservoir to the BCA thereby enhancing its survival in a hostile environment such as soil. For example, it is well known that the widely used BCA *Trichoderma* spp. must not be applied in the stage of spores (conidia) if not supported by a suitable carrier. This is due to the high sensitivity to soil fungistasis showed by these asexual reproductive structures (Pan et al. 2006). Hence, the application of *Trichoderma*-based formulations can fail if spores (even at the stage of early germination) are applied to the soil without an adequate nutrients supply (Yang et al. 2011). A number of examples in which OA+BCA combinations performed better than single OA treatments are available. For instance, Zhao et al. (2011) developed different formulations using as a carrier an organic fertilizer supplemented with different BCAs (see in Table 22.1). The carrier did not show any disease suppressive effect by itself but in combination with the BCAs resulted in a suitable formulation that effectively controlled Fusarium wilt caused by *Fusarium oxysporum* f. sp. *melonis* in melon.

The use of organic-based carriers in OA+BCA control strategy has two main beneficial outcomes. On the one hand, recycling organic material (i.e., pruning remains) may help farmers to deal with waste derived from their activity. For instance, this is an urgent need in some Mediterranean countries in the case of olive (*Olea europaea* L.) mill waste management, an important by-product from olive oil industry activity (Papasotiriou et al. 2013). On the other hand, some organic-based carriers such as specific composts from agriculture wastes have been demonstrated to be effective on its own in the control of a number of soil-borne pathogens (Trillas et al. 2006). For instance, Papasotiriou et al. (2013) have demonstrated that the use of olive mill waste compost reduced *V. dahliae* microsclerotia germination as well as the number of hyphae per germinated microsclerotium *in planta*. Likewise, Alfano et al. (2011) have shown that the use of composted olive mill waste has in vivo suppressive effect against *Fusarium oxysporum* f. sp. *lycopersici* and *Pythium ultimum* [the causal agents of Fusarium wilt and damping off on tomato (*Solanum lycopersicum* Mill) seedling, respectively]. Both suppression by competition (nutrients and/or space) and antagonistic effect due to microorganisms inhabiting the compost are likely involved in the suppressive effect.

22.6.2 Combining OAs with BCAs

Disease management strategies are obviously focused on the improvement of the crop's health. However, application of OA+BCA combinations can provide additional beneficial effects to the crop (i.e., better plant development, enhanced yield, plant growth, etc.). This is a consequence of the fertilizing properties of OAs, which can release chemical substances with similar or better outcomes than synthetic fertilizers (Ding et al. 2013). Furthermore, it is well known that some BCAs have the capability to promote plant growth by means of a number of direct mechanisms (Lugtenberg and Kamilova 2009). The interested reader can consult excellent reviews on this topic (i.e., Kaewchai et al. 2009; Tailor and Joshi 2014).

A number of studies dealing with the use of OA+BCA combinations and their effects on the plant growth, crop yield, and/or on the soil microbial community structure, besides its effectiveness against pathogens, are available (Table 22.1). Nevertheless, we would like to differentiate between two types of OA+BCA combinations depending on whether they are applied as joint formulations (i.e., blended and/or composted mixtures prior to application, marked in Table 22.1) or as individual treatments that are subsequently applied (either at the same time or not) upon introduction in the target crop/field. *Trichoderma* spp. and a number of bacterial genera are, once again, the most widely used BCAs in this control strategy.

For instance, Bhadauria et al. (2012) reported that application of *T. harzianum* (as seed treatment) plus soil treatment with neem (*Azadirachta indica* A. Juss.) cake was an effective treatment to reduce Fusarium wilt incidence (*Fusarium solani* f. sp. *melongenae*) in brinjal (eggplant, *Solanum melongena* L.) plants. Moreover, this combined treatment reduced the amount of pathogen propagules and did not produce unwanted residues what makes it an excellent eco-friendly strategy for the management of this disease. Likewise, the addition of *T. harzianum* to compost (see Table 22.1) improved the biocontrol effectiveness and induced changes in the biotic (e.g., changes in bacterial community composition) and abiotic (pH modification) characteristics of this AO (Blaya et al. 2013). Jayaraj et al. (2007) used different OAs (farmyard manure, leaf compost, poultry manure, press mud, vermicompost, and neem cake) combined with *P. fluorescens* to control damping-off (*Pythium aphanidermatum*) in tomato. In this case, OAs were incorporated into soil prior to planting while the BCA was applied as seed treatment using a formulation (see Table 22.1). Results showed an enhancement of *P. fluorescens* rhizosphere population as well as a reduction of the disease incidence caused by this oomycete.

Taking into account the expected advantages of mixing BCAs (combination of complementary modes of action) mentioned above, Liu et al. (2013) developed a bioorganic fertilizer using an OA as a carrier (see Table 22.1). They observed better suppression of the bacterial pathogen *Ralstonia solanacearum* in tobacco (*Nicotiana tabacum* L.) plants pot experiments when a formulation containing two BCAs were applied in combination with compost (see Table 22.1). In addition to the enhanced disease suppressive effect, they also found increased plant growth probably due to a synergistic effect derived from the combination of BCAs with the compost. Considering the benefits achieved by the combination OAs and BCAs, a progressive substitution of chemically based fungicides seems to be a practicable strategy (De Ceuster and Hoitink 1999).

22.7 Can OA+BCA Combinations Be a Feasible Disease Control Approach in Woody Plants?

Trees and woody crops are of utmost importance for the life of the planet. For instance, forests cover around 31 % of the world's land surface (Food and Agriculture Organization of the United Nations 2010b), providing many important goods (e.g., wood, paper, etc.) and playing essential roles in processes such as nutrients and water cycling and storage. Trees are also crucial to prevent soil erosion, to mitigate the effects of climate change acting as carbon dioxide sink, and to support microbial, animal, and plant biodiversity in many areas. Therefore, the health of forests and woody agro-ecosystems is of particular relevance.

Many soil-borne pathogens affect woody plants causing serious constraints in economically relevant tree crops and forestry. Among them, species of the genera

Fusarium, *Verticillium*, *Phytophthora*, *Pythium*, *Armillaria*, *Rosellinia*, or *Heterobasidion* can be highlighted as extremely damaging (García-Jiménez et al. 2010). The utilization of BCAs to control these pathogens when affecting woody plants has been investigated in a number of pathosystems (see Pliego and Cazorla 2012, and references therein). The same accounts for the use of OAs although to a lesser extent (Noble and Conventry 2005). Remarkably, however, a search in the literature reveals that, to the best of our knowledge, the combination of BCAs and OAs as a disease control strategy has been implemented in woody plants at a negligible level compared to that in arable crops or seedlings (Table 22.1). A number of reasons could explain why biocontrol strategies in general, and BCA +OA combinations in particular, have been less (or seldom) applied in these particular agro-ecosystems. Thus, it is plausible to think that factors such as large biomass, anatomy, longevity, and/or particularities of tree crops and forests management make it more difficult to develop effective biological control measures against diseases affecting woody plants. For instance, regarding to soil-borne pathogens, large root systems of trees can undergo repeated infection events from pathogen's propagules present in soil. Infection events can then take place either in the same season or in successive ones that contribute to complicate the application of effective biocontrol strategies, including OA+BCA combinations. Pliego and Cazorla (2012) have particularly stressed that the large root systems developed by trees greatly hamper the effectiveness of BCA treatments. Likewise, López-Escudero and Mercado-Blanco (2011) have emphasized the difficulty to control *V. dahliae* in olive because of the pathogen's location within the vascular system, a site always difficult to be reached by chemical or biological treatments. Nevertheless, and in spite of these difficulties, biocontrol measures are feasible for woody plants. For instance, application of BCAs can be done with seedlings, in pots under controlled conditions, and/or during the nursery propagation stage. Thus, Vitullo et al. (2013) focused on pot-growing olive plants at nursery conditions with the aim to guarantee the production of healthy plants. These authors achieved positive results in the control of *V. dahliae* by mixing *Bacillus amyloliquefaciens* and *Burkholderia cepacia* with olive mill waste. However, the important step forward yet to be taken is the application of biocontrol strategies (including OA+BCA with the advantages discussed above) at large scale and under field conditions (tree orchards, forests). The relevant question still to be answered is whether application of BCAs, OAs, and/or OA+BCAs combinations can be done in an economically efficient way considering the particularities of trees (and woody plants in general).

Disease control measures that can be implemented together with OA+BCA combinations (see above) have to confront the idiosyncrasies of woody plants as well, and their potential success can be reduced compared to when they are applied to herbaceous crops. For instance, it is known that efficiency of soil solarization decreases at deep soil layers (López-Herrera et al. 2003). Thus, deep root systems usually developed by trees are less accessible to physically-, chemically- and/or biologically based disease control measures.

A promising alternative to be used in woody plants are endophytic microorganisms adapted to colonize and endure for long periods of time within plant tissues.

Among the agro-biotechnological applications that bacterial and fungal endophytes pose, their potential as BCAs are yet insufficiently explored (Mercado-Blanco and Lugtenberg 2014). However, effective control of Verticillium wilt of olive has been achieved in nursery-propagated plants by the olive root endophyte *P. fluorescens* PICF7 (Prieto et al. 2009) or against poplar canker (caused by three pathogens viz. *Cytospora chrysosperma*, *Phomopsis macrospora*, and *Fusicoccum aesculi*) by using the endophyte *Bacillus pumillus* (Ren et al. 2013). Considering the advantages discussed above, the use of AO (endophytic)+BCA combinations may constitute and interesting approach to be used in the control of diseases affecting woody plants.

22.8 Conclusions

The growing public concern about the undesirable effects derived from an overzealous use of agrochemicals, mainly fungicides and herbicides, has encouraged the search for more environmentally friendly plant disease control alternatives. Chemical inputs have caused, among other side effects, the development of plant pathogen resistance and hazard to animal and human health. For a number of reasons, many plant pathologists have devoted their research efforts to seek novel alternatives for the effective control of phytopathogens that, in addition, aim to diminish the risk of undesirable effects. The implementation of IDM strategies encompassing, among others, measures such as the combined use of BCA and OA likely constitutes the best option towards the success in plant disease management. It must be emphasized that the application of any soil-borne pathogen control method, either individually or combined with other(s), may result in major changes affecting not only the structure and physical–chemical characteristics of the soil but also the indigenous microbiota residing therein. These changes can have a profound influence on the pathogen control process, even determining the success or failure of the strategy used. Obviously, the introduction of OA, BCA, or OA+BCA combinations into a given agro-ecosystem also provokes major changes (Fig. 22.1), which should be studied and understood in detail.

A crucial step for the success of biocontrol strategies is the way the BCA is applied or delivered. Indeed, the choice of the most appropriate carrier when developing a BCA-based formulation is of utmost importance. The carrier should not only serve as nutrients supply but also be a proper support enabling microorganisms to have long shelf lives and to cope with the adverse, highly competing conditions they have to face soon after they are released into the target site (soil, rhizosphere, seeds, etc.). The development of OA-based carriers constitutes an excellent approach because they can simultaneously enhance the survival rate of the BCA, antagonize the target pathogen, and act as plant fertilizers.

To our knowledge, studies combining BCAs and OAs to control diseases of woody plants are scant. Several factors may explain this circumstance and have been briefly presented in this chapter. Nevertheless, the combination of OA and

BCA emerges as an interesting approach yet to be explored. BCAs displaying endophytic lifestyle also offer a number of advantages (e.g., adaptation to live within the plant tissue, plant growth promotion, etc.) to be exploited as well. Promising results have been obtained from these environmentally friendly tools under controlled conditions (i.e., greenhouse, nursery-production stage). The challenge now is to better understand and exploit the benefits of combining them as well as to develop correct strategies for their efficient use in agro-ecosystems and forestry.

References

Abeysinghe S (2009) Use of nonpathogenic *Fusarium oxysporum* and rhizobacteria for suppression of Fusarium root and stem rot of *Cucumis sativus* caused by *Fusarium oxysporum* f. sp. *radicis-cucumerinum*. Arch Phytopathol Plant Prot 42:73–82

Abo-Elyousr KAM, Hashem M, Ali EH (2009) Integrated control of cotton root rot disease by mixing fungal biocontrol agents and resistance inducers. Crop Prot 28:295–301

Agrios G (2005) Plant pathology, 5th edn. Elsevier, New York

Alabouvette C, Steinberg C (2006) The soil as a reservoir for antagonists to plant diseases. In: Eilenberg J, Hokkanen HMT (eds) An ecological and societal approach to biological control. Springer, Dordrecht, pp 123–144

Alfano G, Lustrato G, Lima G, Vitullo D, Ranalli G (2011) Characterization of composted olive mill wastes to predict potential plant disease suppressiveness. Biol Control 59:199–207

Antonopoulos DF, Tjamos SE, Antoniou PP, Rafeletos P, Tjamos EC (2008) Effect of *Paenibacillus alvei*, strain K165, on the germination of *Verticillium dahliae* microsclerotia *in planta*. Biol Control 46:166–170

Atkinson CJ, Fitzgerald JD, Hipps NA (2010) Potential mechanisms for achieving agricultural benefits from biochar application to temperate soils: a review. Plant Soil 337:1–18

Avila Miranda ME, Herrera Estrella A, Peña Cabriales JJ (2006) Colonization of the rhizosphere, rhizoplane and endorhiza of garlic (*Allium sativum* L.) by strains of *Trichoderma harzianum* and their capacity to control allium white-rot under field conditions. Soil Biol Biochem 38:1823–1830

Bailey KL, Lazarovits G (2003) Suppressing soil-borne diseases with residue management and organic amendments. Soil Tillage Res 72:169–180

Bashan Y, de-Bashan LE, Prabhu SR, Hernandez JP (2014) Advances in plant growth-promoting bacterial inoculant technology: formulations and practical perspectives (1998–2013). Plant Soil 378:1–33

Bhadauria BP, Singh Y, Puri S, Singh PK (2012) Ecofriendly management of *Fusarium* wilt of Brinjal. Ecol Environ Conserv 18:1049–1052

Blaya J, López-Mondéjar R, Lloret E, Pascual JA, Ros M (2013) Changes induced by *Trichoderma harzianum* in suppressive compost controlling Fusarium wilt. Pestic Biochem Physiol 107:112–119

Borrego-Benjumea A, Basallote-Ureba MJ, Abbasi PA, Lazarovits G, Melero-Vara JM (2014) Effects of incubation temperature on the organic amendment-mediated control of *Fusarium* wilt of tomato. Ann Appl Biol 164:453–463

Boukaew S, Klinmanee C, Prasertsan P (2013) Potential for the integration of biological and chemical control of sheath blight disease caused by *Rhizoctonia solani* on rice. Worl J Microbiol Biotechnol 29:1885–1893

Boulter JI, Trevors JT, Boland GJ (2002) Microbial studies of compost: bacterial identification, and their potential for turfgrass pathogen suppression. World J Microbiol Biotechnol 18:661–671

Brent K, Hollomon DW (2007) Fungicide resistance in crop pathogens: how can it be managed? Fungicide Resistance Action Committee Monograph No. 1. Croplife International, Belgium

Cao Y, Zhang Z, Ling N, Yuan Y, Zheng X, Shen B, Shen Q (2011) *Bacillus subtilis* SQR 9 can control Fusarium wilt in cucumber by colonizing plant roots. Biol Fertil Soils 47:495–506

Codex Alimentarius (2013) Pesticide residues in food and feed FAO and WHO http://www.codexalimentarius.net. Accessed 25 Feb 2014

Compant S, Brader G, Muzammil S, Sessitsch A, Lebrihi A, Mathieu F (2013) Use of beneficial bacteria and their secondary metabolites to control grapevine pathogen diseases. BioControl 58:435–455

Cook RJ (1988) Biological control and holistic plant-health care in agriculture. Am J Alternative Agric 3:51–62

Cook RJ, Baker KF (1983) Developmental history of biological control of plant pathogens. In: Cook RJ, Baker KF (eds) The nature and practice of biological control of plant pathogens. The American Phytopathological Society, St. Paul, MN, pp 30–56

Daniel R (2005) The metagenomics of soil. Nat Rev Microbiol 3:470–478

Davis JG, Whiting D (2014) Choosing a soil amendment. In: Gardening series/basics. Colorado State University Extension, Fact sheet No. 7235. http://www.ext.colostate.edu/pubs/garden/07235.html. Accessed 20 Feb 2014

De Ceuster TJJ, Hoitink AJ (1999) Prospects for composts and biocontrol agents as substitutes for methyl bromide in biological control of plant diseases. Compost Sci Util 7:6–15

Decaëns T (2010) Macroecological patterns in soil communities. Glob Ecol Biogeogr 19:287–302

Deepak (2011) Soil amendments, plant extracts and plant products for integrated disease management in agricultural crops: a review. Afr J Agric Res 6:6790–6797

Ding C, Shen Q, Zhang R, Chen W (2013) Evaluation of rhizosphere bacteria and derived bio-organic fertilizers as potential biocontrol agents against bacterial wilt (*Ralstonia solanacearum*) of potato. Plant Soil 366:453–466

Doan TT, Bouvierc C, Bettarelc Y, Bouvier T, Henry-des-Tureaux T, Janeau JL, Lamballe P, Van Nguyen B, Jouquet P (2014) Influence of buffalo manure, compost, vermicompost and biochar amendments on bacterial and viral communities in soil and adjacent aquatic systems. Appl Soil Ecol 73:78–86

Domínguez P, Miranda L, Soria C, de los Santos B, Chamorro M, Romero F, Daugovish O, López-Aranda JM, Medina JJ (2014) Soil biosolarization for sustainable strawberry production. Agron Sustain Dev 34:821–829

El-Hassan SA, Gowen SR (2006) Formulation and delivery of the bacterial antagonist *Bacillus subtilis* for management of lentil vascular wilt caused by *Fusarium oxysporum* f. sp. *lentis*. J Phytopathol 154:148–155

Fiers M, Edel-Hermann V, Chatot C, Le Hingrat Y, Alabouvette C, Steinberg C (2012) Potato soil-borne diseases. A review. Agron Sustain Dev 32:93–132

Food and Agriculture Organization of the United Nations (2010a) Database on commercially available organic fertilizers and water-retaining products. http://www.fao.org/ag/agp/orgfert/intro.htm. Accesses 29 Jan 2014

Food and Agriculture Organization of the United Nations (2010b) Global forest resources assessment 2010: global tables. http://foris.fao.org/static/data/fra2010/FRA2010Globaltables EnJune29.xls. Accesses 14 Oct 2014

Fungicide Resistance Action Committee (2013) List of plant pathogenic organisms resistant to disease control agents. www.frac.info. Accessed 29 Jan 2014

García-Jiménez J, Monte E, Trapero A (2010) Los hongos y oomicetos fitopatógenos. In: Jiménez Díaz R, Montesinos Segui E (eds) Enfermedades de las plantas causadas por hongos y oomicetos. Naturaleza y control integrado, Phytoma, pp 23–50

Ghorbani R, Wilcockson S, Koocheki A, Leifert C (2008) Soil management for sustainable crop disease control: a review. Environ Chem Lett 6:149–162

Giotis C, Markelou E, Theodoropoulou A, Toufexi E, Hodson R, Shotton P, Shiel R, Cooper J, Leifert C (2009) Effect of soil amendments and biological control agents (BCAs) on soil-borne root diseases caused by *Pyrenochaeta lycopersici* and *Verticillium albo-atrum* in organic greenhouse tomato production systems. Eur J Plant Pathol 123:387–400

Goicoechea N (2009) To what extent are soil amendments useful to control Verticillium wilt? Pest Manag Sci 65:831–839

Gu Y, Zhang X, Tu S, Lindström K (2009) Soil microbial biomass, crop yields, and bacterial community structure as affected by long-term fertilizer treatments under wheat-rice cropping. Eur J Soil Biol 45:239–246

Haggag WM, Saber MSM (2000) Evaluation of three compost as multi-strain carriers for biofertilizer and biocontrol agents against *Fusarium* wilt disease of some legume plants. Arab J Biotechnol 3:133–144

Hamblin A (1995) The concept of agricultural sustainability. In: Andrews JH, Tommerup I (eds) Advances in plant pathology, vol 11. Academy, New York, pp 1–19

Hannan MA, Hasan MM, Hossain I, Rahman SME, Ismail AM, Oh DH (2012) Integrated management of foot rot of lentil using biocontrol agents under field condition. J Microbiol Biotechnol 22:883–888

Hayat R, Ali S, Amara U, Khalid R, Ahmed I (2010) Soil beneficial bacteria and their role in plant growth promotion: a review. Ann Microbiol 60:579–598

Hirooka T, Ishii H (2013) Chemical control of plant diseases. J Gen Plant Pathol 79:390–401

Huang X, Zhang N, Yong X, Yang X, Shen Q (2012) Biocontrol of *Rhizoctonia solani* damping-off disease in cucumber with *Bacillus pumilus* SQR-N43. Microbiol Res 167:135–143

Ismail Y, McCormick S, Hijri M (2013) The arbuscular mycorrhizal fungus, *Glomus irregulare*, controls the mycotoxin production of *Fusarium sambucinum* in the pathogenesis of potato. FEMS Microbiol Lett 348:46–51

Israel S, Mawar R, Lodha S (2005) Soil solarisation, amendments and bio-control agents for the control of *Macrophomina phaseolina* and *Fusarium oxysporum* f. sp. *cumini* in aridisols. Ann Appl Biol 146:481–491

Jacobsen CS, Hjelmsø MH (2014) Agricultural soils, pesticides and microbial diversity. Curr Opin Biotechnol 27:15–20

Jaiswal AK, Elad Y, Graber ER, Frenkel O (2014) *Rhizoctonia solani* suppression and plant growth promotion in cucumber as affected by biochar pyrolysis temperature, feedstock and concentration. Soil Biol Biochem 69:110–118

Jayaraj J, Parthasarathi P, Adhakrishnan NV (2007) Characterization of a *Pseudomonas fluorescens* strain from tomato rhizosphere and its use for integrated management of tomato damping-off. BioControl 52:683–702

Joshi D, Hooda KS, Bhatt JC (2009) Integration of soil solarization with bio-fumigation and *Trichoderma* spp. for management of damping-off in tomato (*Lycopersicon esculentum*) in the mid altitude region of north-western Himalayas. Indian J Agric Sci 79:754–757

Jung T, Colquhoun IJ, Hardy GESJ (2013) New insights into the survival strategy of the invasive soilborne pathogen *Phytophthora cinnamomi* in different natural ecosystems in Western Australia. For Pathol 43:266–288

Kaewchai S, Soytong K, Hyde KD (2009) Mycofungicides and fungal biofertilizers. Fungal Divers 38:25–50

Kavroulakis N, Ntougias S, Besi MI, Katsou P, Damaskinou A, Ehaliotis C, Zervakis GI, Papadopoulou KK (2010) Antagonistic bacteria of composted agro-industrial residues exhibit antibiosis against soil-borne fungal plant pathogens and protection of tomato plants from *Fusarium oxysporum* f. sp. *radicis-lycopersici*. Plant Soil 333:233–247

Kawaguchi A, Inoue K (2012) New antagonistic strains of non-pathogenic *Agrobacterium vitis* to control grapevine crown gall. J Phytopathol 160:509–518

Kim IY, Pusey PL, Zhao Y, Korban SS, Choi H, Kim KK (2012) Controlled release of *Pantoea agglomerans* E325 for biocontrol of fire blight disease of apple. J Control Release 161:109–115
Lang J, Hu J, Ran W, Xu Y, Shen Q (2012) Control of cotton *Verticillium* wilt and fungal diversity of rhizosphere soils by bio-organic fertilizer. Biol Fertil Soils 48:191–203
Larkin RP (2008) Relative effects of biological amendments and crop rotations on soil microbial communities and soilborne diseases of potato. Soil Biol Biochem 40:1341–1351
Latha P, Anand T, Prakasam V, Jonathan EI, Paramathma M, Samiyappan R (2011) Combining *Pseudomonas*, *Bacillus* and *Trichoderma* strains with organic amendments and micronutrient to enhance suppression of collar and root rot disease in physic nut. Appl Soil Ecol 49:215–223
Lelièvre L, Groh M, Angebault C, Maherault AC, Didier E, Bougnoux ME (2013) Azole resistant *Aspergillus fumigatus*: an emerging problem. Med Mal Infect 43:139–145
Liu M, Hu F, Chen X, Huang Q, Jiao J, Zhang B, Li H (2009) Organic amendments with reduced chemical fertilizer promote soil microbial development and nutrient availability in a subtropical paddy field: the influence of quantity, type and application time of organic amendments. Appl Soil Ecol 42:166–175
Liu Y, Shi J, Feng Y, Yang X, Li X, Shen Q (2013) Tobacco bacterial wilt can be biologically controlled by the application of antagonistic strains in combination with organic fertilizer. Biol Fertil Soils 49:447–464
López-Escudero FJ, Blanco-López MA (1999) First report of transmission of *Verticillium dahliae* by infested manure in olive orchards in Andalucía (Southern Spain). Plant Dis 83:1178
López-Escudero FJ, Mercado-Blanco J (2011) *Verticillium* wilt of olive: a case study to implement an integrated management strategy to control a soil-borne pathogen. Plant Soil 344:1–50
López-Herrera CJ, Perez-Jiménez RM, Basallote-Ureba MJ, Zea Bonilla T, Melero-Vara JM (2003) Effect of soil solarization on the control of Phytophthora root rot in avocado. Plant Pathol 46:329–340
Lugtenberg B, Kamilova F (2009) Plant-growth-promoting rhizobacteria. Annu Rev Microbiol 63:541–556
Luo J, Ran W, Hu J, Yang X, Xu Y, Shen Q (2010) Application of bio-organic fertilizer significantly affected fungal diversity of soils. Soil Sci Soc Am J 74:2039–2048
Mazzola M (2002) Mechanisms of natural soil suppressiveness to soilborne diseases. Anton Leeuw J Microbiol 81:557–564
Melero-Vara JM, López-Herrera CJ, Prados-Ligero AM, Vela-Delgado MD, Navas-Becerra JA, Basallote-Ureba MJ (2011) Effects of soil amendment with poultry manure on carnation *Fusarium* wilt in greenhouses in southwest Spain. Crop Prot 30:970–976
Mercado-Blanco J, Bakker AHM (2007) Interactions between plants and beneficial *Pseudomonas* spp.: exploiting bacterial traits for crop protection. Antonie Van Leeuwenhoek 92:367–389
Mercado-Blanco J, Lugtenberg BJJ (2014) Biotechnological applications of bacterial endophytes. Curr Biotechnol 3:60–75
Milgroom MG, Cortesi P (2004) Biological control of chestnut blight with hypovirulence: a critical analysis. Annu Rev Phytopathol 42:311–338
Moreira AC, Domingos AC, Fontes AM, Semedo J, Melo E, Machado H, Reis M, Horta M, Cravador A (2007) Evaluation of cork and holm oak seedling viability to *Phytophthora cinnamomi* infection treated with compost and mycorrhizae fungi. IOBC/wprs Bull 57:73–76
Nakkeeran S, Fernando WGD, Siddiqui ZA (2006) Plant growth promoting rhizobacteria formulations and its scope in commercialization for the management of pests and disease. In: Siddqui ZA (ed) PGPR: biocontrol and biofertilization. Springer, Dordrecht, pp 257–296
Narayanasami P (2013) Mechanisms of action of fungal biological control agents. In: Nayaranassami P (ed) Biological management of diseases of crops, vol 1, Characteristics of biological control agents. Springer, Dordrecht, pp 99–200
Neeraj, Singh K (2011) Organic amendments to soil inoculated arbuscular mycorrhizal fungi and *Pseudomonas fluorescens* treatments reduce the development of root-rot disease and enhance the yield of *Phaseolus vulgaris* L. Eur J Soil Biol 47:288–295

Noble R (2011) Risks and benefits of soil amendment with composts in relation to plant pathogens. Australas Plant Pathol 40:157–167
Noble R, Conventry E (2005) Suppression of soil-borne plant diseases with composts: a review. Biocontrol Sci Technol 15:3–20
Ongley ED (ed) (1996) Control of water pollution from agriculture – FAO irrigation and drainage paper 55. FAO, Roma
Pan W, Mu C, Jiang X, Tian Y, Zhu C (2006) Chlamydospore and conidia of *Trichoderma* and soil fungistasis. Chin J Biol Control 22:87–91
Papasotiriou FG, Varypatakis KG, Christofi N, Tjamos SE, Paplomatas EJ (2013) Olive mill wastes: a source of resistance for plants against *Verticillium dahliae* and a reservoir of biocontrol agents. Biol Control 67:51–60
Pegg GF, Brady BL (2002) *Verticillium* wilts. CAB International, Wallingford
Pliego C, Cazorla FM (2012) Biocontrol of tree root diseases. In: Bruijn FJ (ed) Molecular microbial ecology of the rhizosphere, vol 2. Wiley Blackwell, New Jersey, pp 655–663
Pokharel R (2011) Soil solarization, an alternative to soil fumigants. Crop series soils, Colorado State University, Fact Sheet n° 0.505
Porras M, Barrau C, Romero F (2007) Effects of soil solarization and *Trichoderma* on strawberry production. Crop Prot 26:782–787
Prashar P, Kapoor N, Sachdeva S (2014) Rhizosphere: its structure, bacterial diversity and significance. Rev Environ Sci Biotechnol 13:63–77
Prieto P, Navarro-Raya C, Valverde-Corredor A, Amyotte SG, Dobinson KF, Mercado-Blanco J (2009) Colonization process of olive tissues by *Verticillium dahliae* and its *in planta* interaction with the biocontrol root endophyte *Pseudomonas fluorescens* PICF7. Microb Biotechnol 2:499–511
Raaijmakers JM, Paulitz TC, Steinberg C, Alabouvette C, Moënne-Loccoz Y (2009) The rhizosphere: a playground and battlefield for soilborne pathogens and beneficial microorganisms. Plant Soil 321:341–361
Ranganathswamy M, Patibanda AK, Nageshwara Rao G (2013) Evaluation of toxicity of agrochemicals on *Trichoderma* isolates in vitro. J Mycopathol Res 51:289–293
Ren JH, Li H, Wang YF, Ye JR, Yan AQ, Wu XQ (2013) Biocontrol potential of an endophytic *Bacillus pumilus* JK-SX001 against poplar canker. Biol Control 67:421–430
Roberts DP, Lohrke SM, Meyer SLF, Buyer JS, Bowers JH, Baker CJ, Lie W, de Souza JT, Lewis JA, Chung S (2005) Biocontrol agents applied individually and in combination for suppression of soilborne diseases of cucumber. Crop Prot 24:141–155
Ruano-Rosa D, López-Herrera CJ (2009) Evaluation of *Trichoderma* spp. as biocontrol agents against avocado white root rot. Biol Control 51:66–77
Ruano-Rosa D, Cazorla FM, Bonilla N, Martín-Pérez R, de Vicente A, López-Herrera CJ (2014) Biological control of avocado white root rot with combined applications of *Trichoderma* spp. and rhizobacteria. Eur J Plant Pathol 138:751–762
Soil Science Glossary Terms Committee (2008) Glossary of soil science terms 2008. American Society of Agronomy, USA
Sparks TC (2013) Insecticide discovery: an evaluation and analysis. Pestic Biochem Physiol 107:8–17
Suárez-Estrella F, Arcos-Nievas MA, López MJ, Vargas-García MC, Moreno J (2013) Biological control of plant pathogens by microorganisms isolated from agro-industrial composts. Biol Control 67:509–515
Tailor JA, Joshi BH (2014) Harnessing plant growth promoting rhizobacteria beyond nature: a review. J Plant Nutr 37:1534–1571
Trabelsi D, Mhamdi R (2013) Microbial inoculants and their impact on soil microbial communities: a review. Biomed Res Int. doi:10.1155/2013/863240
Trillas MI, Casanova E, Cotxarrera L, Ordovás J, Borrero C, Avilés M (2006) Composts from agricultural waste and the *Trichoderma asperellum* strain T-34 suppress *Rhizoctonia solani* in cucumber seedlings. Biol Control 39:32–38

Tupe SG, Chaudhary PM, Deshpande SR, Deshpande MV (2014) Development of novel molecules for the control of plant pathogenic fungi in agriculture. In: Kharwar RN, Upadhyay RS, Dubey NK, Raghuwanshi R (eds) Microbial diversity and biotechnology in food security. Springer, New Delhi, pp 315–325

Varma A, Chincholkar S (eds) (2007) Microbial siderophores. Springer, New York

Vitullo D, Altieri R, Esposito A, Nigro F, Ferrara M, Alfano G, Ranalli G, De Cicco V, Lima G (2013) Suppressive biomasses and antagonist bacteria for an eco-compatible control of *Verticillium dahliae* on nursery-grown olive plants. Int J Environ Sci Technol 10:209–220

Wang B, Yuan J, Zhang J, Shen Z, Zhang M, Li R, Ruan Y, Shen Q (2013) Effects of novel bioorganic fertilizer produced by *Bacillus amyloliquefaciens* W19 on antagonism of *Fusarium* wilt of banana. Biol Fertil Soils 49:435–446

Wei Z, Yang X, Yin S, Shen Q, Ran W, Xu Y (2011) Efficacy of *Bacillus*-fortified organic fertiliser in controlling bacterial wilt of tomato in the field. Appl Soil Ecol 48:152–159

Weiland JE (2014) *Pythium* species and isolate diversity influence inhibition by the biological control agent *Streptomyces lydicus*. Plant Dis 98:653–659

Woo SL, Ruocco M, Vinale F, Nigro M, Marra R, Lombardi N, Pascale A, Lanzuise S, Manganiello G, Lorito M (2014) *Trichoderma*-based products and their widespread use in agriculture. Open Mycol J 8:71–126

Wu HS, Yang XN, Fan JQ, Miao WG, Ling N, Xu YC, Huang QW, Shen QR (2009) Suppression of *Fusarium* wilt of watermelon by a bio-organic fertilizer containing combinations of antagonistic microorganisms. Biocontrol 54:287–300

Xu XM, Jeffries P, Pautasso M, Jeger MJ (2011) Combined use of biocontrol agents to manage plant diseases in theory and practice. Phytopathology 101:1024–1031

Yang X, Chen L, Yong X, Shen Q (2011) Formulations can affect rhizosphere colonization and biocontrol efficiency of *Trichoderma harzianum* SQR-T037 against Fusarium wilt of cucumbers. Biol Fertil Soils 47:239–248

Yao S, Merwin IA, Abawi GS, Thies JE (2006) Soil fumigation and compost amendment alter soil microbial community composition but do not improve tree growth or yield in an apple replant site. Soil Biol Biochem 38:587–599

Yoom MY, Cha B, Kim JC (2013) Recent trends in studies on botanical fungicides in agriculture. Plant Pathol J 29:1–9

Zaidi NW, Singh US (2013) *Trichoderma* in plant health management. In: Mukherjee PK, Horwitz BA, Singh US, Mukherjee M, Schmoll M (eds) *Trichoderma*: biology and applications. CABI, Boston, MA, pp 230–246

Zhang QC, Shamsi IH, Xu DT, Wang GH, Lin XY, Jilani G, Hussain N, Chaudhry AN (2012) Chemical fertilizer and organic manure inputs in soil exhibit a vice versa pattern of microbial community structure. Appl Soil Ecol 57:1–8

Zhao Q, Dong C, Yang X, Mei X, Ran W, Shen Q, Xu Y (2011) Biocontrol of *Fusarium* wilt disease for *Cucumis melo* melon using bio-organic fertilizer. Appl Soil Ecol 47:67–75

Chapter 23
Combining Application of Vermiwash and Arbuscular Mycorrhizal Fungi for Effective Plant Disease Suppression

Mohammad Haneef Khan, M.K. Meghvansi, Rajeev Gupta, K.K. Chaudhary, Kamal Prasad, Sazada Siddiqui, Vijay Veer, and Ajit Varma

23.1 Introduction

Disease suppression is a biological process which challenges the performance of the pathogen. It can be achieved by either curbing the propagation of the pathogen population or by neutralizing the pathogen derived harmful effects (viz., chemicals)

M.H. Khan (✉) • R. Gupta • V. Veer
Defence Research Laboratory, Defence R&D Organisation, Post Bag 2, Tezpur 784001, Assam, India
e-mail: haneef.itrc@gmail.com; rajeev_drlt@rediffmail.com; vijayveer50@yahoo.com

M.K. Meghvansi
Ministry of Defence, Defence R&D Organisation, Defence Research Laboratory, Post Bag 2, Tezpur 784001, Assam, India
e-mail: mk_meghvansi@yahoo.co.in

K.K. Chaudhary
Department of Biotechnology, Institute School of Life Sciences, Jaipur National University, Jaipur, Rajasthan, India
e-mail: kamal22055@yahoo.co.in

K. Prasad
Symbiosis Sciences Pvt Ltd., Sector-37, Gurgaon 122001, India
e-mail: kamalprsd27@gmail.com

S. Siddiqui
Biology Department, College of Science, King Khalid University, Abha, PO Box 10255 61321, Saudi Arabia
e-mail: guddank@rediffmail.com

A. Varma
Amity Institute of Microbial Technology, Amity University Uttar Pradesh, E-3 Block, 4th Floor, Sector 125, Noida, UP 201303, India
e-mail: ajitvarma@amity.edu

M.K. Meghvansi, A. Varma (eds.), *Organic Amendments and Soil Suppressiveness in Plant Disease Management*, Soil Biology 46,
DOI 10.1007/978-3-319-23075-7_23

to the host. However, it can be argued that by suppressing the pathogen's disease causing strategy, nature may provide chance to evolve themselves or other mechanisms of them to cause disease to the host. Therefore, it is arduous to identify any superior strategy against the disease. We can only measure their relative merits based on the observations in nature (Gordon and Leveau 2010). Nevertheless, disease suppressive agents should not only maintain the nature's rule of biodiversity by not ceasing the life of the pathogen, it should also keep the host unaffected. Maintenance of plant's health has been a major drawback of chemical agents for disease control which made environmentalist, agriculturists to think beyond chemicals and environmental friendly approach. An apotheosis is banished methyl bromide which is a most efficient fumigant and soil disinfectant, but caused and increased environmental burden of toxicity. In addition, pathogens are developing resistance against them and, therefore, it requires a surrogate as plant pathogen suppressing agent (Martin 2003). Hence, there is a dire need for an economical, efficient environmentally benevolent replacement for sustainable agriculture practices (Bonanomi et al. 2007).

Vermiwash (compost extract or worm tea or compost tea) is one of the by-products of vermicompost. It is an organic fertilizer decoction obtained from the units of vermiculture/vermicompost as drainage. Arbuscular mycorrhizal (AM) fungi, forming the order Glomales of the Glomeromycota, occur on the roots of 80 % of vascular flowering plant species (Smith and Read 2008), but they are obligate biotrophs and cannot be cultured without the plant. Mycorrhizal fungi facilitate nutrient and water uptake from soil. Vermiwash and symbiotic organisms such as AM fungi which are conventionally used as fertilizer supplement have, in recent decades been considered as potential disease management agents also. Nevertheless, use of vermiwash and AM fungi has been mainly focused on managing soil-borne phytopathogens. Role of foliar spray of vermiwash and AM fungi in managing foliar plant diseases has received little attention of the researchers. Further, still there are many unknowns with regard to the underlying mechanisms of disease suppression by the vermiwash and AM fungi. Given the complexity of the plant disease, changes in behavior of the phytopathogens under the influence of diverse environmental conditions, it is evident that no single component strategy for disease management is effective and sustainable. This has led to the concept of integrating all the promising disease management strategies such as combined application of AM fungi and compost/vermiwash in a holistic way. During past few decades, a considerable amount of knowledge on various management strategies against plant diseases such as good agronomic practices, use of botanicals, chemical fungicides, and biocontrol agents has been generated by researchers working in different parts of the world. However, there is dearth of information in the current scientific literature on combined use of vermiwash foliar spray and AM fungi for managing plant diseases. In view of the above, this chapter highlights the potential of individual and combined approach of vermiwash and AM fungi with a particular emphasis on understanding the possible underlying molecular mechanisms involved in the suppression of plant diseases.

23.2 Compost/Vermiwash, AM Fungi, and Their Effect on Plant Disease Suppression

23.2.1 Effect of Compost/Vermiwash on Plant Disease Suppression

Vermiwash can be used both as foliar spray and in the root zone of the plants (Meghvansi et al. 2012). It is also called Vermi-Tea or Vermi-liquid. It was found to develop resistance to diseases in plants and was beneficial in nurseries, lawns, and orchards (Meghvansi et al. 2012). As a foliar spray, it was reported to have yielded good results, especially initiating flowering and long lasting inflorescence of *Anthuriums* (Rao 2005). It could also be used as a liquid fertilizer applied to the rhizosphere (Rao 2005). Compost teas are reported to control plant pathogens through different mechanisms. The most reported factor influencing the efficacy of compost teas in inhibiting the development of plant disease is their microbial content. The microorganisms present in the tea may act as pathogen antagonists through their ability to compete for space and nutrients, to destroy pathogens by parasitism, to produce antimicrobial compounds, or to induce systemic resistance in plants (Mehta et al. 2014). Other work hypothesized that the physico-chemical properties of the compost teas, namely nutrients and organic molecules such as humic or phenolic compounds (Siddiqui et al. 2008), may protect the plant against disease through improved nutritional status, direct toxicity toward the pathogen, or induced systemic resistance (Fig. 23.1). The potential parameters that affect the efficacy of compost teas are two-fold: the target pathosystem (pathogen and host plant) and the preparation methodologies of the teas (aeration, compost type, nutrient additives, duration of fermentation, etc.) (Scheuerell and Mahaffee 2002). Therefore, vermiwash has the potential and got plenty of approaches to downgrade the pathogen's ability of causing disease. Since the first report of suppressive composts (Hoitink et al. 1977) several examples with wide array of pathosystems and composts, obtained from a large variety of raw materials and utilization of different technologies, have been published. Indisputably, suppression of plant pathogens by compost is a globally known phenomenon. However, its practical applicability is still limited primarily owing to inconsistencies in performance. Although several published reports suggested that the compost is effective in controlling the diseases in more than 50 % cases (Bonanomi et al. 2010), yet there is also apprehension that there is a risk of promoting or introducing disease causing pathogens via compost (Hadar and Papadopoulou 2012). In addition, after reviewing large number of positive cases of disease suppression, Noble and Coventry (2005) concluded that the compost's effect is relatively lesser and inconsistent when it is applied in the field, as compared with results from artificial media. In addition, there are meager reports on effect of compost/vermiwash on foliar phytopathogens.

Fig. 23.1 Vermiwash follows different strategies/approaches against plant pathogens

23.2.2 Effect of AM Fungi on Plant Disease Suppression

AM fungi provide an effective alternative method of disease suppression, especially for those pathogens which affect belowground plant organs. AM fungi have an enormous potential for use as biocontrol agents for soil and root-borne diseases. These fungi compete for nutrients and release certain compounds which suppress the growth of soil or root-borne plant pathogens. It also adopts various other strategies to indirectly suppress plant disease such as change in root morphology by increased lignification, abiotic stress alleviation, alteration in microbial community in the rhizosphere, and chemical composition in plant tissues like antifungal chitinase, isoflavonoids, etc. (Pal and Gardener 2006). Specifically, ectomycorrhizae may involve physical barrier of the fungal hyphae around the plant root along with antibiosis (Pal and Gardener 2006). The mycorrhizas also incite both systemic acquired resistance (SAR) and induced systemic resistance (ISR) (Wasternack and Hause 2013; Khan et al. 2010) and are involved in priming of several defense mechanisms required for plants against diseases. Although, these responses have mostly being seen against soil-borne pathogens, yet our knowledge of foliar disease suppression by mycorrhizal fungi is still limited. We know pathogen does not attack only at the root but also at the aboveground part of the plant. Hence, the AM fungi performance against foliar plant pathogens should be evaluated under varied conditions and hosts.

23.2.3 Combined Effect of Compost/Vermiwash and AM Fungi on Plant Disease Suppression

The combination of different methods to manage the disease, with ecological approaches, proposed almost 15 years ago (Ristaino and Johnston 1999), is still considered valid. This is the reason why plant disease control may sometimes be achieved by a single measure (vermiwash or AM fungi), but the long-term reduction of disease losses generally requires the combination of more than one control measures. An integrated disease-control program, based on knowledge of pathogen biology and diseases most likely to occur in an area, is the most effective and efficient means of controlling pathogens in the long run. Combination of AM fungi and organic amendments particularly vermicompost against soil-borne phytopathogens has been studied. AM fungi colonization of rice (Kale et al. 1992) and sorghum (Cavender et al. 2003) was found to increase significantly with vermicompost applications, and AM fungi colonization has also been correlated with traditional compost applications (Tarkalson et al. 1998), although this relationship has not been studied in depth. Therefore, it is necessary to discuss and throw light on more than one control measure for sustainable and more effective plant disease suppression strategy.

23.3 The Underlying Possible Mechanisms by Plant Disease Suppressive Agents

23.3.1 Vermiwash

Although several studies have demonstrated the role and their possible mechanisms of composts in disease suppression (Yu et al. 2011; Termorshuizen et al. 2006), yet studies on its usage as a liquid extract and their underlying mechanism are limited.

The compost generates an environment of competition among microbes for nutrient. By incorporating compost, the microbes present in it increase the diversity and, hence, the competition for nutrients among the native and incorporated microbes. Competition is the common phenomena for general disease suppression mechanism. Therefore, the level of disease suppressiveness is typically related to the level of active microbial biomass in a soil (Edwards 2004). The larger the soil's active microbial biomass, the greater the soil's capacity to use nutrients leading to lowering of the nutrient availability to pathogens. In other words, when most soil nutrients are tied up in microbial bodies, the competition for readily available mineral nutrients gets a higher level (Edwards 2004). Therefore, it can be stated that competition for limited nutrients is a key for general suppression. This type of strategy is mainly observed against *Fusarium oxysporum* f. sp. *radicis-lycopersici*, *Pyrenochaeta lycopersici*, *Pythium ultimum*, and *Rhizoctonia solani*. This effect was observed with a marked increase in the siderophores producers which reduces the level of iron required for germination of pathogens. A well-known example of siderophore producer is fluorescent *Pseudomonas*, which competes with the *Fusarium* germination and propagation (Mehta et al. 2014).

Antibiosis is another approach for plant disease suppression by compost, which is mediated by microbes via secondary metabolites, lysis, enzymatic activity, or other substances (Fravel 1988). Zwittermicin A and kanosamine, products of *B. cereus* UW85, are well-known antibiotics used against oomycetes like *Phytophthora* (Milner et al. 1996). *Gl. Virens* that produces Gliotoxin, an antibiotic produced by the microbes of composted mineral soil, has been found effective against the control of damping-off of zinnia seedlings (*Zinnia elegans*) caused by *Py. ultimum* and *R. solani* (Lumsden et al. 1992).

Hyperparasitism is also a type of antagonism where a microorganism directly kills a pathogen by attacking it (Heydari and Pessarakli 2010). In this attack, nonpathogenic microbes parasitize or lyse the mycelium, oospores, hyphae, or sclerotia of several pathogenic fungi such as *Pythium*, *Phytophthora*, *Verticillium*, *Rhizoctonia*, and *Sclerotinia* (Diánez et al. 2005). A very well-known example of hyperparsitism is suppression of *R. solani* by *Trichoderma harzianum* (Chet and Baker 1980), which is frequently found in composts (Kuter et al. 1983). *Acremonium alternatum*, *Acrodontium crateriforme*, *Ampelomyces quisqualis*, *Cladosporium oxysporum*, *Gl. virens*, *Humicola fuscoatra*, and *Verticillium chlamydosporium* also have the capacity to parasitize powdery mildew pathogens and *Ph. capsici* oospores (Sutherland and Papavizas 2008).

Another important factor for disease suppression by composts is induced systemic resistance (ISR). ISR develops, in case of soil-borne pathogens, when the rhizosphere is inoculated with a weakly virulent pathogen. After the initiation of systemic resistance by weak pathogen, the plant develops the capacity for future effective response to a more virulent pathogen (Pharand et al. 2002; Zhang et al. 1998). Zhang et al. (1996) found that compost induced resistance in cucumber to both pythium root rot and anthracnose caused by *Colletotrichum orbiculare* and that this phenomenon was negated by sterilization. They reported that the effect of compost on peroxidase activity in cucumber was more pronounced after plant infection. Similarly, high glucanase activity was found in *Arabidopsis thaliana* and cucumber plants grown in compost after infection, compared with that in plants grown in peat (Zhang et al. 1998). They concluded that compost induced systemic acquired resistance in a different way from its induction by pathogens or salicylic acid. These findings suggest that the microflora in the compost had an effect on these PR proteins in both plant types, but that much of the activation resulted from infection by the pathogen. Further in two preliminary tests, the expression levels of the PR proteins, PR-Q, chitinase1, and peroxidase were not elevated when melon plants were grown in the suppressive compost, compared with their levels in plants grown in conducive peat in the absence of FOM (Yogev et al. 2010). These results indicate the lacunae of consistent results for suppressive compost against variety of crops. Therefore, the application of suppressive composts and its mechanism needs validation on variety of crops. In other systems, PR-Q and peroxidase were found to be upregulated in transgenic tobacco plants treated with suppressive compost expressing viral movement proteins (Hofius et al. 2001) and in marrow (*Cucurbita pepo* L.) plants infected with cucumber mosaic virus (CMV) (Tecsi et al. 1996), respectively. Despite having several application based studies, the less known mechanism behind the disease suppressive property of OA needs attention. Further, the nutrients in the OA applied in the rhizosphere region of soil may percolate down especially in the rainy season or higher application of water or in crops with high irrigation. Therefore to overcome this problem, compost extracts, a liquid form of compost, were developed. These are much easier to handle for applying to the crops than solid composts, which are bulky and heavy and need soil incorporation. These extracts are increasingly being applied, as soil drenches or soil and foliar sprays, to enhance plant growth and control plant diseases and pests (Simsek-Ersahin 2011). Since last decade, the utilization of vermicompost extracts/teas as bio-control agents has accelerated (Simsek-Ersahin 2011). The consistency and performance of compost have been observed several times in wide variety of conditions belonging to the rhizosphere. However, they have been tested on very few occasions against foliar plant pathogens. Therefore, studies for foliar spray of compost against foliar plant pathogen need validation and, therefore, these types of study are warranted. This also indicates that single measure may have immediate or for a time-being effect on plant diseases, but for long-term effect single approach is less favorable.

23.3.2 AM Fungi

It has been established that AM fungi provide several benefits to the plants, mainly the increase in nutrient uptake (Smith and Read 2008). Despite this, there is still an ambiguity that the AMF has any direct involvement in the host's defense signaling against phytopathogens. However, there are several reports mentioning the indirect functions contributing to intensify the plant defense responses. The mycorrhizal fungi protect plant roots from diseases in several ways. Improved phosphorus uptake in the host plant has commonly been associated with mycorrhizal fungi (Meghvansi and Mahna 2009). When plants are not deprived of nutrients, they are better able to tolerate or resist disease-causing organisms. Protection from the pathogen *F. oxysporum* was shown in a field study using a cool-season annual grass and mycorrhizal fungi. In this study, the disease was suppressed in mycorrhizae-colonized grass inoculated with the pathogen (Newsham et al. 1995). In field studies with eggplant, fruit numbers went from an average of 3.5 per plant to an average of 5.8 per plant when inoculated with *Gigaspora margarita* mycorrhizal fungi. Average fruit weight per plant increased from 258 to 437 g. A lower incidence of verticillium wilt was also realized in the mycorrhizal plants (Matsubara et al. 1995). In a study conducted by Tabin et al. (2009), mycorrhizal inoculation not only reduced the percentage of damping-off disease of *Aquilaria agallocha* seedlings caused by the pathogenic fungus (*Py. aphanidermatum*) but also significantly increased host plant height, total biomass, and dry matter. The effects of arbuscular mycorrhizal fungi *G. mosseae*, *G. fasciculatum*, and *Rh. leguminosarum* biovar *phaseoli* were examined on the pathosystem of *Sclerotinia sclerotiorum* (Lib.) de Bary (Ss) and common bean by Aysan and Demir (2009). Treatments of single inoculations of AMF and *Rh. leguminosarum* isolates reduced disease severity by 10.3–24.1 %. The mechanism for disease suppression by AM fungi is described below.

As stated earlier that AM fungi intensify plant defense response against pathogen by anatomical alterations in the root system (Wehner et al. 2010), microbial changes in the rhizosphere and enhancing the attenuated plant defense responses by altering the host's signaling pathways (Pozo and Azcón-Aguilar 2007). This is accomplished primarily through modulation in Jasmonic acid (JA) and salicylic acid (SA) dependent pathways (Pozo and Azcón-Aguilar 2007). Furthermore, the AMF is likely to have role in induction of hydrolytic enzymes (Pozo et al. 1999), enhanced levels of Pathogenesis-related (PR) proteins, accrual of phytoalexins (Larose et al. 2002), callose deposition (Cordier et al. 1998), and reactive oxygen species generation (Salzer et al. 1999). Hence, there are several reports exemplifying the potential of AMF in reducing the severity and incidence of plant disease since a long time. During AMF's colonization, a strong genetic shift occurs which leads to the enhancement of signaling pathways of plant defense response against phytopathogen. After having symbiotic relationship with its host, AMF possibly enhances genes encoded products having antimicrobial activity. For instance, induction of *Medicago truncatula* genes TC104515 (6659-fold), TC101060, and TC98064 was observed in the roots colonized with *Glomus intraradices* (Liu et al. 2007). These genes were predicted to encode cysteine rich proteins that display antifungal activity (Terras et al. 1995). Their function is to elicit the hypersensitivity response with the matching resistance gene (de Wit 1992). This response is mediated by reactive oxygen species (ROS) produced early in the plant–pathogen interaction (Levine et al. 1994). Accumulation of reactive oxygen species (ROS) in the mycorrhized plants has also been observed (Pozo and Azcón-Aguilar 2007). However, TC104515 transcripts were detected only in roots colonized with *G. intraradices* and not in roots colonized with *G. versiforme* or *Gi. gigantea*. This gene was also not expressed in *M. truncatula*/*G. mosseae* roots (Hohnjec et al. 2005). So, there is a considerable variation in the genetic shifts of different plant-related defense genes colonized with diverse AMF species, which needs further exploration. In an experiment, a group of genes was identified through differential expression in shoots of AMF-colonized plant, showing striking similarities with defense/stress signaling genes and ACRE genes (Liu et al. 2007). The ACRE genes were previously known to respond instantaneously in tomato upon infection of *Cladosporium fulvum* and suggested to be involved in the initial development of defense signaling (Durrant et al. 2000). Based on the split-root analyses, some of ACRE genes including two WRKY-type transcription factors and a TOLL-type protein showed a greater increase in transcripts in the non-colonized roots and shoots of the mycorrhizal plants (Liu et al. 2007). On the contrary, leaves of mycorrhizal plants infected with the phytopathogens *Botrytis cinerea* or tobacco mosaic virus showed no significant improvement in incidence and severity of necrotic lesions than those of nonmycorrhizal ones (Shaul et al. 1999). Further investigation also revealed the induction of PR-1 and PR-3 expression was observed in the leaves of both non-mycorrhizal and mycorrhizal plants. Although accretion and mRNA steady-state levels of these proteins were lower, their appearances were delayed in the leaves of the mycorrhizal plants. They concluded that prior infection of AMF than pathogen attack is required. These evidences are strongly in favor of AMF triggered localized and systemic priming of plants.

One of the most studied phytohormones in the AMF–plant interactions is Jasmonic acid (JA) which is known to regulate the accommodation of AMF and the nutrient provided by it within the plant root cells. Increase in the endogenous JA levels in arbusculated cells of plant upon phytopathogen attack has also been reported (Hause et al. 2002; Meixner et al. 2005). Like other pathogens, SA also recognizes AMF as pathogen and acts against it by delaying its colonization or in some cases suppresses its growth (Fig. 23.2). Nevertheless, enhanced SA level was found in mycorrhizal defective (myc-) mutants in response to AMF (García-Garrido and Ocampo 2002). A study conducted by our research group (Unpublished) demonstrated that *G. mosseae*-colonized plant seemed to follow a relatively broad-spectrum strategy against cercospora leaf spot disease suppression as was evident by significantly expressed genes related to different activities such as cell death, carotenoid, salicylic acid, and systemic acquired resistance. This result does align with suggestion of Pozo and Azcón-Aguilar (2007) that AM fungi activate the priming effect of the plant. Nevertheless, the studies related to this topic are still scarce. Pozo and Azcón-Aguilar (2007) suggested that AM fungi incite a priming effect on the defense system of the colonizing plant. The priming effect means AM fungi is a nonpathogenic fungal organism, which activates the defense system of the plants by colonizing it. Thus, priming effect helps plant to retaliate against any kind of pathogen, viz., be it soil-borne or foliar pathogen. However, the priming effect can defend any pathogen attack only if AM fungi are colonized before pathogen attack. We know that in nature this is not the case. The pathogen will not wait for AM fungi to get colonized, instead pathogen may infect at every stage of plant's life. So, in that case, how AM fungi will help their partner to defend the attack of already infected pathogen? These are the few queries of which answers are still not known.

23.3.3 Combined Approaches

The individual effect of AMF and organic amendments as discussed above has shown that they have potential to suppress the phytopathogens, but their combined effect can be beneficial or deleterious for plant health. However, there are limited studies establishing the combined effect of AM fungi and vermiwash synergistically and enhancing the plant growth even more compared to their individual effect. Khan et al. (2014) demonstrated the foliar application of vermiwash and AM Fungi inoculation in the soil improves the plant growth and nutrient uptake. In addition, it

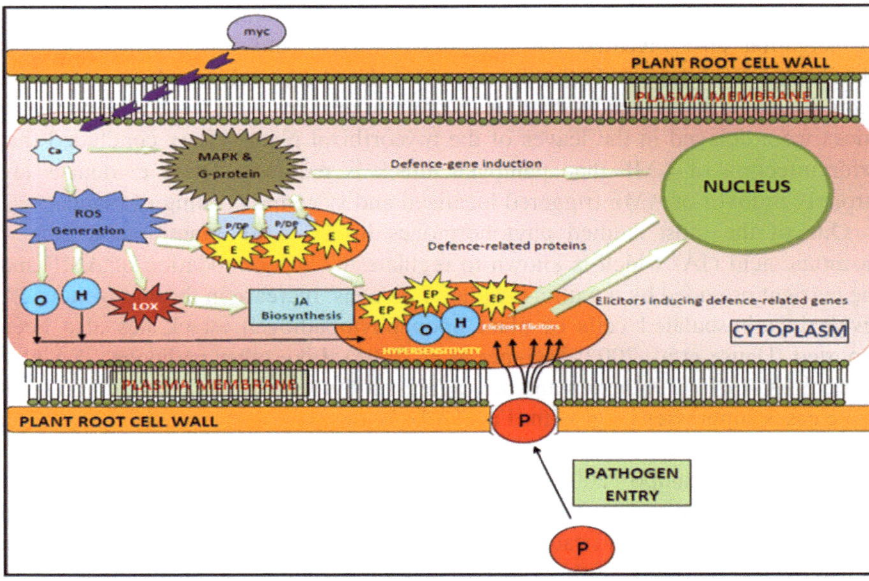

Fig. 23.2 Schematic representation of AMF-induced defense signaling in plant's cell. The myc (myc factor) from AMF triggers calcium dependent downstream processes ($[Ca^{2+}]_{cyt}$; abbreviated as Ca) which includes induction of ROS generation, and MAPK and G-protein alterations. ROS includes O_2^- (abbreviated as O) and H_2O_2 (abbreviated as H). ROS also induces LOX, which leads to JA biosynthesis. Antioxidant enzymes (E) such as SOD, POD, catalase, and APX which play an important role in ROS metabolism gets phosphorylated (EP) through MAPK and G-protein. MAPK and G-protein also triggers plant's defense genes. As pathogen enters, it either secretes some elicitors or by damaging cell wall caused by the pathogen that triggers plant's defense genes. These defense-related genes encoding proteins attack on pathogens and try to neutralize them. Whereas, antioxidant enzymes and ROS act constitutively on the pathogen infected site and initiate hypersensitivity reaction which leads to the apoptosis of the infective cells. (*Source*: Khan et al. 2010; J Phytol 2: 53–69, with permission)

was found that vermiwash spray influences the nutrient stoichiometry and growth by contributing more N to the plant colonized with AM fungi. However, there were only two AM fungi tested in this experiment; there is scope for other AM fungi to provide better results. Perner et al. (2007) observed that the addition of compost in combination with mycorrhizal inoculation can improve nutrient status and flower development of plants grown on peat-based substrates. Labidi et al. (2007) also suggested that the effect of compost addition on growth of the AM fungal biomass could be one way to improve survival of planted seedlings in arid regions. In a field experiment undertaken by Caravaca et al. (2002) to evaluate the effect of mycorrhizal inoculation with *G. intraradices* and added composted residue on the establishment of *Pistacia lentiscus* L. seedlings in a semiarid area showed that after 1 year of plantation, the plant height of *P. lentiscus* seedlings increased by 106 % with respect to the control. Again, Caravaca et al. (2006) observed that combined treatment involving the addition of a medium dose of amendment (100 mg C kg^{-1}

soil) and the mycorrhizal inoculation with *G. intraradices* or *G. deserticola* produced an additive effect on the plant growth with respect to the treatments applied individually (about 77 % greater than plants grown in the amended soil and about 63 % greater than inoculated plants). Maji et al. (2013) conducted an experiment wherein they observed the response of foliar disease of Mulberry variety S-1635 including pseudo CLS disease caused by *Pseudocercospora mori* under organic versus conventional farming system for 2 years (2007–2009). In this study, they applied following doses: FYM (20 tons/ha/year) and NPK 336:180:112 kg/ha/year in five split doses (recommended package), Vermicompost (30 tons/ha/year) in five split doses, Vermicompost (30 tons/ha/year in five split doses) + twice foliar spray of vermiwash@600 L/ha/crop, Vermicompost (25 tons/ha/year in five split doses) + green manure (*Crotalaria juncea*), Vermicompost (20 tons/ha/year) + green manure + recommended dose of bacterial and fungal biofertilizer (*Azotobacter chroococcum* @ 20 kg/ha/year and arbuscular mycorrhizal fungi (AMF) @ 80 kg/ha/year), T7–Vermicompost (15 tons/ha/year) + biofertlizers (*A. chroococcum* 20 kg/ha/year and AMF 80 kg/ha/year) + NPK: 168:90:56 in five split doses. Based on these results, Maji et al. (2013) suggested that the application of balanced organic and inorganic fertilizers helps in enrichment of soil beneficial mycoflora and nutrient supply for a healthy plant growth which may bring forth resistance to diseases. The issue raised by Maji et al. (2013) is of utmost importance as the balance between AM fungi and other foliar defense elicitor, viz., vermiwash, as discussed earlier, is warranted. This is because, in almost all the mentioned studies pertaining to the plant receiving combined effect of compost/ vermiwash and AM fungi also observed that colonization of AM fungi got reduced upon application of foliar spray of vermiwash. This influence of vermiwash on AM fungal colonization may further influence the plant defense system activated by AM fungi. Khan et al. (2014) demonstrated that foliar spray of vermiwash on *Ca. assamicum* colonized *Rhizophagus irregulare* showed lesser colonization than that of *G. mosseae*. de Román et al. (2011) also showed that acibenzolar-*S*-methyl (ASM), chemical elicitor, impairs the AM fungal root colonization. This indicates that the negative influence is likely due to alterations in defense status of the plant rather than to changes in resource allocation patterns. However, they also suggested that the AM association may activate the plant defense mechanisms and overcome such effects. This may be an answer to our question being asked earlier that what happens when AM fungi colonizes after that pathogen infection, which usually occurs in nature. However, it needs both extensive and intensive studies to get established. Nevertheless, this suggests that the priming effect incited by the AM fungi in this condition will be less in the plant. However, there is a scope for vermiwash to partly incite and induce the plant defense system on behalf of AM fungi. The synergistic effect of AM fungi and foliar spray of vermiwash on *C. tezpurensis*-infected *Ca. assamicum* was observed in a study conducted by our group (Unpublished) targeting 22 genes pertaining to plant defense system and other physiological parameters. This study showed early induction of almost all defense genes which are shown to be lately induced in their individual treatments.

This demonstrates that there is a compensation or make-over effect of combined measures of vermiwash and AM fungi.

23.4 Conclusion

Over the past many years, researchers, agriculturists, and farmers went for chemical approaches to control the plant diseases. However, their side effects came under scanner lately demanding for a sustainable replacement. This replacement should be eco-friendly which advocates the diversity law of nature by only suppressing but ceasing the pathogen population and their propagation. To achieve this aim, single measures such as compost/vermiwash and AM fungi were found to be effective. However recently, the lesser impact of these single measures came into light wherein, they cannot provide sustainable disease suppression/management. Therefore now, there is a dire need for the multiple approaches, just like multiple drug therapy (MDT) for tuberculosis kind of deadly disease, in the suppression of plant disease until the mentioned community come up with an alternative. It is clear from the potencies of vermiwash/compost and AM fungi that they follow or help the plant to adopt various types of strategies to counter the attack of the pathogen. In addition, they are eco-friendly and also help in enhancement of plant growth. Therefore, further research on their combined effect on multi-climatic, multi-locational, and large-scale field trials is necessary to come up with concrete evidences of their potential to provide an all-round protection to the plant. Further, ecological studies for their both negative and positive effect on environment and, physiological and in-depth molecular studies are warranted to understand the underlying mechanisms of disease suppression more precisely.

Acknowledgements MHK, MKM, and RG thank Director, Defence Research Laboratory, Defence R&D Organisation, Tezpur, for providing finance and necessary lab support to conduct our research work.

References

Aysan E, Demir S (2009) Using arbuscular mycorrhizal fungi and *Rhizobium leguminosarum*, biovar*phaseoli* against *Sclerotinia sclerotiorum* (Lib.) de bary in the common bean *Phaseolus vulgaris* L.). Plant Pathol J 8:74–78

Bonanomi G, Antignani V, Pane C, Scala F (2007) Suppression of soilborne fungal diseases with organic amendments. J Plant Pathol 89:311–324

Bonanomi G, Antignani V, Capodilupo M, Scala F (2010) Identifying the characteristics of organic soil amendments that suppress soil-borne plant diseases. Soil Biol Biochem 42: 136–144

Caravaca F, Bareab JM, Roldána A (2002) Synergistic influence of an arbuscular mycorrhizal fungus and organic amendment on *Pistacia lentiscus* L. seedlings afforested in a degraded semiarid soil. Soil Biol Biochem 34:1139–1145

Caravaca F, Tortosa G, Carrasco L, Cegarra J, Roldán A (2006) Interaction between AM fungi and a liquid organic amendment with respect to enhancement to the performance to the leguminous shrub *Retama sphaerocarpa*. Biol Fertil Soils 43:30–38

Cavender ND, Atiyeh RM, Knee M (2003) Vermicompost stimulates mycorrhizal colonization of roots of *Sorghum bicolor* at the expense of plant growth. Pedobiologia 47:85–89

Chet I, Baker R (1980) Induction of suppressiveness to *Rhizoctonia solani* in soil. Phytopathology 70:994–998

Cordier C, Pozo MJ, Barea JM, Gianinazzi S, Gianinazzi-Pearson V (1998) Cell defense responses associated with localized and systemic resistance to *Phytophthora parasitica* induced in tomato by an arbuscular mycorrhizal fungus. Mol Plant Microbe Interact 11:1017–1028

de Román M, Fernández I, Wyatt T, Sahrawy M, Heil M, Pozo MJ (2011) Elicitation of foliar resistance mechanisms transiently impairs root association with arbuscular mycorrhizal fungi. J Ecol 99:36–45

de Wit PJGM (1992) Molecular characterization of gene-for-gene systems in plant-fungus interactions and the application of avirulence genes in control of plant pathogens. Annu Rev Phytopathol 30:391–418

Diánez F, Santos M, Tello JC (2005) Suppression of soilborne pathogens by compost: suppressive effects of grape marc compost on phytopathogenic oomycetes. Acta Hortic 697:441–460

Durrant WE, Rowland O, Piedras P, Hammond-Kosack KE, Jones JDG (2000) cDNA-AFLP reveals a striking overlap in race-specific resistance and wound response gene expression profiles. Plant Cell 12:963–977

Edwards CA (ed) (2004) Earthworm ecology. Lewis Publishers/CRC Press, Boca Raton, FL

Fravel D (1988) Role of antibiosis in the biocontrol of plant diseases. Annu Rev Phytopathol 26:75–91

García-Garrido JM, Ocampo JA (2002) Regulation of the plant defence response in arbuscular mycorrhizal symbiosis. J Exp Bot 53:1377–1386

Gordon TR, Leveau JHJ (2010) Plant pathology: a story about biology. Annu Rev Phytopathol 48: 293–309

Hadar Y, Papadopoulou KK (2012) Suppressive composts: microbial ecology links between abiotic environments and healthy plants. Annu Rev Phytopathol 50:133–153

Hause B, Maier W, Miersch O, Kramell R, Strack D (2002) Induction of jasmonate biosynthesis in arbuscular mycorrhizal barley roots. Plant Physiol 130:1213–1220

Heydari A, Pessarakli M (2010) A review on biological control of fungal plant pathogens using microbial antagonists. J Biol Sci 10:273–290

Hofius D, Herbers K, Melzer M, Omid A, Tacke E, Wolf S, Sonnewald U (2001) Evidence for expression level-dependent modulation of carbohydrate status and viral resistance by the potato leaf roll virus movement protein in transgenic tobacco plants. Plant J 28:529–543

Hohnjec N, Vieweg MF, Puhler A, Becker A, Kuster H (2005) Overlaps in the transcriptional profiles of *Medicago truncatula* roots inoculated with two different *Glomus* fungi provide insights into the genetic program activated during arbuscular mycorrhiza. Plant Physiol 137: 1283–1301

Hoitink HAJ, Van Doren DM, Schmitthenner AF (1977) Suppression of *Phytophthora cinnamomi* in a composted hardwood bark potting medium. Phytopathology 67:561–565

Kale RD, Mallesh BC, Kubra B, Bagyaraj DJ (1992) Influence of vermicompost application on the available macronutrients and selected microbial-populations in a paddy field. Soil Biol Biochem 24:1317–1320

Khan MH, Meghvansi MK, Panwar V, Gogoi HK, Singh L (2010) Arbuscular mycorrhizal fungi-induced signaling in plant defence against phytopathogens. J Phytol 2:53–69

Khan MH, Meghvansi MK, Gupta R, Veer V, Singh L, Kalita MC (2014) Foliar spray with vermiwash modifies the arbuscular mycorrhizal dependency and nutrient stoichiometry of bhut jolokia (*Capsicum assamicum*). PLoS One 9(3), e92318

Kuter GA, Nelson EB, Hoitink HAJ, Madden LV (1983) Fungal populations in container media amended with composted hardwood bark suppressive and conducive to *Rhizoctonia* damping-off. Phytopathology 73:1450–1456

Labidi S, Nasr H, Zouaghi M, Wallander H (2007) Effects of compost addition on extra-radical growth of arbuscular mycorrhizal fungi in *Acacia tortilis* ssp. *raddianasavanna* in a pre-Saharan area. Appl Soil Ecol 35:184–192

Larose G, Chenevert MP, Gagne S, Piche VH (2002) Flavonoid levels in roots of *Medicago sativa* are modulated by the developmental stage of the symbiosis and the root colonizing arbuscular mycorrhizal fungus. J Plant Physiol 159:1329–1339

Levine A, Tenhaken R, Dixon R, Lamb CJ (1994) H_2O_2 from the oxidative burst orchestrates the plant hypersensitive disease resistance response. Cell 79:583–593

Liu J, Maldonado-Mendoza I, Lopez-Meyer M, Cheung F, Town CD, Harrison MJ (2007) Arbuscular mycorrhizal symbiosis is accompanied by local and systemic alterations in gene expression and an increase in disease resistance in the shoots. Plant J 50:529–544

Lumsden RD, Locke JC, Adkins ST, Walter JF, Ridout CJ (1992) Isolation and localization of the antibiotic gliotoxin produced by *Gliocladium virens* from alginate prill in soil and soilless media. Phytopathology 82:230–235

Maji MD, Setua GC, Ghosh A (2013) Evaluation of severity of foliar diseases of mulberry under organic versus conventional farming systems. J Plant Prot Sci 5:55–58

Martin FN (2003) Development of alternative strategies for management of soilborne pathogens currently controlled with methyl bromide. Annu Rev Phytopathol 41:325–350

Matsubara Y-I, Tamura H, Harada T (1995) Growth enhancement and verticillium wilt control by vesicular-arbuscular mycorrhizal fungus inoculation in eggplant. J Jpn Hortic Soc 64:555–561

Meghvansi MK, Mahna SK (2009) Evaluating the symbiotic potential of *Glomus intraradices* and *Bradyrhizobium japonicum* in vertisol with two soybean cultivars. American-Eurasian J Agron 2:21–25

Meghvansi MK, Khan MH, Gupta R, Gogoi HK, Singh L (2012) Vegetative and yield attributes of okra and naga chilli as affected by foliar sprays of vermiwash on acidic soil. J Crop Improv 26:520–531

Mehta CM, Palni U, Franke-Whittle IH, Sharma AK (2014) Compost: its role, mechanism and impact on reducing soil-borne plant diseases. Waste Manag 34:607–622

Meixner C, Ludwig-Müller J, Miersch O, Gresshoff P, Staehelin C, Vierheilig H (2005) Lack of mycorrhizal autoregulation and phytohormonal changes in the supernodulating soybean mutants 1007. Planta J 222:709–715

Milner JL, Silo-Suh LA, Lee JC, He H, Clardy J, Handelsman J (1996) Production of kanosamine by *Bacillus cereus* UW85. Appl Environ Microbiol 62:3061–3065

Newsham KK, Fitter AH, Watkinson AR (1995) Arbuscular mycorrhiza protect an annual grass from root pathogenic fungi in the field. J Ecol 83:991–1000

Noble R, Coventry E (2005) Suppression of soil-borne plant diseases with composts: a review. Biocontrol Sci Technol 15:3–20

Pal KK, Gardener BM (2006) Biological control of plant pathogens. Plant Health Instructor. doi:10.1094/PHI-A-2006-1117-02

Perner H, Schwarz D, Bruns C, Mäder P, George E (2007) Effect of arbuscular mycorrhizal colonization and two levels of compost supply on nutrient uptake and flowering of pelargonium plants. Mycorrhiza 17:469–474

Pharand B, Carisse O, Benhamou N (2002) Cytological aspects of compost mediated induced resistance against fusarium crown and root rot in tomato. Phytopathology 92:424–438

Pozo MJ, Azcón-Aguilar C (2007) Unraveling mycorrhiza-induced resistance. Curr Opin Plant Biol 10:393–398

Pozo MJ, Dumas-Gaudot E, Barea JM (1999) β-1,3-glucanase activities in tomato roots inoculated with arbuscular mycorrhizal fungi and/or *Phytophthora parasitica* and their possible involvement in bioprotection. Plant Sci 141:149–157

Rao BRC (2005) Vermicomposting. IEC CELL-KUDCEMP, Mysore

Ristaino JB, Johnston SA (1999) Ecologically based approaches to management of phytophthora blight on bell pepper. Plant Dis 83:1080–1089

Salzer P, Corbiere H, Boller T (1999) Hydrogen peroxide accumulation in *Medicago truncatula* roots colonized by the arbuscular mycorrhiza-forming fungus *Glomus mosseae*. Planta 208: 319–325

Scheuerell S, Mahaffee W (2002) Compost tea: principles and prospects for plant disease control. Compost Sci Util 10:313–338

Shaul O, Galili S, Volpin H, Ginzberg I, Elad Y, Chet I, Kapulnik Y (1999) Mycorrhiza-induced changes in disease severity and pr protein expression in tobacco leaves. Mol Plant Microbe Interact 12:1000–1007

Siddiqui Y, Meon S, Ismail MR, Ali A (2008) *Trichoderma*-fortified compost extracts for the control of choanephora wet rot in okra production. Crop Prot 27:385–390

Simsek-Ersahin Y (2011) The use of vermicompost products to control plant disease and pests. In: Karaca A (ed) Biology of earthworms, soil biology, vol 24. Springer, Berlin, pp 191–213

Smith SE, Read DJ (2008) Mycorrhizal symbiosis. Academic, London

Sutherland ED, Papavizas GC (2008) Evaluation of oospore hyperparasites for the control of phythophthora crown rot of pepper. J Phytopathol 31:33–39

Tabin T, Arunachalam A, Shrivastava K, Arunachalam K (2009) Effect of arbuscular mycorrhizal fungi on damping-off disease in *Aquilaria agallocha* Roxb. seedlings. Trop Ecol 50:243–248

Tarkalson DD, Jolley VD, Robbins CW, Terry RE (1998) Mycorrhizal colonization and nutrient uptake of dry bean in manure and compost manure treated subsoil and untreated topsoil and subsoil. J Plant Nutr 21:1867–1878

Tecsi LI, Smith AM, Maule AJ, Leegood RC (1996) A spatial analysis of physiological changes associated with infection of cotyledons of marrow plants with cucumber mosaic virus. Plant Physiol 111:975–985

Termorshuizen AJ, van Rijn E, van der Gaag DJ, Alabouvette C, Chen Y, Lagerlöf J, Malandrakis AA, Paplomatas EJ, Rämert B, Ryckeboer J, Steinberg C, Zmora-Nahum S (2006) Suppressiveness of 18 composts against 7 pathosystems, variability in pathogen response. Soil Biol Biochem 38:2461–2477

Terras FR, Eggermont K, Kovaleva V, Raikhel NV, Osborn RW, Kester A, Rees SB, Torrekens S, Van Leuven F, Vanderleyden J (1995) Small cysteine-rich antifungal proteins from radish: their role in host defense. Plant Cell 7:573–588

Wasternack C, Hause B (2013) Jasmonates: biosynthesis, perception, signal transduction and action in plant stress response, growth and development. An update to the 2007 review in Annals of Botany. Ann Bot Lond 111:1021–1058

Wehner J, Antunes PM, Powell JR, Mazukatow J, Rillg MC (2010) Plant pathogen protection by arbuscular mycorrhizas: a role of fungal diversity? Pedobiologia 53:197–201

Yogev A, Raviv M, Hader Y, Cohen R, Wolf S, Gil L, Katan J (2010) Induced resistance as putative component of compost suppressiveness. Biol Control 54:46–51

Yu Z, Zeng GM, Chen YN, Zhang JC, Yu Y, Li H, Liu ZF, Tang L (2011) Effects of inoculation with *Phanerochaete chrysosporium* on remediation of pentachlorophenol contaminated soil waste by composting. Process Biochem 46:1285–1291

Zhang W, Dick WA, Hoitink HAJ (1996) Compost-induced systemic acquired resistance in cucumber to *Pythium* root rot and anthracnose. Phytopathology 86:1066–1070

Zhang W, Han DY, Dick WA, Davis KR, Hoitink HAJ (1998) Compost and compost water extract-induced systemic acquired resistance in cucumber and Arabidopsis. Phytopathology 88:450–455

Chapter 24
Organic Amendments and Soil Suppressiveness: Results with Vegetable and Ornamental Crops

Massimo Pugliese, Giovanna Gilardi, Angelo Garibaldi, and Maria Lodovica Gullino

24.1 Introduction

Vegetable and ornamental crops are high-value production systems, economically important worldwide, facing severe limitations in the use of chemicals and continuous innovations and adaptations to climate change and new diseases. Many new crops and varieties were introduced in the last decades, together with changes in the horticultural industry and in the food market. Potted plants are partially replacing cut aromatic and ornamental plants, while new products such as ready-to-eat processed salads are requesting improved growing techniques and new production areas. Rapid changes in the production systems are influencing disease development and their management. Together with the phase-out of methyl bromide and the regulatory constraints for the use of soil fumigants, growers are facing also new diseases as a consequence of the introduction of new cultivars and crops and the intensification of the production systems.

This review will focus on the use of organic amendments, compost in particular, and soil suppressiveness for the management of diseases of vegetable and ornamental crops.

M. Pugliese • M.L. Gullino (✉)
Agroinnova, University of Torino, Largo Paolo Braccini 2, Grugliasco, Torino, Italy

DISAFA, University of Torino, Largo Paolo Braccini 2, Grugliasco, Torino, Italy
e-mail: marialodovica.gullino@unito.it

G. Gilardi • A. Garibaldi
Agroinnova, University of Torino, Largo Paolo Braccini 2, Grugliasco, Torino, Italy

M.K. Meghvansi, A. Varma (eds.), *Organic Amendments and Soil Suppressiveness in Plant Disease Management*, Soil Biology 46,
DOI 10.1007/978-3-319-23075-7_24

24.2 Soil Suppressiveness and Organic Amendments

Soil suppressiveness is considered a complex system in which soil, microflora and plants play the main role. Suppressive soils or substrates are those in which the disease development is naturally controlled, even in the presence of a virulent pathogen, a susceptible plant host, and with good environmental conditions for the development of the disease. Both biotic and abiotic elements are considered to be important for the suppression of plant diseases, but the microbial activity is considered as a key element. All natural soils have a general disease suppression compared to the same pasteurised soil, and it is directly related to the amount of microbial activity. In cropping systems, due to soil cultivation and management, a specific suppression is concerned, where an individual or group of microorganisms, selected for their antagonistic activity, is directly responsible for disease suppression.

The application of organic amendments is a strategy commonly used in traditional agricultural systems for providing nutrients to the crops and for improving soil fertility. Several chemical and biological changes in the soil are associated with the incorporation of amendments and correlated to the control of soilborne diseases, with a good potential for their management thus reducing chemical inputs (Bailey and Lazarovits 2003; Bonanomi et al. 2007; Bonilla et al. 2012). However, a widespread use of organic amendments for disease control is still not being achieved, due to many factors such as the type of amendments, the lack of standardisation, the inconsistency in their efficacy and the complexity in their use. In most cases, the application rates effective under controlled conditions are too high for field crops; in others prior crop management practices do not allow a proper use of amendments. A gap between good results observed in laboratory and greenhouse compared to few promising results in the field is still relevant today, as mechanisms of action are largely unknown and risk avoidance is too much limited compared to other disease control strategies. Some studies indicate that the effectiveness of organic amendments is variable and, in some cases, can enhance severity of some diseases (Mazzola 2007). Organic amendments include manure, crop and food residues, compost, organic fertilisers, etc. Their use can help to control soilborne pathogens in vegetable and ornamental crops, especially when applied in conjunction with other management practices and considering a system approach. The aim is to maintain the soil's stability and resilience and to promote a self-regulation and self-balance of the agro-ecosystem. Such an approach is very interesting in the case of organic farming, where the use of amendments, in combination with mulching and other cultural practices, is effective against many soilborne pathogens.

Amendments can be applied together with other methods, like soil solarisation, anaerobic soil disinfestation and soil fumigations, to reduce the density of pathogens. When added to soil, amendments such as cow or poultry manure and cruciferous residues are subjected to microbial degradation that results in the generation of both toxic and volatile compounds directly affecting soilborne pathogens propagules or indirectly increasing microbial antagonistic activity in the soil. Positive effects of solarisation integrated with organic amendment have been observed for several soilborne fungi (*Rhizoctonia solani*, *Pythium* spp., *Fusarium oxysporum*, *Verticillium* spp., *Sclerotium rolfsii*), nematodes and also many weeds (Gamliel 2000; Mattner et al. 2008). However, released toxic compounds may result in phytotoxic effects on crops and some limitations to practical applications. In other cases they are applied and integrated with agronomical strategies, like the use of resistant grafted plants, in order to delay the root infections and provide additional times for the establishment of disease-suppressive microbial communities in the rhizosphere. The application of organic amendments can further promote the re-establishment of a more balanced and suppressive soil microflora, when combined with cultural practices like no-tillage and soil mulching. Furthermore, the development of plant disease is reduced thanks to the good root systems growing in a soil rich in organic matter and managed accurately (Chellemi 2010).

Among organic amendments, composts and *Brassica* pellets are considered those more promising. The use of *Brassica* species as green manure is considered a type of biofumigation that, involving the release of volatile compounds such as thiocyanates and nitriles, control multiple soilborne pathogens (Larkin and Griffin 2007; Handiseni et al. 2012). Studies carried out under greenhouse conditions showed improved control of *Colletotrichum coccodes* of tomato by mixing into the soil *Brassica carinata* dried pellets (Table 24.1; Gilardi et al. 2014a). The use of organic amendments, such as *B. juncea* green manure, provided a positive effect on eggplant grafted onto *Solanum torvum* partially resistant to *Verticillium dahliae* eggplant (Garibaldi et al. 2010). The combination of green manure with soil solarisation is also very effective and reduces the period of time for the soil covered with plastic films. Under simulated conditions of optimal and suboptimal temperature, it is possible to control Fusarium wilt of lettuce, rocket and basil with biofumigation, using *Brassica carinata* pellet, combined, respectively, with 7 and 14 days of soil solarisation (Garibaldi et al. 2010; Gilardi et al. 2014b). Field trials

Table 24.1 Incidence of *C. coccodes* expressed as percentage of infected roots on tomato cv. Arawak, grafted or not-grafted, in a naturally infested soil, with or without the addition of *Brassica* pellets, and the effect on yield [adapted from Gilardi et al. (2014a)]

Rootstocks	Biofumigation	Training system	% of roots affected by the attacks of *C. coccodes*		Total yield (g/plant)
–[a]	No	1 branch	35.9	C[b]	4837.7 a
Beaufort	Yes	1 branch	21.3	ab	6821.3 de
Beaufort	Yes	2 branches	23.4	abc	7051.5 de
Arnold	Yes	1 branch	14.1	a	6556.7 c
Arnold	Yes	2 branches	14.1	a	6915.4 d
–	Yes	1 branch	24.4	abc	4991.6 a

[a]Not-grafted Arawak plants served as control
[b]Means of the same column, followed by the same letter, do not significantly differ following Tukey's test ($P < 0.05$)

using Brassicaceae seed meal formulations demonstrated to be an effective tool for the management of apple tree replant diseases (Mazzola and Brown 2010). However, some studies indicate that the effectiveness of *Brassica* residues is variable and, in some cases, disease severity can be enhanced (Lu et al. 2010).

24.3 Compost

Compost is the material derived from the decomposition of organic material such as recycled plant waste, biosolids, fish or other organic materials. Composting is a process which turns biomass into compost with the use of oxygen and certain microorganisms. Increasing the opportunities to use compost in agriculture and in particular in horticulture as a (potting) substrate for plants would contribute to the recycling of wastes and to reducing the use of non-renewable fertilisers.

24.3.1 Compost Quality and Use in Agriculture

Quality aspects of compost are of most importance in order to assure a proper use in agriculture. Compost quality refers to the overall state of the material with regard to physical, chemical and biological characteristics. These parameters are indicators of the ultimate impact of the compost on the environment. In particular, the most important parameters from the point of view of environment protection standards, public health and the soil are those related to pathogens, inorganic and organic potentially toxic compounds (heavy metals) and stability. Within the EU, standards on the use and quality of compost exist in most Member States, while there is not yet a comprehensive European Community legislation. Moreover, common analysis is not enough to assess compost quality according to specific uses, such as for potting mixes, vegetable and ornamental crops, soil-less systems and suppressing plant diseases. Consequently, it is important to define and use also agronomical tests to assess compost quality, and compost suppressiveness to plant pathogens is also a key point for high-quality compost to be taken into consideration.

Farmers' willingness to use compost is strictly connected to various quality aspects of compost. Compost is commonly used as a soil amendment to increase organic matter content and fertility by improving physical, chemical and biological soil conditions (Hoitink and Fahy 1986). The nutritive value of composts and their potential to enhance soil quality makes them ideal for agriculture but may unnecessarily increase the heavy metal content of the soil when applied at high dosages (Ramos and López-Acevedo 2004). Composts have the advantage to significantly increase soil organic matter (SOM) contents, a key soil quality indicator that is on the contrary declining in many regions of the world (Bellamy et al. 2005). Additional benefits of compost addition to soil are promotion of soil biological activity, reduction of erosion losses, decrease of bulk density, improvement of structural stability, nutrient availability and plant uptake and increase of water holding capacity (Shiralipour et al. 1992; Tejada and Gonzalez 2007). Crop growth or yield is usually increased by compost amendments in the field. Compost is also interesting as a peat substitute, in particular after recent increasing concern of the environmental impact of peat extraction and the damage of peat lands and natural habitats by the horticulture industry that lead to the adoption of alternative substrates (Silva et al. 2007). Also in field horticulture, there are great market opportunities for compost, although its use on leafy vegetables is unlikely due to the potential for microbiological contamination by human pathogens, especially in the case of municipal solid waste compost (Farrell and Jones 2009).

24.3.2 *Compost Suppressiveness*

The use of compost as a peat substitute to control root pathogens in Italy was first suggested in 1988 (Garibaldi 1988). The suppressive capacity of compost against soilborne pathogens has been demonstrated in several studies, and, consequently, the use of disease-suppressive compost can reduce crop losses caused by soilborne diseases and benefit growers (Hoitink and Fahy 1986; Hoitink and Boehm 1999; Noble and Coventry 2005; Pugliese et al. 2007; Hadar 2011). Compost showed to be the most suppressive material, with more than 50 % of cases showing effective disease control, compared to other amendments such as crop residues and peat (Bonanomi et al. 2007). In field trials compost showed, in most experiments, to be suppressive with an application rate of at least 15 tons/ha. Compost prepared from cannery wastes was able to suppress anthracnose caused by *Colletotrichum coccodes* and bacterial spot caused by *Xanthomonas campestris* pv. *vesicatoria* on tomato in soil (Abbasi et al. 2002). Lower applications, like 4 tons/ha, have also been reported to be sufficient for reducing dry root rot of bean caused by *Macrophomina phaseolina* (Lodha et al. 2002). In other cases, repetition for five consecutive years of compost at 10 tons/ha was necessary to suppress damping off of cucumber and lettuce caused by *Pythium ultimum* and *Rhizoctonia solani* (Fuchs 1995). Suppressive effect of compost is generally proportional to the inclusion rate in soil, like in the case of damping off of cress by *P. ultimum* and wilt of flax by *Fusarium oxysporum* f. sp. *lini* (Fuchs 1995; Serra-Wittling et al. 1996), but not always. Application of compost suppressed root rot of chile peppers caused by *Phytophthora capsici* when applied at 48 tons/ha but at higher rates (72 tons/ha), promoted the disease, probably by increasing soil salinity (Dickerson 1999), and suppressed damping off caused by *R. solani*. However, disease promotion of root rot of bean caused by *R. solani* on soil amended with dairy manure compost has also been observed (Voland and Epstein 1994). In the case of vascular diseases caused by *Fusarium* species and root rots and damping off caused by *Pythium* species, amending soil with compost generally suppressed or did not affect the diseases (Noble and Coventry 2005). Different results can be obtained by different composts on the same pathosystem. For example, verticillium wilt of potato caused by

V. dahliae was promoted by dairy manure compost but suppressed by vegetable waste compost (Noble 2011). Soil type and conditions, like texture, pH and moisture, can also influence suppressiveness to soilborne pathogens (Bruehl 1975). Coventry et al. (2005) found that vegetable waste compost was ineffective against *Sclerotium cepivorum* in a silt soil but suppressive on the same pathogen, causal agent of *Allium* white rot, in sandy loam and peat soils.

In container experiments using soil or sand, compost derived from green wastes and/or dairy cow manure generally showed a suppressive effect on *Pythium* species and *Rhizoctonia solani*, but results did not necessarily translate into the field (Noble and Coventry 2005). Compost equally suppressed white rot of onion caused by *Sclerotium cepivorum* in pot tests and in the field (Coventry et al. 2005). In other experiments composts suppressing *Phytophthora* on citrus seedlings in pot experiments were ineffective in field trials with the same soils (Widmer et al. 1998). Compost suppressiveness also showed to be dependent on the type of wastes used for preparation. For example, bark compost suppressed *Pythium* root rot, while grape marc showed neutral or promoting effects to disease (Erhart et al. 1999), and vermicomposted animal manure suppressed infection of tomato seedlings caused by *Phytophthora nicotianae*, but not root and stem rot of cucumber caused by *Fusarium oxysporum* f. sp. *radicis-cucumerinum* (Kannangara et al. 2000; Szczech and Smolinska 2001).

Low rates of compost in growing media are generally indicated, in order to avoid negative growth effects and phytotoxicity caused by high pH and electrical conductivity and other phytotoxic compounds present in composts (Sullivan and Miller 2001). However, it is generally necessary to include at least 20 % v/v of compost in containers in order to observe a suppressive effect. Lower rates are successfully applied for few specific cases, like *Ralstonia solanacearum* and *Rhizoctonia solani* (Voland and Epstein 1994; Islam and Toyota 2004). Cases of increase of disease severity caused by composts used in containers have also been reported. A 50 % spruce bark compost increased black root rot caused by *Thielaviopsis basicola* in poinsettias and Fusarium wilt of cyclamen, compared to a peat substrate (Krebs 1990). Highly saline composts were reported to enhance *Pythium* and *Phytophthora* diseases, while composts with higher nitrogen or ammonium content enhance Fusarium wilts (Hoitink et al. 2001). Among soilborne pathogens, *Rhizoctonia solani* is considered to be the most difficult one to be controlled with compost (Scheuerell et al. 2005; Bonanomi et al. 2007). Variability also depends on the pathosystem. A compost from wood chips and horse manure stimulated disease caused by *Rhizoctonia solani* on cauliflower but suppressed it on pine (Termorshuizen et al. 2006). Success or failure of compost for disease control depends on the nature of the raw materials from which the compost was prepared, on the composting process used and on the maturity and quality of the compost (Termorshuizen et al. 2006). Composting temperatures are important also for the eradication of plant pathogens and nematodes and the sanitisation of compost (Noble and Roberts, 2004). Fortifying composts with beneficial microorganisms is one possible factor that can help in the success of compost, increasing the efficacy and reliability of disease control (De Clercq et al. 2004).

24.3.3 Mechanisms of Action of Disease Suppression

Disease suppressiveness depends on soil or substrate properties, including both abiotic and biotic parameters (Mazzola 2004; Janvier et al. 2007). Regarding the influence of physicochemical properties of suppressive soils and substrates towards diseases, soils with higher pH showed to be more suppressive towards Fusarium wilts (Höper et al. 1995) but conducive for nematodes (Rimé et al. 2003). Acidic pH reduce incidence of potato scab caused by *Streptomyces scabies* (Lacey and Wilson 2001) or enhance suppression of take-all of wheat with *Trichoderma koningii* (Duffy et al. 1997). Concerning the N content of soil, a positive association was found on the suppressiveness towards *Pseudomonas syringae* on bean and cucumber (Rotenberg et al. 2005), *Fusarium* spp. on asparagus (Hamel et al. 2005), *Gaeumanomyces graminis* var. *tritici* and *Rhizoctonia solani* on wheat (Pankhurst et al. 2002) and ectoparasitic nematodes (Rimé et al. 2003). The form of N, either NO_3 or NH_4, is also important (Janvier et al. 2007), and NH_3 or HNO_2 showed to be able to kill microsclerotia of *Verticillium dahliae* in several soils (Tenuta and Lazarovits 2004). Higher C content showed to reduce incidence of *Pythium* damping off of tomato and *Fusarium solani* f. sp. *pisi* on pea and *Fusarium culmorum* on barley but to positively affect *Thielaviopsis basicola* (Oyarzun et al. 1998; van Bruggen and Semenov 1999; Rasmussen et al. 2002).

Other physicochemical characteristics are also important, like soil texture, cations and oligoelements. Suppressiveness to Fusarium wilts of flax and Armillaria root disease on lodgepole pine was found to be reduced in sandy soils (Höper et al. 1995; Mallett and Maynard 1998). Higher clay content was associated with less *Gaeumanomyces graminis* var. *tritici* on wheat after treatment with *Trichoderma koningii* (Duffy et al. 1997). No correlation on Fusarium wilt of banana (Dominguez et al. 2001) and *Fusarium* root rot of asparagus (Hamel et al. 2005) were found between soil texture and suppressiveness instead. Higher levels of Mg and K were found to reduce incidence of fungal disease (Duffy et al. 1997; Peng et al. 1999) and suppressiveness of nematodes (Rimé et al. 2003), providing contrasting results depending on the pathogen. Al, Fe, Na or Zn contents generally reduced disease levels (Oyarzun et al. 1998). After analysing 28 physical and chemical properties of ten soils, Ownley et al. (2003) found that 16 soil properties were correlated with disease suppression and proposed a model including six key soil properties ($N\text{–}NO_3$, CEC, Fe, % silt, soil pH and zinc) to explain the variance in take-all disease of wheat treated with phenazine-producing *Pseudomonas fluorescens*. In the case of suppressive composts, higher rates of CaO, MgO, K_2O and $N\text{–}NH_4$ and a higher CEC showed to suppress *Rhizoctonia solani* more than the control soil (Pérez-Piqueres et al. 2006). A loss in the disease-suppressive effect of composts following sterilisation or heat treatments has been demonstrated in several papers (Hoitink et al. 1997; Cotxarrera et al. 2002; Reuveni et al. 2002; Chen and Nelson 2008; Pugliese et al. 2011). A declining of microbial activity after long periods of maturation and, consequently, a reduction of disease suppression have been also reported (Zmora-Nahum et al. 2008).

Also the use of water extracts from composts showed to suppress several soilborne pathogens (El-Masry et al. 2002), indicating a predominant biological component rather than chemical or physical in the suppressive effect. Compost acts as a food source and shelter for the antagonists that compete with plant pathogens or parasitise them, for those beneficials that produce antibiotics and for those microorganisms that induce resistance in plants: high-quality compost should contain disease-suppressive microorganisms (Noble and Coventry 2005; Hadar 2011).

According to Hoitink and Boehm (1999), the following biological mechanisms are involved in compost suppressiveness:

(a) Competition for nutrients by beneficial microorganisms
(b) Parasitism against pathogens by beneficial microorganisms
(c) Antibiotic production by beneficial microorganisms
(d) Activation of disease-resistance genes in plants by microorganisms (induced systemic resistance)
(e) Improved plant nutrition and vigour, leading to enhanced disease resistance

The mode of actions (a), (d) and (e) generally occurs when disease suppressiveness is not accompanied by a reduction in soilborne pathogen inoculums (Lumsden et al. 1983; Lievens et al. 2001).

Bacteria belonging to genera *Bacillus* spp., *Enterobacter* spp., *Pseudomonas* spp., *Streptomyces* spp., *Penicillium* spp. as well as several *Trichoderma* spp. isolates and other fungi have been identified as biocontrol agents (BCAs) in compost-amended substrates (Chen et al. 1987; Boehm et al. 1993; Hoitink et al. 1997; Boulter et al. 2002; Pugliese et al. 2008). The isolation from roots of eggplants grown in compost of strains of *Pseudomonas fluorescens* and of *Fusarium oxysporum* controlling Verticillium wilt and the presence of microbial species that interact at rhizosphere level and suppress the disease of plants germinated in compost indicate that suppression is related to microorganisms, rather than to the growing substrate (Malandraki et al. 2007; Chen and Nelson 2008). Microorganisms, selected from a compost suppressive against Fusarium wilts, controlled *Fusarium oxysporum* and few other soilborne diseases like *Phytophthora nicotianae* and *Rhizoctonia solani* (Table 24.2; Pugliese et al. 2008). The addition of such microorganisms and BCAs might be considered a good strategy to increase compost suppressiveness and to partially restore disease suppressiveness of steam-sterilised compost (Table 24.3; Pugliese et al. 2011).

The presence of toxic or volatile compounds in compost, sometimes correlated with changes to the physical properties of the growing medium or soil or to soil pH and conductivity, is another possible mechanism (Noble 2011), suggesting compost use as alternative to chemical fumigants for managing soilborne pathogens, also integrated with soil solarisation (Katan 2000). Immature composts release volatile compounds containing sulphur, organic acids and ammonia that may be responsible for disease suppression (Scheuerell et al. 2005; Coventry et al. 2006). Phytotoxic compounds produced by soil microorganisms after application of farmyard

Table 24.2 Activity of microorganisms isolated from a suppressive compost against soilborne pathogens [adapted from Pugliese et al. (2008)]

		% of disease control		
Microorganism	Pathogen	*F. oxysporum* f. sp. *basilici*/basil	*Phytophthora nicotianae*/tomato	*Rhizoctonia solani*/bean
K5	Yes	69 ab[a]	28 bc	13 cd
K6	Yes	56 abc	0 c	15 cd
K7	Yes	64 ab	0 c	22 bc
E12	Yes	0 c	25 bc	14 cd
E15	Yes	0 c	31 bc	1 d
E19	Yes	10 bc	0 c	49 b
B3	Yes	16 bc	73 a	11 cd
B17	Yes	10 bc	82 a	29 bc
–	Yes	0 c	0 c	0 d
–	No	100 a	100 a	100 a

[a]Tukey's HSD test ($P < 0.05$)

Table 24.3 Effect of *Trichoderma* spp. added to a substrate made by compost and peat on the suppression of *Rhizoctonia solani* on bean [adapted from Pugliese et al. (2011)]

Substrate mix (% v/v)					
Compost	Peat	Antagonist (dosage)	Pathogen	Disease suppressiveness (%)[a]	Biomass (%)[a]
40	60	*T. harzianum* T-22 (4 g l^{-1})	Yes	40 b[b]	92 a
40	60	*T. viride* TV1 (4 g l^{-1})	Yes	−59 d	60 bc
40	60	*T. harzianum* ICC012 +*T. viride* ICC080 (2 g l^{-1})	Yes	−33 cd	53 c
40	60	Inoculated control	Yes	−46 d	63 bc
40	60	Control	No	100 a	115 a
0	100	Inoculated control	Yes	0[c] c	75[c] bc
0	100	Control	No	100 a	100 a

[a]Values represent the means of at least two bioassays
[b]Different letters represent significant differences between treatments according to Tukey's HSD test ($P < 0.05$). Negative figures indicate significant disease aggravation as compared to peat control
[c]The level of disease and biomass in the peat control is, respectively, 52 % of alive plants and 27.75 g

compost were found to suppress apple replant diseases (Gur et al. 1998). Investigating a wide range of biological and chemical characteristics of composts and compost-peat mixtures in relation to plant disease suppression, Termorshuizen et al. (2006) demonstrated that only pH increase resulting from compost amendment showed a consistent relationship with the suppression of some diseases, such as *Fusarium oxysporum*, but that there is no single factor conferring suppressiveness to composts.

Several approaches were used to monitor compost suppressiveness, microbial activity and related effects after organic amendment application to soil and substrates, including analysis of phospholipid fatty acids (PLFAs), enzymatic activities and DNA-based techniques (Noble and Coventry 2005). Overall, enzymatic and microbiological parameters, rather than chemical ones, are considered much more informative for predicting suppressiveness (Bonanomi et al. 2010).

Hydrolysis of fluorescein diacetate (FDA) and dehydrogenase activity have been suggested as indicators for damping off and root rot diseases (Chen et al. 1988; Scheuerell et al. 2005; Giotis et al. 2009), but the technique has not been found to be consistently reliable for predicting compost suppressiveness in other pathosystems (Erhart et al. 1999; Termorshuizen et al. 2006; Rotenberg et al. 2007). Factors like microbial community composition, decomposition time, amendment quality and pathosystem tested may interact with each other and make it difficult to identify specific indicators for disease suppression. According to Bonanomi et al. (2010), the response of pathogen populations is a reliable feature only for pathogens with a limited saprophytic ability (e.g. *Thielaviopsis basicola* and *Verticillium dahliae*) and for some organic matter types (e.g. crop residues and organic wastes with C/N lower than 15). The most useful parameters to predict disease suppression were FDA activity, substrate respiration, microbial biomass, total culturable bacteria, fluorescent pseudomonads and *Trichoderma* populations. Specific indicators have been indicated only for some pathogens. For instance, suppressiveness in peat substrate amended with compost may be predicted by total extractable carbon, *O*-aryl C and C/N ratio for *Pythium ultimum*; by alkyl/*O*-alkyl ratio, *N*-acetyl-glucosaminidase and chitobiosidase enzymatic activities for *Rhizoctonia solani*; and by electrical conductivity for *Sclerotinia minor* (Pane et al. 2011). DNA-based techniques such as analysis of terminal restriction fragment length polymorphisms (T-RFLPs) and denaturing gradient gel electrophoresis (DGGE) showed correlations between microbial diversity of compost-amended substrates and their suppressiveness to bean root rot, cucumber root rot caused by *Pythium aphanidermatum* and southern blight caused by *Sclerotium rolfsii* (Postma et al. 2005; Liu et al. 2007; Rotenberg et al. 2007).

24.4 Conclusions

Control of soilborne diseases with organic amendments must be viewed not as a stand-alone management approach but rather part of a system approach where several aspects of the impact of crop production practices on resident soil microbial communities are addressed. Organic amendments like *Brassica* manure are of particular interest for field crops, combined with soil solarisation, including fruit tree replant diseases, but not for other *Brassica* crops and vegetables like cabbage, cauliflower, broccoli, radish and wild rocket. Compost suppressiveness can be used both for potted plants and for field crops, combined with other management strategies like soil solarisation and grafting. Induced resistance by compost has also been observed and consequently used for the control of other pathogens or pests. However, quality standards are required in order to avoid phytotoxicity effects on plants and reduce the variability in the control of diseases. New approaches to monitor how microbial community structures in soil change as a result of organic amendment may lead to a better understanding of which changes in microbial communities are responsible for conferring the disease-suppressive effects. This may eventually lead to improved and more reliable disease control resulting from organic amendment of soil, sand or peat, both in container crops in greenhouses and in the field.

Acknowledgements Work carried out within the EU Project "Reducing mineral fertilizers and chemicals use in agriculture by recycling treated organic waste as compost and biochar products" (Collaborative project REFERTIL, FP7/2007–2013, grant agreement n° 289785).

References

Abbasi PA, Al-Dhamani J, Shain F, Hoitink HAJ, Miller SA (2002) Effect of compost amendments on disease severity and yield of tomato in conventional and organic production systems. Plant Dis 86:156–161

Bailey KL, Lazarovits G (2003) Suppressing soil-borne diseases with residue management and organic amendments. Soil Tillage Res 72:169–180

Bellamy PH, Loveland PJ, Bradley RI, Lark RM, Kirk GJD (2005) Carbon losses from all soils across England and Wales 1978–2003. Nature 437:245–248

Boehm M, Madden LV, Hoitink HAJ (1993) Effect of organic matter decomposition level on bacterial species diversity and composition in relationship to pythium damping-off severity. Appl Environ Microbiol 59:4171–4179

Bonanomi G, Antignani V, Pane C, Scala F (2007) Suppression of soilborne fungal diseases with organic amendments. J Plant Pathol 89:311–324

Bonanomi G, Antignani V, Capodilupo M, Scala F (2010) Identifying the characteristics of organic soil amendments that suppress soilborne plant diseases. Soil Biol Biochem 42:136–144

Bonilla N, Gutierrez-Barranquero JA, de Vicente A, Cazorla FM (2012) Enhancing soil quality and plant health through suppressive organic amendments. Diversity 4:475–491

Boulter JI, Trevors JT, Boland GJ (2002) Microbial studies of compost: bacterial identification, and their potential for turfgrass pathogen suppression. World J Microbiol Biotechnol 18:661–671

Bruehl GW (1975) Biology and control of soil-borne plant pathogens. The American Phytopathological Society, St. Paul, MN, p 216

Chellemi DO (2010) Back to the future: total system management (organic, sustainable). In: Gisi U, Chet I, Gullino ML (eds) Recent developments in management of plant diseases. Plant pathology in the 21st century. Springer Science, Dordrecht, pp 285–292

Chen MH, Nelson EB (2008) Seed-colonizing microbes from municipal biosolids compost suppress *Pythium ultimum* damping-off on different plant species. Phytopathology 98:1012–1018

Chen W, Hoitink HAJ, Schmitthenner AF (1987) Factors affecting suppression of Pythium damping-off in container media amended with composts. Phytopathology 77:755–760

Chen W, Hoitink HAJ, Schmitthenner AF, Tuovinen OH (1988) The role of microbial activity in suppression of damping-off caused by *Pythium ultimum*. Phytopathology 78:314–322

Cotxarrera L, Trillas-Gay MI, Steinberg C, Alabouvette C (2002) Use of sewage sludge compost and *Trichoderma asperellum* isolates to suppress Fusarium wilt of tomato. Soil Biol Biochem 34:467–476

Coventry E, Noble R, Mead A, Whipps JM (2005) Suppression of Allium white rot (*Sclerotium cepivorum*) in different soils using vegetable wastes. Eur J Plant Pathol 111:101–112

Coventry E, Noble R, Mead A, Marin FR, Perez JA, Whipps JM (2006) Allium white rot suppression with composts and *Trichoderma viride* in relation to sclerotia viability. Phytopathology 96:1009–1020

De Clercq D, Vandesteen L, Coosemans J, Ryckeboer J (2004) Use of compost as suppressor of plant diseases. In: Lens P, Hamelers B, Hoitink H, Bidlingmaier W (eds) Resource recovery and reuse in organic solid waste management. IWA Publishing, London, pp 317–337

Dickerson GW (1999) Damping-off and root rot. BioCycle 40:62–63

Dominguez J, Negrin MA, Rodriguez CM (2001) Aggregate water-stability, particle-size and soil solution properties in conducive and suppressive soils to fusarium wilt of banana from Canary Islands (Spain). Soil Biol Biochem 33:449–455

Duffy BK, Ownley BH, Weller DM (1997) Soil chemical and physical properties associated with suppression of take-all of wheat by *Trichoderma koningii*. Phytopathology 87:1118–1124

El-Masry MH, Khalil AI, Hassouna MS, Ibrahim HAH (2002) In situ and in vivo suppressive effect of agricultural composts and their water extracts on some phytopathogenic fungi. World J Microbiol Biotechnol 18:551–558

Erhart E, Burian K, Hartl W, Stich K (1999) Suppression of Pythium ultimum by biowaste composts in relation to compost microbial biomass, activity and content of phenolic compounds. J Phytopathol 147:299–305

Farrell M, Jones DL (2009) Critical evaluation of municipal solid waste composting and potential compost markets. Bioresour Technol 100:4301–4310

Fuchs JG (1995) Practical use of quality compost for plant health and vitality improvement. In: Insam H, Riddech N, Klammer S (eds) Microbiology of composting. Springer, Berlin, pp 435–444

Gamliel A (2000) Soil amendments: a non chemical approach to the management of soilborne pest. Acta Hortic 532:39–47

Garibaldi A (1988) Research on substrates suppressive to *Fusarium oxysporum* and *Rhizoctonia solani*. Acta Hortic 221:271–277

Garibaldi A, Gilardi G, Clematis F, Gullino ML, Lazzeri L, Malaguti L (2010) Effect of green *Brassica* manure and *Brassica* defatted seed meals in combination with grafting and soil solarization against Verticillium wilt of eggplant and Fusarium wilt of lettuce and basil. Acta Hortic 883:295–302

Gilardi G, Colla P, Pugliese M, Baudino M, Gullino ML, Garibaldi A (2014a) Control of *Colletotrichum coccodes* on tomato by grafting and soil amendments. J Phytopathol 162:116–123

Gilardi G, Demarchi S, Gullino ML, Garibaldi A (2014b) Effect of simulated soil solarization and organic amendments on *Fusarium* wilt of rocket and basil under controlled conditions. J Phytopathol 162. doi:10.1111/jph.12223

Giotis C, Markelou E, Theodoropoulou A, Toufexi E, Hodson R, Shotton P, Shiel R, Cooper J, Leifert C (2009) Effect of soil amendments and biological control agents (BCAs) on soil-borne root diseases caused by *Pyrenochaeta lycopersici* and *Verticillium albo-atrum* in organic greenhouse tomato production systems. Eur J Plant Pathol 123:387–400

Gur A, Luzzati J, Katan J (1998) Alternatives for soil fumigation in combating apple replant disease. Acta Hortic 477:107–113

Hadar Y (2011) Suppressive compost: when plant pathology met microbial ecology. Phytoparasitica 39:311–314

Hamel C, Vujanovic V, Jeannotte R, Nakano-Hylander A, St-Arnaud M (2005) Negative feedback on a perennial crop: Fusarium crown and root rot of asparagus is related to changes in soil microbial community structure. Plant Soil 268:75–87

Handiseni M, Brown J, Zemetra R, Mazzola M (2012) Use of Brassicaceous seed meals to improve seedling emergence of tomato and pepper in *Pythium ultimum* infested soils. Arch Phytopathol Plant Prot 45:1204–1209

Hoitink HAJ, Boehm MJ (1999) Biocontrol within the context of soil microbial communities: a soil-dependent phenomenon. Annu Rev Phytopathol 37:427–446

Hoitink HAJ, Fahy PC (1986) Basis for the control of soilborne plant pathogens with composts. Annu Rev Phytopathol 24:93–114

Hoitink HAJ, Stone AG, Han DY (1997) Suppression of plant disease by composts. HortSci 32:184–187

Hoitink HAJ, Krause MS, Han DY (2001) Spectrum and mechanisms of plant disease control with composts. In: Stofella PJ, Kahn BA (eds) Compost utilization in horticultural cropping systems. Lewis Publishers, Boca Raton, pp 263–274

Höper H, Steinberg C, Alabouvette C (1995) Involvement of clay type and pH in the mechanisms of soil suppressiveness to Fusarium wilt of flax. Soil Biol Biochem 27:955–967

Islam TMD, Toyota K (2004) Suppression of bacterial wilt of tomato by *Ralstonia solanacearum* by incorporation of composts in soil and possible mechanisms. Microbes Environ 19:53–60

Janvier C, Villeneuve F, Alabouvette C, Edel-Hermann V, Mateille T, Steinberg C (2007) Soil health through soil disease suppression: which strategy from descriptors to indicators? Soil Biol Biochem 39:1–23

Kannangara T, Utkhede RS, Paul JW, Punja ZW (2000) Effects of mesophilic and thermophilic composts on suppression of Fusarium root and stem rot of greenhouse cucumber. Can J Microbiol 46:1021–1028

Katan J (2000) Physical and cultural methods for the management of soil-borne pathogens. Crop Prot 19:725–731

Krebs E (1990) Rinden-Kultursubstrate und Schadpilze. Deutscher Gartenbau 44:2874–2877

Lacey MJ, Wilson CR (2001) Relationship of common scab incidence of potatoes grown in Tasmanian ferrosol soils with pH, exchangeable cations and other chemical properties of those soils. J Phytopathol 149:679–683

Larkin RP, Griffin TS (2007) Control of soilborne potato diseases using Brassica green manures. Crop Prot 26:1067–1077

Lievens B, Vaes K, Coosemans J, Ryckeboer J (2001) Systemic resistance induced in cucumber against Pythium root rot by source separated household waste and yard trimmings composts. Compost Sci Util 9:221–229

Liu B, Gumpertz ML, Hu S, Beagle Ristaino J (2007) Long-term effects of organic and synthetic soil fertility amendments on soil microbial communities and the development of southern blight. Soil Biol Biochem 39:2302–2316

Lodha S, Sharma SK, Aggarwal RK (2002) Inactivation of *Macrophomina phaseolina* propagules during composting and effect of composts on dry root rot severity and on seed yield of clusterbean. Eur J Plant Pathol 108:253–261

Lu P, Gilardi G, Gullino ML, Garibaldi A (2010) Biofumigation with *Brassica* plants and its effect on the inoculum potential of *Fusarium* yellows of Brassica crops. Eur J Plant Pathol 126:387–402

Lumsden RD, Lewis JA, Milner PD (1983) Effect of composted sewage sludge on several soilborne pathogens and diseases. Phytopathology 73:1543–1548

Malandraki I, Tjamos SE, Pantelides IS, Paplomatas EJ (2007) Thermal inactivation of compost suppressiveness implicates possible biological factors in disease management. Biol Control 44:180–187

Mallett KI, Maynard DG (1998) Armillaria root disease, stand characteristics, and soil properties in young lodgepole pine. Forest Ecol Manag 105:37–44

Mattner SW, Porter IJ, Gounder RK, Shanks AL, Wren DJ, Alle D (2008) Factors that impact on the ability of biofumigants to suppress fungal pathogens and weed of strawberry. Crop Prot 27:1165–1173

Mazzola M (2004) Assessment and management of soil microbial community structure for disease suppression. Annu Rev Phytopathol 42:35–59
Mazzola M (2007) Manipulation of rhizosphere bacterial communities to induce suppressive soils. J Nematol 39:213–220
Mazzola M, Brown J (2010) Efficacy of brassicaceous seed meal formulations for the control of apple replant disease in organic and conventional orchard production systems. Plant Dis 94:835–842
Noble R (2011) Risks and benefits of soil amendment with composts in relation to plant pathogens. Australas Plant Pathol 40:157–167
Noble R, Coventry E (2005) Suppression of soil-borne plant diseases with composts: a review. Biocontrol Sci Technol 15:3–20
Noble R, Roberts SJ (2004) Eradication of plant pathogens and nematodes during composting: a review. Plant Pathol 53:548–568
Ownley BH, Duffy BK, Weller DM (2003) Identification and manipulation of soil properties to improve the biological control performance of phenazine-producing *Pseudomonas fluorescens*. Appl Environ Microbiol 69:3333–3343
Oyarzun PJ, Gerlagh M, Zadoks JC (1998) Factors associated with soil receptivity to some fungal root rot pathogens of peas. Appl Soil Ecol 10:151–169
Pane C, Spaccini R, Piccolo A, Scala F, Bonanomi G (2011) Compost amendments enhance peat suppressiveness to *Pythium ultimum*, *Rhizoctonia solani* and *Sclerotinia minor*. Biol Control 56:115–124
Pankhurst CE, McDonald HJ, Hawke BG, Kirkby CA (2002) Effect of tillage and stubble management on chemical and microbiological properties and the development of suppression towards cereal root disease in soils from two sites in NSW, Australia. Soil Biol Biochem 34:833–840
Peng HX, Sivasithamparam K, Turner DW (1999) Chlamydospore germination and fusarium wilt of banana plantlets in suppressive and conducive soils are affected by physical and chemical factors. Soil Biol Biochem 31:1363–1374
Pérez-Piqueres A, Edel-Hermann V, Alabouvette C, Steinberg C (2006) Response of soil microbial communities to compost amendments. Soil Biol Biochem 38:460–470
Postma J, Bart PJ, Geraats R, van Elsas JD (2005) Characterization of the microbial community involved in the suppression of *Pythium aphanidermatum* in cucumber grown on rockwool. Phytopathology 95:808–818
Pugliese M, Garibaldi A, Gullino ML (2007) The use of compost in horticulture for controlling soil-borne pathogens. Phytopathology 97:s95
Pugliese M, Liu BP, Gullino ML, Garibaldi A (2008) Selection of antagonists from compost to control soil-borne pathogens. J Plant Dis Protect 115:220–228
Pugliese M, Liu BP, Gullino ML, Garibaldi A (2011) Microbial enrichment of compost with biological control agents to enhance suppressiveness to four soil-borne diseases in greenhouse. J Plant Dis Protect 118:45–50
Ramos MC, López-Acevedo M (2004) Zinc levels in vineyard soils from the alt penedès-anoia region (ne spain) after compost application. Adv Environ Res 8:687–696
Rasmussen PH, Knudsen IMB, Elmholt S, Jensen DF (2002) Relationship between soil cellulolytic activity and suppression of seedling blight of barley in arable soils. Appl Soil Ecol 19:91–96
Reuveni R, Raviv M, Krasnovsky A, Freiman L, Medina S, Bar A, Orion D (2002) Compost induces protection against *Fusarium oxysporum* in sweet basil. Crop Prot 21:583–587
Rimé D, Nazaret S, Gourbiere F, Cadet P, Moenne-Loccoz Y (2003) Comparison of sandy soils suppressive or conducive to ectoparasitic nematode damage on sugarcane. Phytopathology 93:1437–1444
Rotenberg D, Cooperband L, Stone A (2005) Dynamic relationships between soil properties and foliar disease as affected by annual additions of organic amendment to a sandy-soil vegetable production system. Soil Biol Biochem 37:1343–1357

Rotenberg D, Jimenez Wells A, Chapman EJ, Whitfield AE, Goodman RM, Cooperbrand LR (2007) Soil properties associated with organic matter-mediated suppression of bean root rot in field soil amended with fresh and composted paper mill residuals. Soil Biol Biochem 39:2936–2948

Scheuerell SJ, Sullivan DM, Mahaffee WF (2005) Suppression of seedling damping-off caused by *Pythium ultimum*, *P. irregulare*, and *Rhizoctonia solani* in container media amended with a diverse range of Pacific Northwest compost sources. Phytopathology 95:306–315

Serra-Wittling C, Houot S, Alabouvette C (1996) Increased soil suppressiveness to Fusarium wilt of flax after addition of municipal solid waste compost. Soil Biol Biochem 28:1207–1214

Shiralipour A, McConnell DB, Smith WH (1992) Uses and benefits of MSW compost: a review and an assessment. Biomass Bioenerg 3:267–279

Silva MTB, Menduina AM, Seijo YC, Viquiera FDF (2007) Assessment of municipal solid waste compost quality using standardized methods before preparation of plant growth media. Waste Manag Res 25:99–108

Sullivan DM, Miller RO (2001) Compost quality attributes, measurements, and variability. In: Stofella PJ, Kahn BA (eds) Compost utilization in horticultural cropping systems. Lewis Publishers, Boca Raton, pp 95–120

Szczech M, Smolinska U (2001) Comparison of suppressiveness of vermicomposts produced from animal manures and sewage sludge against *Phytophthora nicotianae* Breda de Haan var. *nicotianae*. J Phytopathol 149:77–82

Tejada M, Gonzalez JL (2007) Influence of organic amendments on soil structure and soil loss under simulated rain. Soil Tillage Res 93:197–205

Tenuta M, Lazarovits G (2004) Soil properties associated with the variable effectiveness of meat and bone meal to kill microsclerotia of *Verticillium dahliae*. Appl Soil Ecol 25:219–236

Termorshuizen AJ, van Rijn E, van der Gaag DJ, Alabouvette C, Chen Y, Lagerlöf J, Malandrakis AA, Paplomatas EJ, Rämert B, Ryckeboer J, Steinberg C, Zmora-Nahum S (2006) Suppressiveness of 18 composts against 7 pathosystems: variability in pathogen response. Soil Biol Biochem 38:2461–2477

van Bruggen AHC, Semenov AM (1999) A new approach to the search for indicators of root disease suppression. Australas Plant Pathol 28:4–10

Voland RP, Epstein AH (1994) Development of suppressiveness to diseases caused by *Rhizoctonia solani* in soils amended with composted and noncomposted manure. Plant Dis 78:461–466

Widmer TL, Graham JH, Mitchell DJ (1998) Composted municipal waste reduces infection of citrus seedlings by *Phytophthora nicotianae*. Plant Dis 82:683–688

Zmora-Nahum S, Danon M, Hadar Y, Chen Y (2008) Compost curing reduces suppression of plant diseases. Compost Sci Util 16:250–256

Chapter 25
Effect of NaCl on Tolerance to *Fusarium* Crown Rot and Symbiosis-Specific Changes in Free Amino Acids in Mycorrhizal Asparagus

Yoh-ichi Matsubara, Jia Liu, and Tomohiro Okada

25.1 Introduction

Asparagus decline is a serious and increasing threat in asparagus-producing regions all over the world (Reid et al. 2001; Hamel et al. 2005; Wong and Jeffries 2006; Knaflewski et al. 2008). It is supposed to be caused by the contribution of both biotic (disease) factors (Wong and Jeffries 2006; Knaflewski et al. 2008) and abiotic (allelopathy, etc.) factors (Yong 1984; Miller et al. 1991; Lake et al. 1993). As biotic factors, the most common phenomenon is *Fusarium* crown and root rot, caused by *Fusarium proliferatum* (Fp), *F. oxysporum* f. sp. *asparagi* (Foa), *F. redolens*, etc. (Reid et al. 2002; Wong and Jeffries 2006; Knaflewski et al. 2008). In Japan, Nahiyan et al. (2011) demonstrated that Fp and Foa are dominant *Fusarium* species in asparagus decline fields by PCR-SSCP analysis. However, the diseases are still difficult to control because no resistant cultivar or disinfesting method has been developed. On the other hand, biological control of *Fusarium* disease was tried by inoculation with nonpathogenic isolates of the *Fusarium* species (Blok et al. 1997; Elmer 2004). However, the method is not enough to control and has no growth-promoting effect.

Arbuscular mycorrhizal fungi (AMF) are ubiquitous soil inhabitants and form a symbiotic relationship with roots of most of the terrestrial plants. AMF promote host plant growth by enhancing phosphorus uptake through symbiosis (Marschner and Dell 1994) and hence an alternative to high inputs of fertilizers and pesticides in sustainable crop production systems. Previously, the author reported tolerance to *Fusarium* root rot caused by Foa in mycorrhizal asparagus (cv. Mary Washington

Y. Matsubara (✉)
Faculty of Applied Biological Sciences, Gifu University, Gifu 501-1193, Japan
e-mail: ymatsu@gifu-u.ac.jp

J. Liu • T. Okada
The United Graduate School of Agricultural Science, Gifu University, Gifu 501-1193, Japan

M.K. Meghvansi, A. Varma (eds.), *Organic Amendments and Soil Suppressiveness in Plant Disease Management*, Soil Biology 46,
DOI 10.1007/978-3-319-23075-7_25

500 W) plants (Matsubara et al. 2003). On the other hand, chemical control of *Fusarium* disease was also tried by the treatment of sodium chloride (Elmer 1992; Reid et al. 2001); however, many points remain unclear about the mechanisms of disease reduction in sodium chloride (NaCl) and AMF-treated asparagus plants.

As for the changes in amino acid constituents related to disease tolerance in mycorrhizal plants, Baltruschat and Schonbeck (1975) demonstrated that the propagation of *Thielaviopsis basicola* was inhibited by the increase of arginine and citrulline in mycorrhizal tobacco plants. In addition, some reports mentioned that the free amino acid level in plants changes through AMF colonization. Sood (2003) and Fattah and Mohamedin (2000) reported that increases in the contents of free amino acids occurred in mycorrhizal tomato and sorghum plants, respectively. On the other hand, Rolin et al. (2001) reported that AMF colonization decreased total amino acid levels in mycorrhizal leek plants. However, it has been unclear how the contents of free amino acid change through AMF and NaCl in asparagus plants and how the changes are associated with disease tolerance.

In this study, suppression of *Fusarium* crown rot and the changes in free amino acid contents in mycorrhizal asparagus plants with NaCl treatment were investigated in order to clarify the mechanisms of disease tolerance.

25.2 Protocol

25.2.1 Inoculation of AMF

Seeds of asparagus (*Asparagus officinalis* L., 'Welcome') were inoculated with two AMF species [*Glomus* sp. R10 (Gr) and *Gigaspora margarita* (GM), supplied by Idemitsu Kosan Co., Ltd. for Gr and Central Glass Co., Ltd. for GM] according to Matsubara et al. (2003). The inoculated plants (AMF+) and the non-inoculated control plants (AMF−) were raised in autoclaved commercial soil and administered by mixed fertilizer (N:P:K = 13:11:13, 0.5 g per plant). Forty plants per plot with three replications were irrigated as regularly and grown in a greenhouse.

25.2.2 Treatment of Sodium Chloride

Treatment of sodium chloride (NaCl) was carried out according to the method of Reid et al. (2001). From 8 weeks after AMF inoculation, NaCl (50, 100 mM, w/v) was added (10 ml/plant, NaCl+) to bed soil once a week until Foa inoculation (16 weeks after AMF inoculation). Non-NaCl-added (NaCl−) plants were treated with distilled water.

25.2.3 *Inoculation of Fusarium* proliferatum

Two isolates of *F. proliferatum* (Fp:N1-31, SUF1207) were grown on potato-dextrose agar media. The conidia were harvested in potato sucrose liquid media and incubated at 25 °C in the dark for 7 days. The conidial suspension was sieved and the concentrations adjusted to 10^6 conidia per ml. Sixteen weeks after AMF inoculation, each plant was inoculated by 50 ml of the conidial suspension onto the roots.

25.2.4 *Estimation of Symptoms of Fusarium Crown Rot*

Ten weeks after inoculation of Fp, the symptoms of *Fusarium* crown rot were rated to 6 degrees as follows: 0, no symptom; frequency of diseased storage roots in a root system—1, less than 20 %; 2, 20–40 %; 3, 40–60 %; 4, 60–80 %; 5, 80–100 %.

25.2.5 *Evaluation of AMF Colonization Level*

Sixteen weeks after AMF inoculation, roots of asparagus were preserved with 70 % ethanol and stained according to Phillips and Hayman (1970). The rate of AMF colonization in 1-cm segments of lateral roots (abbreviated RFCSL) was calculated. Hence, RFCSL expresses the percentage of 1-cm AMF-colonized segments to the total 1-cm segments of all lateral roots; the number of total segments was approx. 30 per plant. Average colonization was calculated from the values of five plants.

25.2.6 *Determination of Free Amino Acids in Plants*

Sixteen weeks after AMF inoculation, plants were sampled and partitioned into shoots and storage roots from ten plants, and all samplers were frozen in liquid nitrogen. The samples for free amino acid analysis were collected from ten plants as follows: shoots (approx. 1 cm long from the base) and storage roots (approx. 1 cm from the crown). Free amino acids in each 200-mg weighed samples were extracted at 0 °C in 2 mL 0.2 N perchloric acid solution mixed with 1 mL 0.25 μM D,L-norleucine as an internal standard. Extracts were centrifuged at 14,000 rpm at 4 °C, and pH was adjusted to 4.0 with $KHCO_3$. Then, the extracts (20 μL in each time) were filtrated by a GL Chromatodisc (GL science Co., Ltd., Tokyo, Japan). Free amino acid concentrations (41 constituents) were measured using an automatic amino acid analyzer (JLC-500, JEOL Co., Ltd., Tokyo, Japan) using ninhydrin.

25.2.7 Statistical Analysis

Mean values were separated by *t*-test for dry weight and free amino acid contents at $P \leq 0.05$. All analyses were performed using statistical analysis software (SSRI, Tokyo, Japan).

25.3 Salient Observations

Sixteen weeks after AMF inoculation, AMF+ (Gr and GM) plants had higher dry weight of shoots than AMF− plants in NaCl− plots, regardless of the fungal spices (Fig. 25.1). In NaCl+ plots, dry weight of shoots in AMF+NaCl+ and roots in Gr +NaCl 50 increased compared to AMF−NaCl+; no significant difference occurred in dry weight of shoots and roots by NaCl treatment in control plants.

AMF colonization was confirmed in all the inoculated plants, and no colonization occurred in AMF− plants. The colonization levels reached more than 60 % in all the plots 16 weeks after AMF inoculation; no difference appeared between Gr and GM (Fig. 25.2).

As for disease incidence, AMF− plants showed 100 % incidence and highest severity in the two Fp isolates (Fig. 25.3). However, Gr+ and GM+ plants with or

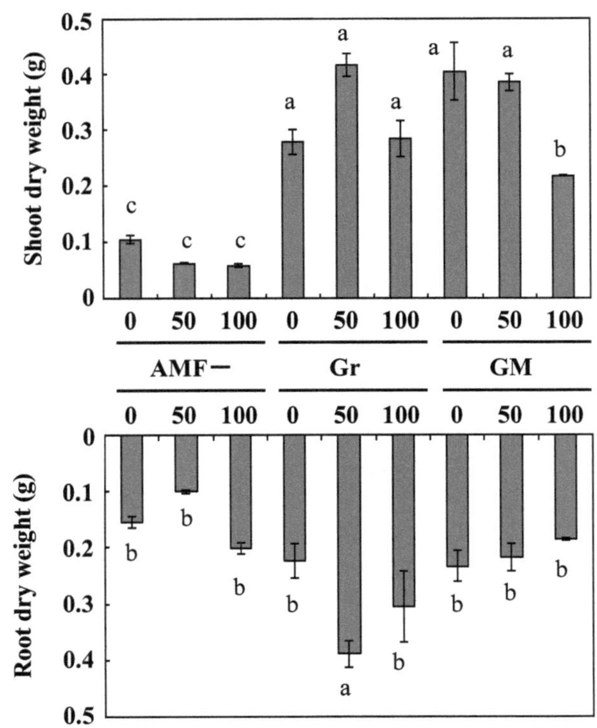

Fig. 25.1 Influence of NaCl and AMF on dry weight of asparagus plants 16 weeks after AMF inoculation. 0, water; 50, 50 mM NaCl; 100, 100 mM NaCl; AMF−, non-AMF inoculated; Gr, *Glomus* sp. R10; GM, *Gigaspora margarita*. *Bars* represent standard errors ($n = 10$). Column denoted by different *letters* indicate significant difference according to Tukey's multiple range test ($P \leq 0.05$)

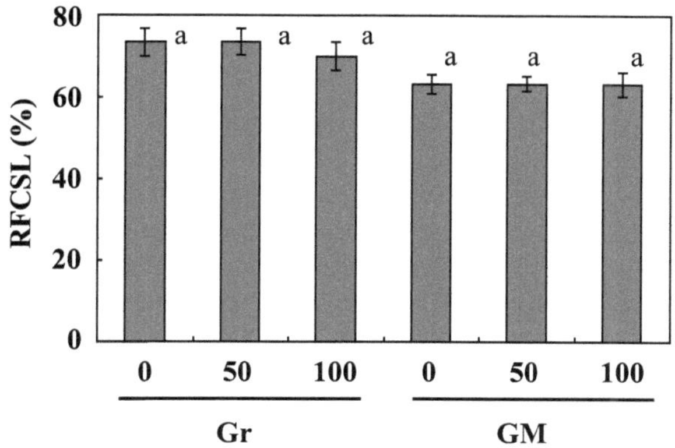

Fig. 25.2 Influence of NaCl on AMF colonization level (RFCSL) 16 weeks after AMF inoculation. 0, 50, 100, Gr, GM, see Fig. 25.1. *Bars* represent standard errors ($n = 5$). Column denoted by different *letters* indicate significant difference according to Tukey's multiple range test ($P \leq 0.05$)

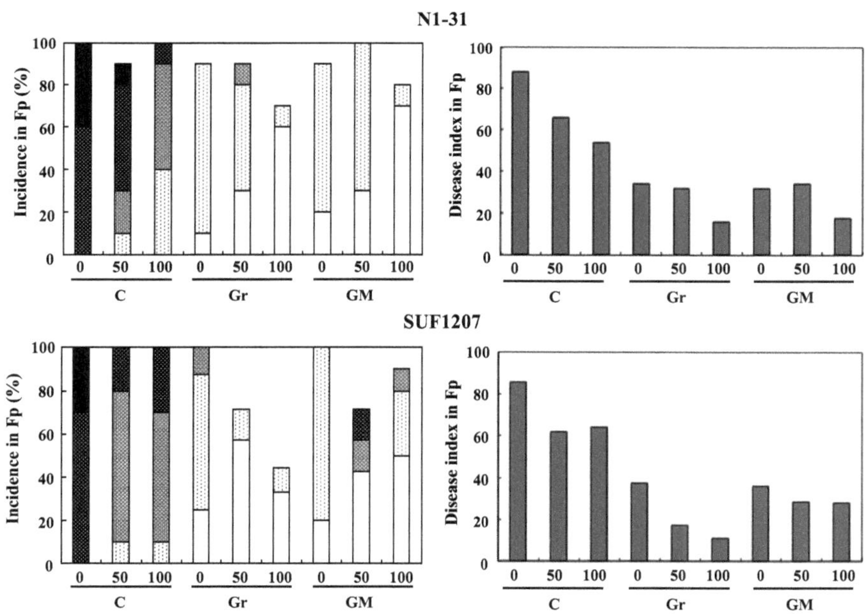

Fig. 25.3 Disease incidence and index in NaCl and mycorrhizal asparagus plants 10 weeks after Fp (*Fusarium proliferatum*; N1-31, SUF1207) inoculation. 0, water; 50, 50 mM NaCl; 100, 100 mM NaCl. C, control; Gr, *Glomus* sp. R10; GM, *Gigaspora margarita*. Ratio of diseased storage roots. *Square* 0–20; *dotted square* 20–40; *checked square* 40–60; *dark checked square* 60–80; *filled square* 80–100 (%)

without NaCl showed lower incidence and severity than AMF− plants in the two isolates. In this case, synergistic effect in disease suppression occurred in some of the AMF plants treated with NaCl.

Sixteen weeks after Gr inoculation, the increase in 11 constituents of amino acids in shoots and 18 in roots occurred in AMF plants, and in addition, maximal increase in six constituents of shoots (asparagine, alanine, GABA, threonine, phenylalanine, lysine) and four of roots (GABA, threonine, citrulline, glycine) occurred in AMF+NaCl plants compared to control (Figs. 25.4 and 25.5). +NaCl plots without AMF showed an increase in 11 constituents in shoots and eight in roots.

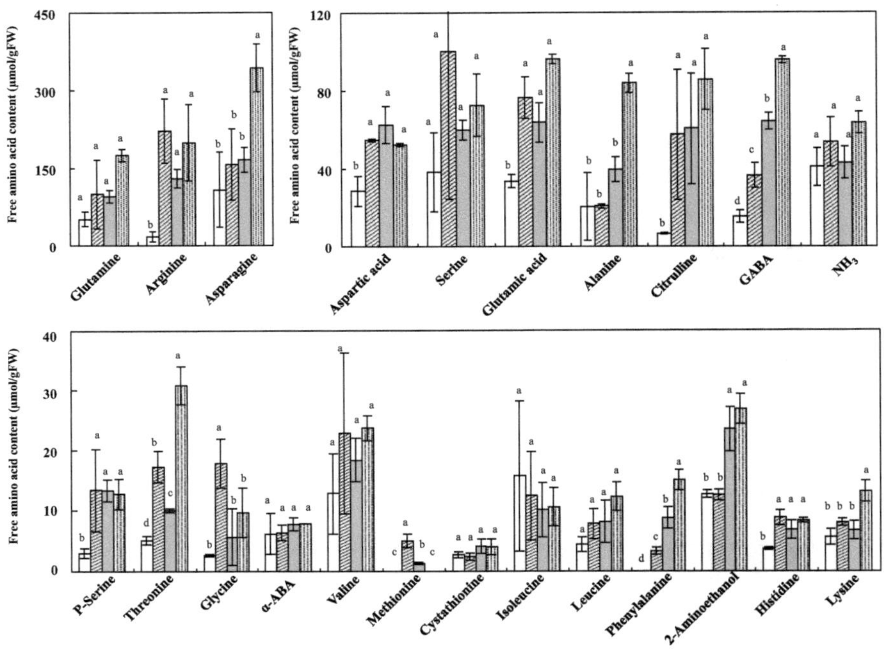

Fig. 25.4 Influence of NaCl and AMF on free amino acid content in asparagus shoots 16 weeks after AMF inoculation. *Bars* represent SE ($n = 10$). Column denoted by different *letters* indicate significant difference according to Tukey's multiple test ($P \leqq 0.05$). *Square* control; *stripped square* 100 mM NaCl; *shaded square Glomus* sp. R10 (Gr); *dotted square* Gr + 100 mM NaCl

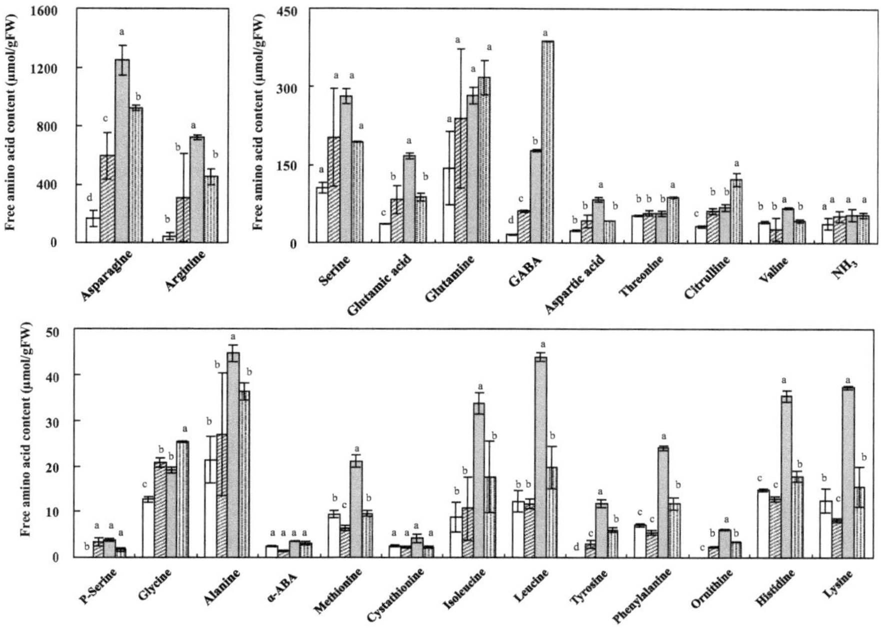

Fig. 25.5 Influence of NaCl and AMF on free amino acid content in asparagus roots 16 weeks after AMF inoculation. *Bars* represent SE ($n = 10$). Column denoted by different *letters* indicate significant difference according to Tukey's multiple test ($P \leqq 0.05$). *Square* control; *stripped square* 100 mM NaCl; *shaded square Glomus* sp. R10 (Gr); *dotted square* Gr + 100 mM NaCl

25.4 Conclusion

In this study, dry weight of shoots increased in AMF+NaCl− plants compared to AMF−NaCl− plants. In addition, AMF+NaCl+ plants showed higher dry weight of shoots than AMF−NaCl+ plants. From these findings, growth promotion effect through symbiosis appeared in mycorrhizal asparagus plants in both NaCl+ and NaCl− condition. Porras-Soriano et al. (2009) reported that dry weight of shoots and roots increased in mycorrhizal olive plants compared to control plants under NaCl treatment. They also mentioned that no significant difference occurred in AMF colonization levels by NaCl treatment, which supposed that reduction of salt stress appeared in mycorrhizal plants. Our results partially agreed with the findings and suggest that AMF could induce growth-promoting effect in host plants under NaCl treatment. In addition, it is expected that AMF might induce alleviation of salt stress to horticultural plants. Recently, salt stress is used for increasing functional constituencies, such as sugar and amino acids; however, salt stress resulted in growth reduction and the decrease in yield and fruit size in tomato (Kitano et al. 2008). In our results, growth-promoting effect under NaCl treatment appeared in mycorrhizal asparagus plants, and several amino acid contents increased in mycorrhizal plots. From these facts, AMF might lead the potential to enhance plant growth and increase functional constituents in host plants under NaCl treatment.

Matsubara et al. (2003) reported that mycorrhizal asparagus ('MW500W') plants showed lower incidence and severity of *Fusarium* root rot compared to control. In addition, NaCl-treated asparagus plants had lower severity of *Fusarium* root rot symptom than non-NaCl plants (Reid et al. 2001; Elmer 2004). Most of our results in 'Welcome' with Fp agreed with those findings, and additionally, synergisic effects on alleviation of *Fusarium* crown rot symptom by using AMF and NaCl were confirmed. In this study, NaCl treatment was carried out according to Reid et al. (2001), and NaCl 50 showed better results than NaCl 100. However, it is necessary to investigate sustainable method of NaCl treatment including chemical property of soil for inducing growth enhancement and disease suppression under field condition. In our results, AMF promoted the growth of asparagus plants 16 weeks after AMF and 10 weeks after Fp (data not shown) inoculation. In addition, both the incidence and severity of symptoms in Fp were alleviated by pre-colonization with Gr and GM. Ozgonen and Erkilic (2007) reported that growth promotion and reduction of *Phytophthora capsici* had no correlation with the mycorrhizal colonization level in peppers. Lozano et al. (1996) reported that alleviation of drought showed no correlation with the mycorrhizal colonization level in lettuce. In our results, Gr+NaCl showed relatively lower symptoms of *Fusarium* crown rot than GM in the two Fp isolates, with no significant difference in colonization level between the two species. Thus, the colonization level might have less association with the reduction of *Fusarium* crown rot in this study.

In the present study, AMF promoted the growth of asparagus plants, and the severity of symptoms in Fp was alleviated by pre-colonization with AMF. Baltruschat and Schonbeck (1975) demonstrated that in tobacco plants, an increase in both arginine and citrulline occurred in mycorrhizal plants, which inhibited the propagation of *Thielaviopsis basicola*. Starratt and Lazarovits (1999) reported low levels of the herbicide trifluralin-induced resistance to Fusarium wilt and elevated levels of free amino acids in melon seedlings. In this study, the increase in several free amino acids through mycorrhizal symbiosis and NaCl in asparagus plants was confirmed. From these findings, suppression of *Fusarium* crown rot in this study is closely associated with increase in free amino acids. On the other hand, Dehne and Schonbeck (1979) reported that the lignification in the endodermis and the stele enhanced by AMF colonization suppressed Fusarium wilt in tomato plants. Matsubara et al. (2003) reported that pectic substances in asparagus roots increased by AMF colonization, and they supposed that the resulting rigidity of root tissue suppressed *Fusarium* infection. Thus, some physiological and histological factors may be associated with disease tolerance in mycorrhizal plants.

On the other hand, Pozo et al. (2002) reported that in tomato plants with a split root system, tolerance to *Phytophthora parasitica* appeared in both non-AMF inoculated roots and inoculated roots in AMF plants, so that induced systemic disease resistance was recognized. In this study, several free amino acids increased in shoots, where no colonization occurred. From these facts, we will estimate the induced systemic resistance in mycorrhizal asparagus plants with split root system, and further work is required to determine whether the changes in free amino acid contents have a direct or indirect relationship to the induced systemic resistance.

Our results suggest that AMF could inhibit symptoms of *Fusarium* crown rot in asparagus plants, and synergistic effect of disease suppression could be expected by the combination use of AMF and NaCl. This proposal seeks to develop a sustainable practice to manage the disease and improve plant health, thus contributing to an improvement in asparagus decline.

References

Baltruschat H, Schonbeck F (1975) The influence of endotrophic mycorrhiza on the infestation of tobacco by *Thielaviopsis basicola*. Phytopathol Z 84:172–188

Blok WJ, Zwankhuizen MJ, Bollen GJ (1997) Biological control of *Fusarium oxysporum* f. sp. *asparagi* by applying non-pathogenic isolates of *F. oxysporum*. Biocontrol Sci Technol 7:527–541

Dehne HW, Schonbeck F (1979) The influence of endotrophic mycorrhiza on plant diseases. II. Phenol metabolism and lignification. Phytopathol Z 95:210–216

Elmer WH (1992) Suppression of Fusarium crown and root rot of asparagus with sodium chloride. Phytopathology 82:97–104

Elmer WH (2004) Combining nonpathogenic strains of *Fusarium oxysporum* with sodium chloride to suppress Fusarium crown rot of asparagus in replanted fields. Plant Pathol 53:751–758

Fattah GM, Mohamedin AH (2000) Interactions between a vesicular-arbuscular mycorrhizal fungus (*Glomus intraradices*) and *Streptomyces coelicolor* and their effects on sorghum plants grown in soil amended with chitin of brawn scales. Biol Fertil Soils 32:401–409

Hamel C, Vujanovic V, Nakano-Hylander A, Jeannotte R, St-Arnaud M (2005) Factors associated with Fusarium crown and root rot of asparagus outbreaks in Quebec. Phytopathology 95:867–873

Kitano M, Hidaka K, Zushi K, Araki T (2008) Production of value-added vegetable by applying environmental stresses to roots in soil-less culture. J SHITA 20:210–218 (In Japanese with English abstract)

Knaflewski M, Golinski P, Kostecki M, Waskiewicz A, Weber Z (2008) Mycotoxins and mycotoxin-producing fungi occurring in asparagus spears. Acta Hortic 776:183–189

Lake RJ, Falloon PG, Cook DWM (1993) Replant problem and chemical components of asparagus roots. N Z J Crop Hortic Sci 21:53–58

Lozano JM, Azcon R, Palma JM (1996) Superoxide dismutase activity in arbuscular mycorrhizal *Lactuca sativa* plants subjected to drought stress. New Phytol 134:327–333

Marschner H, Dell B (1994) Nutrient uptake in mycorrhizal symbiosis. Plant Soil 159:89–102

Matsubara Y, Hasegawa N, Ohba N (2003) Relation between fiber and pectic substance in root tissue and tolerance to Fusarium root rot in asparagus plants infected with arbuscular mycorrhizal fungus. J Jpn Soc Hortic Sci 72:275–280

Miller HG, Ikawa M, Peirce LC (1991) Caffeic acid identified as an inhibitory compound in asparagus root filtrate. HortSci 26:1525–1527

Nahiyan ASM, Boyer LR, Jeffries P, Matsubara Y (2011) PCR-SSCP analysis of *Fusarium* diversity in asparagus decline in Japan. Eur J Plant Pathol 130:197–203

Ozgonen H, Erkilic A (2007) Growth enhancement and Phytophthora blight (*Phytophthora capsici* Leonian) control by arbuscular mycorrhizal fungal inoculation in pepper. Crop Prot 26:1682–1688

Phillips JM, Hayman DS (1970) Improved procedures for clearing roots and staining parasitic and vesicular-arbuscular mycorrhizal fungi for rapid assessment of infection. Trans Br Mycol Soc 55:158–163

Porras-Soriano A, Soriano-Martin ML, Porras-Piedra A, Azcon R (2009) Arbuscular mycorrhizal fungi increased growth, nutrient uptake and tolerance to salinity in olive trees under nursery conditions. J Plant Physiol 166:1350–1359

Pozo MJ, Cordier C, Gaudot ED, Barea JM, Aguilar CA (2002) Localized versus systemic effect of arbuscular mycorrhizal fungi on defence responses to *Phytophthora* infection in tomato plants. J Exp Bot 53:525–534

Reid TC, Hausbeck MK, Kizilkaya K (2001) Effects of sodium chloride on commercial asparagus and of alternative forms of chloride salt on Fusarium crown and root rot. Plant Dis 85:1271–1275

Reid TC, Hausbeck MK, Kizilkaya K (2002) Use of fungicides and biological controls in the suppression of Fusarium crown and root rot of asparagus under green house and growth chamber conditions. Plant Dis 86:493–498

Rolin D, Pfeffer PE, Douds DD, Farrell HM, Shachar Y (2001) Arbuscular mycorrhizal symbiosis and phosphorus nutrition: effects on amino acid production and turnover in leek. Symbiosis 30:1–14

Sood SG (2003) Chemotactic response of plant-growth-promoting bacteria towards roots of vesicular-arbuscular mycorrhizal tomato plants. Microbiol Ecol 45:219–227

Starratt AN, Lazarovits G (1999) Herbicide-induced disease resistance and associated increases in free amino acid levels in melon plants. Can J Plant Pathol 21:33–36

Wong JY, Jeffries P (2006) Diversity of pathogenic *Fusarium* populations associated with asparagus roots in decline soils in Spain and the UK. Plant Pathol 55:331–342

Yong CC (1984) Autointoxication in root exudates of *Asparagus officinalis* L. Plant Soil 82:247–253

Index

M.K. Meghvansi, A. Varma (eds.), *Organic Amendments and Soil Suppressiveness in Plant Disease Management*, Soil Biology 46,
DOI 10.1007/978-3-319-23075-7

R

MIX
Papier aus verantwortungsvollen Quellen
Paper from responsible sources
FSC® C105338

If you have any concerns about our products,
you can contact us on
ProductSafety@springernature.com

In case Publisher is established outside the EU,
the EU authorized representative is:
Springer Nature Customer Service Center GmbH
Europaplatz 3, 69115 Heidelberg, Germany

Printed by Libri Plureos GmbH
in Hamburg, Germany